Design of Linear Multivariable Feedback Control Systems

Joseph J. Bongiorno Jr. · Kiheon Park

Design of Linear Multivariable Feedback Control Systems

The Wiener–Hopf Approach using Transforms and Spectral Factorization

Springer

Joseph J. Bongiorno Jr.
Department of Electrical and Computer Engineering, NYU Tandon School of Engineering
Polytechnic Institute, New York University
Brooklyn, NY, USA

Kiheon Park
College of Information and Communication Engineering
Sungkyunkwan University
Suwon-si, Korea (Republic of)

ISBN 978-3-030-44355-9 ISBN 978-3-030-44356-6 (eBook)
https://doi.org/10.1007/978-3-030-44356-6

Mathematics Subject Classification (2010): 93-01, 93C05, 93C15, 93C35, 93C55, 93C57, 93C62, 93D09, 93D15, 93E20, 47A68

This Springer imprint is published by the registered company Springer Nature Switzerland AG
The registered company address is: Gewerbestrasse 11, 6330 Cham, Switzerland

This book is dedicated to Dante C. Youla our mentor and colleague for his inspiration, encouragement, and collaboration.

Preface

This book is based on some of the research done by the authors and their colleague, Dante C. Youla, over a period that spanned almost 50 years beginning with Youla (1961) and continuing most recently with Park and Bongiorno (2009). Specifically, their work on the design of analog optimal time-invariant linear multivariable feedback control systems using transforms and the Wiener–Hopf methodology is presented within a unified framework for both the analog and the digital case. The authors are especially grateful to Dante C. Youla for helping us to resolve a number of mathematical issues connected with the material presented in this book.

The focus is on feedback control systems that can be modeled as an interconnection of subsystems each of which is specified except for one that is chosen by the designer and called the controller. Attention is restricted to those systems in which all subsystems can be modeled with linear constant-coefficient differential equations (analog systems) or to those in which all subsystems can be modeled with linear constant-coefficient difference equations (digital systems). Exogenous inputs that are deterministic and persistent (steps, ramps, etc.) and/or ones that are stochastic are considered. In the book, a quadratic functional is used to measure system performance. The design task is to choose and synthesize the controller transfer matrix so that the desired overall system performance is achieved. The relationship of the works of others to the work presented here is described throughout the book in sections entitled "Historical Perspective and Discussion".

A key first step in the design process is the parameterization of the class of stabilizing controllers, which is the topic of Chap. 2. Larin, Naumenko, and Suntsev (1971, 1972, 1973) were the first to solve this problem. However, this achievement was not recognized at the time and solutions were also provided in Kučera (1974) and Youla, Jabr, and Bongiorno (1976b) with the parameterization in the latter attracting particular notoriety. In order to apply the Wiener–Hopf methodology when tracking of persistent deterministic reference inputs and rejection of persistent deterministic disturbance inputs is required, a refinement of the parameterization of stabilizing controllers is needed. In particular, with these inputs the subset of stabilizing controllers for which the error variables and the controller output variables are stable must be employed. The required parameterization is presented in Chap. 3

and used to obtain the optimal controller with respect to a quadratic performance functional for the standard configuration. In addition, a methodology is described that enables one to trade off optimality for improved stability margin and for reduced sensitivity to plant model uncertainty. Results are also provided for one-degree-of-freedom (1DOF) and three-degree-of-freedom (3DOF) systems. Chapter 3 contains material that first appeared in Park and Bongiorno (2009). Optimal design of decoupled systems is addressed in Chap. 4 which contains results not previously published. The numerical calculation of the optimal controller parameters and realization of the controller are treated in Chap. 5.

With regard to applications, three examples are provided. Example 3.23 concerns the optimal $\mathcal{H}_2$ design of a 1DOF depth and pitch controller for a submerged submarine traveling at 30 knots speed using the linearized model for submarine motion given in Sect. 10.4 of Grimble (2006). Example 3.24 concerns the optimal $\mathcal{H}_2$ design of a 2DOF controller for a Rosenbrock process: i.e., a multivariable process whose transfer matrix exhibits a severe coupling phenomenon (Åström et al. 2002). Specifically, the transfer matrix has a nonminimum phase zero at $s = 1$ and this yields fundamental limitations on control system performance. The design of an optimal decoupling 2DOF $\mathcal{H}_2$ controller is determined in Example 4.10 for the same Rosebrock process so that the price paid in performance for imposing the decoupling requirement could be assessed.

This book is intended for readers familiar with linear systems and matrix theory. In particular, the reader should be familiar with the material in Appendix A and enough matrix theory to follow the presentation in Appendix B. Many examples with solutions are provided which contain important insights, extensions, and/or elaborations on the topics treated in the book. Examples of particular importance are identified with an asterisk.

Brooklyn, USA — Joseph J. Bongiorno, Jr.
Suwon-si, Korea (Republic of) — Kiheon Park

Contents

Chapter 1
Introduction

1.1 Overview

The design of analog and digital feedback control systems is addressed in this book. The focus is on control systems that can be modeled as an interconnection of subsystems each of which is specified except for one, called the controller, which is to be chosen by the designer. Associated with each subsystem is a set of signals that are the elements of a designated vector called the input and which generate responses that are the elements of a designated vector called the output. The input and/or output vectors can contain one or more elements; hence, the subsystems are multivariable ones that include single-input-output (single-variable) subsystems as a special case. Also associated with each subsystem is a transfer matrix, which connects the input vector to the contribution it makes in the output vector. The symbol used to designate the transfer matrix of a subsystem is shown within the block representing it in the system block diagram. The design task is to choose and synthesize the controller transfer matrix so that the desired overall system performance is achieved. In this book, quadratic measures of system performance are used exclusively. The essential elements of complex function theory, transforms, and stochastic processes needed are summarized in Appendix A. The essential elements of matrix theory needed are reviewed in Appendix B.

When dealing with analog systems, attention is restricted to subsystems that can be modeled by a system of ordinary linear differential equations with real constant coefficients. As a consequence, all analog subsystem transfer matrices contain elements that are ratios of polynomials in the Laplace variable s with real coefficients or, equivalently, real rational functions of s. The traditional transform used by engineers for digital systems is the z-transform. It turns out for the problems of interest here that it is more convenient to use the λ-transform, where $\lambda = 1/z$. When dealing with digital systems, attention is restricted to subsystems that can be modeled by a system of ordinary linear differential equations with real constant

J. J. Bongiorno Jr. and K. Park, *Design of Linear Multivariable Feedback Control Systems*, https://doi.org/10.1007/978-3-030-44356-6_1

coefficients. As a consequence, all digital subsystem transfer matrices contain elements that are ratios of polynomials in the variable λ with real coefficients or, equivalently, real rational functions of λ. So it should not be surprising that the methodologies used for analog and digital systems parallel one another.

Sampled-data systems contain both analog and digital elements and represent a greater modeling challenge. Typically in such systems, all subsystems except for the controller are analog. The controller consists of a digital processor operating at a fixed rate followed by a hold circuit. The input to the digital processor is a sequence of sampled measurements and the digital processor output samples are the inputs to the hold circuit. Usually, the output of the hold circuit is held constant during the inter-sample interval at the value of the sample at the beginning of the interval. When attention is restricted to the behavior of the system at the sampling instants, the sampled-data system can be easily modeled as a digital system. When the system behavior in the interval between the samples needs to be taken into account, it is still possible with a more complex digital system model to account for the performance of the sampled-data system (Rosenwasser and Lampe 2006). Hence, all the results for digital systems presented in this book can be applied to sampled-data systems.

1.2 System Models

The block diagram of an analog control system that incorporates most of the features one might ever encounter in practice is shown in Fig. 1.1. All symbols designate Laplace transforms and the sizes of all column vectors representing the inputs and outputs in the system are consistent with the sizes of the subsystem transfer matrices. The subsystem that is to be controlled is referred to as the plant and the vector $d(s)$ represents the environmental disturbances that impact the plant. The vector $v(s)$ represents the variables that are available to control the plant.

Whenever feasible and economical, a sensor capable of measuring some or all of the environmental disturbances impacting the plant is included in the feedforward path as shown. The matrix $L(s)$ is the sensor transfer matrix and the vector $n_l(s)$ accounts for sensor error. The feedforward path and the controller provide a mechanism for modifying the plant input $v(s)$ so as to suppress the impact of the environmental disturbances $d(s)$ on the plant output vector $y(s)$. The same is true for the feedback path that contains a sensor that measures the plant output $y(s)$. The matrix $F(s)$ is the sensor transfer matrix and the vector $n_m(s)$ accounts for sensor error.

Often it is not physically possible or economical to measure the disturbance vector $d(s)$. The task of suppressing the impact of disturbances on system performance in those cases must be accomplished only with the aid of the feedback path. Feedback also plays an essential role in reducing the sensitivity of system performance to changes in the plant transfer matrix $P(s)$ and is essential for the control of unstable plants. The block with the transfer matrix $T_d(s)$ does not represent a

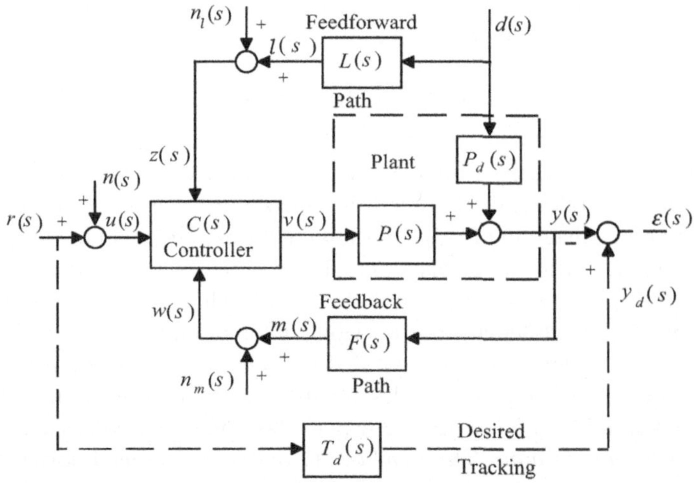

Fig. 1.1 Three-degree-of-freedom system (© 1990 Taylor and Francis Ltd. http://www.tandfonline.com. Reprinted, with permission, from Park and Bongiorno (1990))

component of the physical system. It is simply the mechanism used in the block diagram to define the system tracking error $\varepsilon(s)$. Accordingly, the path in which it lies is indicated with a dashed line. Obviously, $y_d(s)$ represents the desired tracking of the reference input $r(s)$ that can be corrupted by noise $n(s)$ when applied to the controller.

There are three vector inputs to the controller and one can partition the controller transfer matrix accordingly so that

$$v(s) = -C_w(s)w(s) - C_z(s)z(s) + C_u(s)u(s) \tag{1.1}$$

and

$$C(s) = [-C_w(s) - C_z(s) \;\; C_u(s)]. \tag{1.2}$$

The negative signs in (1.2) are introduced to reflect the fact that the feedforward path and feedback path are intended to cancel the undesired components present in the output y. Clearly, there are three submatrices associated with the controller that the designer can choose. For this reason, the system shown in Fig. 1.1 is referred to as a *three-degree-of-freedom* (3DOF) system. The second term on the right-hand side of (1.1) and the second submatrix on the right-hand side of (1.2) disappear when feedforward compensation is not included. The system is then called a *two-degree-of-freedom* (2DOF) system. In some control applications, there are restrictions on the implementation of the controller that correspond to $C_w = C_u$, $C_z = 0$. The controller is then of the form

$$C(s) = C_u(s)[-I \quad I], \tag{1.3}$$

where I denotes the identity matrix. Now there is only one matrix to choose in the design of the controller and the system is called a *one-degree-of-freedom* (1DOF) system. The block diagram of a 3DOF digital control system is identical to the one shown in Fig. 1.1 with the Laplace variable s replaced by the digital transform variable λ. Whenever it is not essential to distinguish between analog and digital systems, the functional dependence on s or λ is often omitted.

All systems of interest here can be divided into two parts: one part is just the controller and the other part includes all the remaining subsystems. The block diagram shown in Fig. 1.2 is a generic way to show all such systems and it is called the *standard configuration*. The symbols G and C represent Laplace transform transfer matrices when the system is analog and λ-transform transfer matrices when the system is digital. The block G is called the *generalized plant* and contains the physical plant and all subsystems other than the controller. The column vector e contains all the exogenous inputs to the system. Specifically, in the standard configuration representation of the 3DOF system shown in Fig. 1.1 the column vector e is an amalgamation of the column vectors r, n, n_l, n_m, and d. The measurements column vector y_m is an amalgamation of the column vectors u, w, and z. For the column vector y_c, one can choose an amalgamation of the controlled variables ε and v since one desires small tracking errors with minimal control effort. One of the earliest papers in which the standard configuration is used to model linear multivariable feedback control systems is Cheng and Pearson (1981), but a similar configuration was also treated in Larin et al. (1971, 1972, 1973).

The standard configuration is a useful representation for establishing fundamental properties of feedback control systems irrespective of the subsystems and their interconnection within the generalized plant. On the other hand, engineering insights are generally obtained from the analysis of specific feedback system implementations. A comprehensive approach covering both methodologies is employed in the sequel.

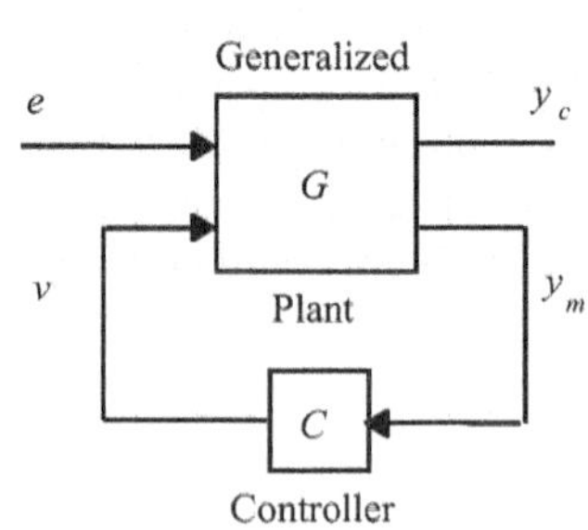

Fig. 1.2 Standard configuration (© 1989 IEEE. Reprinted with permission, from Park and Bongiorno (1989))

An insightful alternative to the standard configuration can be generated once the following observations are made. First, one can always view the exogenous input vector e as an amalgamation of a reference input vector r, the noise vector n corrupting r, and a disturbance vector d that accounts for all other exogenous system inputs. Second, one can view y_c as an amalgamation of an error vector $\varepsilon = y_d - y_t$ and a vector y_s that contains all the system responses that need to be kept small. In particular, y_s should include the signals which need to be kept small so that the linear system model remains valid and minimal control effort is required. The vector y_t represents the system outputs, which need to track $y_d = T_d r$. Third, r and its noise n involve only low power level signals that can only enter the system through the controller. An alternative to the standard configuration in this case is the one shown in Fig. 1.3. It is called the 2DOF *standard configuration* because the controller transfer matrix can be partitioned into two blocks so that

$$v = -C_y y + C_u u \tag{1.4}$$

and

$$C = \begin{bmatrix} -C_y & C_u \end{bmatrix}. \tag{1.5}$$

A key element in the design of many feedback control systems is the requirement that the system tracks certain persistent reference input signals such as steps and ramps. In these cases, the 2DOF standard configuration is a natural representation for the system since it clearly exposes the tracking requirement and is especially useful when decoupling is needed. The system is decoupled when the transfer matrix connecting r to y_t is diagonal. Two of the earliest papers in which the 2DOF standard configuration is used to model linear multivariable feedback control systems are Pernebo (1981a, b).

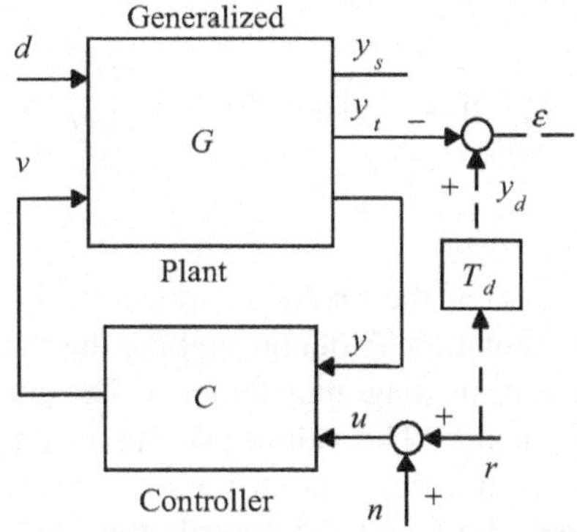

Fig. 1.3 Two-degree-of-freedom standard configuration

1.3 Stability, Signal Models, Performance Measures, and Saturation Constraints

As a consequence of superposition in linear systems, once the controller is selected, each of the output vectors y_c, y_m, and v in the standard configuration is given by the sum of two vectors: one completely determined by the exogenous input e and one completely determined by initial conditions in the system. So it is necessary for satisfactory performance that any contributions from the initial conditions or transients in response to exogenous inputs disappear in time. Then, the output vectors are determined essentially by e. The system is called *stable* in this case and system design can focus on the behavior of the system in response to e; this is the approach adopted here. First, the class of all stabilizing controllers is identified. Then the one within this class is sought for which performance requirements are met. Attention is restricted to exogenous input signals that are deterministic with real rational transforms, or are stochastic with real rational power spectral densities, or are combinations of these. Often an input signal is viewed as the product of a real random variable with a deterministic function. Such signals are called *shape-deterministic* signals and attention is restricted to those for which the deterministic part has a real rational transform. Since only physical inputs which have real rational transforms and since only physical subsystems with real rational transfer matrices are considered, it is assumed throughout whenever not explicitly stated otherwise that all polynomial matrices and all transfer matrices are real (i.e., all coefficients are real).

It is useful in addition to the notion of system stability to define a *stable transform*. A Laplace transform whose finite poles all have negative real parts or a λ-transform whose finite poles all have magnitudes greater than unity are called stable transforms. When the standard configuration is stable, when there are no stochastic inputs, when any persistent deterministic reference inputs are ones for which the controlled variables have zero steady-state value, and when all other deterministic inputs have stable transforms, then the output vector transform y_c is stable. It follows under these conditions that in the digital case (see Appendix A)

$$J = \sum_{k=0}^{\infty} \hat{y}_c'(k)\hat{y}_c(k) = \frac{1}{2\pi j} \oint_{|\lambda|=1} y_{c*}(\lambda)y_c(\lambda)d\lambda/\lambda = \frac{1}{2\pi j} \oint_{|\lambda|=1} \mathrm{Tr}[y_c(\lambda)y_{c*}(\lambda)]d\lambda/\lambda, \tag{1.6}$$

where $\hat{y}_c'(k)$ is the transpose of the inverse λ-transform of $y_c(\lambda)$ and $y_{c*}(\lambda)$ is the transpose of $y_c(1/\lambda)$. The notation Tr denotes taking the trace of a matrix which for a square matrix is the sum of its diagonal elements. The quantity J can be used as a measure of system performance. The validity of this lies in the fact that when any element of $\hat{y}_c(k)$ is nonzero a positive contribution is made to J and the larger the magnitude of this element, the larger the contribution. So the smaller J, the better the performance one can expect since ideally $\hat{y}_c(k) = 0$ is desired. Contrary to standard practice a circumflex over the inverse transform is used to distinguish it from its transform rather than the other way around. This is done to keep the notation simpler because transform quantities are primarily used here. Quadratic performance measures in the form of the last integral in (1.6) are the ones used in this work when designing a controller for digital systems that takes into account transient performance in response to deterministic inputs. Since this integral can also be used as a norm on the Hardy subspace of all real rational stable $y_c(\lambda)$, these quadratic performance measures are also often called $\mathcal{H}_2$ performance measures in the literature.

It is also typical in the analog case for y_c to be strictly proper (i.e., y_c is zero for $s = \infty$). It then follows from Parseval's formula under the same conditions as cited above for the digital case that (see Appendix A)

$$J = \int_0^\infty \hat{y}_c'(t)\hat{y}_c(t)\,dt = \frac{1}{2\pi j}\int_{-j\infty}^{j\infty} y_{c*}(s)y_c(s)ds = \frac{1}{2\pi j}\int_{-j\infty}^{j\infty} \mathrm{Tr}[y_c(s)y_{c*}(s)]ds, \quad (1.7)$$

where $\hat{y}_c'(t)$ is the transpose of the inverse Laplace transform of $y_c(s)$ and $y_{c*}(s)$ is the transpose of $y_c(-s)$. Again J can be used as a measure of system performance. The usefulness of (1.7) as a performance measure lies in the fact that when any element of $\hat{y}_c(t)$ is nonzero over any finite time interval a positive contribution is made to J and the larger the magnitude of this element, the larger the contribution. So the smaller J, the better the performance one can expect since ideally $\hat{y}_c(t) = 0$ is desired. Quadratic performance measures in the form of the last integral in (1.7) are the primary ones used here when designing the controller for analog systems to take into account transient performance in response to deterministic inputs. Since this integral can also be used as a norm on the Hardy subspace of all real rational stable strictly proper $y_c(s)$, these quadratic performance measures are also often called $\mathcal{H}_2$ performance measures in the literature.

When the deterministic inputs are shape-deterministic instead, then $< y_c y_{c*} >$ is used instead of $y_c y_{c*}$ in the integrands where $< \cdot >$ denotes the ensemble average. When only stationary zero-mean stochastic signals are applied to a stable standard configuration, then for digital systems (see Appendix A)

$$< \hat{y}_c'(k)\hat{y}_c(k) > = \frac{1}{2\pi j}\oint_{|\lambda|=1} \mathrm{Tr}[\Phi_{y_c y_c}(\lambda)]d\lambda/\lambda \quad (1.8)$$

can be used as a measure of system performance since it represents the mean square value of the norm of the vector $\hat{y}_c$. The matrix $\Phi_{y_c y_c}(\lambda)$ is the two-sided λ-transform of the covariance matrix $< \hat{y}_c(k+l)\hat{y}'_c(l) >$ and must be free of poles on the unit circle for the integral to exist. The relationship corresponding to (1.8) for analog systems is (see Appendix A)

$$<\hat{y}_c'(t)\hat{y}_c(t)> = \frac{1}{2\pi j}\int_{-j\infty}^{j\infty} \mathrm{Tr}[\Phi_{y_cy_c}(s)]ds. \tag{1.9}$$

The matrix $\Phi_{y_cy_c}(s)$ is the two-sided Laplace transform of the covariance matrix $<\hat{y}_c(t+\tau)\hat{y}_c'(\tau)>$. For the integral to exist, $s\,\Phi_{y_cy_c}(s)$ must be strictly proper and $\Phi_{y_cy_c}(s)$ must be free of poles on the imaginary axis.

It should be emphasized that (1.6) and (1.7) and their counterparts for shape-deterministic inputs are typically used to measure transient performance while (1.8) and (1.9) measure steady-state performance after all transients have disappeared. When a system is subjected to both shape-deterministic and stochastic inputs both measures can be combined into a single performance functional

$$J = \frac{1}{2\pi j}\oint_{|\lambda|=1} \mathrm{Tr}[Q(\lambda)\Phi_{y_c}(\lambda)]d\lambda/\lambda, \tag{1.10}$$

where

$$\Phi_{y_c}(\lambda) = \alpha_1 <y_c(\lambda)y_{c*}(\lambda)> + \alpha_2\Phi_{y_cy_c}(\lambda) \tag{1.11}$$

in the case of digital systems and

$$J = \frac{1}{2\pi j}\int_{-j\infty}^{j\infty} \mathrm{Tr}[Q(s)\Phi_{y_c}(s)]ds, \tag{1.12}$$

where

$$\Phi_{y_c}(s) = \alpha_1 <y_c(s)y_{c*}(s)> + \alpha_2\Phi_{y_cy_c}(s) \tag{1.13}$$

in the case of analog systems. The matrix Q is included to allow for frequency weighting and is chosen so that $Q = Q_*$ and so that it is nonnegative definite on the unit circle in the case of digital systems and is nonnegative definite on the imaginary axis in the case of analog systems. The scalar constants α_1, α_2 are nonnegative and allow for weighting the relative importance of the contribution from deterministic and stochastic components to the controlled vector y_c. Traditionally, $\Phi_{y_cy_c}$ is called a spectral density matrix. In the sequel, matrices that are of the form of the generalized spectral density matrix Φ_{y_c} are also referred to as spectral density matrices and only a single subscript is used for simplicity to distinguish them from the traditional ones.

The subset of all stabilizing digital controllers that yield finite values for (1.10) and the subset of all stabilizing analog controllers that yield finite values for (1.12) are parameterized in this book and the controllers that minimize these performance measures are identified. In addition, a methodology is presented that enables one to trade off optimality for improved stability margin and reduced sensitivity of the performance functional to plant model uncertainty. Moreover, specialized results for the systems shown in Figs. 1.1 and 1.3 are presented. The design of optimal 1DOF and 2DOF decoupled systems is also treated.

A key element in the development of these results is the recognition that the deterministic components of the inputs to the physical process under control must be bounded. For example, the rudder angle of an aircraft or boat is limited. When such a limit is reached, *saturation* is said to have occurred. Clearly, a necessary condition for avoiding saturation is that the deterministic parts of the inputs to the physical plant have *pseudo-stable* transforms. That is, transforms which have no poles in Re $s \geq 0$ for the analog case and no poles in $|\lambda| \leq 1$ for the digital case except possibly for simple poles on the stability boundaries (the finite $s = j\omega$ axis in the analog case or on $|\lambda| = 1$ in the digital case). Otherwise, the physical plant inputs would continue to grow until the saturation level is reached and the assumption of linearity would no longer hold.

For the standard configuration shown in Fig. 1.2, one can write $y_c = y_{c1} + y_{c2}$ for the controlled variables, where y_{c1} represents the deterministic components and y_{c2} represents the stochastic ones. It is particularly convenient to set

$$y_{c1} = \begin{bmatrix} y_{t1} - y_{td1} \\ y_{s1} - y_{sd1} \end{bmatrix} = \begin{bmatrix} \varepsilon_{t1} \\ \varepsilon_{s1} \end{bmatrix}, \tag{1.14}$$

where y_{td1} and y_{sd1} represent the transforms, respectively, of the desired responses of the deterministic tracking signals y_{t1} and the deterministic signals likely to cause saturation y_{s1}. When C is a stabilizing controller and ε_{t1} is stable, asymptotic tracking is assured. When v_1 denotes the deterministic component of the output from C; when as is typical $y_{s1} = G_{ac}v_1$, where G_{ac} accounts for the actuators and any pre-compensation of the plant; and when G_{ac} is stable except possibly for simple poles on the stability boundary; then y_{s1} is the pseudo-stable transform of a bounded temporal function for all stable v_1. Thus, attention is restricted to deterministic exogenous inputs e_1 for which a stabilizing controller C exists such that ε_{t1} and v_1 are stable. When y_{s1} is stable and $y_{sd1} = 0$ or when y_{s1} is pseudo-stable and y_{sd1} is such that any poles of y_{s1} on the stability boundary are not poles of ε_{s1}, then ε_{s1} is also stable. In this case, the performance functional (1.6) with $y_c = y_{c1}$ can be used for the digital case. The performance functional (1.7) can be used for the analog case provided y_{c1} is also strictly proper.

An example in which y_{s1} is pseudo-stable and $y_{t1} = G_p y_{s1} = G_t e_1$ is now considered. When $y_{td1} = e_1 = m_d/s$ and m_d is a constant vector; when the output from a pre-compensator and actuator is $y_{s1} = G_{ac}v_1 = \tilde{G}_{ac}(s)v_1/s$, where $\tilde{G}_{ac}$ is stable and $\tilde{G}_{ac}(0) \neq 0$; when $G_p(s)$ is analytic at $s = 0$ and $G_{po} = G_p(0)$; and when a stabilizing controller is used so that the closed-loop transfer matrix G_t is stable and $G_{to} = G_t(0) = I$; then the error

$$\varepsilon_{t1} = y_{t1} - y_{td1} = (G_t - I)m_d/s \tag{1.15}$$

is stable since

$$\lim_{t\to\infty} \hat{\varepsilon}_{t1}(t) = \lim_{s\to 0} s\,\varepsilon_{t1}(s) = \lim_{s\to 0}(G_t - I)m_d = 0. \tag{1.16}$$

Moreover, since y_{s1} is pseudo-stable and on the $s = j\omega$ axis possesses only a simple pole at $s = 0$, one can find a constant vector m such that

$$y_{s1} = m/s + \tilde{y}_{s1}, \tag{1.17}$$

where $\tilde{y}_{s1}$ is stable. In addition, it follows from

$$\lim_{t\to\infty} \hat{y}_{t1}(t) = \lim_{s\to 0} s\,y_{t1}(s) = \lim_{s\to 0} sG_p y_{s1} = G_{po}\,m = G_{to}\,m_d = m_d \tag{1.18}$$

that the vector m must satisfy $G_{po}\,m = m_d$. In this case, $y_{sd\,1} = m/s$ can be chosen and

$$\varepsilon_{s1} = y_{s1} - y_{sd1} = \tilde{y}_{s1} \tag{1.19}$$

is stable. A similar approach is used in Polyakov (2001) for single-variable sampled-data control systems.

Typically, G_{po} has row rank and there is at least one solution for m given by

$$m = m_o = G'_{po}\left(G_{po}G'_{po}\right)^{-1} m_d. \tag{1.20}$$

Clearly, all solutions are given by $m = m_o + m_h$, where $G_{po}\,m_h = 0$. Hence, $m'_h G'_{po} = 0$ and one gets

$$m'_h m_o = m'_h G'_{po}\left(G_{po}G'_{po}\right)^{-1} m_d = 0. \tag{1.21}$$

Thus,

$$\left\|m^2\right\| = m'm = \left(m'_o + m'_h\right)(m_o + m_h) = \|m_o\|^2 + \|m_h\|^2 \geq \|m_o\|^2. \tag{1.22}$$

So the solution $m = m_o$ is the minimum norm solution and is in keeping with the desire to avoid saturation. Moreover, m_o as given by (1.20) is in terms of given data: G_{po} and m_d.

1.4 Notation and Terminology

The notation and terminology used throughout is collected here for easy reference. All results are obtained working exclusively with rational transforms. The Laplace transform in the complex variable s is used for analog systems and the λ-transform is used for digital systems. The λ-transform is related to the z-transform through the relationship $\lambda = 1/z$. In the few instances that it is necessary to do so, a circumflex is used to distinguish an inverse transform from its transform as in (1.6) and (1.7) for example. Often function arguments are omitted for brevity when no confusion is possible or when the discussion is applicable to both analog and digital systems. Laplace transforms and λ-transforms are distinguished from one another when necessary by showing explicitly the dependence on s or λ. The complex variable p is used generically to represent either s or λ when useful to do so. The conjugate of p is denoted by $\bar{p}$. The real part of p is denoted by Re p and $\mathrm{Re}\, p \geq 0$ is used to denote the finite part of the closed right half p-plane. The notation $|p|$ is used for the magnitude of p and $|p| > 1$ represents the finite part of the open region of the complex p-plane outside the unit circle. A transform is called *good* in the analog case when it is free of poles on $s = j\omega$, ω finite. It is called *good* in the digital case when it is free of poles on $|\lambda| = 1$. It is called *stable* in the analog case when all its finite poles lie in $\mathrm{Re}\, s < 0$, the open left-half s-plane. It is called *stable* in the digital case when all its finite poles lie in $|\lambda| > 1$, the region outside the unit circle. A transform that is not stable is called *unstable*. In general, the finite poles of a transform can be divided into those that lie in the *stable region* ($\mathrm{Re}\, s < 0$ or $|\lambda| > 1$) and those that do not. The former ones are called the *stable poles* and the latter ones are called the *unstable poles*. Transforms that have no poles in Re $s \geq 0$ for the analog case and no poles in $|\lambda| \leq 1$ for the digital case, except possibly for simple poles on the stability boundaries (the finite $s = j\omega$ axis in the analog case or on $|\lambda| = 1$ in the digital case), are called *pseudo-stable*.

A polynomial in s is called *Hurwitz* when all its zeros lie in $\mathrm{Re}\, s < 0$ and a polynomial in λ is called *Schur* when all its zeros lie in $|\lambda| > 1$. A polynomial in p is called *stable* when its reciprocal is stable. Clearly, a polynomial is Hurwitz or Schur iff it is stable. A *monic* polynomial is one whose leading coefficient is unity.

For any matrix G', G^*, det G, and Tr G are used for the transpose, conjugate transpose, determinant, and trace of G, respectively. A diagonal matrix G of order h is denoted by $G = \mathrm{diag}\{g_1, g_2, \ldots, g_h\}$. The identity matrix is the special case of a diagonal matrix with ones on the diagonal and is denoted by I. A positive definite (nonnegative definite) matrix G is indicated by $G > 0$ ($G \geq 0$). A matrix of transforms is called *good* when all its elements are good. It is called *stable* when all its elements are stable. The matrix $G_*(s)$ is the conjugate transpose of $G(-\bar{s})$. That is, $G_*(s) = G^*(-\bar{s})$ which for a real rational matrix reduces to $G_*(s) = G'(-s)$. The matrix $G_*(\lambda)$ is the conjugate transpose of $G(1/\bar{\lambda})$. That is, $G_*(\lambda) = G^*(1/\bar{\lambda})$ which for a real rational matrix reduces to $G_*(\lambda) = G'(1/\lambda)$. It is important to note that G_* has a different interpretation for analog and digital systems, but it is always clear from the context which interpretation needs to be made. A real rational matrix

G in s or λ is called *parahermitian* when $G = G_*$. Any parahermitian matrix $G(s)$ or $G(\lambda)$ which is nonnegative definite on $s = j\omega$ or $|\lambda| = 1$, respectively, is called parahermitian-positive.

A matrix $G(p)$ is called *proper (strictly proper))* when it is finite (zero) at $p = \infty$. Otherwise, it is called *improper*. A square proper matrix whose inverse is also proper is called *biproper*. The notation $G(p) \leq 0(p^{\nu})$ means that no entry in $G(p)$ grows faster than p^{ν} as $p \to \infty$. When $G(p)$ is square and polynomial and the coefficient matrix of the highest power of p is nonsingular, it is called *regular*.. The *rank* of a matrix $G(p)$ is its normal rank: i.e., the highest order of all non-identically zero minors of $G(p)$. The *McMillan degree* $\delta(G\,;\,p_o)$ of $p = p_o$ (finite or infinite) as a pole of the rational matrix $G(p)$ is the largest multiplicity it possesses as a pole of any minor of $G(p)$. The McMillan degree $\delta\,(G)$ of $G(p)$ is the sum of the McMillan degrees of its distinct poles.

In the partial fraction expansion of $G(s)$, the contributions made by all its finite poles in Re $s \leq 0$, Re $s > 0$, and by its poles at s = ∞ are denoted by $\{G(s)\}_+$, $\{G(s)\}_-$, and $\{G(s)\}_\infty$, respectively. Clearly, $\{G(s)\}_+$ is analytic in Re $s > 0$, $\{G(s)\}_-$ is analytic in Re $s \leq 0$, and both are strictly proper. The contribution $\{G(s)\}_\infty$ is polynomial. Similarly, in the partial fraction expansion of $G(\lambda)$, the contributions made by all its finite poles in $|\lambda| \geq 1, |\lambda| < 1$, and by its poles at $\lambda = \infty$ are denoted by $\{G(\lambda)\}_+$, $\{G(\lambda)\}_-$, and $\{G(\lambda)\}_\infty$, respectively. Clearly, $\{G(\lambda)\}_+$ is analytic in $|\lambda| < 1$, $\{G(\lambda)\}_-$ is analytic in $|\lambda| \geq 1$, and both are strictly proper. The contribution $\{G(\lambda)\}_\infty$ is polynomial.

For any parahermitian-positive matrix $G(s)$, it is established in Youla (1961, 2015) that there are factors $\Omega_l(s)$ and $\Omega_r(s)$ analytic together with their respective left and right inverses in Re$\,s > 0$ such that $G(s) = \Omega_l(s)\Omega_{l*}(s) = \Omega_{r*}(s)\Omega_r(s)$. Similarly, for any parahermitian matrix $G(\lambda)$ which is nonnegative definite on $|\lambda| = 1$, one can show as in Appendix C that there are factors $\Omega_l(\lambda)$ and $\Omega_r(\lambda)$ analytic together with their respective left and right inverses in $|\lambda| < 1$ such that $G(\lambda) = \Omega_l(\lambda)\Omega_{l*}(\lambda) = \Omega_{r*}(\lambda)\Omega_r(\lambda)$. The matrix $\Omega_l(\Omega_r)$ is called a left (right) *Wiener–Hopf factor* for G in both cases. Examples of parahermitian-positive matrices are power spectral density matrices. Such matrices are the transforms of ensemble averages and such averages are denoted by $< \cdot >$ as in (1.8) and (1.9) for example.

The *Schur* product $G \circ R$ of two same-size matrices G and R is the matrix whose i-row, j-column entry is $g_{ij}r_{ij}$. The *Kronecker* product $G \otimes R$ is the matrix whose ij-block is given by $g_{ij}R$. The vector vec $G = [g_1' \;\; g_2' \;\; \cdots \;\; g_n']'$ is formed by stacking all the columns of the matrix G. Useful references for the properties of Kronecker matrix products are Brewer (1978) and Horn and Johnson (1991). A useful reference for the properties of Schur matrix products is Horn and Johnson (1985)

Additional notation and terminology needed for Appendices B and C is summarized in Sects. B.2 and C.2.

1.5 Historical Perspective and Commentary

The methodologies employed to design linear time-invariant feedback control systems are generally either time-domain methodologies or frequency-domain methodologies. Time-domain methodologies are based on a state-variable representation for each given subsystem. Frequency-domain methodologies are based on a transfer matrix representation for each subsystem. The design methodologies are also distinguished by the performance measure employed. When tracking accuracy and control effort with the nominal plant model is key, then a quadratic or $\mathcal{H}_2$ performance measure is often used. When robustness or insensitivity to uncertainty or change in the plant model or exogenous inputs is key, then an $\mathcal{H}_\infty$ performance measure is often used. The $\mathcal{H}_\infty$-norm of a real rational stable transfer matrix is the least upper bound for all the singular values of the matrix on the imaginary axis for analog systems or on the unit circle for digital systems.

Examples of books focusing on the time-domain approach with an $\mathcal{H}_2$ performance measure are Kwakernaak and Sivan (1972), Anderson and Moore (1989), and Saberi et al. (1995). In Chen (2000), the time-domain approach with an $\mathcal{H}_\infty$ performance measure is considered and the results are used in practical applications: specifically, the control of the voice-coil-motor actuator of a hard disk drive, the control of a piezoelectric actuator, and the control of a gyro-stabilized mirror targeting system. Examples of books focusing on the time-domain approach and covering designs for both $\mathcal{H}_2$ and $\mathcal{H}_\infty$ performance measures are Chen and Francis (1995) and Zhou et al. (1996). Books focusing on the frequency-domain approach are Kučera (1979), Vidyasagar (1985), Grimble (2006), and Rosenwasser and Lampe (2006). In Kučera (1979), $\mathcal{H}_2$ performance measures are used exclusively; in Vidyasagar (1985), the emphasis is on $\mathcal{H}_\infty$ performance measures; and in Grimble (2006), both $\mathcal{H}_2$ and $\mathcal{H}_\infty$ performance measures are used. Grimble (2006) covers robustness in depth and contains applications of the theory that include designs of aircraft, ship, submarine, and industrial control systems. The $\mathcal{H}_2$ performance measure is used in Rosenwasser and Lampe (2006) and as in Chen and Francis (1995) the focus is on sampled-data systems. In Aliev and Larin (1998), $\mathcal{H}_2$ designs for both time-domain and frequency-domain methodologies are covered with an emphasis on computational algorithms. $\mathcal{H}_2$ designs for both time-domain and frequency-domain methodologies are also covered in Kučera (1991), with an emphasis on the relationships between the two methodologies.

In addition to these analytically oriented design methodologies, a number of graphically oriented frequency-domain design methodologies for multivariable systems along classical lines have been proposed. The inverse Nyquist array of Rosenbrock, the characteristic locus method of MacFarlane, and the quantitative feedback theory of Horowitz are discussed in Maciejowski (1989) along with designs based on $\mathcal{H}_2$ and $\mathcal{H}_\infty$ performance measures. Another alternative to analytically oriented design methodologies is the use of convex programming along the lines taken in Boyd and Barratt (1991) or indirectly through linear matrix inequalities as described in Boyd et al. (1994) and in Skelton and Iwasaki (1995).

In this book attention is focused on an analytical frequency-domain design methodology using an $\mathcal{H}_2$ performance measure.

The earliest efforts to design controllers for feedback systems by minimizing a quadratic or $\mathcal{H}_2$ performance measure are described in Newton et al. (1957). These efforts employ the calculus of variations to generate the integral equation, known as the Wiener–Hopf equation, which the impulse response function of the optimal controller satisfies. Equations of this kind had been treated in Wiener (1949). Transforms, partial fraction expansions, and spectral factorizations are then used to solve the Wiener–Hopf equation and an explicit formula for the controller transfer function is obtained for analog 1DOF single-input-output systems containing stable plants. Since the design methodology presented in this book also makes use of transforms, partial fraction expansions, and spectral factorizations, it is referred to as the Wiener–Hopf design methodology in recognition of the insightful work of these early researchers. In Weston and Bongiorno (1972), the Wiener–Hopf design methodology is extended to analog 1DOF multivariable systems with stable plants by utilizing the variational technique in Bongiorno (1969) and the matrix spectral factorization technique in Youla (1961).

The Wiener–Hopf methodology is applied in Chang (1961) to analog 2DOF single-input-output systems containing stable plants. The restriction to stable plants is needed in the cases just cited because the methodologies involve first finding an optimal stable open-loop controller and then realizing it within a single-loop feedback structure. This methodology leads to pole-zero cancelations in $\mathrm{Re}\, s \geq 0$ when the plant is unstable; hence, the system is unstable. In Newton et al. (1957) and Chang (1961), it is suggested that unstable plants can be handled by first stabilizing the plant with minor-loop feedback. Although this approach allows one to make use of the methodology available for stable plants, it leaves open the question of whether or not the performance one can realize is different among the many possible choices one can make for the minor-loop transfer function. This is especially a problem in the multivariable case and/or when the plant cannot be stabilized with stable minor-loop feedback. That there exist unstable plants that cannot be stabilized with a proper stable minor-loop feedback transfer matrix is made clear in Youla et al. (1974). Also left open is whether the open-loop design can actually be realized with a stable closed-loop feedback system. Indeed, it is pointed out in Aliev and Larin (1998, pp. 66–67) that all stabilization by minor-loop feedback is not equivalent to all stabilizing controllers for the original plant. Obviously, an unequivocal approach is needed for unstable plants.

The first Wiener–Hopf solution for unstable multivariable analog plants is contained in Larin et al. (1971, 1972, 1973). However, this notable achievement for its time went unrecognized by other researchers and similar results were published in Youla et al. (1976b) for analog systems and in Kučera (1979) for digital systems. In these works, all stabilizing controllers are effectively parameterized in terms of a free stable matrix and the optimal $\mathcal{H}_2$ choice for the free matrix parameter is determined. In Kučera (1979), it was shown that all stabilizing controllers can be generated from the parameterized solution set of a Bezout equation. (There is a difference of opinion on whether the Bezout equation should actually be called a

Diophantine equation or an Aryabhatta equation as noted in the preface to Vidyasagar (1985).) Additional commentary on the history and attributes of stabilizing controller parameterizations is provided in Sect. 2.8. Also relevant are the comments in Aliev and Larin (2007) and the reply by Kučera (2007).

The frequency-domain methodology in Kučera (1979) is applicable to 1DOF multivariable digital plants and is called the polynomial equation approach. In contrast with the Wiener–Hopf methodology in which a partial fraction expansion is needed, the polynomial equation approach involves finding solutions to a pair of linear matrix polynomial equations. Only open-loop optimal designs with zero initial conditions are considered in connection with the tracking of deterministic inputs. The stable closed-loop realization of these designs is not addressed for arbitrary initial conditions and with output feedback instead of state-variable feedback. Hence, the results for open-loop designs are of limited value for unstable plants. Optimal $\mathcal{H}_2$ closed-loop designs are given for the regulator problem with stochastic inputs; however, some important restrictions regarding signal and/or system poles on the unit circle are not adequately generalized although specific examples in which such poles can be handled successfully are provided.

The next major step forward in connection with the Wiener–Hopf design method occurs in Youla and Bongiorno (1985) where analog 2DOF multivariable systems are treated. Not only is the optimal $\mathcal{H}_2$ design determined in this paper, but all designs which yield a finite value for the $\mathcal{H}_2$ performance measure are parameterized in terms of two free stable matrices. This important parameterization is extended to analog 3DOF systems in Park and Bongiorno (1990) and to the analog standard configuration in Park and Bongiorno (1989). Digital 2DOF and 3DOF systems are treated in Grimble (1988) using the polynomial equation approach. However, in contrast with the treatments cited above for the analog case, shape-deterministic inputs and plants with poles on the stability boundary are not treated directly. Instead, a limiting procedure is suggested. Moreover, only the optimal solution is given and not all solutions for which the $\mathcal{H}_2$ performance measure is finite. The standard configuration is also treated using a state-variable approach in Doyle et al. (1989) for analog systems and using the polynomial equation approach in Hunt et al. (1994) for digital systems. However, the first of these two papers does not deal directly with optimal designs for persistent inputs that have poles on the imaginary axis and the second paper does not deal fully with the issue of stability. Also, the second paper does not parameterize all designs for which the $\mathcal{H}_2$ performance measure is finite. It does provide a comprehensive chronological account of the publications related to the polynomial equation approach.

The analog 2DOF standard configuration is treated in Corrêa and Da Silveira (1995), Xie et al. (2000), and Park et al. (2002). The first two papers allow persistent deterministic exogenous inputs with poles in $\operatorname{Re} s \geq 0$ where the plant has none and insists on asymptotic tracking with zero steady-state error. In this case, the plant input must have poles in $\operatorname{Re} s \geq 0$ and the $\mathcal{H}_2$-norm of the plant input is not finite. Corrêa and Da Silveira (1995) circumvent this issue by only incorporating the stable part of the plant input into the system performance measure. Xie et al. (2000)

circumvent this issue by only incorporating the $\mathcal{H}_2$-norm of the transfer matrix between the reference input and the plant input into the system performance measure. As a consequence of these approaches, however, plant input saturation can become a problem. In Park et al. (2002) on the other hand, the only persistent deterministic exogenous inputs considered are ones with poles on the imaginary axis that the plant is capable of tracking with a stable plant input. In this case, the $\mathcal{H}_2$-norm of the plant input is finite and can be included in the system performance measure to avoid plant input saturation and, moreover, it is possible to make a connection between state-space and Wiener–Hopf design formulas.

The parameterization of the subset of all stabilizing controllers that yield finite $\mathcal{H}_2$ performance in Park and Bongiorno (1990) led to the consideration of a trade-off between stability margin and nominal performance for 2DOF and 3DOF analog multivariable systems which is described in Bongiorno et al. (1997). Basically, an approximate measure of stability margin is embodied in an $\mathcal{H}_2$ performance measure which is a functional on the free stable matrix parameter. Then this stability margin performance measure is minimized subject to a constraint on the allowed increase in nominal performance that is given by the $\mathcal{H}_2$-norm of the free stable matrix parameter. The extension of this methodology to the standard configuration of Fig. 1.2 is treated in Sect. 3.9. The trade-off problem turned out to be a special case of a more general one treated in Khargonekar and Rotea (1991) and involved obtaining a Wiener–Hopf solution in terms of Kronecker products.

In the papers by Doyle et al. (1989), Khargonekar and Rotea (1991), and Hunt et al. (1994), the standard configuration exogenous input is assumed to be a white noise or simply an indeterminate to indicate an input channel. In this setting, the dynamics of the persistent inputs are absorbed into the generalized plant block. However, in these papers certain stabilizability assumptions are made which lead to a distinct disadvantage: the results are not applicable to 2DOF and 3DOF systems with unstable reference inputs. In Park and Bongiorno (1989), this problem is avoided by keeping the model of the persistent inputs separate from the generalized plant; however, persistent inputs with unstable poles only on the stability boundary can be accommodated. In Corrêa and Da Silveira (1995) and Xie et al. (2000), a generalized 2DOF configuration is considered and unstable persistent inputs are treated. The approach begins with a parameterization of the subset of stabilizing controllers for which the error transform is stable. However, the parameterization does not ensure that actual plant inputs remain bounded with time. That is, the real issue of plant saturation is not specifically respected.[1]

A methodology for overcoming these limitations is presented in Park and Bongiorno (2009), which begins with a parameterization of the subset of stabilizing controllers for which the error transform is stable and the actual plant inputs remain bounded with time. The optimal $\mathcal{H}_2$ solution is then obtained from within the subset of free parameters for which the cost functional is finite. A key element in this

[1] This paragraph is reprinted, with permission, from Park and Bongiorno (2009).

approach which distinguishes it from the earlier works cited above is the explicit recognition and imposition of the requirement that the controller output should be stable for the reasons described in the last few paragraphs of Sect. 1.3. This is the approach followed in Chap. 3 of this book. Additional remarks on Park and Bongiorno (2009) are contained in the reply by Bongiorno and Park (2010) to comments by Aliev and Larin (2010).

The Wiener–Hopf design methodology has also been successful in connection with the design of optimal decoupled linear multivariable systems. This methodology is used in Lee and Bongiorno (1993a, b) for analog 2DOF and 3DOF systems and in both Youla and Bongiorno (2000) and Bongiorno and Youla (2001) for analog 1DOF systems. Optimal tracking under decoupling constraints has also been considered by Corrêa et al. (2001). In Park et al. (2002), the results contained in Lee and Bongiorno (1993a) are extended to the 2DOF standard configuration and computational algorithms based on state-variable representations are also given. An important step in the Wiener–Hopf solution of the optimal decoupling problem is the recognition in Lee and Bongiorno (1993a) that the Schur matrix product plays a key role. Another essential step in the solution of the optimization problem is the parameterization of all closed-loop transfer matrices that can be realized with stabilizing decoupling controllers. Relevant results in this regard for 1DOF systems are contained in Vardulakis (1987), Lin and Hsieh (1993), Linnemann and Wang (1993), and Gómez and Goodwin (2000) for unity-feedback systems. The extension to nonunity-feedback systems is given in Youla and Bongiorno (2000). Relevant results with regard to stabilizing decoupling controllers for 2DOF and 3DOF systems containing plants with square transfer matrices are given in Desoer and Gündes (1986) and the extension to plants with rectangular transfer matrices is given in Lee and Bongiorno (1993b).

In this book, the comprehensive analog $\mathcal{H}_2$ Wiener–Hopf frequency-domain analytical design methodology developed by Youla, Bongiorno, and coworkers for linear time-invariant multivariable feedback control systems is presented in a unified way for analog and digital systems. Moreover, the tracking of persistent exogenous deterministic inputs is explicitly accounted for. The focus is on theoretical insights regarding engineering issues that are not easily obtained from a state-variable approach.

1.6 Examples

Example 1.1 The standard configuration in Fig. 1.2 represents the mathematical relationship

$$\begin{bmatrix} y_c \\ y_m \end{bmatrix} = \begin{bmatrix} G_{11} & G_{12} \\ G_{21} & G_{22} \end{bmatrix} \begin{bmatrix} e \\ v \end{bmatrix} = G \begin{bmatrix} e \\ v \end{bmatrix},$$

where the submatrices of G are compatibly sized with respect to the vectors y_c, y_m, e, and v. (a) Find $G_{y_c e}$, the transfer matrix between e and y_c for the standard configuration. (b) Find the expressions for the G_{ij} that correspond to the 3DOF system shown in Fig. 1.1 when $y_c' = [\varepsilon' \;\; v']$, $y_m' = [w' \;\; z' \;\; u']$, and $e' = [d' \;\; r' \;\; n' \;\; n_m' \;\; n_l']$. (c) Evaluate $G_{y_c e}$ for the 3DOF system. (d) Find explicit formulas which give ε and v in response to d, r, n, n_m, and n_l for the 3DOF system using the results obtained in part (c).

Solution (a) Clearly, $v = Cy_m$ and it follows that

$$y_m = G_{21}e + G_{22}v = G_{21}e + G_{22}Cy_m \Leftrightarrow y_m = (I - G_{22}C)^{-1}G_{21}e.$$

Hence, from

$$y_c = G_{11}e + G_{12}v = G_{11}e + G_{12}Cy_m = [G_{11} + G_{12}C(I - G_{22}C)^{-1}G_{21}]e$$

one gets

$$G_{y_c e} = G_{11} + G_{12}C(I - G_{22}C)^{-1}G_{21}.$$

(b) It is evident from Fig. 1.1 that

$$\left[\begin{array}{c} y_c \\ \hline y_m \end{array}\right] = \left[\begin{array}{c} \varepsilon \\ v \\ \hline w \\ z \\ u \end{array}\right] = \left[\begin{array}{c} T_d r - P_d d - Pv \\ v \\ \hline FP_d d + FPv + n_m \\ Ld + n_l \\ r + n \end{array}\right] = \left[\begin{array}{ccccc|c} -P_d & T_d & 0 & 0 & 0 & -P \\ 0 & 0 & 0 & 0 & 0 & I \\ \hline FP_d & 0 & 0 & I & 0 & FP \\ L & 0 & 0 & 0 & I & 0 \\ 0 & I & I & 0 & 0 & 0 \end{array}\right] \left[\begin{array}{c} e \\ \hline v \end{array}\right].$$

Thus, for the 3DOF system

$$\left[\begin{array}{c|c} G_{11} & G_{12} \\ \hline G_{21} & G_{22} \end{array}\right] = \left[\begin{array}{ccccc|c} -P_d & T_d & 0 & 0 & 0 & -P \\ 0 & 0 & 0 & 0 & 0 & I \\ \hline FP_d & 0 & 0 & I & 0 & FP \\ L & 0 & 0 & 0 & I & 0 \\ 0 & I & I & 0 & 0 & 0 \end{array}\right].$$

(c) Using (1.2) and the G_{ij} obtained in part (b) yields

$$C(I-G_{22}C)^{-1}=(I-CG_{22})^{-1}C=(I+C_wFP)^{-1}[-C_w \quad -C_z \quad C_u].$$

It then follows from part (a) that

$$\begin{aligned}G_{y_ce}&=G_{11}+G_{12}C(I-G_{22}C)^{-1}G_{21}=G_{11}+G_{12}(I+C_wFP)^{-1}[-C_w \quad -C_z \quad C_u]G_{21}\\&=G_{11}+\begin{bmatrix}-P\\I\end{bmatrix}(I+C_wFP)^{-1}[-(C_wFP_d+C_zL) \quad C_u \quad C_u \quad -C_w \quad -C_z].\end{aligned}$$

Hence, with

$$G_{y_ce}=\begin{bmatrix}G_{\varepsilon d} & G_{\varepsilon r} & G_{\varepsilon n} & G_{\varepsilon m} & G_{\varepsilon l}\\G_{vd} & G_{vr} & G_{vn} & G_{vm} & G_{vl}\end{bmatrix}=\begin{bmatrix}G_{\varepsilon e}\\G_{ve}\end{bmatrix}$$

one gets

$$\begin{aligned}&G_{\varepsilon d}=-P_d+P(I+C_wFP)^{-1}(C_wFP_d+C_zL),\\&G_{\varepsilon r}=T_d-P(I+C_wFP)^{-1}C_u,\\&[G_{\varepsilon n} \quad G_{\varepsilon m} \quad G_{\varepsilon l}]=-P(I+C_wFP)^{-1}[C_u \quad -C_w \quad -C_z],\end{aligned}$$

and

$$G_{ve}=(I+C_wFP)^{-1}[-(C_wFP_d+C_zL) \quad C_u \quad C_u \quad -C_w \quad -C_z].$$

(d) From the fact that $y_c=[\varepsilon' \quad v']'=G_{y_ce}e$, it follows that

$$\varepsilon=[I \quad 0]y_c=[I \quad 0]G_{y_ce}e=G_{\varepsilon e}e=\varepsilon_d+\varepsilon_r+\varepsilon_n+\varepsilon_{n_m}+\varepsilon_l,$$

where

$$\begin{aligned}&\varepsilon_d=G_{\varepsilon d}d=[P(I+C_wFP)^{-1}(C_wFP_d+C_zL)-P_d]d,\\&\varepsilon_r=G_{\varepsilon r}r=[T_d-P(I+C_wFP)^{-1}C_u]r,\end{aligned}$$

and

$$\begin{aligned}\varepsilon_n+\varepsilon_m+\varepsilon_l&=[G_{\varepsilon n} \quad G_{\varepsilon m} \quad G_{\varepsilon l}][n' \quad n_m' \quad n_l']'\\&=-P(I+C_wFP)^{-1}(C_un-C_wn_m-C_zn_l).\end{aligned}$$

Similarly,

$$\begin{aligned} v &= [0 \quad I]y_c = [0 \quad I]G_{y_ce}e = G_{ve}e \\ &= (I + C_wFP)^{-1}[-(C_wFP_d + C_zL) \quad C_u \quad C_u \quad -C_w \quad -C_z]e \\ &= -(I + C_wFP)^{-1}[C_wFP_d + C_zL)d - C_u(r+n) + C_wn_m + C_zn_l], \end{aligned}$$

Comment: Results similar to those obtained in part (d) are derived in a simpler and more direct way at the beginning of Sect. 2.6. That is, the determination of a particular result by treating a system as a special case of the standard configuration may not always be the best approach algebraically.

Example 1.2 It is important in this example to replace G in Fig. 1.2 with

$$G = \tilde{G} = \begin{bmatrix} \tilde{G}_{11} & \tilde{G}_{12} \\ \tilde{G}_{21} & \tilde{G}_{22} \end{bmatrix}$$

so that it can be distinguished from the G in Fig. 1.3. With regard to Fig. 1.3, one can write

$$\begin{aligned} y_c &= \begin{bmatrix} y_s \\ y_t \end{bmatrix} = G_{11}d + G_{12}v = \begin{bmatrix} G_{11s} \\ G_{11t} \end{bmatrix} d + \begin{bmatrix} G_{12s} \\ G_{12t} \end{bmatrix} v, \\ y_m &= \begin{bmatrix} y \\ u \end{bmatrix} = \begin{bmatrix} G_{21}d + G_{22}v \\ r + n \end{bmatrix} \end{aligned}$$

and $e' = [d' \quad r' \quad n']$. (a) Find the $\tilde{G}_{ij}$ that corresponds to the 2DOF standard configuration. (b) Find G_{y_ce} using the results obtained in Example 1.1 and evaluate it when $C = [-C_y \quad C_u]$. (c) Find explicit formulas which give $\varepsilon = T_dr - y_t$ and y_s in response to d, r, and n.

Solution (a) For the 2DOF standard configuration,

$$y_c = [G_{11} \quad 0 \quad 0]e + G_{12}v$$

and

$$y_m = \begin{bmatrix} G_{21} & 0 & 0 \\ 0 & I & I \end{bmatrix} e + \begin{bmatrix} G_{22} \\ 0 \end{bmatrix} v.$$

It is now evident that

$$\tilde{G}_{11} = [G_{11} \quad 0 \quad 0], \; \tilde{G}_{12} = G_{12},$$

$$\tilde{G}_{21} = \begin{bmatrix} G_{21} & 0 & 0 \\ 0 & I & I \end{bmatrix}, \ \tilde{G}_{22} = \begin{bmatrix} G_{22} \\ 0 \end{bmatrix}.$$

(b) From the solution of Example 1.1(a),

$$\begin{aligned} G_{y_ce} &= \tilde{G}_{11} + \tilde{G}_{12}C(I - \tilde{G}_{22}C)^{-1}\tilde{G}_{21} = \tilde{G}_{11} + \tilde{G}_{12}(I - C\tilde{G}_{22})^{-1}C\tilde{G}_{21} \\ &= [\,G_{11} \quad 0 \quad 0\,] + G_{12}\left(I - [\,-C_y \quad C_u\,]\begin{bmatrix} G_{22} \\ 0 \end{bmatrix}\right)^{-1}[\,-C_y \quad C_u\,]\tilde{G}_{21} \\ &= [\,G_{11} \quad 0 \quad 0\,] + G_{12}(I + C_yG_{22})^{-1}[\,-C_yG_{21} \quad C_u \quad C_u\,]. \end{aligned}$$

(c) Clearly,

$$\begin{aligned} y_c &= \begin{bmatrix} y_s \\ y_t \end{bmatrix} = G_{y_ce}e \\ &= [G_{11} - G_{12}(I + C_yG_{22})^{-1}C_yG_{21}]d + G_{12}(I + C_yG_{22})^{-1}C_u(r + n). \end{aligned}$$

Now

$$y_s = [\,I \quad 0\,]y_c, \ \varepsilon = T_d r - [\,0 \quad I\,]y_c.$$

Since $G_{11} = [\,G'_{11s} \quad G'_{11t}\,]'$ and $G_{12} = [\,G'_{12s} \quad G'_{12t}\,]'$, one gets

$$y_s = [G_{11s} - G_{12s}(I + C_yG_{22})^{-1}C_yG_{21}]d + G_{12s}(I + C_yG_{22})^{-1}C_u(r + n)$$

and

$$y_t = [G_{11t} - G_{12t}(I + C_yG_{22})^{-1}C_yG_{21}]d + G_{12t}(I + C_yG_{22})^{-1}C_u(r + n).$$

Chapter 2
Stabilizing Controllers, Tracking, and Disturbance Rejection

2.1 Overview

The family of controllers that stabilize the standard configuration is parameterized in terms of a free stable real rational matrix in this chapter. The family of controllers that stabilize the 2DOF standard configuration and the 3DOF system is then derived from this parameterization. Stability is only one of several objectives in feedback control system design, however. Often it is also required that the system steady-state error in response to certain persistent deterministic inputs be zero. In addition, as explained at the end of Sect. 1.3 with regard to saturation, it is important that not only is zero steady-state error realized, but also zero steady-state value at the controller output. Moreover, this is required for the finite value of the quadratic or $\mathcal{H}_2$ performance measures used when they include penalties for both nonzero error and nonzero control effort. There are also practical reasons for insisting that the controller output not contain any persistent deterministic components. Such components cause a persistent drain on the sources that supply power to the actuators providing the physical inputs to the plant.

These steady-state requirements are equivalent to insisting that both the error transform and the controller output transform be stable. The family of persistent deterministic inputs for which this is the case in the 3DOF system is established in this chapter along with a parameterization of the subset of stabilizing controllers that achieve this desired objective. This is accomplished for both tracking of reference inputs and for rejection of disturbance inputs. The focus on the 3DOF system is justified by the insights it provides with regard to engineering and practical considerations: the generalizations possible for the 2DOF standard configuration are treated in Example 2.12 and for the standard configuration are presented in Sect. 3.3.

J. J. Bongiorno Jr. and K. Park, *Design of Linear Multivariable Feedback Control Systems*, https://doi.org/10.1007/978-3-030-44356-6_2

Finally, the subset of all proper stabilizing controllers is parameterized and a tunable realization for all such controllers is described. These results are useful when the controller transfer matrix for which the performance measure is minimized turns out to be proper as is often the case.

In this chapter, the stage is set for Chap. 3 in which the Wiener–Hopf design methodology is employed to make an optimal choice for the free parameters. The presentation of the material in this chapter relies heavily on Appendix B as a prerequisite.

2.2 Polynomial Matrix Descriptions

The *polynomial matrix description* for analog systems was introduced in Rosenbrock (1970) and also treated in Wolovich (1974), Kailath (1980), Callier and Desoer (1982), and Chen (1984). In Antsaklis and Michel (2006), polynomial matrix descriptions for both analog and digital systems are treated. Any system with such a description can be represented in terms of transformed quantities by equations of the form

$$T(p)\xi(p) = U(p)u(p) + i_T(p), \tag{2.1}$$

$$y(p) = V(p)\xi(p) + W(p)u(p) + i_y(p), \tag{2.2}$$

where p is the Laplace transform variable in the case of analog systems and the λ-transform variable in the case of digital systems. The elements of the vector ξ are the transforms of the system variables that completely characterize system behavior. The elements of the vectors $i_T(p)$ and $i_y(p)$ are polynomials in p and account for the contribution from initial conditions in the analog case and from initial data strings in the digital case. The matrices T, U, V, and W are polynomial matrices in p with real coefficients and T is square; u is the transform of the system input vector; and y is the transform of the system output vector. Clearly, a unique solution for ξ exists iff (if, and only if) $\det T \not\equiv 0$. In this case, the system is called *nondegenerate*. Attention is restricted to nondegenerate systems and this is to be implicitly understood throughout.

The uniqueness of ξ, however, is not enough to establish the inverse transform $\hat{\xi}$: one must also invoke causality (see Appendix A). For analog systems, the causality requirement is met by specifying that the domain of existence of the Laplace transform is the half-plane to the right of all finite poles of $\xi(s)$. For this case, one then gets by expanding $\xi(s)$ into the sum of a vector with polynomial elements and a vector with strictly proper elements that the inverse Laplace transforms of the strictly proper elements are zero for $t < 0$. The inverse transform of the polynomial elements contains impulses, doublets, and/or impulses of a higher order.

For digital systems and the λ-transform, causality considerations require that one choose the circular contour for the inversion integral with a sufficiently small radius so that $\xi(\lambda)$ is analytic on and within this contour. This is possible for arbitrary $i_T(\lambda)$ only if det $T(0) \neq 0$. Otherwise, $T^{-1}(\lambda)$ has a pole at $\lambda = 0$. When $u(\lambda)$ is analytic at $\lambda = 0$, det $T(0) \neq 0$ is also sufficient for the analyticity of $\xi(\lambda)$ at $\lambda = 0$. Hence, for digital systems det $T(0) \neq 0$ is assumed which is more stringent than det $T(\lambda) \not\equiv 0$.

The polynomial $\Delta = \mu \det T$, where μ is a constant selected to make Δ monic, is called the *characteristic polynomial* of the system. For a nondegenerate system, it easily follows that

$$\xi = T^{-1}Uu + T^{-1}i_T \tag{2.3}$$

and

$$y = Gu + VT^{-1}i_T + i_y, \tag{2.4}$$

where

$$G = VT^{-1}U + W \tag{2.5}$$

is the system *transfer matrix* whose elements are real rational functions of p.

Definition 2.1 A system is called stable when with $u \equiv 0$, ξ is a stable transform for all i_T.

Lemma 2.1 *A system is stable iff its characteristic polynomial is stable or, alternatively, Δ is Hurwitz in the analog case or Schur in the digital case.*

Proof With $u \equiv 0$, $\xi = T^{-1}i_T$ is stable for any polynomial vector i_T iff T^{-1} is stable. Since T is a polynomial matrix, T^{-1} is stable iff $\Delta = \mu \det T$ is stable.

It is important to note that when T^{-1} is stable, it follows from (2.5) that G is stable and $VT^{-1}i_T$ is stable. Hence, the transfer matrix of a stable system is stable and the output y is stable for all stable u since i_y is polynomial. However, a stable transfer matrix does not imply a stable system: cancelation of unstable poles of T^{-1} by the polynomial matrices U and V is possible.

As shown in Appendix B, Sect. B.11, the real rational transfer matrix $G(p)$ can be expressed, respectively, in terms of left and right coprime polynomial matrix pairs $A(p)$, $B(p)$ and $B_1(p)$, $A_1(p)$ that have real coefficients. Specifically, one can write

$$G = A^{-1}B = B_1A_1^{-1} \tag{2.6}$$

and, with η and η_1 constants chosen to make ψ_G monic,

$$\psi_G = \eta \det A = \eta_1 \det A_1 \tag{2.7}$$

is the *characteristic denominator* for G. The representations of G in (2.6) are called *coprime polynomial matrix fraction descriptions*. For these descriptions, when p_o is a zero of ψ_G of multiplicity m_o, then p_o is a pole of G with McMillan degree m_o (see Sect. B.6). That is,

$$\delta(G;p_o) = m_o. \tag{2.8}$$

It is clear from (2.5) since U, V, and W are polynomial matrices that p_o must be a zero of $\det T$ with at least the same multiplicity m_o. Moreover, it is possible for $\det T$ to have zeros, where ψ_G does not. Specifically,

$$h_G = \Delta/\psi_G \tag{2.9}$$

is a polynomial. The zeros of h_G are called the *hidden poles* of the system.

2.3 Stability Analysis of the Standard Configuration

The standard configuration (Fig. 1.2) treated here consists of an interconnection of a generalized plant represented by the polynomial matrix description

$$T_G\xi_G = U_1 e + U_2 v + i_{T_G} \tag{2.10}$$

$$y_c = V_1\xi_G + W_{11}e + W_{12}v + i_c \tag{2.11}$$

$$y_m = V_2\xi_G + W_{21}e + W_{22}v + i_m \tag{2.12}$$

with a controller represented by the polynomial matrix description

$$T_C\xi_C = U_C y_m + i_{T_C} \tag{2.13}$$

$$v = V_C\xi_C + W_C y_m + i_v. \tag{2.14}$$

Using (2.12) in (2.13) and (2.14) gives

$$T_C\xi_C = U_C V_2\xi_G + U_C W_{21}e + U_C W_{22}v + U_C i_m + i_{T_C} \tag{2.15}$$

$$v = V_C\xi_C + W_C V_2\xi_G + W_C W_{21}e + W_C W_{22}v + W_C i_m + i_v. \tag{2.16}$$

Equations (2.10) (2.15), and (2.16) are compactly described by the polynomial matrix description

$$\begin{bmatrix} T_G & 0 & -U_2 \\ -U_C V_2 & T_C & -U_C W_{22} \\ -W_C V_2 & -V_C & (I - W_C W_{22}) \end{bmatrix} \begin{bmatrix} \xi_G \\ \xi_C \\ v \end{bmatrix} = \begin{bmatrix} U_1 \\ U_C W_{21} \\ W_C W_{21} \end{bmatrix} e + \begin{bmatrix} i_{T_G} \\ U_C i_m + i_{T_C} \\ W_C i_m + i_v \end{bmatrix}. \tag{2.17}$$

Equation (2.17) together with (2.11) is a complete polynomial matrix description for the standard configuration. Hence, the system is stable iff $\det T$ is a stable polynomial, where

$$T = \begin{bmatrix} T_G & 0 & -U_2 \\ -U_C V_2 & T_C & -U_C W_{22} \\ -W_C V_2 & -V_C & (I - W_C W_{22}) \end{bmatrix}. \tag{2.18}$$

It is easy to confirm that the transfer matrix from input v to output y_m for the generalized plant is

$$G_{22} = V_2 T_G^{-1} U_2 + W_{22} \tag{2.19}$$

and that the transfer matrix from input y_m to output v for the controller is

$$C = V_C T_C^{-1} U_C + W_C. \tag{2.20}$$

Multiplying T on the right by the block triangular matrix

$$T_o = \begin{bmatrix} I & 0 & T_G^{-1} U_2 \\ 0 & I & T_C^{-1} U_C G_{22} \\ 0 & 0 & I \end{bmatrix} \tag{2.21}$$

then gives

$$\hat{T} = T T_o = \begin{bmatrix} T_G & 0 & 0 \\ -U_c V_2 & T_C & 0 \\ -W_C V_2 & -V_C & (I - C G_{22}) \end{bmatrix}. \tag{2.22}$$

Since

$$\det \hat{T} = (\det T)(\det T_o) = \det T, \tag{2.23}$$

it immediately follows that

$$\det T = (\det T_G)(\det T_C)\det(I - CG_{22}). \tag{2.24}$$

Hence, the standard configuration is stable iff $\det T$ is Hurwitz in the analog case and Schur in the digital case or, generically, iff $\det T$ is a stable polynomial.

It is important to recognize that CG_{22} is the transfer matrix of the loop closed on the generalized plant by the controller. The transfer matrices C and G_{22} are real rational matrices since the polynomial matrices in the polynomial matrix description for the generalized plant and controller have real coefficients. In addition, to within multiplicative constants, $\det T_G$ is the characteristic polynomial of the generalized plant and $\det T_C$ is the characteristic polynomial of the controller. It is also insightful to introduce the left and right coprime polynomial matrix fraction descriptions

$$C = A_C^{-1} B_C \tag{2.25}$$

and

$$G_{22} = -B_1 A_1^{-1}. \tag{2.26}$$

(The coprime polynomial matrix fraction description $A^{-1}B = B_1 A_1^{-1}$ is used to represent G, $-G_{22}$, or $-P_{22}$ in different sections of this book. It is always made clear which is intended so that no confusion arises.) Since C and G_{22} are real rational matrices, the polynomial matrices in (2.25) and (2.26) have real coefficients.

It now follows from (2.24) that

$$\det T = \left(\frac{\det T_G}{\det A_1}\right)\left(\frac{\det T_C}{\det A_C}\right)\det\left(A_C A_1 + B_C B_1\right). \tag{2.27}$$

Clearly,

$$\left(\frac{\det T_C}{\det A_C}\right) = \eta_C\left(\frac{\Delta_C}{\psi_C}\right) = \eta_C h_C, \tag{2.28}$$

where h_C is a polynomial whose zeros account for any hidden poles in the controller and η_C is a constant to account for the fact that Δ_C and ψ_C are monic polynomials.

In order to make further progress, one needs to recognize that the transfer matrix of the generalized plant follows from (2.10) through (2.12) with i_{T_G}, i_c, and i_m zero. Specifically,

$$y = \begin{bmatrix} y_c \\ y_m \end{bmatrix} = G \begin{bmatrix} e \\ v \end{bmatrix} = Gu, \tag{2.29}$$

where

$$G = \begin{bmatrix} G_{11} & G_{12} \\ G_{21} & G_{22} \end{bmatrix} = \begin{bmatrix} V_1 \\ V_2 \end{bmatrix} T_G^{-1} [U_1 \quad U_2] + \begin{bmatrix} W_{11} & W_{12} \\ W_{21} & W_{22} \end{bmatrix}. \tag{2.30}$$

Hence, by the same reasoning that led to (2.9),

$$\left(\frac{\det T_G}{\det A_1} \right) = \eta_G \left(\frac{\Delta_G}{\psi_{G_{22}}} \right) = \eta_G \left(\frac{h_G \psi_G}{\psi_{G_{22}}} \right), \tag{2.31}$$

where η_G is an appropriate constant; the polynomial h_G accounts for the hidden poles of the generalized plant; and ψ_G and $\psi_{G_{22}}$ are the characteristic denominators of the generalized plant transfer matrix G and its submatrix G_{22}, respectively. Since G_{22} is a submatrix of G, every pole of G_{22} has a McMillan degree which is less or equal to its McMillan degree as a pole of G (see Sect. B.7). That is,

$$h_{22} = \left(\frac{\psi_G}{\psi_{G_{22}}} \right) \tag{2.32}$$

is a polynomial and it follows from (2.27) and (2.28) that

$$\det T = \eta_C \eta_G h_C h_G h_{22} \det(A_C A_1 + B_C B_1). \tag{2.33}$$

Now the right-hand side of (2.33) is a product of polynomial factors and the choice of controller has no impact on h_G or h_{22}. So if either one of these polynomials or, equivalently, their product is not stable, the standard configuration is unstable no matter what one chooses for the controller. This leads immediately to the notion of admissibility.

Definition 2.2 The generalized plant in the standard configuration is said to be *admissible*

(i) when Δ_G/ψ_G and $\psi_G/\psi_{G_{22}}$ are stable polynomials or, equivalently,
(ii) their product $\Delta_G/\psi_{G_{22}}$ is a stable polynomial.

The two equivalent ways of defining admissibility in Definition 2.2 lead to different insights, which are otherwise not so easily derived from one another. These insights are particularly important when the stability and optimization of the 3DOF system is treated. Remarkably, it is possible to characterize all stabilizing controllers for an admissible generalized plant and this is done next using polynomial matrices with real coefficients.

It is first clear from (2.33) that the controller must be synthesized with no unstable hidden poles or, equivalently, h_C must be a stable polynomial. Next one must choose a left coprime real polynomial matrix pair A_C, B_C so that

$$A_C A_1 + B_C B_1 = L, \tag{2.34}$$

where L is a real polynomial matrix whose determinant is a stable polynomial. Since the pair B_1, A_1 is right coprime, there exists (see Sect. B.10) a real polynomial matrix pair X_1, Y_1 such that

$$X_1 A_1 + Y_1 B_1 = I. \tag{2.35}$$

Hence,

$$L X_1 A_1 + L Y_1 B_1 = L \tag{2.36}$$

and all polynomial matrix solution pairs to (2.34) are of the form

$$A_C = L X_1 + A_{C_o} \tag{2.37}$$

and

$$B_C = L Y_1 + B_{C_o}, \tag{2.38}$$

where A_{C_o} and B_{C_o} are real polynomial matrices satisfying

$$A_{C_o} A_1 + B_{C_o} B_1 = 0 \tag{2.39}$$

or

$$A_{C_o} = -B_{C_o} B_1 A_1^{-1} = B_{C_o} G_{22}. \tag{2.40}$$

One can now introduce the left coprime matrix fraction description

$$G_{22} = -A^{-1} B \tag{2.41}$$

for which there exist (see Lemma B.5) real polynomial matrices X, Y such that

$$A X + B Y = I. \tag{2.42}$$

Then from (2.40)

$$A_{C_o} = -B_{C_o}A^{-1}B \tag{2.43}$$

must be true. Premultiplying (2.42) by $B_{C_o}A^{-1}$ and using (2.43) gives

$$B_{C_o}X - A_{C_o}Y = B_{C_o}A^{-1}. \tag{2.44}$$

Since the left-hand side of (2.44) is a real polynomial matrix,

$$M = B_{C_o}A^{-1} \tag{2.45}$$

must be a real polynomial matrix or

$$B_{C_o} = MA, \tag{2.46}$$

where M is a real polynomial matrix. Using (2.46) in (2.43) immediately yields

$$A_{C_o} = -MB \tag{2.47}$$

which is also a real polynomial matrix as required. That is, all real polynomial matrix solution pairs (2.37) and (2.38) of (2.34) must be of the form

$$A_C = LX_1 - MB \tag{2.48}$$

and

$$B_C = LY_1 + MA. \tag{2.49}$$

Conversely, it easily follows from

$$BA_1 = AB_1 \tag{2.50}$$

that the pair (2.48) (2.49) is a solution to (2.34) for any real polynomial matrix M. However, the solution pair needed is one which is left coprime. It is now established that all such pairs are obtained from (2.48) and (2.49) by choosing for L, M a left coprime polynomial matrix pair.

It follows from Appendix B, Sect. B.10, that A_C, B_C is left coprime iff $[A_C \quad B_C]$ has row rank for all p. That is, A_C, B_C is left coprime iff for every p the only solution to

$$a[A_C \quad B_C] = a[(LX_1 - MB) \quad (LY_1 + MA)] = 0 \tag{2.51}$$

is $a = 0$. A useful identity which helps in the identification of the solution of (2.51) is obtained when (2.48) and (2.49) are substituted into (2.34). The result is

$$(LX_1 - MB)A_1 + (LY_1 + MA)B_1 = L. \tag{2.52}$$

Multiplying (2.52) on the left by a and using (2.51) then gives

$$aL = a[(LX_1 - MB) \quad (LY_1 + MA)]\begin{bmatrix} A_1 \\ B_1 \end{bmatrix} = 0 \tag{2.53}$$

Hence, (2.51) is equivalent to

$$[a(LX_1 - MB) \quad a(LY_1 + MA)] = aM[-B \quad A] = 0 \tag{2.54}$$

and it follows that

$$aM[-B \quad A]\begin{bmatrix} -Y \\ X \end{bmatrix} = aM(AX + BY) = aM = 0. \tag{2.55}$$

Combining (2.53) and (2.55) yields

$$a[L \quad M] = 0. \tag{2.56}$$

That is, any solution a of (2.51) must be a solution of (2.56). The converse is also true: if a satisfies (2.56), then $aL = 0$ and $aM = 0$ and it easily follows that a satisfies (2.51). So the pair A_C, B_C given by (2.48) and (2.49) is left coprime iff L, M is left coprime. One additional constraint on the pair L, M remains. The representation (2.25) requires that

$$\det A_C = \det (LX_1 - MB) \not\equiv 0. \tag{2.57}$$

It is now possible to state the key theorem concerning the stability of the standard configuration.

Theorem 2.1 *A standard configuration with given polynomial matrix description for the generalized plant and for the controller is stable iff the generalized plant is admissible, the controller is free of unstable hidden poles, and the controller transfer matrix can be written in the form*

$$C = (X_1 - KB)^{-1}(Y_1 + KA), \tag{2.58}$$

where K is a stable real rational matrix such that

$$\det(X_1 - KB) \not\equiv 0. \tag{2.59}$$

Proof The necessity for admissibility and no unstable hidden poles in the controller has already been pointed out. From (2.48) and (2.49), it follows that the controller transfer matrix must have the form

$$C = A_C^{-1}B_C = (LX_1 - MB)^{-1}(LY_1 + MA) = (X_1 - KB)^{-1}(Y_1 + KA), \tag{2.60}$$

where L is such that $\det L$ is a stable polynomial and $K = L^{-1}M$. Hence, K must be a stable real rational matrix. Moreover,

$$\det(LX_1 - MB) = (\det L)\det(X_1 - KB). \tag{2.61}$$

Hence, Eq. (2.59) is equivalent to (2.57) since $\det L \not\equiv 0$. The sufficiency of the stated conditions easily follows from the fact that any stable real rational matrix K satisfying (2.59) can be written as $L^{-1}M$, where L, M is left coprime; $\det L$ is a stable polynomial; and both L and M have real coefficients. Then one can substitute $K = L^{-1}M$ into (2.58) and (2.59) to confirm that (2.57) is satisfied and to make the identifications (2.48) and (2.49). It then follows from (2.33) that $\det T$ is Hurwitz in the analog case and Schur in the digital case or, equivalently, the system is stable.

There are a number of important additional observations to make before closing this section. The first concerns an alternative form to (2.58) for the controller transfer matrix. The second concerns an algebraic result relating to the requirement in connection with admissibility that $\psi_G/\psi_{G_{22}}$ be a stable polynomial. The third concerns showing that the stability results presented here hold irrespective of the polynomial matrix fraction description chosen for G_{22}. The presentation of these observations relies on an important identity, the generalized Bezout identity, assumed throughout this book and described in Lemma B.5 of Appendix B for the coprime polynomial matrix fraction descriptions of a rational matrix. Specifically, the real polynomial matrices in (2.26), (2.35), (2.41), and (2.42) can be chosen so that

$$\begin{bmatrix} A & B \\ Y_1 & -X_1 \end{bmatrix}\begin{bmatrix} X & B_1 \\ Y & -A_1 \end{bmatrix} = \begin{bmatrix} X & B_1 \\ Y & -A_1 \end{bmatrix}\begin{bmatrix} A & B \\ Y_1 & -X_1 \end{bmatrix} = \begin{bmatrix} I & 0 \\ 0 & I \end{bmatrix}. \tag{2.62}$$

It is now possible to make the first observation and it is stated in the form of

Lemma 2.2 *The controller transfer matrix given by (2.58) satisfies the identity*

$$C = (X_1 - KB)^{-1}(Y_1 + KA) = (Y + A_1K)(X - B_1K)^{-1}. \tag{2.63}$$

Proof From the second matrix product in (2.62), one gets

$$YA = A_1Y_1 \tag{2.64}$$

and

$$A_1X_1 + YB = I. \tag{2.65}$$

Solving (2.64) for Y_1, substituting the result into (2.58), and making use of (2.65) then leads to

$$C = [I - (Y + A_1K)B]^{-1}(Y + A_1K)A. \tag{2.66}$$

Invoking the easily verified matrix identity

$$(I - MN)^{-1}M = M(I - NM)^{-1} \tag{2.67}$$

now gives

$$C = (Y + A_1K)[I - B(Y + A_1K)]^{-1}A. \tag{2.68}$$

However,

$$I - B(Y + A_1K) = AX - BA_1K = AX - AB_1K = A(X - B_1K) \tag{2.69}$$

and (2.68) reduces to the final expression for C in (2.63) which completes the proof of the Lemma.

Attention is now directed to the second observation to be made. Specifically,

Lemma 2.3 *For the generalized plant transfer matrix in the standard configuration, $\psi_G/\psi_{G_{22}}$ is a stable polynomial iff AG_{21}, $G_{12}A_1$, and $G_{11} + G_{12}A_1Y_1G_{21}$ are stable.*

This Lemma plays an important role in the next chapter where its significance is best appreciated. Before giving the proof of the Lemma, the basis for its conjecture is presented. From (2.29), (2.30), and the fact that $v = Cy_m$ in Fig. 1.2, one gets

$$y_c = G_{11}e + G_{12}Cy_m \tag{2.70}$$

and

$$y_m = G_{21}e + G_{22}Cy_m. \tag{2.71}$$

Solving (2.71) for y_m and substituting the result into (2.70) gives $y_c = \tilde{G}e$, where

$$\tilde{G} = G_{11} + G_{12}C(I - G_{22}C)^{-1}G_{21} \tag{2.72}$$

is the transfer matrix for the standard configuration. Using the matrix identity (2.67) and the expression (2.58) for C then leads to

$$\tilde{G} = G_{11} + G_{12}[I + (X_1 - KB)^{-1}(Y_1 + KA)B_1A_1^{-1}]^{-1}(X_1 - KB)^{-1}(Y_1 + KA)G_{21} \tag{2.73}$$

or

$$\tilde{G} = G_{11} + G_{12}A_1[(X_1 - KB)A_1 + (Y_1 + KA)B_1]^{-1}(Y_1 + KA)G_{21}. \tag{2.74}$$

Using (2.35) and (2.50), one gets finally that

$$\tilde{G} = G_{11} + G_{12}A_1Y_1G_{21} + G_{12}A_1KAG_{21} = \tilde{G}_{11} + G_{12}A_1KAG_{21}. \tag{2.75}$$

The standard configuration and $\tilde{G}$ are stable for every choice of stable K when the generalized plant is admissible and the controller has no unstable hidden poles. In this case then the choice $K = 0$ indicates that $\tilde{G}_{11} = G_{11} + G_{12}A_1Y_1G_{21}$ must be stable. It is also true when the standard configuration is stable that the transfer matrices from e to y_m and from e to v must be stable. The former transfer matrix is given by

$$\tilde{G}_m = (X - B_1K)AG_{21} \tag{2.76}$$

and the latter by

$$\tilde{G}_v = (Y + A_1 K) A G_{21}. \tag{2.77}$$

Hence,

$$A\tilde{G}_m + B\tilde{G}_v = AG_{21} \tag{2.78}$$

must be stable. Since the stability of the polynomial $\psi_G/\psi_{G_{22}}$ is equivalent to the stability of the polynomial $\psi_{G'}\Big/\psi_{G'_{22}}$, it follows as shown in the proof below that $G_{12}A_1$ must also be stable. Moreover, the only aspect of the admissibility of the generalized plant that impacts on $\tilde{G}$ is the stability of the polynomial $\psi_G/\psi_{G_{22}}$, and it should now be clear why one might conjecture Lemma 2.3 for which a direct algebraic proof is now given. In the proof, extensive use is made of several properties (see Sect. B.7) of the McMillan degree.

Proof The polynomial $\psi_G/\psi_{G_{22}}$ is stable iff the unstable poles of G_{22} and G coincide and have identical McMillan degrees. Since

$$G_2 = \begin{bmatrix} G_{21} & G_{22} \end{bmatrix} \tag{2.79}$$

is a submatrix of G, the McMillan degree of any pole of G_2 is equal to or lower than its McMillan degree as a pole of G. Similarly, since G_{22} is a submatrix of G_2, the McMillan degree of any pole of G_{22} is equal to or lower than its McMillan degree as a pole of G_2. But the unstable poles of G_{22} and G coincide and have identical McMillan degrees. Hence, the same is true for G_2 and G_{22}. That is, $\psi_{G_2}/\psi_{G_{22}}$ must be a stable polynomial.

Now one can always write $G_2 = A_2^{-1}B_2$ where A_2, B_2, is a left coprime polynomial matrix pair. Premultiplying (2.79) by A_2 then gives that

$$B_2 = \begin{bmatrix} A_2G_{21} & A_2G_{22} \end{bmatrix} = \begin{bmatrix} A_2G_{21} & -A_2A^{-1}B \end{bmatrix}. \tag{2.80}$$

Hence, $A_2A^{-1}B$ must be a stable matrix. But this is the case iff $H_2 = A_2A^{-1}$ is stable since A, B is left coprime and A_2 is a polynomial matrix (see Appendix B, Lemma B.8). Moreover,

$$(\det H_2)^{-1} = \left(\frac{\det A}{\det A_2}\right) = \eta\psi_{G_{22}}\psi_{G_2}^{-1} \tag{2.81}$$

is stable since η is simply a constant to account for the fact that $\det A$ and $\det A_2$ may not be monic. So

$$H_2^{-1} = \left(\frac{\text{adj } H_2}{\det H_2}\right) = AA_2^{-1} \tag{2.82}$$

must be stable. Premultiplying (2.79) by A then gives that

$$AA_2^{-1}B_2 = [AG_{21} \quad AG_{22}] = [AG_{21} \quad -B] \tag{2.83}$$

must be a stable matrix. Thus, AG_{21} must be stable. Clearly, these same arguments can be applied to

$$G' = \begin{bmatrix} G'_{11} & G'_{21} \\ G'_{12} & G'_{22} \end{bmatrix} = \begin{bmatrix} G'_{11} & G'_{21} \\ G'_{12} & -(A'_1)^{-1}B'_1 \end{bmatrix} \tag{2.84}$$

to conclude that $A'_1G'_{12}$ or, equivalently, $G_{12}A_1$ must be stable.

Now with regard to $\tilde{G}_{11}$ it is natural to post-multiply G by

$$\Gamma = \begin{bmatrix} I & 0 \\ A_1Y_1G_{21} & I \end{bmatrix} \tag{2.85}$$

to get

$$G_o = G\Gamma = \begin{bmatrix} \tilde{G}_{11} & G_{12} \\ \tilde{G}_{21} & G_{22} \end{bmatrix}, \tag{2.86}$$

where

$$\tilde{G}_{21} = G_{21} + G_{22}A_1Y_1G_{21}. \tag{2.87}$$

The matrix

$$\Gamma_{21} = A_1Y_1G_{21} = A_1Y_1A^{-1}AG_{21} \tag{2.88}$$

is stable because AG_{21} must be stable and because it follows from the fact that

$$A_1Y_1A^{-1}B = YB \tag{2.89}$$

is stable that $A_1 Y_1 A^{-1}$ is stable (A, B is left coprime and $A_1 Y_1$ is a polynomial matrix). So Γ and

$$\Gamma^{-1} = \begin{bmatrix} I & 0 \\ -A_1 Y_1 G_{21} & I \end{bmatrix} \tag{2.90}$$

are both stable and one gets from (2.86) that the unstable poles of G and G_o coincide and have identical McMillan degrees (see Example B.3). Another important observation to make is that (2.87) is equivalent to

$$\tilde{G}_{21} = (I - B_1 Y_1) G_{21} = (I - B_1 Y_1) A^{-1} A G_{21}. \tag{2.91}$$

Now $(I - B_1 Y_1) A^{-1}$ stable follows from

$$B_1 X_1 = B_1 (I - Y_1 B_1) A_1^{-1} = (I - B_1 Y_1) B_1 A_1^{-1} = (I - B_1 Y_1) A^{-1} B \tag{2.92}$$

since $B_1 X_1$ is stable, $(I - B_1 Y_1)$ is polynomial, and A, B is left coprime. This fact and the fact that $A G_{21}$ is stable yields that $\tilde{G}_{21}$ must be stable. So from

$$G_o = \begin{bmatrix} 0 & 0 \\ \tilde{G}_{21} & 0 \end{bmatrix} + \begin{bmatrix} \tilde{G}_{11} & G_{12} \\ 0 & G_{22} \end{bmatrix} = G_a + G_b \tag{2.93}$$

it follows that the unstable poles of G, G_o, and

$$G_b = \begin{bmatrix} \tilde{G}_{11} & G_{12} \\ 0 & G_{22} \end{bmatrix} \tag{2.94}$$

coincide and have identical McMillan degrees. Since the same is required of G and G_{22}, the unstable poles of G_b and G_{22} must coincide and have identical McMillan degrees. Now it is always possible to construct a submatrix of G_b given by

$$S_b = \begin{bmatrix} g & h' \\ 0 & S_{22} \end{bmatrix}, \tag{2.95}$$

where g is one of the elements of $\tilde{G}_{11}$, h' contains elements from G_{12} which are in the same row of G_b as g, and S_{22} is a square submatrix of G_{22} whose columns are the same as those from which the elements of h' are taken. Then

$$\det S_b = g \det S_{22} \tag{2.96}$$

is a minor of G_b. The matrix S_{22} can always be selected so that $\det S_{22}$ possesses any particular unstable pole of G_{22} with multiplicity equal to its McMillan degree as a pole of G_{22}. Since S_b is a submatrix of G_b and it is required that G_b and G_{22} possess this same pole with equal McMillan degree it follows from (2.96) that g must be stable. Now g can be any one of the elements of $\tilde{G}_{11}$. Hence, $\tilde{G}_{11}$ must be stable. This completes the proof for the necessity of AG_{21}, $G_{12}A_1$, and $\tilde{G}_{11}$ stable in Lemma 2.3.

With regard to sufficiency, it is first noted that when AG_{21} is stable one can again conclude as done immediately following (2.94) that the unstable poles of G and G_{22} coincide and have identical McMillan degrees when the same is true of G_b and G_{22}. When $\tilde{G}_{11}$ is also stable, it then follows from

$$G_b = \begin{bmatrix} \tilde{G}_{11} & 0 \\ 0 & 0 \end{bmatrix} + \begin{bmatrix} 0 & G_{12} \\ 0 & G_{22} \end{bmatrix} \tag{2.97}$$

that all one needs to show is that the poles of

$$G_c = \begin{bmatrix} G_{12} \\ G_{22} \end{bmatrix} \tag{2.98}$$

and G_{22} coincide and have identical McMillan degrees. When $G_{12}A_1$ is stable, it follows from (2.98) that

$$G_d = G_c A_1 = \begin{bmatrix} G_{12}A_1 \\ G_{22}A_1 \end{bmatrix} = \begin{bmatrix} G_{12}A_1 \\ -B_1 \end{bmatrix} \tag{2.99}$$

is stable. Moreover, from

$$G_c = G_d A_1^{-1} \tag{2.100}$$

one gets for every unstable pole p_o of G_c that

$$\delta(G_c; p_o) \leq \delta(G_d; p_o) + \delta(A_1^{-1}; p_o). \tag{2.101}$$

Now

$$\delta(G_d; p_o) = 0 \tag{2.102}$$

since G_d is stable and

$$\delta(A_1^{-1}; p_o) = \delta(-B_1 A_1^{-1}; p_o) = \delta(G_{22}; p_o) \tag{2.103}$$

since B_1, A_1 are right coprime (see Lemma B.7 and Remark B.8). Thus,

$$\delta(G_c; p_o) \leq \delta(G_{22}; p_o). \tag{2.104}$$

On the other hand, since G_{22} is a submatrix of G_c,

$$\delta(G_c; p_o) \geq \delta(G_{22}; p_o). \tag{2.105}$$

Hence,

$$\delta(G_c; p_o) = \delta(G_{22}; p_o) \tag{2.106}$$

for every unstable pole p_o of G_c and, therefore, the unstable poles of G_c and G_{22} coincide and have identical McMillan degrees. This completes the proof of Lemma 2.3.

The final observation to be made concerns the fact that Theorem 2.1 holds irrespective of the coprime polynomial matrix fraction descriptions used for G_{22}. This is important because such descriptions are only unique to within the choice of certain arbitrary unimodular matrices. Indeed, as established in Lemma B.6 of Appendix B,

$$-G_{22} = A^{-1}B = B_1A_1^{-1} \tag{2.107}$$

and

$$-G_{22} = \tilde{A}^{-1}\tilde{B} = \tilde{B}_1\tilde{A}_1^{-1} \tag{2.108}$$

are both coprime polynomial matrix fraction descriptions of $-G_{22}$ iff

$$\tilde{A} = NA, \tilde{B} = NB \tag{2.109}$$

and

$$\tilde{A}_1 = A_1N_1, \tilde{B}_1 = B_1N_1, \tag{2.110}$$

where N and N_1 are unimodular polynomial matrices. Moreover, there exist polynomial matrices $\tilde{X}_1$, $\tilde{Y}_1$ such that

$$\tilde{X}_1\tilde{A}_1 + \tilde{Y}_1\tilde{B}_1 = I. \tag{2.111}$$

Premultiplying (2.111) by N_1 then leads to

$$N_1\tilde{X}_1A_1 + N_1\tilde{Y}_1B_1 = I. \tag{2.112}$$

Subtracting (2.35) from (2.112) now yields

$$(N_1\tilde{X}_1 - X_1)A_1 + (N_1\tilde{Y}_1 - Y_1)B_1 = 0. \tag{2.113}$$

Equation (2.113) is identical in form to (2.39) and so the same reasoning that led to (2.46) and (2.47) gives

$$(N_1\tilde{X}_1 - X_1) = -MB, (N_1\tilde{Y}_1 - Y_1) = MA, \tag{2.114}$$

where M is an arbitrary polynomial matrix. Now when one uses (2.108) and (2.111) in Theorem 2.1, the formula for the stabilizing controller transfer matrix becomes

$$\tilde{C} = (\tilde{X}_1 - \tilde{K}\tilde{B})^{-1}(\tilde{Y}_1 + \tilde{K}\tilde{A}). \tag{2.115}$$

Using (2.109) and the solutions for $\tilde{X}_1$ and $\tilde{Y}_1$ from (2.114) in (2.115) and setting

$$\tilde{K} = N_1^{-1}(K - M)N^{-1} \tag{2.116}$$

leads to

$$\tilde{C} = (X_1 - KB)^{-1}(Y_1 + KA) \tag{2.117}$$

which is identical in form to (2.58). Moreover, since N and N_1 are unimodular, for every stable K one gets a stable $\tilde{K}$ in (2.116) and conversely for every stable $\tilde{K}$ in (2.116) the solution for K is

$$K = M + N_1\tilde{K}N \tag{2.118}$$

a stable matrix. So the set of all stabilizing controllers given by (2.58) as K ranges over all stable real rational matrices is identical to the set of all controllers given by (2.115) as $\tilde{K}$ ranges over all stable real rational matrices. Because of the importance of this observation, it is incorporated into

Lemma 2.4 *The set of all stabilizing controllers given by (2.58) as K ranges over all stable real rational matrices is invariant with respect to the choice of the real coprime polynomial matrix fraction descriptions used for* G_{22}.

2.4 Stability Analysis of the 2DOF Standard Configuration

The 2DOF standard configuration is shown in Fig. 1.3. It is important to recall that the block T_d is not an actual component in the system and, therefore, plays no role with regard to stability of the interconnected physical components. Moreover, the physical system can be represented by the standard configuration in Fig. 1.2 when the identifications

$$e = \begin{bmatrix} d' & r' & n' \end{bmatrix}', \tag{2.119}$$

$$y_c = \begin{bmatrix} y_s' & y_t' \end{bmatrix}', \tag{2.120}$$

and

$$y_m = \begin{bmatrix} y' & u' \end{bmatrix}' \tag{2.121}$$

are made. In particular, when

$$T_G\xi_G = U_1 d + U_2 v + i_{T_G}, \tag{2.122}$$

$$\begin{bmatrix} y_s' & y_t' \end{bmatrix}' = V_1\xi_G + W_{11} d + W_{12} v + i_c, \tag{2.123}$$

and

$$y = V_2\xi_G + W_{21} d + W_{22} v + i_y \tag{2.124}$$

is a polynomial matrix description for the generalized plant in Fig. 1.3, it follows that

$$T_G\xi_G = \begin{bmatrix} U_1 & 0 & 0 \end{bmatrix} e + U_2 v + i_{T_G}, \tag{2.125}$$

$$y_c = V_1\xi_G + \begin{bmatrix} W_{11} & 0 & 0 \end{bmatrix} e + W_{12} v + i_c, \tag{2.126}$$

and

$$y_m = \begin{bmatrix} V_2 \\ 0 \end{bmatrix}\xi_G + \begin{bmatrix} W_{21} & 0 & 0 \\ 0 & I & I \end{bmatrix} e + \begin{bmatrix} W_{22} \\ 0 \end{bmatrix} v + \begin{bmatrix} i_y \\ 0 \end{bmatrix} \tag{2.127}$$

is the polynomial matrix description of the generalized plant in Fig. 1.2 for the 2DOF standard configuration. Hence, it is clear from a comparison of (2.10) through (2.12) with (2.125) through (2.127) that the stability results obtained for the standard configuration apply here when

$$\tilde{U}_1 = [\, U_1 \quad 0 \quad 0\,], \tag{2.128}$$

$$\tilde{W}_{11} = [\, W_{11} \quad 0 \quad 0\,], \tag{2.129}$$

$$\tilde{V}_2 = [\, V_2' \quad 0'\,]', \tag{2.130}$$

$$\tilde{W}_{21} = \begin{bmatrix} W_{21} & 0 & 0 \\ 0 & I & I \end{bmatrix}, \tag{2.131}$$

and

$$\tilde{W}_{22} = [\, W_{22}' \quad 0'\,]' \tag{2.132}$$

are used instead of U_1, W_{11}, V_2, W_{21}, and W_{22} in (2.10) through (2.12). It then follows that in place of the G_{ij} in (2.30) one gets

$$\tilde{G}_{11} = V_1 T_G^{-1} \tilde{U}_1 + \tilde{W}_{11} = [\, G_{11} \quad 0 \quad 0\,], \tag{2.133}$$

$$\tilde{G}_{12} = V_1 T_G^{-1} U_2 + W_{12} = G_{12}, \tag{2.134}$$

$$\tilde{G}_{21} = \tilde{V}_2 T_G^{-1} \tilde{U}_1 + \tilde{W}_{21} = \begin{bmatrix} G_{21} & 0 & 0 \\ 0 & I & I \end{bmatrix}, \tag{2.135}$$

and

$$\tilde{G}_{22} = \tilde{V}_2 T_G^{-1} U_2 + \tilde{W}_{22} = \begin{bmatrix} G_{22} \\ 0 \end{bmatrix}, \tag{2.136}$$

where

$$G_{11} = V_1 T_G^{-1} U_1 + W_{11}, \tag{2.137}$$

$$G_{21} = V_2 T_G^{-1} U_1 + W_{21}, \tag{2.138}$$

and

$$G_{22} = V_2 T_G^{-1} U_2 + W_{22}. \tag{2.139}$$

Hopefully, there will be no confusion between the G_{ij} introduced here and those used in Sect. 2.3. This risk is taken in order to avoid a profusion of symbols. In fact, the G_{ij} introduced here are transfer matrices associated with (2.122) through (2.124), the polynomial matrix description of the generalized plant in the 2DOF standard configuration. The G_{ij} introduced in Sect. 2.3 are transfer matrices associated with (2.10) through (2.12), the polynomial matrix description of the generalized plant in the standard configuration; hence, the symbol $\tilde{G}_{ij}$ is used here to represent them (see Problem 1.2). It is especially important to recognize that G_{22} represents the transfer matrix between v and y_m in the standard configuration of Sect. 2.3 and between v and y in the 2DOF standard configuration treated here. The two are connected by (2.136).

It is clear from (2.136) and Example B.2 that $\delta(\tilde{G}_{22};p_o) = \delta(G_{22};p_o)$ for any pole p_o; hence,

$$\psi_{\tilde{G}_{22}} = \psi_{G_{22}}. \tag{2.140}$$

It also follows from

$$\tilde{G} = \begin{bmatrix} \tilde{G}_{11} & \tilde{G}_{12} \\ \tilde{G}_{21} & \tilde{G}_{22} \end{bmatrix} = \begin{bmatrix} G_{11} & 0 & 0 & G_{12} \\ G_{21} & 0 & 0 & G_{22} \\ 0 & I & I & 0 \end{bmatrix} \tag{2.141}$$

that

$$\tilde{G}\Gamma = \begin{bmatrix} G_{11} & G_{12} & 0 & 0 \\ G_{21} & G_{22} & 0 & 0 \\ 0 & 0 & I & I \end{bmatrix} = \begin{bmatrix} G & 0 & 0 \\ 0 & I & I \end{bmatrix}, \tag{2.142}$$

where

$$\Gamma = \begin{bmatrix} I & 0 & 0 & 0 \\ 0 & 0 & I & 0 \\ 0 & 0 & 0 & I \\ 0 & I & 0 & 0 \end{bmatrix} \tag{2.143}$$

is nonsingular. Hence, when Example B.2 and Property B.5 are recalled, it is easy to see that

$$\psi_{\tilde{G}} = \psi_{\tilde{G}\Gamma} = \psi_G. \tag{2.144}$$

Thus, $\psi_{\tilde{G}_{22}}\psi_{\tilde{G}}^{-1}$ is stable iff $\psi_{G_{22}}\psi_G^{-1}$ is stable. Also, from (2.31),

$$h_{\tilde{G}} = \frac{\Delta_{\tilde{G}}}{\psi_{\tilde{G}}} = \frac{\eta \det T_G}{\psi_G} = \frac{\Delta_G}{\psi_G} = h_G, \tag{2.145}$$

where η is a constant chosen so that $\eta \det T_G$ is monic. In analogy with Definition 2.2, it is now natural to introduce

Definition 2.3 The generalized plant in the 2DOF standard configuration is said to be *admissible*

(i) when Δ_G/ψ_G and $\psi_G/\psi_{G_{22}}$ are stable polynomials or, equivalently,
(ii) their product $\Delta_G/\psi_{G_{22}}$ is a stable polynomial.

Further progress requires determination of the connection between coprime polynomial matrix fraction descriptions for $\tilde{G}_{22}$ and G_{22}. Specifically, one can always introduce

$$G_{22} = -A^{-1}B = -B_1A_1^{-1} \tag{2.146}$$

along with the equations

$$AX + BY = I \tag{2.147}$$

and

$$X_1A_1 + Y_1B_1 = I. \tag{2.148}$$

It then follows from (2.136) that

$$\tilde{G}_{22} = -\begin{bmatrix} A & 0 \\ 0 & I \end{bmatrix}^{-1} \begin{bmatrix} B \\ 0 \end{bmatrix} = -\begin{bmatrix} B_1 \\ 0 \end{bmatrix} A_1^{-1} \tag{2.149}$$

and, moreover,

$$\begin{bmatrix} A & 0 \\ 0 & I \end{bmatrix}\begin{bmatrix} X & 0 \\ 0 & I \end{bmatrix} + \begin{bmatrix} B \\ 0 \end{bmatrix}[Y \quad 0] = \begin{bmatrix} I & 0 \\ 0 & I \end{bmatrix} \tag{2.150}$$

and

$$X_1A_1 + [Y_1 \quad 0]\begin{bmatrix} B_1 \\ 0 \end{bmatrix} = I. \tag{2.151}$$

Comparing (2.149) through (2.151) with (2.26), (2.35), (2.41), and (2.42) now indicates that A, B, X, Y, A_1, B_1, X_1, and Y_1 in the latter group are replaced by

$$\tilde{A} = \begin{bmatrix} A & 0 \\ 0 & I \end{bmatrix}, \quad \tilde{B} = \begin{bmatrix} B \\ 0 \end{bmatrix} \tag{2.152}$$

$$\tilde{X} = \begin{bmatrix} X & 0 \\ 0 & I \end{bmatrix}, \quad \tilde{Y} = [\,Y \quad 0\,] \tag{2.153}$$

$$\tilde{A}_1 = A_1, \tilde{B}_1 = [\,B_1' \quad 0'\,]' \tag{2.154}$$

and

$$\tilde{X}_1 = X_1, \tilde{Y}_1 = [\,Y_1 \quad 0\,]\,. \tag{2.155}$$

Using these replacements and $\tilde{K}$ for K in (2.58) then gives

$$C = \left(X_1 - \tilde{K}\begin{bmatrix} B \\ 0 \end{bmatrix}\right)^{-1}\left([\,Y_1 \quad 0\,] + \tilde{K}\begin{bmatrix} A & 0 \\ 0 & I \end{bmatrix}\right). \tag{2.156}$$

It follows with

$$\tilde{K} = [\,K \quad H\,] \tag{2.157}$$

that

$$C = (X_1 - KB)^{-1}[\,(Y_1 + KA) \quad H\,]. \tag{2.158}$$

Comparing (1.5) and (2.158) immediately yields

$$C_u = (X_1 - KB)^{-1}H \tag{2.159}$$

and

$$C_y = -\,(X_1 - KB)^{-1}(Y_1 + KA). \tag{2.160}$$

So in analogy with Theorem 2.1 one can state without any additional proof

Theorem 2.2 *A 2DOF standard configuration with given polynomial matrix description for the generalized plant and for the controller is stable iff the generalized plant is admissible, the controller is free of unstable hidden poles, and the controller transfer matrix can be written in the form*

$$C = [\,-C_y \quad C_u\,] = (X_1 - KB)^{-1}[\,(Y_1 + KA) \quad H\,], \tag{2.161}$$

where K and H are stable real rational matrices and K is such that

$$\det(X_1 - KB) \not\equiv 0. \tag{2.162}$$

It is important to recognize that since the polynomial matrices in (2.146) through (2.148) are assumed chosen so that the generalized Bezout identity (2.62) is satisfied, one gets from (2.152) through (2.155) that

$$\begin{bmatrix} \tilde{A} & \tilde{B} \\ \tilde{Y}_1 & -\tilde{X}_1 \end{bmatrix}\begin{bmatrix} \tilde{X} & \tilde{B}_1 \\ \tilde{Y} & -\tilde{A}_1 \end{bmatrix} = \begin{bmatrix} \tilde{X} & \tilde{B}_1 \\ \tilde{Y} & -\tilde{A}_1 \end{bmatrix}\begin{bmatrix} \tilde{A} & \tilde{B} \\ \tilde{Y}_1 & -\tilde{X}_1 \end{bmatrix} = \begin{bmatrix} I & 0 \\ 0 & I \end{bmatrix}. \tag{2.163}$$

Indeed, that the first of the matrix products is an identity matrix easily follows once it is recognized that

$$\tilde{Y}_1\tilde{X} - \tilde{X}_1\tilde{Y} = [Y_1 \quad 0]\begin{bmatrix} X & 0 \\ 0 & I \end{bmatrix} - X_1[Y \quad 0] = [(Y_1X - X_1Y) \quad 0] = 0. \tag{2.164}$$

That the second matrix product is an identity matrix is an immediate consequence of the fact that $MN = I$ implies $NM = I$ for any two square matrices M, N. Hence, with the appropriate replacements, Lemmas 2.2, 2.3, and 2.4 are also applicable to the 2DOF standard configuration. The alternative formula for C that one can get from (2.63) is, however, not very useful in comparison with the one given by (2.161); therefore, it is not presented here.

The replacements for the three key matrices in Lemma 2.3 are

$$\tilde{A}\tilde{G}_{21} = \begin{bmatrix} A & 0 \\ 0 & I \end{bmatrix}\begin{bmatrix} G_{21} & 0 & 0 \\ 0 & I & I \end{bmatrix} = \begin{bmatrix} AG_{21} & 0 & 0 \\ 0 & I & I \end{bmatrix}, \tag{2.165}$$

$$\tilde{G}_{12}\tilde{A}_1 = G_{12}A_1, \tag{2.166}$$

and

$$\begin{aligned} \tilde{G}_{11} + \tilde{G}_{12}\tilde{A}_1\tilde{Y}_1\tilde{G}_{21} &= [G_{11} \quad 0 \quad 0] + G_{12}A_1[Y_1 \quad 0]\begin{bmatrix} G_{21} & 0 & 0 \\ 0 & I & I \end{bmatrix} \\ &= [(G_{11} + G_{12}A_1Y_1G_{21}) \quad 0 \quad 0]. \end{aligned} \tag{2.167}$$

Clearly, the left-hand sides of (2.165) through (2.167) are stable iff AG_{21}, $G_{12}A_1$, and $(G_{11} + G_{12}A_1Y_1G_{21})$ are stable. Thus,

Lemma 2.5 *For the generalized plant transfer matrix in the 2DOF standard configuration,* $\psi_G/\psi_{G_{22}}$ *is a stable polynomial iff* AG_{21}, $G_{12}A_1$, *and* $G_{11} + G_{12}A_1Y_1G_{21}$ *are stable.*

It should also be clear that the appropriate replacement for Lemma 2.4 is

Lemma 2.6 *The set of all stabilizing controllers given by (2.161) as* K *and* H *range over all stable real rational matrices is invariant with respect to the choice of coprime polynomial matrix fraction descriptions used for* G_{22}.

2.5 Stability Analysis of the 3DOF System

Most feedback control system implementations have the structure of the 3DOF system or some simplification of it (i.e., the 1DOF or the 2DOF system). So it is of interest and insightful to specialize the results for the 2DOF standard configuration to that of the 3DOF system. When the 3DOF system shown in Fig. 1.1 is viewed as a particular implementation of the 2DOF standard configuration shown in Fig. 1.3, the generalized plant takes the form shown in Fig. 2.1.

The blocks P_a and P_p, with

$$P = P_p P_a, \tag{2.168}$$

are not shown explicitly in Fig. 1.1, but are included here so that the outputs from the actuator block P_a can be associated with the vector y_s: these outputs are typically the internal variables of the physical plant that need to be kept small in order to minimize the control effort required and in order to assure that the linear model remains valid. The physical plant output y in Fig. 1.1 is designated by y_t in Fig. 2.1 since it is in fact the output that is to be tracked. Moreover, this is best done to avoid confusion with the measurement vector

$$y = [w' \quad z']'. \tag{2.169}$$

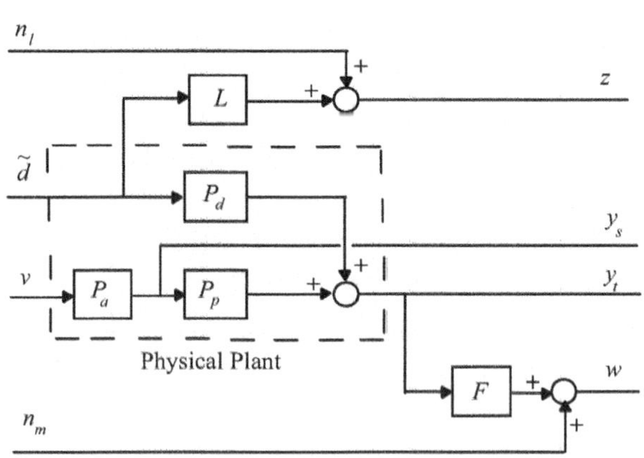

Fig. 2.1 Generalized plant for the 3DOF system

Similarly, the disturbance input d is denoted by $\tilde{d}$ in Fig. 2.1 so as to avoid confusion with the exogenous input vector

$$d = [\tilde{d}' \quad n_l' \quad n_m']' \tag{2.170}$$

in Fig. 1.3.

The objective here is to find the expressions for T_G and G so that Definition 2.3, Theorem 2.2, and Lemma 2.5 can be specialized to the 3DOF system. Since the quantities sought do not in any way depend on the initial conditions or initial data strings, for simplicity these are taken to be zero. The first step is to introduce a polynomial matrix description for each subsystem in Fig. 2.1. Specifically,

$$T_a \xi_a = U_a v \tag{2.171}$$

$$y_s = V_a \xi_a + W_a v \tag{2.172}$$

is used to represent the actuators and

$$T_p \xi_p = [U_{p_1} \quad 0 \quad 0] d + U_{p_2} y_s \tag{2.173}$$

$$y_t = V_p \xi_p + [W_{p_1} \quad 0 \quad 0] d + W_{p_2} y_s \tag{2.174}$$

is used to model the balance of the physical plant, where

$$P_a = V_a T_a^{-1} U_a + W_a, \tag{2.175}$$

$$P_d = V_p T_p^{-1} U_{p_1} + W_{p_1}, \tag{2.176}$$

and

$$P_p = V_p T_p^{-1} U_{p_2} + W_{p_2}. \tag{2.177}$$

Similarly,

$$T_l \xi_l = [U_l \quad 0 \quad 0] d \tag{2.178}$$

$$z = V_l \xi_l + [W_l \quad I \quad 0] d \tag{2.179}$$

with

$$L = V_l T_l^{-1} U_l + W_l \tag{2.180}$$

is used for the feedforward transducer and

$$T_f \xi_f = U_f y_t \tag{2.181}$$

$$w = V_f \xi_f + W_f y_t + [0 \quad 0 \quad I] d \tag{2.182}$$

with

$$F = V_f T_f^{-1} U_f + W_f \tag{2.183}$$

is used for the feedback transducer.

Using (2.172) in (2.173) and (2.174) yields

$$T_p \xi_p - U_{p_2} V_a \xi_a = [U_{p_1} \quad 0 \quad 0] d + U_{p_2} W_a v \tag{2.184}$$

$$y_t = V_p \xi_p + W_{p_2} V_a \xi_a + [W_{p_1} \quad 0 \quad 0] d + W_{p_2} W_a v\,. \tag{2.185}$$

Using (2.185) in (2.181) and (2.182) then gives

$$T_f \xi_f - U_f V_p \xi_p - U_f W_{p_2} V_a \xi_a = U_f [W_{p_1} \quad 0 \quad 0] d + U_f W_{p_2} W_a v, \tag{2.186}$$

$$w = V_f \xi_f + W_f V_p \xi_p + W_f W_{p_2} V_a \xi_a + [W_f W_{p_1} \quad 0 \quad I] d + W_f W_{p_2} W_a v\,. \tag{2.187}$$

Hence, for the generalized plant in Fig. 2.1, Eq. (2.122) holds with

$$\xi_G = [\xi_a' \quad \xi_p' \quad \xi_f' \quad \xi_l']' \tag{2.188}$$

and

$$T_G = \begin{bmatrix} T_a & 0 & 0 & 0 \\ -U_{p_2} V_a & T_p & 0 & 0 \\ -U_f W_{p_2} V_a & -U_f V_p & T_f & 0 \\ 0 & 0 & 0 & T_l \end{bmatrix}. \tag{2.189}$$

The remaining coefficient matrices in (2.122) through (2.124) can also be identified from the above equations, but this is not necessary to complete the task at hand and is omitted for brevity. Clearly, for the 3DOF system,

$$\Delta_G = \eta \det T_G = \eta (\det T_a)(\det T_p)(\det T_f)(\det T_l). \tag{2.190}$$

The remaining item to be determined is the generalized plant transfer matrix. This can be done directly from Fig. 2.1. Clearly,

$$\begin{bmatrix} y_s \\ y_t \end{bmatrix} = \begin{bmatrix} 0 & 0 & 0 \\ P_d & 0 & 0 \end{bmatrix} d + \begin{bmatrix} P_a \\ P \end{bmatrix} v \tag{2.191}$$

and

$$y = \begin{bmatrix} w \\ z \end{bmatrix} = \begin{bmatrix} FP_d & 0 & I \\ L & I & 0 \end{bmatrix} d + \begin{bmatrix} FP \\ 0 \end{bmatrix} v \tag{2.192}$$

So

$$G = \left[\begin{array}{c|c} G_{11} & G_{12} \\ \hline G_{21} & G_{22} \end{array}\right] = \left[\begin{array}{ccc|c} 0 & 0 & 0 & P_a \\ P_d & 0 & 0 & P \\ \hline FP_d & 0 & I & FP \\ L & I & 0 & 0 \end{array}\right] \tag{2.193}$$

and

$$\psi_{G_{22}} = \psi_{FP}. \tag{2.194}$$

Hence, in place of Definition 2.3 one has

Definition 2.4 The 3DOF system is said to be *admissible (i)* when

$$h_G = \frac{\Delta_G}{\psi_G} = \frac{\eta(\det T_a)(\det T_p)(\det T_f)(\det T_l)}{\psi_G} \tag{2.195}$$

and ψ_G/ψ_{FP} are stable polynomials or, equivalently, *(ii)* their product

$$\frac{\Delta_G}{\psi_{FP}} = \frac{\eta(\det T_a)(\det T_p)(\det T_f)(\det T_l)}{\psi_{FP}} \tag{2.196}$$

is a stable polynomial.

Now one can introduce coprime polynomial matrix fraction descriptions for FP satisfying the generalized Bezout identity (2.62) such that

$$FP = A^{-1}B = B_1 A_1^{-1}. \tag{2.197}$$

Then

$$G_{22}=\begin{bmatrix}FP\\0\end{bmatrix}=-\begin{bmatrix}A&0\\0&I\end{bmatrix}^{-1}\begin{bmatrix}-B\\0\end{bmatrix}=-\begin{bmatrix}-B_1\\0\end{bmatrix}A_1^{-1},\tag{2.198}$$

$$\begin{bmatrix}A&0\\0&I\end{bmatrix}\begin{bmatrix}X&0\\0&I\end{bmatrix}+\begin{bmatrix}-B\\0\end{bmatrix}[-Y\quad 0]=\begin{bmatrix}I&0\\0&I\end{bmatrix},\tag{2.199}$$

and

$$X_1A_1+[-Y_1\quad 0]\begin{bmatrix}-B_1\\0\end{bmatrix}=I.\tag{2.200}$$

It is clear from a comparison of (2.198) through (2.200) with (2.146) through (2.148) that the results sought for the 3DOF system can be obtained by replacing A, B, X, Y, A_1, B_1, X_1, and Y_1 in the results previously obtained for the 2DOF standard configuration with

$$\tilde{A}=\begin{bmatrix}A&0\\0&I\end{bmatrix},\ \tilde{B}=\begin{bmatrix}-B\\0\end{bmatrix}\tag{2.201}$$

$$\tilde{X}=\begin{bmatrix}X&0\\0&I\end{bmatrix},\ \tilde{Y}=[-Y\quad 0]\tag{2.202}$$

$$\tilde{A}_1=A_1,\ \tilde{B}_1=\begin{bmatrix}-B_1\\0\end{bmatrix}\tag{2.203}$$

and

$$\tilde{X}_1=X_1,\tilde{Y}_1=[-Y_1\quad 0].\tag{2.204}$$

In particular, when these replacements are made in (2.161), when K is partitioned compatibly as

$$K=-[K_1\quad H_2],\tag{2.205}$$

and when H is replaced with H_1, one gets

$$C=\left(X_1+[K_1\quad H_2]\begin{bmatrix}-B\\0\end{bmatrix}\right)^{-1}\left[\left([-Y_1\quad 0]-[K_1\quad H_2]\begin{bmatrix}A&0\\0&I\end{bmatrix}\right)\quad H_1\right]\tag{2.206}$$

which reduces to

$$C = (X_1 - K_1 B)^{-1}\left[-(Y_1 + K_1 A) \quad -H_2 \quad H_1\right]. \tag{2.207}$$

So the equivalent to Theorem 2.2 for the 3DOF system is

Theorem 2.3 *An admissible 3DOF system is stable iff the controller is free of unstable hidden poles and the controller transfer matrix can be written in the form*

$$C = \left[-C_w \quad -C_z \quad C_u\right] = (X_1 - K_1 B)^{-1}\left[-(Y_1 + K_1 A) \quad -H_2 \quad H_1\right], \tag{2.208}$$

where K_1, H_1, and H_2 are stable real rational matrices and K_1 is such that

$$\det(X_1 - K_1 B) \not\equiv 0. \tag{2.209}$$

Attention is turned next to Lemma 2.5. For the 3DOF system, the three key matrices become

$$\tilde{A}G_{21} = \begin{bmatrix} A & 0 \\ 0 & I \end{bmatrix}\begin{bmatrix} FP_d & 0 & I \\ L & I & 0 \end{bmatrix} = \begin{bmatrix} AFP_d & 0 & A \\ L & I & 0 \end{bmatrix}, \tag{2.210}$$

$$G_{12}\tilde{A}_1 = \begin{bmatrix} P_a \\ P \end{bmatrix} A_1 = \begin{bmatrix} P_a A_1 \\ PA_1 \end{bmatrix}, \tag{2.211}$$

and

$$\begin{aligned} G_{11} + G_{12}\tilde{A}_1\tilde{Y}_1 G_{21} &= \begin{bmatrix} 0 & 0 & 0 \\ P_d & 0 & 0 \end{bmatrix} - \begin{bmatrix} P_a \\ P \end{bmatrix} A_1 \left[Y_1 FP_d \quad 0 \quad Y_1\right] \\ &= \begin{bmatrix} -P_a A_1 Y_1 FP_d & 0 & -P_a A_1 Y_1 \\ (P_d - PA_1 Y_1 FP_d) & 0 & -PA_1 Y_1 \end{bmatrix}. \end{aligned} \tag{2.212}$$

So since A and Y_1 are polynomial matrices one gets in place of Lemma 2.5

Lemma 2.7 *For the 3DOF system ψ_G/ψ_{FP} is a stable polynomial iff AFP_d,L, $P_a A_1$, PA_1, $P_a A_1 Y_1 FP_d$, and $(I - PA_1 Y_1 F)P_d$ are all stable.*

Another fundamental result with regard to admissibility can be obtained from (2.196) using the polynomial matrix description of the physical plant contained in (2.171) and (2.184), and the physical plant transfer matrix which is easily determined from Fig. 2.1. Specifically,

$$\begin{bmatrix} T_a & 0 \\ -U_{p_2} V_a & T_p \end{bmatrix}\begin{bmatrix} \xi_a \\ \xi_p \end{bmatrix} = \begin{bmatrix} 0 & U_a \\ U_{p_1} & U_{p_2} W_a \end{bmatrix}\begin{bmatrix} \tilde{d} \\ v \end{bmatrix} \tag{2.213}$$

and

$$\begin{bmatrix} y_s \\ y_t \end{bmatrix} = \begin{bmatrix} 0 & P_a \\ P_d & P \end{bmatrix} \begin{bmatrix} \tilde{d} \\ v \end{bmatrix}. \tag{2.214}$$

Hence, the characteristic polynomial for the physical plant is

$$\Delta_p = \eta_p \det \begin{bmatrix} T_a & 0 \\ -U_{p_2} V_a & T_p \end{bmatrix} = \eta_p (\det T_a)(\det T_p), \tag{2.215}$$

where η_p is the constant needed for Δ_p to be monic, and the transfer matrix for the physical plant is

$$P_o = \begin{bmatrix} 0 & P_a \\ P_d & P \end{bmatrix}. \tag{2.216}$$

It follows that

$$\Delta_p = h_o \psi_{P_o}, \tag{2.217}$$

where ψ_{P_o} is the characteristic denominator of P_o and h_o is the monic polynomial accounting for the hidden poles of the physical plant. Moreover, since P is a submatrix of P_o,

$$\psi_{P_o} = h_1 \psi_P, \tag{2.218}$$

where h_1 is a monic polynomial. It should also be clear that

$$\Delta_f = \eta_f \det T_f = h_f \psi_F \tag{2.219}$$

and

$$\Delta_l = \eta_l \det T_l = h_l \psi_L, \tag{2.220}$$

where h_f and h_l are the monic polynomials which account for the hidden poles of the feedback and feedforward transducers, respectively, and ψ_F and ψ_L are the characteristic denominators of F and L, respectively. Thus, (2.196) becomes

$$\frac{\Delta_G}{\psi_{FP}} = \frac{\Delta_p \Delta_f \Delta_l}{\psi_{FP}} = \frac{h \psi_P \psi_F \psi_L}{\psi_{FP}}, \tag{2.221}$$

where

$$h = h_o h_1 h_f h_l. \tag{2.222}$$

Now when p_o is a finite pole of FP, it follows that

$$\delta(FP; p_o) \leq \delta(F; p_o) + \delta(P; p_o). \tag{2.223}$$

Hence, ψ_{FP} divides into $\psi_F \psi_P$ without remainder and (2.221) is a stable polynomial iff h, ψ_L, and $\psi_F \psi_P / \psi_{FP}$ are all stable polynomials. Clearly, h is a stable polynomial iff h_o, h_1, h_f, and h_l are all stable polynomials. That is, iff the physical plant, the feedback transducer, and the feedforward transducer, are all free of unstable hidden poles and, in addition, h_1 is a stable polynomial. The polynomial h_1 can be determined from (2.218):

$$h_1 = \psi_{P_o} / \psi_P. \tag{2.224}$$

Finally, it is noted that ψ_L is a stable polynomial iff L is stable. It now follows from *(ii)* of Definition 2.4 without the need for any additional proof that

Lemma 2.8 *The 3DOF system is admissible iff (i) the feedback transducer, the feedforward transducer, and the physical plant are free of unstable hidden poles and (ii) L is stable and both ψ_{P_o}/ψ_P and $\psi_F \psi_P / \psi_{FP}$ are stable polynomials.*

Lemmas 2.7 and 2.8 establish important properties that hold for any admissible 3DOF system. These properties play a key role in the formulation of the Wiener–Hopf design methodology for such systems and are essential in connection with Theorem 2.3. It should also be emphasized that as in earlier cases the results for 3DOF systems do not depend on the choice of coprime polynomial matrix fraction descriptions used for *FP*. Specifically, in place of Lemma 2.6 one has

Lemma 2.9 *The set of all stabilizing controllers given by (2.208) as K_1, H_1, and H_2 range over all stable real rational matrices is invariant with respect to the choice of coprime polynomial matrix fraction descriptions used for FP.*

2.6 Tracking and Disturbance Rejection in the 3DOF System

The tracking and disturbance rejection capabilities of the 3DOF system shown in Fig. 1.1 are investigated in this section. It is assumed that the system is admissible and the controller is a stabilizing one as described in Theorem 2.3 with transfer matrix given by (2.208). The reference noise vector n and the transducer noise vectors n_l and n_m are typically stochastic in nature and, therefore, are devoid of any deterministic or shape-deterministic components. The elements of the disturbance input vector d ($\tilde{d}$ in Fig. 2.1) and the reference input vector r, however, often contain deterministic or shape-deterministic components that are transforms of such signals as steps, ramps, and exponentially growing functions of time. In this case, d and r possess unstable poles. The objective here is to identify the class of inputs d and r of this type for which the tracking error ε and the controller output v have stable transforms and to parameterize the subset of stabilizing controllers that achieve this objective.

Definition 2.5 The reference input r is called *trackable* if, when it is the only exogenous input, there exists a stabilizing controller for which both the error transform and the controller output transform are stable.

Definition 2.6 The disturbance input d is called *rejectable* if, when it is the only exogenous input, there exists a stabilizing controller for which both the error transform and the controller output transform are stable.

The reason for insisting on the stability of v is a consequence of the following considerations. The use of a quadratic performance measure that includes a penalty on the y_s shown in Fig. 2.1 requires that $P_a v$ be stable. So any unstable poles of v would have to be canceled by zeros of P_a. Since for admissibility any actuator poles and/or zeros in the unstable region (Re $s \geq 0$ for analog systems, $|\lambda| \leq 1$ for digital systems) must appear in $P = P_p P_a$ as well, and since such poles and zeros typically degrade the performance achievable, it is customary to pick actuators for which P_a is stable and maintains full column rank in the unstable region. Indeed, $P_a = I$ is often used to model the actuators. In these cases, stability of y_s is assured iff v is stable. There are also mathematical benefits from insisting on v stable: with this restriction, the task of parameterizing the subset of all stabilizing controllers for which ε is stable is made more transparent than in Cheng and Pearson (1981), Pernebo (1981), and Saeks and Murray (1981).

Since a stabilizing controller is assumed, the contributions to ε and v from initial conditions in the analog case or from initial data strings in the digital case are stable. So there is no loss in generality in ignoring the contribution from these sources. That is, one can work exclusively with exogenous inputs and transfer matrices. From Fig. 1.1 and (1.1), it follows that

$$v = -C_w(m + n_m) - C_L(Ld + n_l) + C_u(r + n), \tag{2.225}$$

where

$$m = F(Pv + P_d d). \tag{2.226}$$

Eliminating m yields

$$v = -(R_w F P_d + R_z L)d + R_u r + v_n, \tag{2.227}$$

where

$$[-R_w \quad -R_z \quad R_u] = (I + C_w F P)^{-1}[-C_w \quad -C_z \quad C_u] \tag{2.228}$$

and

$$v_n = -R_w n_m - R_z n_l + R_u n. \tag{2.229}$$

It now immediately follows from

$$\varepsilon = T_d r - P v - P_d d \tag{2.230}$$

in Fig. 1.1 that the error transform is given by

$$\varepsilon = (T_d - P R_u) r - (P_d - P R_w F P_d - P R_z L)d - P v_n. \tag{2.231}$$

When the parameterizations of C_u, C_w, and C_z given in (2.208) for a stabilizing controller are used and when (2.62) and (2.197) are recalled, the closed-loop transfer matrices R_u, R_w, and R_z simplify significantly. Specifically, one gets

$$(I + C_w F P)^{-1} = [I + (X_1 - K_1 B)^{-1}(Y_1 + K_1 A) B_1 A_1^{-1}]^{-1} = A_1 (X_1 - K_1 B) \tag{2.232}$$

and it easily follows from (2.228) that

$$[-R_w \quad -R_z \quad R_u] = [-A_1(Y_1 + K_1 A) \quad -A_1 H_2 \quad A_1 H_1]. \tag{2.233}$$

Since cancelations between unstable terms in (2.227) or in (2.231) cannot be counted on in practice, the contributions to v and ε from d and r must be stable separately. In addition, the contributions from r depend only on R_u, which is determined by the free stable real rational matrix H_1 and the contributions from d depend only on R_w and R_z which are determined, respectively, by the free stable real rational matrices K_1 and H_2. An important consequence of this separation of the design parameters is that one can address the issues of tracking error in response to reference inputs independently from the issue of disturbance rejection. Initially, attention is focused on the contributions from r given by

$$v_r = R_u r = A_1 H_1 r \tag{2.234}$$

and

$$\varepsilon_r = (T_d - PR_u)r = (T_d - PA_1 H_1)r. \tag{2.235}$$

The first observation to make is that admissibility together with Lemma 2.7 requires that PA_1 be stable. In addition, stability requires that H_1 be stable. So PA_1H_1 is stable and there is no good reason to choose T_d to be anything but a stable matrix. In this case, $T_d - PA_1H_1$ and A_1H_1 are both stable. With

$$r = A_r^{-1} b_r \tag{2.236}$$

a left coprime real polynomial matrix fraction description for r, it then follows that v_r and ε_r are stable iff the real rational matrices

$$\Psi = A_1 H_1 A_r^{-1} \tag{2.237}$$

and

$$\Phi = (T_d - PA_1 H_1)A_r^{-1} = T_d A_r^{-1} - P\Psi \tag{2.238}$$

are both stable. So the question of interest is whether or not one can choose a stable H_1 so that Φ and Ψ are stable? Clearly, the answer is yes iff there exist stable solution pairs Φ, Ψ satisfying

$$\Phi + P\Psi = T_d A_r^{-1} \tag{2.239}$$

and among such solution pairs there is at least one for which the solution for H_1 in (2.237) is a stable matrix. Actually, under reasonable assumptions, it is established in the sequel that one can exhibit all stable solution pairs Φ, Ψ and for every such pair the solution for H_1 in (2.237) is a stable matrix.

The first step is the introduction of the coprime polynomial matrix fraction descriptions

$$P = A_p^{-1} B_p = B_{p_1} A_{p_1}^{-1}, \tag{2.240}$$

where

$$\begin{bmatrix} A_p & B_p \\ Y_{p_1} & -X_{p_1} \end{bmatrix} \begin{bmatrix} X_p & B_{p_1} \\ Y_p & -A_{p_1} \end{bmatrix} = \begin{bmatrix} X_p & B_{p_1} \\ Y_p & -A_{p_1} \end{bmatrix} \begin{bmatrix} A_p & B_p \\ Y_{p_1} & -X_{p_1} \end{bmatrix} = \begin{bmatrix} I & 0 \\ 0 & I \end{bmatrix}. \tag{2.241}$$

Then (2.239) yields

$$A_p\Phi + B_p\Psi = A_pT_dA_r^{-1}. \tag{2.242}$$

So if Φ and Ψ are both stable, it is necessary that

$$\tilde{T}_d = A_pT_dA_r^{-1} \tag{2.243}$$

be a stable matrix. Given this is the case, the family of all stable solution pairs Φ, Ψ can be parameterized with the aid of

$$A_pX_p + B_pY_p = I \tag{2.244}$$

which is a consequence of (2.241). Indeed, it should be evident that all stable solution pairs Φ, Ψ are of the form

$$\Phi = X_p\tilde{T}_d + \Phi_o \tag{2.245}$$

and

$$\Psi = Y_p\tilde{T}_d + \Psi_o, \tag{2.246}$$

where Φ_o and Ψ_o are stable matrices satisfying

$$\Phi_o = -A_p^{-1}B_p\Psi_o = -B_{p_1}A_{p_1}^{-1}\Psi_o. \tag{2.247}$$

Since B_{p_1}, A_{p_1} is right coprime and Ψ_o must be stable, Φ_o is stable iff

$$K_r = A_{p_1}^{-1}\Psi_o \tag{2.248}$$

is stable. That is,

$$\Phi = X_p\tilde{T}_d - B_{p_1}K_r \tag{2.249}$$

and

$$\Psi = Y_p\tilde{T}_d + A_{p_1}K_r \tag{2.250}$$

parameterizes all stable real rational solution pairs Φ, Ψ in terms of the arbitrary stable real rational matrix K_r.

Thus, one can choose a stable H_1 for which both Ψ and Φ are stable iff one can choose a stable K_r for which

$$H_1 = A_1^{-1}\Psi A_r = A_1^{-1}Y_pA_pT_d + A_1^{-1}A_{p_1}K_rA_r \tag{2.251}$$

is stable. Now, from (2.241)

$$Y_pA_p = A_{p_1}Y_{p_1}; \tag{2.252}$$

hence,

$$H_1 = A_1^{-1}A_{p_1}(Y_{p_1}T_d + K_rA_r). \tag{2.253}$$

Clearly, when $A_1^{-1}A_{p_1}$ is stable, H_1 is stable for all stable K_r since T_d is stable and since Y_{p_1} and A_r are polynomial matrices. In this regard the following Lemma plays an important role.

Lemma 2.10 *When $\psi_{FP}(\psi_F\psi_P)^{-1}$ is stable, $A_1^{-1}A_{p_1}$ is stable iff F is stable.*

Proof When $A_1^{-1}A_{p_1}$ is stable,

$$\det(A_1^{-1}A_{p_1}) = \frac{\det A_{p_1}}{\det A_1} = \eta\left(\frac{\psi_F\psi_P}{\psi_{FP}}\right)\psi_F^{-1} \tag{2.254}$$

must be stable where η is the appropriate constant needed because the ψ's are monic polynomials. Hence,

$$\psi_F^{-1} = \eta^{-1}\psi_{FP}(\psi_F\psi_P)^{-1}\det(A_1^{-1}A_{p_1}) \tag{2.255}$$

must be stable or, equivalently, F must be stable. Conversely, when F is stable

$$A_1^{-1}A_{p_1} = (X_1 + Y_1FP)A_{p_1} = X_1A_{p_1} + Y_1FB_{p_1} \tag{2.256}$$

is stable. This completes the proof.

Since in almost all practical cases the feedback transducer transfer matrix F is stable, there is ample justification to make this hypothesis. In this circumstance, it follows from Lemmas 2.8 and 2.10 that (2.253) parameterizes all stable H_1 for which ε_r and v_r are stable and one has with no additional proof needed

Theorem 2.4 *When (1) $r = A_r^{-1}b_r$ is a left coprime real polynomial matrix fraction description for the reference input and (2) T_d and F are stable, the reference input is trackable with a stabilizable 3DOF system iff $A_pT_dA_r^{-1}$ is stable. Moreover, all stabilizing controllers for which r is trackable are given by (2.208) with*

$$H_1 = A_1^{-1} A_{p_1} (Y_{p_1} T_d + K_r A_r), \tag{2.257}$$

where K_r is a stable real rational matrix.

Remark 2.1 Theorem 2.4 is primarily of interest when A_r^{-1} is unstable. When A_r^{-1} is stable, r is stable and it follows from (2.234) and (2.235) that v_r and ε_r are stable for all stable H_1. That is, the desired tracking capability is realized in this case without the need to impose any additional restriction on H_1 beyond that required by system stability.

Remark 2.2 The equations that characterize ε and v in the 2DOF system obtained when the feedforward path in Fig. 1.1 is removed and the controller submatrix C_z is deleted correspond to setting $R_z = 0$ in (2.227) and (2.231). Clearly, this has no impact on v_r and ε_r so Theorem 2.4 is also applicable to the 2DOF system.

Remark 2.3 Since A_r, b_r is left coprime, $A_p T_d A_r^{-1}$ is stable iff

$$A_p T_d A_r^{-1} b_r = A_p T_d r \tag{2.258}$$

is stable or, when T_d is square and possesses an inverse, iff

$$r = T_d^{-1} A_p^{-1} r_o, \tag{2.259}$$

where r_o is a stable real rational vector. In this case, (2.259) parameterizes the family of all real rational reference inputs that can be tracked with zero steady-state error and zero steady-state controller output under Theorem 2.4. Typically, T_d^{-1} is also stable; then the unstable poles of r must be a subset of the unstable poles of A_p^{-1} which are the unstable poles of the plant transfer matrix P.

Application of Theorem 2.4 requires finding coprime polynomial matrix fraction descriptions for both FP and P. By imposing a somewhat stronger assumption on F and T_d, however, one can avoid the need of finding a coprime polynomial matrix fraction description for P and obtain a simpler formula in place of (2.257). Indeed, it is next established that

Theorem 2.5 *When (1) $r = A_r^{-1} b_r$ is a left coprime real polynomial matrix fraction description for the reference input and (2) T_d, T_d^{-1}, and $(T_d F - I)P$ are stable, the reference input is trackable with a stabilizable 3DOF system iff AA_r^{-1} is stable. Moreover, all stabilizing controllers for which r is trackable are given by (2.208) with*

$$H_1 = Y_1 + K_r A_r, \tag{2.260}$$

where K_r is a stable real rational matrix.

Proof Equation (2.238) can be rewritten as

$$\Phi = T_d A_r^{-1} - T_d F P \Psi + (T_d F - I) P \Psi. \tag{2.261}$$

The last term on the right-hand side of (2.261) is stable when Ψ is stable. Hence, Φ is stable in this case iff $T_d(A_r^{-1} - FP\Psi)$ is stable or, since T_d and T_d^{-1} are stable, $(A_r^{-1} - FP\Psi)$ is stable. That is, there exists a stable solution pair Φ, Ψ to (2.261) iff there exists a stable solution pair $\tilde{\Phi}, \Psi$ to

$$\tilde{\Phi} = A_r^{-1} - FP\Psi = A_r^{-1} - A^{-1} B \Psi \tag{2.262}$$

or, equivalently,

$$A\tilde{\Phi} + B\Psi = AA_r^{-1}. \tag{2.263}$$

A necessary and sufficient condition for this to be the case is that AA_r^{-1} be stable. One can now apply the same reasoning as in connection with Theorem 2.4 to get that all stable real rational solution pairs $\tilde{\Phi}, \Psi$ are given by

$$\Psi = YAA_r^{-1} + A_1 K_r \tag{2.264}$$

and

$$\tilde{\Phi} = XAA_r^{-1} - B_1 K_r. \tag{2.265}$$

It then follows from (2.237) that

$$H_1 = A_1^{-1} \Psi A_r = A_1^{-1} YA + K_r A_r = Y_1 + K_r A_r \tag{2.266}$$

and the proof is complete.

Remark 2.4 Obviously, Remarks 2.1 and 2.2 apply here as well.

Remark 2.5 The assumption that T_d, T_d^{-1}, and

$$\Gamma = (T_d F - I)P = T_d FP - P \tag{2.267}$$

are stable is a stronger assumption than just assuming that T_d and F are stable. This can be seen from the following reasoning. When T_d, T_d^{-1}, and Γ are stable, one gets from (2.267) for any unstable pole located at p_o that

$$\delta(FP;p_o) = \delta(T_d FP;p_o) = \delta(\Gamma + P;p_o) = \delta(P;p_o). \tag{2.268}$$

That is, $\psi_P \psi_{FP}^{-1}$ is stable. Now when the system is also admissible, one has from Lemma 2.8 that $\psi_{FP}(\psi_F \psi_P)^{-1}$ is stable. So

$$\psi_F^{-1} = [\psi_P \psi_{FP}^{-1}][\psi_{FP}(\psi_F \psi_P)^{-1}] \tag{2.269}$$

is stable or, equivalently, F must be stable. However, admissibility and a stable F and T_d are not enough to guarantee a stable Γ and T_d^{-1}.

Remark 2.6 It has already been pointed out that one benefit of the stronger Assumption (2) in Theorem 2.5 is that a coprime polynomial matrix fraction description for P is not needed and a simpler parameterization of H_1 is obtained. There are other benefits as well. It easily follows from (2.267) that

$$\Gamma A_1 = T_d B_1 - P A_1. \tag{2.270}$$

Hence, PA_1 is stable and this is consistent with one of the requirements needed for admissibility in Lemma 2.7. Assumption (2) in Theorem 2.5 also turns out to be significant with regard to tracking when 1DOF systems are treated (see Examples 2.13 and 3.12).

Remark 2.7 Since A_r, b_r is left coprime, AA_r^{-1} is stable iff

$$AA_r^{-1} b_r = Ar \tag{2.271}$$

is stable. This is the case iff

$$r = A^{-1} r_o, \tag{2.272}$$

where r_o is a stable real rational vector. Equation (2.272) parameterizes the family of all real rational reference inputs that can be tracked with zero steady-state error and zero steady-state controller output under Theorem 2.5. Since (2.268) holds, it follows from (2.197) that the unstable poles of A^{-1} and P are the same. It is then clear from (2.272) that the unstable poles of r must be a subset of the unstable poles of the plant transfer matrix P.

Attention is now turned to the contribution to v and the contribution to ε on account of d in (2.227) and (2.231), respectively. These contributions are given by

$$v_d = -A_1 H d \tag{2.273}$$

and

$$\varepsilon_d = -(P_d - PA_1 H)d, \tag{2.274}$$

where, when (2.233) is recalled,

$$H = (Y_1 + K_1 A)FP_d + H_2 L. \tag{2.275}$$

Using (2.64) one gets

$$A_1 H = (Y + A_1 K_1)AFP_d + A_1 H_2 L \tag{2.276}$$

and

$$P_d - PA_1 H = (I - PA_1 Y_1 F)P_d - PA_1(K_1 AFP_d + H_2 L) \tag{2.277}$$

which are clearly stable because of admissibility and Lemma 2.7. With

$$d = A_d^{-1} b_d \tag{2.278}$$

a left coprime real polynomial matrix fraction description for d, it then follows that v_d and ε_d are stable iff the real rational matrices

$$\Psi = A_1 H A_d^{-1} \tag{2.279}$$

and

$$\Phi = (P_d - PA_1 H)A_d^{-1} = P_d A_d^{-1} - P\Psi \tag{2.280}$$

are both stable. These two equations are identical in form to (2.237) and (2.238) with P_d replacing T_d and H replacing H_1; however, here H need not be stable, only K_1 and H_2 in H must be stable. It still follows in similar fashion, however, that

$$\tilde{P}_d = A_p P_d A_d^{-1} \tag{2.281}$$

must be a stable matrix and all stable real rational solution pairs Φ, Ψ of (2.280) are given by

$$\Phi = X_p \tilde{P}_d - B_{p_1} K_m \tag{2.282}$$

and

$$\Psi = Y_p \tilde{P}_d + A_{p_1} K_m, \tag{2.283}$$

where K_m is an arbitrary stable real rational matrix. Hence, in an admissible 3DOF system it is possible for both v_d and ε_d to be stable iff one can choose stable K_1, H_2, and K_m so that $\det(X_1 - K_1 B) \not\equiv 0$ and so that

$$\Psi = A_1 H A_d^{-1} = Y_p \tilde{P}_d + A_{p_1} K_m \tag{2.284}$$

or, equivalently,

$$A_1(K_1 Q_d + H_2 L A_d^{-1}) - A_{p_1} K_m = Y_p \tilde{P}_d - YAFP_d A_d^{-1}, \tag{2.285}$$

where

$$Q_d = AFP_d A_d^{-1} = AFA_p^{-1} \tilde{P}_d. \tag{2.286}$$

Now it is again consistent with practice to assume F is stable in which case it follows from $AFP = AFA_p^{-1} B_p = B$, a stable matrix, that AFA_p^{-1} and, therefore, Q_d are stable matrices.

Equation (2.285) is equivalent to

$$H_2 L A_d^{-1} - \tilde{K}_m = \Gamma_d - K_1 Q_d, \tag{2.287}$$

where

$$\Gamma_d = A_1^{-1}(Y_p \tilde{P}_d - YAFP_d A_d^{-1}) = (A_1^{-1} Y_p - Y_1 F A_p^{-1}) \tilde{P}_d \tag{2.288}$$

and

$$\tilde{K}_m = A_1^{-1} A_{p_1} K_m. \tag{2.289}$$

Using $A_1^{-1} = X_1 + Y_1 FP$ and $A_p^{-1} = X_p + PY_p$ in (2.288) yields

$$\Gamma_d = (X_1 Y_p - Y_1 F X_p) \tilde{P}_d. \tag{2.290}$$

Clearly, when F is stable, Γ_d is stable. Moreover, one also has from Lemma 2.10 that $A_1^{-1} A_{p_1}$ is stable. In addition, admissibility together with Lemma 2.7 requires that $PA_1 = B_{p_1} A_{p_1}^{-1} A_1$ be stable or, equivalently, $A_{p_1}^{-1} A_1$ be stable. Hence, $\tilde{K}_m$ is stable iff K_m is stable and the problem of finding stable real rational matrices K_1, H_2, and K_m that satisfy (2.285) has been reduced to one of finding stable real rational matrices K_1, H_2, and $\tilde{K}_m$ that satisfy (2.287).

The real coprime polynomial matrix fraction description

$$LA_d^{-1} = A_g^{-1}B_g \tag{2.291}$$

satisfying

$$A_gX_g + B_gY_g = I \tag{2.292}$$

is now introduced. Then (2.287) yields

$$H_2A_g^{-1}B_g = \Gamma_d - K_1Q_d + \tilde{K}_m; \tag{2.293}$$

hence, the left-hand side must be a stable matrix. Since a stable H_2 is required and the pair A_g, B_g is left coprime, $H_2A_g^{-1}B_g$ is stable iff $H_2A_g^{-1}$ is stable or, equivalently,

$$H_2 = K_gA_g, \tag{2.294}$$

where K_g is a stable matrix. Then (2.293) is satisfied for any choice of stable K_1 when

$$\tilde{K}_m = K_gB_g + K_1Q_d - \Gamma_d. \tag{2.295}$$

That is, v_d and ε_d are stable under the assumptions here iff

$$[K_1 \quad H_2] = [K_1 \quad K_gA_g] \tag{2.296}$$

with K_1 and K_g arbitrary stable real rational matrices. It is of course essential to establish that there is at least one choice of K_1 for which $\det(X_1 - K_1B) \not\equiv 0$. Obviously, for the choice $K_1 = 0$ this is the case since one can always construct X_1 so that $\det X_1 \not\equiv 0$ (Appendix B, Remark B.6); so there is at least one stabilizing 3DOF controller described in Theorem 2.3 for which d is rejectable. It is now possible to state without additional proof that

Theorem 2.6 *When (1) $d = A_d^{-1}b_d$ is a left coprime real polynomial matrix fraction description for the disturbance input, (2) the real polynomial matrix pair A_g, B_g satisfies (2.291) and (2.292), and (3) the feedback sensor transfer matrix F is a stable matrix, then the disturbance input d is rejectable with a stabilized 3DOF system iff $A_pP_dA_d^{-1}$ is stable. Moreover, all stabilizing controllers for which d is rejectable are given by (2.208) with*

$$[K_1 \quad H_2] = [K_1 \quad K_gA_g], \tag{2.297}$$

where K_1 and K_g are stable real rational matrices belonging to the nonempty set of such matrices for which $\det(X_1 - K_1 B) \not\equiv 0$.

Remark 2.8 Equation (2.297) parameterizes in terms of arbitrary stable matrices K_1 and K_g all stable K_1, H_2 for which all pairs v_d, ε_d given by (2.273), (2.274) are stable. Although this parameterization may include cases in which a particular choice of stable K_1 yields $\det(X_1 - K_1 B) \equiv 0$, it can be demonstrated that in most cases of practical interest this possibility does not occur. In analog systems, for example, it is typical that K_1 is chosen so that the sensitivity matrix satisfies

$$S = (I + C_w FP)^{-1} \to I, \quad s \to \infty. \tag{2.298}$$

It immediately follows from (2.232) since $\det A_1 \not\equiv 0$ that

$$\det(X_1 - K_1 B) = \det S / \det A_1 \not\equiv 0. \tag{2.299}$$

In digital systems it is usually the case that $\det A_1(\lambda) \neq 0$, $B_1(\lambda) = 0$, and $B(\lambda) = 0$ at $\lambda = 0$. Also, since a stable $K_1(\lambda)$ is analytic in $|\lambda| \leq 1$, $K_1(\lambda)$ is finite at $\lambda = 0$. In this case then,

$$X_1(0) - K_1(0)B(0) = X_1(0) \tag{2.300}$$

and

$$X_1(0)A_1(0) + Y_1(0)B_1(0) = X_1(0)A_1(0) = I. \tag{2.301}$$

It follows that at $\lambda = 0$

$$\det(X_1 - K_1 B) = \det X_1 = 1/\det A_1 \neq 0; \tag{2.302}$$

therefore, $\det(X_1 - K_1 B) \not\equiv 0$ for any stable K_1.

Remark 2.9 Theorem 2.6 is primarily of interest when A_d^{-1} is unstable. Otherwise, d is stable and it easily follows from (2.276) and (2.277) being stable that v_d and ε_d are stable. That is, the desired disturbance rejection capability is realized in this case without the need to impose any additional restrictions on K_1 and H_2 beyond that required for system stability.

Remark 2.10 The equation that characterizes v_d and the one that characterizes ε_d in the 2DOF system are obtained when the feedforward path in Fig. 1.1 is removed and the controller submatrix C_z is deleted; these equations are the same as those obtained by setting $H_2 = 0$ in (2.275) and (2.276). Equation (2.287) is then replaced with

$$\tilde{K}_m = K_1 Q_d - \Gamma_d \tag{2.303}$$

which gives a stable $\tilde{K}_m$ for any stable K_1. It now easily follows that Theorem 2.6 holds for the 2DOF system except that all stabilizing controllers for which d is rejectable are now given by

$$C = \begin{bmatrix} -C_w & C_u \end{bmatrix} = (X_1 - K_1 B)^{-1} \begin{bmatrix} -(Y_1 + K_1 A) & H_1 \end{bmatrix}, \tag{2.304}$$

where K_1 and H_1 are stable real rational matrices and K_1 is such that

$$\det(X_1 - K_1 B) \not\equiv 0. \tag{2.305}$$

That is, a rejectable d is rejectable in a 2DOF system with any stabilizing controller; no additional constraints on K_1 or H_1 are needed.

Remark 2.11 Clearly, when $\tilde{P}_d = A_p P_d A_d^{-1}$ is stable, then $A_p P_d = \tilde{P}_d A_d$ and $A_p P_d d = \tilde{P}_d b_d$ are stable. Conversely, when $A_p P_d$ and $A_p P_d d = A_p P_d A_d^{-1} b_d$ are stable, it follows from A_d, b_d left coprime that $A_p P_d A_d^{-1}$ is stable. That is, $A_p P_d A_d^{-1}$ is stable iff $A_p P_d$ and $A_p P_d d$ are stable. Hence, $P_d d = A_p^{-1} d_o$, where d_o is a stable vector and $P_d d$ cannot have any unstable poles that are not poles of the plant. Moreover, when $A_p P_d d$ is stable and $A_p P_d$ has a stable left inverse given by $(A_p P_d)^{\#}$, then $(A_p P_d)^{\#} A_p P_d d = d$ must be stable and Remark 2.9 applies.

It is again of interest to determine conditions under which the need to find a coprime polynomial matrix fraction description for P can be avoided. These conditions are embodied in

Theorem 2.7 *When (1) $d = A_d^{-1} b_d$ is a left coprime real polynomial matrix fraction description for the disturbance input, (2) the real polynomial matrix pair A_g, B_g satisfies (2.291) and (2.292), and (3) T_d, T_d^{-1}, $A T_d A^{-1}$, and $(T_d F - I)P$ are stable matrices, the disturbance input d is rejectable with a stabilized 3DOF system iff $A P_d A_d^{-1}$ is stable. Moreover, all stabilizing controllers for which d is rejectable are given by (2.208) with*

$$\begin{bmatrix} K_1 & H_2 \end{bmatrix} = \begin{bmatrix} K_1 & K_g A_g \end{bmatrix}, \tag{2.306}$$

where K_1 and K_g are stable real rational matrices belonging to the nonempty set of such matrices for which

$$\det(X_1 - K_1 B) \not\equiv 0. \tag{2.307}$$

Proof One again obtains (2.280) or

$$\Phi + P\Psi = \Phi + (T_dF + I - T_dF)P\Psi = P_dA_d^{-1}. \tag{2.308}$$

Multiplying (2.308) on the left by T_d^{-1} and introducing

$$\tilde{\Phi} = T_d^{-1}[\Phi - (T_dF - I)P\Psi] = T_d^{-1}P_dA_d^{-1} - FP\Psi \tag{2.309}$$

gives

$$\tilde{\Phi} + FP\Psi = T_d^{-1}P_dA_d^{-1}. \tag{2.310}$$

Under (3) in the assumptions, it is clear from (2.309) when Ψ is stable that $\tilde{\Phi}$ is stable iff Φ is stable. So a stable solution pair Φ, Ψ exists for (2.308) iff a stable solution pair $\tilde{\Phi}, \Psi$ exists for (2.310).

The coprime polynomial matrix fraction descriptions (2.197) that satisfy the generalized Bezout identity (2.62) can now be introduced to get from (2.310)

$$A\tilde{\Phi} + B\Psi = \tilde{P}_{d_o}, \tag{2.311}$$

where

$$\tilde{P}_{d_o} = AT_d^{-1}P_dA_d^{-1} = (AT_d^{-1}A^{-1})(AP_dA_d^{-1}). \tag{2.312}$$

It is evident from (2.311) that $\tilde{P}_{d_o}$ must be stable. Since T_d^{-1} and $S_d = AT_dA^{-1}$ are assumed stable, it follows that $AT_d^{-1}A^{-1} = S_d^{-1} = \det(T_d^{-1})\ \text{adj}\ (S_d)$ is stable. Hence, $\tilde{P}_{d_o}$ is stable iff $AP_dA_d^{-1}$ is stable and the necessity of this condition is clear.

Invoking similar arguments to the ones used to get (2.284) now yields

$$\Psi = A_1HA_d^{-1} = Y\tilde{P}_{d_o} + A_1K_m \tag{2.313}$$

or in view of (2.275), (2.286), and (2.291)

$$H_2A_g^{-1}B_g = \Gamma_{d_o} - K_1Q_d + K_m, \tag{2.314}$$

where

$$\Gamma_{d_o} = -Y_1(FT_d - I)A^{-1}\tilde{P}_{d_o} = -Y_1T_d^{-1}(T_dF - I)A_p^{-1}(A_pA^{-1})AP_dA_d^{-1}. \tag{2.315}$$

It is next established that Γ_{d_o} is stable by showing that $(T_dF - I)A_p^{-1}$ and (A_pA^{-1}) are stable. In accordance with Remark 2.5, the assumption that $(T_dF - I)P$ is stable forces F to be stable. From $(T_dF - I)P = (T_dF - I)A_p^{-1}B_p$ it then follows that $(T_dF - I)A_p^{-1}$ must be stable. Also,

$$A(T_dF - I)P = AT_dFP - AP = AT_dA^{-1}B - AP \tag{2.316}$$

must be stable. Since AT_dA^{-1} is stable, it follows that $AP = AA_p^{-1}B_p$ must be stable or, equivalently, AA_p^{-1} must be stable. Moreover,

$$1\Big/\det(AA_p^{-1}) = \det A_p/\det A = \eta(\psi_F\psi_P/\psi_{FP})\psi_F^{-1} \tag{2.317}$$

must be stable because F is stable and $\psi_F\psi_P/\psi_{FP}$ is polynomial. Hence, $A_pA^{-1} = (AA_p^{-1})^{-1}$ must be stable and Γ_{d_o} is indeed stable. Thus, the problem has been reduced to finding stable real rational matrices K_1, H_2, and K_m that satisfy (2.314). Clearly, $H_2A_g^{-1}B_g$ must be stable and this is the case iff

$$H_2 = K_gA_g, \tag{2.318}$$

where K_g is a stable matrix. Then (2.314) is satisfied for any choice of stable K_1 when

$$K_m = K_gB_g + K_1Q_d - \Gamma_{d_o}. \tag{2.319}$$

That is, v_d and ε_d are stable under the assumptions here iff

$$[\,K_1 \quad H_2\,] = [\,K_1 \quad K_gA_g\,] \tag{2.320}$$

with K_1 and K_g arbitrary stable real rational matrices. It is again essential to establish that there is at least one choice for which $\det(X_1 - K_1B) \not\equiv 0$. Clearly, for $K_1 = 0$,

$$\det(X_1 - K_1B) = \det X_1 \not\equiv 0 \tag{2.321}$$

since one can always construct X_1 so that $\det X_1 \not\equiv 0$ (Appendix B, Remark B.6); this shows that there is at least one stabilizing 3DOF controller described in Theorem 2.3 for which d is rejectable. This completes the proof of Theorem 2.7.

Remark 2.12 It should be apparent that with only minor adjustments Remarks 2.8 through 2.11 are applicable here as well.

Remark 2.13 When the contributions to v and ε from inputs $r = A_r^{-1}b_r$, $d = 0$ or from $r = 0$, $d = A_d^{-1}b_d$ are stable, it follows from superposition that the contributions to v and ε from inputs $r = A_r^{-1}b_r$, $d = A_d^{-1}b_d$ are stable. So Theorems 2.4 through 2.7 are equally useful when both reference inputs and disturbance inputs are present simultaneously as in the practical case. It should also be clear that the controller parameterizations given in Theorems 2.4 through 2.7 do not depend explicitly on b_r or b_d. Hence, v and ε are stable no matter what the polynomial vectors b_r and b_d are; so in this regard the controllers are robust. More specifically, it should be clear that v_r and ε_r are stable for all possible polynomial vectors b_r, whether coprime to A_r or not, iff the matrices Ψ and Φ given by (2.237) and (2.238) are stable. Again when T_d and F are stable this is the case iff the conditions of Theorem 2.4 are satisfied; that is iff $A_pT_dA_r^{-1}$ and K_r are stable and H_1 is given by (2.257). Similarly, v_d and ε_d are stable for all possible polynomial vectors b_d, whether coprime to A_d or not, iff the matrices Ψ and Φ given by (2.279) and (2.280) are stable. Again when F is stable this is the case iff the conditions of Theorem 2.6 are satisfied; that is iff $A_pP_dA_d^{-1}$ and K_d are stable and $[K_1 \quad H_2]$ is given by (2.297). Clearly, analogous observations also apply in connection with Theorems 2.5 and 2.7.

Remark 2.14 Practical requirements often demand that v_r and ε_r both be stable when any one of the elements of r is not zero while all of the other elements are. That is, one might desire instead of $v_r = A_1H_1r$, $\varepsilon_r = (T_d - PA_1H_1)r$ stable that $v_r = A_1H_1D_r$, $\varepsilon_r = (T_d - PA_1H_1)D_r$ be stable, where $D_r = \text{diag}\ \{r_1, r_2, \ldots, r_N\}$ and $r = [r_1 \quad r_2 \ldots r_N]'$. In this case, a coprime polynomial matrix fraction description $D_r = A_r^{-1}B_r$ can be introduced and one can easily replicate the appropriately modified and nearly identical replacements for Theorems 2.4 and 2.5. Similarly, in connection with disturbance inputs $d = [d_1 \quad d_2 \ldots d_M]'$ one might require instead of $v_d = -A_1Hd$, $\varepsilon_d = -(P_d - PA_1H)d$ stable that $v_d = -A_1HD_d$, $\varepsilon_d = -(P_d - PA_1H)D_d$ be stable, where $D_d = \text{diag}\ \{d_1, d_2, \ldots, d_M\}$. In this case, a coprime polynomial matrix fraction description $D_d = A_d^{-1}B_d$ can be introduced and one can easily replicate the appropriately modified and nearly identical replacements for Theorems 2.6 and 2.7.

Example 2.1 This example offers additional insights and illustrates the application of the theory. An analog plant with transfer matrix

$$P_p = \frac{1}{s^2(s-2)}\begin{bmatrix} (s-1) & (s-1) \\ 1 & 2 \end{bmatrix} \tag{2.322}$$

is considered in a 3DOF system with

$$L = F = P_a = T_d = I \tag{2.323}$$

and $P_d = P_p$. It is assumed that the feedback transducer, the feedforward transducer, and the physical plant are free of unstable hidden poles. So this 3DOF system is admissible provided (*ii*) of Lemma 2.8 is satisfied. Obviously, $L = I$ is stable. Since $F = I$, it follows that $\psi_F = 1$ and $\psi_{FP} = \psi_P$. Hence, $(\psi_F \psi_P / \psi_{FP}) = 1$ is a stable polynomial. Moreover,

$$P_o = \begin{bmatrix} 0 & P_a \\ P_d & P \end{bmatrix} = \begin{bmatrix} 0 & P_a \\ P_d & P_p P_a \end{bmatrix} = \begin{bmatrix} I & 0 \\ 0 & P_p \end{bmatrix} \begin{bmatrix} 0 & I \\ I & I \end{bmatrix}. \tag{2.324}$$

Now, the second matrix in the product at the far right-hand side of (2.324) is nonsingular and the characteristic denominator of the first is $\psi_{P_p} = \psi_P$. It therefore follows that $\psi_{P_o} = \psi_P$ and $(\psi_{P_o}/\psi_P) = 1$ is a stable polynomial; hence, the system is indeed admissible.

Clearly,

$$P = P_a P_p = P_p = A_p^{-1} B_p = B_{p_1} A_{p_1}^{-1} \tag{2.325}$$

when

$$A_p = A_{p_1} = s^2(s-2)I \tag{2.326}$$

and

$$B_p = B_{p_1} = \begin{bmatrix} (s-1) & (s-1) \\ 1 & 2 \end{bmatrix}. \tag{2.327}$$

Moreover, since $\det B_p(0) \neq 0$ and $\det B_p(2) \neq 0$, rank $[A_p \quad B_p] = 2$ for all s; so A_p, B_p is left coprime. Similarly, B_{p_1}, A_{p_1} is right coprime.

Consistent with Remark 2.3, the reference input

$$r = A_r^{-1} b_r = \frac{1}{s^2(s-2)} \begin{bmatrix} (s-1) \\ 1 \end{bmatrix} \tag{2.328}$$

should be trackable. Now one cannot choose $A_r = s^2(s-2)I$ because in this case no matter what choice for b_r is made rank $[A_r \quad b_r] = 2$ for all s is impossible. However, with the alternative choice

$$A_r = \begin{bmatrix} 1 & (1-s) \\ 0 & s^2(s-2) \end{bmatrix}, \quad b_r = \begin{bmatrix} 0 \\ 1 \end{bmatrix} \tag{2.329}$$

one gets

$$\text{rank}\begin{bmatrix} A_r & b_r \end{bmatrix} = \text{rank}\begin{bmatrix} 1 & (1-s) & 0 \\ 0 & s^2(s-2) & 1 \end{bmatrix} = 2 \tag{2.330}$$

for all s or, equivalently, the pair in (2.329) is left coprime. So according to Theorem 2.4, the reference input (2.328) is indeed trackable when the A_r of (2.329) is used in (2.257) since in this case $A_p T_d A_r^{-1}$ is stable.

In order to complete the determination of the H_1 given by (2.257), the matrices A_1 and Y_{p_1} have to be found. Since $FP = P = P_p$ in this example, one can take

$$A = A_1 = A_p = A_{p_1}, B = B_1 = B_p = B_{p_1}. \tag{2.331}$$

Hence, $A_1^{-1}A_{p_1} = I$. The determination of Y_{p_1} requires also the determination of X_{p_1}, X_p, and Y_p so that (2.241) is satisfied. Actually, it is established next that one can use any solution pair X_{p_1}, Y_{p_1} to

$$X_{p_1}A_{p_1} + Y_{p_1}B_{p_1} = I. \tag{2.332}$$

Since A_p, B_p is left coprime, there exist matrices $\tilde{X}_p, \tilde{Y}_p$ such that

$$A_p\tilde{X}_p + B_p\tilde{Y}_p = I. \tag{2.333}$$

Then

$$X_p = \tilde{X}_p - B_{p_1}M, \quad Y_p = \tilde{Y}_p + A_{p_1}M \tag{2.334}$$

gives

$$AX_p + BY_p = I \tag{2.335}$$

for any polynomial matrix M which can always be chosen so that

$$Y_{p_1}X_p - X_{p_1}Y_p = Y_{p_1}\tilde{X}_p - X_{p_1}\tilde{Y}_p - M = 0 \tag{2.336}$$

assuring (2.241) is satisfied. So it only remains to find a solution pair X_{p_1}, Y_{p_1} for (2.332). This can be accomplished by assuming X_{p_1} is a constant matrix, setting $Y_{p_1} = s^2 Y_{p_{12}} + sY_{p_{11}} + Y_{p_{10}}$, and finding the unknown constant matrices X_{p_1}, $Y_{p_{12}}$, $Y_{p_{11}}$, and $Y_{p_{10}}$ so that (2.332) is satisfied. The results obtained yield

$$X_{p_1} = \begin{bmatrix} -2 & -2 \\ 1 & 1 \end{bmatrix} \tag{2.337}$$

and

$$Y_{p_1} = \begin{bmatrix} 2(s^2 - s - 1) & -1 \\ -(s^2 - s - 1) & 1 \end{bmatrix}. \tag{2.338}$$

Hence, from (2.257)

$$H_1 = \begin{bmatrix} 2(s^2 - s - 1) & -1 \\ -(s^2 - s - 1) & 1 \end{bmatrix} + K_r \begin{bmatrix} 1 & (1 - s) \\ 0 & s^2(s - 2) \end{bmatrix}. \tag{2.339}$$

It is important to note that at $s = 0$ or at $s = 2$ no restriction is placed on the first column of the product $K_r A_r$ when A_r is given by (2.329). On the other hand, if instead of tracking (2.328) it was required that

$$r = A_r^{-1} b_r = \frac{1}{s^2(s-2)} \begin{bmatrix} b_1 \\ b_2 \end{bmatrix} \tag{2.340}$$

be tracked for all b_1, b_2 then in accordance with Remark 2.13 one would have to use $A_r = s^2(s-2)I$ instead. In this case, $K_r A_r = 0$ at $s = 0$ or at $s = 2$ for any stable K_r. So by requiring more inputs of the form (2.340) to be tracked instead of the single one given by (2.328), design freedom is sacrificed. It is important for this reason to limit the class of inputs to be tracked to include only those truly expected. For example, when $r = A_r^{-1} b_r = b_r / h(s)$ and $h(s)$ is a strict Hurwitz polynomial, then the choice $A_r = h(s)I$ places no restriction on the product $K_r A_r$ because $K_r = \tilde{K}_r / h$ can be chosen with $\tilde{K}_r$ any stable real rational matrix.

Attention is now turned to the issue of disturbance rejection. In accordance with Theorem 2.6 and Remark 2.11, since $A_p P_d = A_p P_p = B_p$ is stable, any input d such that

$$A_p P_d d = B_p d = d_o \tag{2.341}$$

is a stable vector is rejectable. In particular,

$$d = B_p^{-1} d_o = B_p^{-1} [\alpha \quad \beta]' = \frac{1}{(s-1)} \begin{bmatrix} 2 & (1 - s) \\ -1 & (s - 1) \end{bmatrix} \begin{bmatrix} \alpha \\ \beta \end{bmatrix}, \tag{2.342}$$

where α, β are polynomials is rejectable. Moreover, when $\alpha(1) \neq 0$ is assumed, it follows that

$$\text{rank}\,[B_p \quad d_o] = \text{rank} \begin{bmatrix} (s-1) & (s-1) & \alpha \\ 1 & 2 & \beta \end{bmatrix} = 2 \tag{2.343}$$

for all s. Hence, with $A_d = B_p$ and $b_d = [\alpha \quad \beta]'$, $d = A_d^{-1} b_d$ is a left coprime polynomial matrix fraction description for d. Then

$$LA_d^{-1} = B_p^{-1} \tag{2.344}$$

and one can choose in accordance with (2.291) and (2.292)

$$A_g = B_p, B_g = I \tag{2.345}$$

and

$$X_g = 0, Y_g = I. \tag{2.346}$$

It is of interest to note from (2.290) and (2.336) that

$$\Gamma_d = (X_{p_1} Y_p - Y_{p_1} X_p) = 0 \tag{2.347}$$

since on account of (2.331) one can take $X_1 = X_{p_1}$, $Y_1 = Y_{p_1}$. So Γ_d is stable as expected. Finally, (2.297) yields

$$[K_1 \quad H_2] = [K_1 \quad K_g A_g] = [K_1 \quad K_g B_p] \tag{2.348}$$

and, since $\det X_1 = \det X_{p_1} \equiv 0$ for the X_{p_1} given by (2.337), it remains to show that at least one stable K_1 can be chosen so that

$$\det (X_1 - K_1 B) = \det (X_{p_1} - K_1 B_p) \not\equiv 0. \tag{2.349}$$

Indeed, the choice $K_1 = -I$ yields

$$\det (X_1 - K_1 B) = \det (X_{p_1} + B_p) = \det \begin{bmatrix} (s-3) & (s-3) \\ 2 & 3 \end{bmatrix} = (s-3) \not\equiv 0. \tag{2.350}$$

2.7 Parameterization and Realization of Proper Stabilizing Controllers

Two approaches have been taken in the literature with regard to the parameterization of stabilizing controllers. The one adopted in the earlier sections of this chapter relied on the use of coprime polynomial matrix fraction descriptions as in Youla et al. (1976b). The other approach introduced in Desoer et al. (1980) makes use of proper stable coprime rational matrix fraction descriptions instead, motivated by practical considerations which often dictate that the controller transfer matrix be proper. However, to insist on proper stabilizing controllers at the outset for analog systems excludes the possibility of using proportional–integral–derivative (PID) controllers which is common in practice when the exogenous input signal spectra permit. (In the February 2006 issue, vol. 26, no. 1, of the *IEEE Control System Magazine* the extensive use of PID controllers in industrial applications is reported on.) Moreover, when H_2 designs are being considered, one should allow the design methodology to dictate whether or not a proper controller is needed or is best. Also, the use of proper stable coprime rational matrix fractions involves selecting an arbitrary set of stable poles for the proper rational matrices used and these poles must all cancel out in the end anyway. In addition, the use of λ transforms for digital systems removes the need to impose properness on the digital controller transfer matrix: the issue of causality is addressed instead with the requirement that det $(X_1 - K_1 B) \neq 0$ at $\lambda = 0$. For these reasons, the temptation to follow the preference of others (Vidyasagar 1985; Gündes and Desoer 1990; and Zhou et al. 1996) to use proper stable coprime real rational matrix fractions was rejected here. Moreover, in the analog case even when it may be desirable to parameterize all proper stabilizing controllers, this can be done in terms of coprime polynomial matrix fraction descriptions of the plant as shown in the sequel.

Attention is focused on the standard configuration of Fig. 1.2 for generality and on the case in which G_{22} is strictly proper for simplicity. It is of course assumed that the generalized plant is admissible and that the controller is free of unstable hidden poles. The objective is the parameterization of the stable matrix K in (2.63) so that the controller transfer matrix C belongs to the subset of all stabilizing proper controllers. In this regard, it is easier to work with

$$R = C(I - G_{22}C)^{-1} = (I - CG_{22})^{-1}C \tag{2.351}$$

instead of directly with C. Clearly, R is proper when C is proper since $R \to C$ when $G_{22}C \to 0$. Conversely, from

$$C = (I + RG_{22})^{-1}R = R(I + G_{22}R)^{-1}, \tag{2.352}$$

C is proper when R is proper since $C \to R$ when $G_{22}R \to 0$. That is, when G_{22} is strictly proper C is proper iff R is proper; hence, parameterizing all proper C is equivalent to parameterizing all proper R. It is noted before proceeding that the requirement that det $(X_1 - KB) \not\equiv 0$ assures that the inverses in (2.352) exist. This fact follows from

$$I + RG_{22} = I + (I - CG_{22})^{-1}CG_{22} = (I - CG_{22})^{-1} = A_1(X_1 - KB) \tag{2.353}$$

which yields

$$\det(I + G_{22}R) = \det(I + RG_{22}) = (\det A_1)\det(X_1 - BK) \not\equiv 0. \tag{2.354}$$

For the C given by (2.63) one gets

$$R = (Y + A_1K)A = A_1(Y_1 + KA). \tag{2.355}$$

It is convenient at this point to introduce two stable real rational matrices Λ and Ω which have the property that Λ^{-1} and Ω^{-1} are stable matrices and, moreover, $A_1\Lambda^{-1}$ and $\Omega^{-1}A$ together with their inverses are proper matrices. [The structure of the matrix R in (45) of Park and Bongiorno (1989) motivated the introduction of the matrices Λ and Ω.] There are a number of ways in which to construct such matrices. When the polynomial matrix A is row reduced and the polynomial matrix A_1 is column reduced (see Kailath 1980), diagonal matrices Λ and Ω can be chosen with stable polynomials for the diagonal elements. Then, the matrix $A_1\Lambda^{-1}$ and its inverse are proper matrices when the degree of the diagonal element in the j-column of Λ is chosen equal to the column degree of the j-column of A_1 for each possible j and the matrix $\Omega^{-1}A$ and its inverse are proper matrices when the degree of the diagonal element in the i-row of Ω is chosen equal to the row degree of the i-row of A for each possible i. It should also be noted for analog systems that in Antsaklis (1989) the matrices Π and $\tilde{\Pi}$ are acceptable choices for Λ^{-1} and Ω^{-1}, respectively. Another way to construct the matrices Λ and Ω for analog systems which works even if A is not row reduced and A_1 is not column reduced is given in Park et al. (1999). A description of this methodology is presented later.

Introducing the transformation

$$K = \Lambda^{-1}Q\Omega^{-1} \tag{2.356}$$

into the first expression for R in (2.355) yields

$$R = A_1\Lambda^{-1}(\Lambda A_1^{-1}Y\Omega + Q)\Omega^{-1}A. \tag{2.357}$$

Since Λ, Ω, and each of their inverses are stable matrices, it follows from (2.356) that K is stable iff Q is stable. Another important observation to make is that one can always use the polynomial division algorithm if needed to write

$$\Lambda A_1^{-1}Y\Omega = \{\Lambda A_1^{-1}Y\Omega\}_{\infty} + \{\Lambda A_1^{-1}Y\Omega\}_{sp}, \tag{2.358}$$

where the first term on the right-hand side is a polynomial matrix and the second is a strictly proper rational matrix. One can then make the substitution

$$Q = Z - \{\Lambda A_1^{-1} Y\Omega\}_\infty; \tag{2.359}$$

clearly, Q is stable iff Z is stable. Hence, for any stabilizing controller C

$$R = A_1\Lambda^{-1}(\{\Lambda A_1^{-1} Y\Omega\}_{sp} + Z)\Omega^{-1}A, \tag{2.360}$$

where Z is a stable real rational matrix. Moreover, since $A_1\Lambda^{-1}$ and $\Omega^{-1}A$ are biproper matrices (i.e., they are proper together with their inverses), R is proper iff Z is proper. Hence, the controllers described by (2.63) are proper stabilizing controllers iff

$$K = \Lambda^{-1}(Z - \{\Lambda A_1^{-1} Y\Omega\}_\infty)\Omega^{-1}, \tag{2.361}$$

where Z is a proper stable real rational matrix such that

$$\chi = \det(X_1 - KB) \not\equiv 0. \tag{2.362}$$

It is established next that (2.362) is automatically satisfied and imposes no additional constraints on Z. Substituting (2.361) into (2.362) and using (2.358) one easily gets after (2.65) is recalled that $\chi = \chi_o / \det A_1$, where

$$\chi_o = \det[I - A_1\Lambda^{-1}(\{\Lambda A_1^{-1} Y\Omega\}_{sp} + Z)\Omega^{-1}B]. \tag{2.363}$$

Hence, $\chi \not\equiv 0$ iff $\chi_o \not\equiv 0$. Now from $\Omega^{-1}B = -\Omega^{-1}AG_{22}$ it follows that $\Omega^{-1}B$ is strictly proper since $\Omega^{-1}A$ is proper and G_{22} is strictly proper. Since $\chi_o \to 1$ when

$$A_1\Lambda^{-1}(\{\Lambda A_1^{-1} Y\Omega\}_{sp} + Z)\Omega^{-1}B \to 0, \tag{2.364}$$

it readily follows that (2.362) is automatically satisfied for any proper Z. That is

Theorem 2.8 *When the generalized plant in the standard configuration is admissible, G_{22} is strictly proper, and the controller is free of unstable hidden poles, then all proper stabilizing controller transfer matrices are given by*

$$C = (X_1 - KB)^{-1}(Y_1 + KA) = (Y + A_1K)(X - B_1K)^{-1} \tag{2.365}$$

with

$$K = \Lambda^{-1}(Z - \{\Lambda A_1^{-1} Y\Omega\}_\infty)\Omega^{-1}, \tag{2.366}$$

where Z is a proper stable real rational matrix and where the real rational matrices Λ and Ω are stable together with their inverses and selected so that $A_1\Lambda^{-1}$ and $\Omega^{-1}A$ are biproper. It is always possible to construct the matrices Λ and Ω.

Remark 2.15 Substituting (2.366) into (2.365) leads to

$$C = (\tilde{X}_1 - Z\tilde{B})^{-1}(\tilde{Y}_1 + Z\tilde{A}) = (\tilde{Y} + \tilde{A}_1 Z)(\tilde{X} - \tilde{B}_1 Z)^{-1} \tag{2.367}$$

where

$$\tilde{A} = \Omega^{-1}A, \tilde{B} = \Omega^{-1}B, \tilde{A}_1 = A_1\Lambda^{-1}, \tilde{B}_1 = B_1\Lambda^{-1}, \tag{2.368}$$

$$\tilde{X} = X\Omega + B_1\Lambda^{-1}\{\Lambda A_1^{-1}Y\Omega\}_\infty = A^{-1}\Omega - B_1\Lambda^{-1}\{\Lambda A_1^{-1}Y\Omega\}_{sp}, \tag{2.369}$$

$$\tilde{Y} = Y\Omega - A_1\Lambda^{-1}\{\Lambda A_1^{-1}Y\Omega\}_\infty = A_1\Lambda^{-1}\{\Lambda A_1^{-1}Y\Omega\}_{sp}, \tag{2.370}$$

$$\tilde{X}_1 = \Lambda X_1 + \{\Lambda A_1^{-1}Y\Omega\}_\infty\Omega^{-1}B = \Lambda A_1^{-1} - \{\Lambda A_1^{-1}Y\Omega\}_{sp}\Omega^{-1}B, \tag{2.371}$$

and

$$\tilde{Y}_1 = \Lambda Y_1 - \{\Lambda A_1^{-1}Y\Omega\}_\infty\Omega^{-1}A = \{\Lambda A_1^{-1}Y\Omega\}_{sp}\Omega^{-1}A \tag{2.372}$$

are all stable proper real rational matrices that satisfy (see Example 2.15)

$$\begin{bmatrix} \tilde{A} & \tilde{B} \\ \tilde{Y}_1 & -\tilde{X}_1 \end{bmatrix}\begin{bmatrix} \tilde{X} & \tilde{B}_1 \\ \tilde{Y} & -\tilde{A}_1 \end{bmatrix} = \begin{bmatrix} \tilde{X} & \tilde{B}_1 \\ \tilde{Y} & -\tilde{A}_1 \end{bmatrix}\begin{bmatrix} \tilde{A} & \tilde{B} \\ \tilde{Y}_1 & -\tilde{X}_1 \end{bmatrix} = \begin{bmatrix} I & 0 \\ 0 & I \end{bmatrix}. \tag{2.373}$$

Actually, $\tilde{B}$, $\tilde{B}_1$, $\tilde{Y}$, and $\tilde{Y}_1$ are strictly proper and $\tilde{X}$ and $\tilde{X}_1$ are biproper. This corresponds to the parameterization of stabilizing controllers derived using proper stable coprime real rational matrix fractions in the literature cited previously in this section.

Remark 2.16 It is easy to see from (2.351) and (2.352) when G_{22} is proper that R is strictly proper iff C is strictly proper. Hence, Theorem 2.8 holds for proper G_{22} except that now all strictly proper stabilizing controller transfer matrices are given by (2.365) and (2.366), where Z is a strictly proper stable real rational matrix.

Remark 2.17 It should be clear from (2.366) that K is stable iff Z is stable. Hence, (2.367) is an alternative parameterization of all stabilizing controller transfer matrices in terms of a stable real rational matrix Z. When G_{22} is proper, but not strictly proper, it still follows from (2.368) through (2.372) that $\tilde{X}$ and $\tilde{X}_1$ are biproper; now, however, $\tilde{B}$ and $\tilde{B}_1$ are proper, but not strictly proper. So it is no longer automatically true for the subset of all proper stable Z that $(\tilde{X}_1 - Z\tilde{B})$ and $(\tilde{X} - \tilde{B}_1 Z)$ together with their inverses are proper. That is, it is no longer assured without additional restrictions that C is proper when Z is proper or that it is even necessary that Z be proper for proper C. However, when one imposes well-posedness of the feedback loop as the proponents of proper stable coprime rational matrix fraction descriptions do (e.g., see Zhou, Doyle, and Glover, Sect. 5.2, 1996) then this ambiguity disappears. Well-posedness of the feedback loop requires that not only is C a proper stabilizing controller transfer matrix, but one for which all closed-loop transfer matrices are proper including $(I - G_{22}C)^{-1}$. One then gets from the first expression for C in (2.367) that $Z = (\tilde{X}_1 C - \tilde{Y}_1)(\tilde{A} + \tilde{B}C)^{-1} = (\tilde{X}_1 C - \tilde{Y}_1)(I - G_{22}C)^{-1}\tilde{A}^{-1}$ must be proper. Moreover, from $(\tilde{X}_1 - Z\tilde{B})\tilde{A}_1 + (\tilde{Y}_1 + Z\tilde{A})\tilde{B}_1 = I$, it follows that $(\tilde{X}_1 - Z\tilde{B})^{-1} = \tilde{A}_1 + C\tilde{B}_1$ must be proper when C is proper. So, from the fact that $\tilde{X}_1$ is biproper and $\det(\tilde{X}_1 - Z\tilde{B}) = (\det \tilde{X}_1)\det(I - Z\tilde{B}\tilde{X}_1^{-1})$, the first expression for C in (2.367) is proper iff $\det[I - Z(\infty)\tilde{B}(\infty)\tilde{X}_1^{-1}(\infty)] \neq 0$. That is, when G_{22} is proper, the first expression for C in (2.367) parameterizes all proper stabilizing controller transfer matrices for which the feedback loop is well-posed in terms of a proper stable real rational matrix Z for which $\det[I - Z(\infty)\tilde{B}(\infty)\tilde{X}_1^{-1}(\infty)] \neq 0$. In similar fashion, one can establish when G_{22} is proper that the second expression for C in (2.367) parameterizes all proper stabilizing controllers for which the feedback loop is well-posed in terms of a proper stable real rational matrix Z for which $\det[I - \tilde{X}^{-1}(\infty)\tilde{B}_1(\infty)Z(\infty)] \neq 0$.

The procedure given in Park et al. (1999) for constructing the matrices Λ and Ω for analog systems is especially relevant in connection with the Wiener–Hopf methodology; hence, the essential ideas are presented here. Basically, Λ and Ω are chosen so that

$$A_{1*}A_1 + B_{1*}B_1 = \Lambda_*\Lambda \tag{2.374}$$

and

$$AA_* + BB_* = \Omega\Omega_* \tag{2.375}$$

where Λ is a right Wiener–Hopf factor and Ω is a left Wiener–Hopf factor (Appendix C). The matrices Λ and Ω are polynomial matrices because the left-hand sides of the equations are polynomial matrices (see Corollary 2 of Theorem 2 in Youla 1961). Hence, Λ and Ω are stable. Moreover, Λ^{-1} and Ω^{-1} are analytic in $\mathrm{Re}\, s > 0$. In addition, since A_1, B_1 is right coprime and A, B is left coprime, $[A_1' \quad B_1']'$ has full column rank and $[A \quad B]$ has full row rank at any point on the finite $s = j\omega$ axis. It then follows that on the finite imaginary axis

$$\alpha^*[A_{1*} \quad B_{1*}]\begin{bmatrix} A_1 \\ B_1 \end{bmatrix}\alpha = \alpha^*\Lambda_*\Lambda\alpha = \|\Lambda\alpha\|^2 \neq 0 \tag{2.376}$$

and

$$\beta^*[A \quad B]\begin{bmatrix} A_* \\ B_* \end{bmatrix}\beta = \beta^*\Omega\Omega_*\beta = \|\Omega^*\beta\|^2 \neq 0 \tag{2.377}$$

for all nonzero constant vectors α, β. That is, $\det \Lambda \neq 0$ and $\det \Omega \neq 0$ on $s = j\omega$, ω finite; therefore, Λ^{-1} and Ω^{-1} are stable matrices.

One also has from (2.374) that

$$I + G_{22*}G_{22} = A_{1*}^{-1}\Lambda_*\Lambda A_1^{-1} \tag{2.378}$$

and

$$\Lambda_*^{-1}A_{1*}A_1\Lambda^{-1} + \Lambda_*^{-1}B_{1*}B_1\Lambda^{-1} = I. \tag{2.379}$$

Recalling that G_{22} is strictly proper, setting $s = j\omega$, and taking the trace of each of these equations leads to the fact that each element of $\Lambda(j\omega)A_1^{-1}(j\omega)$ and each element of $A_1(j\omega)\Lambda^{-1}(j\omega)$ must have a finite magnitude as $\omega \to \infty$. Hence, ΛA_1^{-1} and $A_1\Lambda^{-1}$ are proper. A similar development beginning with (2.346) would confirm that $\Omega^{-1}A$ and $A^{-1}\Omega$ are proper.

An interesting question concerning the controller transfer matrix given by (2.367) is whether or not it can be realized as an interconnection with one separate block for Z. Then all other blocks in the realization would be fixed by the given data and any later change in the choice of the design parameter Z would require only a change in the one block realizing it. Specifically, a tunable realization of the controller in the form shown in Fig. 2.2 is desired.

Straightforward analysis yields $v = Cy_m$, where

$$C = G_{vy} + G_{vz}Z(I - G_{zz}Z)^{-1}G_{zy}. \tag{2.380}$$

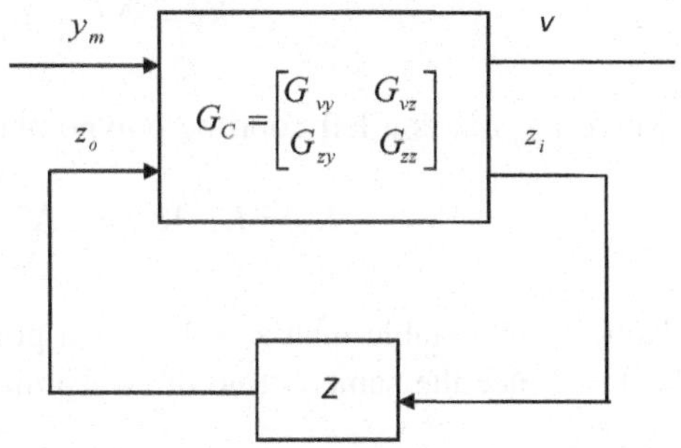

Fig. 2.2 Tunable controller realization

Since $\tilde{A}_1$ is square, it follows from (2.367) that Z has the same number of rows as C; hence, G_{vz} is square. Assuming it is also invertible and using the identity $Z(I - G_{zz}Z)^{-1} = (I - ZG_{zz})^{-1}Z$ then gives

$$C = G_{vy} + (G_{vz}^{-1} - ZG_{zz}G_{vz}^{-1})^{-1}ZG_{zy} \tag{2.381}$$

or

$$C = (G_{vz}^{-1} - ZG_{zz}G_{vz}^{-1})^{-1}[G_{vz}^{-1}G_{vy} + Z(G_{zy} - G_{zz}G_{vz}^{-1}G_{vy})]. \tag{2.382}$$

Equation (2.382) has the same form as the first expression for C in (2.367); it is now clear that one can choose

$$G_{vz} = \tilde{X}_1^{-1}, \quad G_{zz} = \tilde{B}\tilde{X}_1^{-1}, \quad G_{vy} = \tilde{X}_1^{-1}\tilde{Y}_1, \tag{2.383}$$

and

$$G_{zy} = \tilde{A} + \tilde{B}\tilde{X}_1^{-1}\tilde{Y}_1. \tag{2.384}$$

The assumption that G_{vz} is invertible is indeed sustained since $\tilde{X}_1$ is biproper. It is also important to recognize that the controller subsystem transfer matrix

$$G_C = \begin{bmatrix} G_{vy} & G_{vz} \\ G_{zy} & G_{zz} \end{bmatrix} = \begin{bmatrix} I & 0 \\ \tilde{B} & I \end{bmatrix} \begin{bmatrix} \tilde{X}_1^{-1} & 0 \\ 0 & I \end{bmatrix} \begin{bmatrix} \tilde{Y}_1 & I \\ \tilde{A} & 0 \end{bmatrix} \tag{2.385}$$

is proper and, therefore has a state-variable realization. Moreover, for any unstable pole p_o, it follows from (2.385) that $\delta(p_o; G_C) \leq \delta(p_o; \tilde{X}_1^{-1})$. On the other hand, since $G_{vz} = \tilde{X}_1^{-1}$ is a submatrix of G_C, one also has $\delta(p_o; G_C) \geq \delta(p_o; \tilde{X}_1^{-1})$. Thus, $\delta(p_o; G_C) = \delta(p_o; \tilde{X}_1^{-1})$. From (2.371),

$$\tilde{X}_1 = \Lambda L_o^{-1}(L_oX_1 - M_oB), \tag{2.386}$$

where L_o, M_o is a left coprime polynomial matrix pair satisfying

$$K_o = L_o^{-1}M_o = -\Lambda^{-1}\{\Lambda A_1^{-1}Y\Omega\}_{\infty}\Omega^{-1}. \tag{2.387}$$

Now K_o is a stable matrix and L_o is a polynomial matrix; hence, L_o and L_o^{-1} are stable. Since the same is true of Λ, it follows that

$$\delta(p_o; G_C) = \delta(p_o; (L_oX_1 - M_oB)^{-1}L_o\Lambda^{-1}) = \delta(p_o; (L_oX_1 - M_oB)^{-1}) = m_o, \tag{2.388}$$

where m_o is the multiplicity of p_o as a zero of

$$\psi_o = \eta_o \det (L_oX_1 - M_oB), \tag{2.389}$$

the characteristic denominator of $(L_oX_1 - M_oB)^{-1}$. These facts are used in the sequel to establish the conditions under which the tunable controller realization will have no unstable hidden poles.

In order to address the issue of unstable hidden poles, the characteristic polynomial for the tunable controller realization is needed. Clearly, the controller realization shown in Fig. 2.2 has the same form as the standard configuration shown in Fig. 1.2. It then follows by analogy with (2.24) that the characteristic polynomial for the controller is given by

$$\Delta_C = \Delta_G\Delta_Z \det (I - ZG_{zz}) = \frac{\Delta_G\Delta_Z\det (\tilde{X}_1 - Z\Omega^{-1}B)}{\det \tilde{X}_1} \tag{2.390}$$

or

$$\Delta_C = \frac{\eta_o\Delta_G\Delta_Z(\det L_o)\det (\tilde{X}_1 - Z\Omega^{-1}B)}{\psi_o \det \Lambda}, \tag{2.391}$$

where Δ_G and Δ_Z are the characteristic polynomials for the subsystems G_C and Z, respectively.

Recalling (2.361) yields

$$\tilde{X}_1 - Z\Omega^{-1}B = \Lambda(X_1 - KB) = \Lambda L^{-1}(LX_1 - MB) \tag{2.392}$$

when $K = L^{-1}M$ is a left coprime polynomial matrix fraction description for K. It also follows from (2.365) that

$$C = (LX_1 - MB)^{-1}(LY_1 + MA). \tag{2.393}$$

It has already been established that (2.393) is a left coprime polynomial matrix fraction description for C: see (2.51) through (2.57). Hence,

$$\psi_C = \eta_C \det (LX_1 - MB) \tag{2.394}$$

is the characteristic denominator for the tunable controller realization. Using (2.392) and (2.394) in (2.391) gives

$$\Delta_C = \left(\frac{\eta_o}{\eta_C}\right)\left(\frac{\det L_o}{\det L}\right)\left(\frac{\Delta_G}{\psi_o}\right)\Delta_Z\psi_C. \tag{2.395}$$

Now η_o and η_C are simply the scaling constants that make ψ_o and ψ_C monic polynomials. The polynomials $\det L_o$ and $\det L$ are stable. So there are no unstable hidden poles iff there are no zeros of Δ_C/ψ_C in the unstable region, Re $s \geq 0$ for analog systems or $|\lambda| \leq 1$ for digital systems. Clearly, this is the case iff

$$\frac{\Delta_G\Delta_Z}{\psi_o} = \frac{h_G h_Z \psi_G \psi_Z}{\psi_o} \tag{2.396}$$

has no zeros in the unstable region. The polynomials h_G and h_Z account for the hidden poles and ψ_G and ψ_Z are the characteristic denominators of G_C and Z, respectively. On account of (2.388) and (2.389), it follows that every zero of ψ_G in the unstable region is canceled exactly, multiplicity included, by a corresponding zero of ψ_o. Moreover, ψ_Z is stable because Z is stable. Hence, there are no unstable hidden poles iff h_G and h_Z are stable polynomials. That is, there are no unstable hidden poles iff the block G_C and the block Z are realized without unstable hidden poles. This is of course assured when minimal state-variable realizations of G_C and Z are employed.

2.8 Historical Perspective and Commentary

The use of polynomial matrix descriptions and of coprime polynomial matrix fractions in modeling linear systems was popularized by Rosenbrock (1970). Early advances in the characterization and parameterization of stabilizing multivariable controllers were made in Larin et al. (1971, 1972), Kučera (1974), and Youla et al. (1976b) with the parameterization of all stabilizing controllers in the latter attracting particular notoriety. Specific results for 2DOF and 3DOF systems are given in Youla and Bongiorno (1985) and in Park and Bongiorno (1990), respectively; a parameterization of all stabilizing controllers for the standard configuration is employed in Park and Bongiorno (1989). A comprehensive review and comparison of these parameterizations of all stabilizing controllers is given in Aliev and Larin (2008) including ones cited next which involve real rational matrix fractions.

The alternative parameterization of stabilizing controllers in terms of stable coprime real rational matrix fractions was initiated for 1DOF systems in Desoer et al. (1980) and applied to 2DOF systems in Desoer and Gustafson (1984) and in

Vidyasagar (1985). This approach was followed by Nett (1986) to treat stability for the most general of configurations. In all of these works, stability is considered from an input–output viewpoint. That is, instead of defining stability in terms of the response to arbitrary initial conditions as done here, stability is defined in terms of all possible closed-loop transfer matrices with auxiliary inputs injected. Such an approach was not followed in this chapter since more fundamentally stability is connected to system response to initially stored energy or, equivalently, initial conditions. It then follows for stabilized systems that all closed-loop transfer matrices of interest are stable—the definition of stability in these papers. That is, stability with respect to initial conditions assures stable closed-loop transfer matrices, but the converse is not automatically true on account of the possibility of unstable hidden poles. It is appropriate to point out however, that Lemma 2.3 corresponds to Lemma 31 of Nett (1986) where the complete set of conditions of this kind first appeared.

Of the early papers on tracking and disturbance rejection (or regulation), the one by Saeks and Murray (1981) is closest conceptually with the material presented in Sect. 2.6. Here, however, stable controller output transforms in addition to stable error transforms are required on the one hand while on the other hand no restriction that the controller be proper is imposed. In this way, the parameterization of the subset of stabilizing controllers for which tracking and disturbance rejection are achieved is made more pertinent and more transparent. The Wiener–Hopf methodology in which an $\mathcal{H}_2$ performance measure that includes penalties for large plant inputs and system errors then determines whether or not an analog controller should be proper. The question of causality in digital systems is not linked to proper controllers when λ-transforms are used and so there is no need to impose such a condition in this case.

The structure of the tunable controller realization for proper stabilizing controllers first appeared in Doyle et al. (1989).

2.9 Additional Examples

Example 2.2 The generalized plant in the standard configuration of Fig. 1.2 can typically be viewed as an interconnection of nondegenerate dynamic subsystems as depicted in Fig. 2.3. Each subsystem G_q, $q = 1 \to N$, has the polynomial matrix description

$$T_q(p)\xi_q(p) = U_q(p)u_q(p) + i_{T_q}(p)$$
$$y_q(p) = V_q(p)\xi_q(p) + W_q(p)u_q(p) + i_{y_q}(p)$$

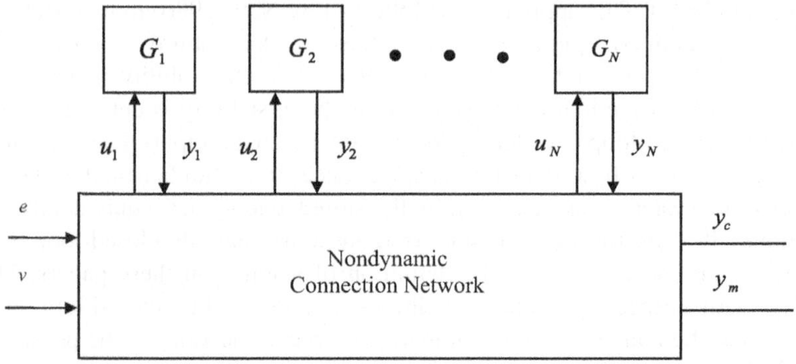

Fig. 2.3 Generalized plant configuration for Example 2.2 (© 1989 IEEE. Reprinted with permission, from Park and Bongiorno (1989))

and the associated transfer matrix

$$G_q(p) = V_q(p)T_q^{-1}(p)U_q(p) + W_q(p).$$

The dynamical subsystems can be compactly described in terms of the block diagonal matrix

$$G = \text{diag }\{G_1, G_2, \ldots, G_N\}$$

and the vectors

$$\xi = [\xi_1' \; \xi_2' \cdots \xi_N']', u = [u_1' \; u_2' \cdots u_N']', \text{and}\, y = [y_1' \, y_2' \cdots y_N']'.$$

The connection network is characterized by a constant matrix M according to

$$\begin{bmatrix} u \\ y_c \\ y_m \end{bmatrix} = \begin{bmatrix} M_{11} & M_{12} & M_{13} \\ M_{21} & M_{22} & M_{23} \\ M_{31} & M_{32} & M_{33} \end{bmatrix} \begin{bmatrix} e \\ v \\ y \end{bmatrix} = M \begin{bmatrix} e \\ v \\ y \end{bmatrix}.$$

It is also convenient to introduce

$$T = \text{diag }\{T_1, T_2, \ldots, T_N\}, U = \text{diag }\{U_1, U_2, \ldots, U_N\},$$
$$V = \text{diag }\{V_1, V_2, \ldots, V_N\}, W = \text{diag }\{W_1, W_2, \ldots, W_N\},$$
$$i_T = [i_{T_1}' \; i_{T_2}' \cdots i_{T_N}']' \text{ and } i_y = [i_{Y_1}' \; i_{Y_2}' \cdots i_{Y_N}']'.$$

(a) Establish that the characteristic polynomial for the generalized plant is given by

$$\Delta_G = \eta\,(\det T)\det(I - GM_{13}),$$

where η is a constant chosen so that Δ_G is monic.

(b) Let $\tilde{G}$ denote the transfer matrix from $[e'\,v']'$ to $[y_c'\,y_m']'$ in Fig. 2.3. That is, $\tilde{G}$ is the transfer matrix of the generalized plant. Show that the appropriate expression to use for G_{22} in (2.24) is

$$\tilde{G}_{22} = M_{32} + M_{33}(I - GM_{13})^{-1}GM_{12}.$$

(c) It is important to recognize that the formula for the characteristic polynomial Δ_G applies to any set of interconnected dynamical systems and the matrix M_{13} is determined by the feedback from the vector of subsystem outputs y to the vector of subsystem inputs u through the connection network. For example, the standard configuration in Fig. 1.2 is a special case of an interconnection with $N = 2$, $G_1 = \tilde{G}$, $G_2 = C$, $u_1 = [e'v']'$, $y_1 = [y_c'\,y_m']'$, $u_2 = [0\,I]y_1 = y_m$, $y_2 = v$. Then from

$$\left[\begin{array}{c} u_1 \\ \hline u_2 \end{array}\right] = \left[\begin{array}{c} e \\ v \\ \hline y_m \end{array}\right] = \left[\begin{array}{c} I \\ 0 \\ \hline 0 \end{array}\right] e + \left[\begin{array}{cc|c} 0 & 0 & 0 \\ 0 & 0 & I \\ \hline 0 & I & 0 \end{array}\right] \left[\begin{array}{c} y_1 \\ \hline y_2 \end{array}\right]$$

show that one gets (2.24).

Solution (a) Clearly,

$$\begin{aligned} T\xi &= Uu + i_T \\ y &= V + Wu + i_y. \end{aligned}$$

Using

$$u = [M_{11} \quad M_{12}] \begin{bmatrix} e \\ v \end{bmatrix} + M_{13}y$$

then gives

$$\begin{aligned} T\xi &= U[M_{11} \quad M_{12}] \begin{bmatrix} e \\ v \end{bmatrix} + UM_{13}y + i_T \\ y &= V\xi + W[M_{11} \quad M_{12}] \begin{bmatrix} e \\ v \end{bmatrix} + WM_{13}y + i_y \end{aligned}$$

or

$$\begin{bmatrix} T & -UM_{13} \\ -V & (I-WM_{13}) \end{bmatrix} \begin{bmatrix} \xi \\ y \end{bmatrix} = \begin{bmatrix} U \\ W \end{bmatrix} [M_{11} \quad M_{12}] \begin{bmatrix} e \\ v \end{bmatrix} + \begin{bmatrix} i_T \\ i_y \end{bmatrix}.$$

In addition one has

$$\begin{bmatrix} y_c \\ y_m \end{bmatrix} = \begin{bmatrix} M_{23} \\ M_{33} \end{bmatrix} [0 \quad I] \begin{bmatrix} \xi \\ y \end{bmatrix} + \begin{bmatrix} M_{21} & M_{22} \\ M_{31} & M_{32} \end{bmatrix} \begin{bmatrix} e \\ v \end{bmatrix}.$$

This pair of equations is of the form

$$T_G \begin{bmatrix} \xi \\ y \end{bmatrix} = U_G \begin{bmatrix} e \\ v \end{bmatrix} + \begin{bmatrix} i_T \\ i_y \end{bmatrix}$$
$$\begin{bmatrix} y_c \\ y_m \end{bmatrix} = V_G \begin{bmatrix} \xi \\ y \end{bmatrix} + W_G \begin{bmatrix} e \\ v \end{bmatrix}$$

and represents a polynomial matrix description for the generalized plant. Hence,

$$\begin{aligned} \Delta_G = \eta \det T_G = \eta \det \begin{bmatrix} T & -UM_{13} \\ -V & (I-WM_{13}) \end{bmatrix} &= \eta \det \left(\begin{bmatrix} I & 0 \\ VT^{-1} & I \end{bmatrix} \begin{bmatrix} T & -UM_{13} \\ -V & (I-WM_{13}) \end{bmatrix} \right) \\ = \eta \det \begin{bmatrix} T & -UM_{13} \\ 0 & (I-GM_{13}) \end{bmatrix} &= \eta \, (\det T) \det\, (I-GM_{13}) \end{aligned}.$$

Note that the generalized plant is nondegenerate iff det $(I - GM_{13}) \not\equiv 0$

(b) One has with $i_T = 0$ and $i_y = 0$ that

$$y = Gu = G[M_{11} \quad M_{12}] \begin{bmatrix} e \\ v \end{bmatrix} + GM_{13}y.$$

It is assumed that the generalized plant is nondegenerate so that

$$y = (I - GM_{13})^{-1} G [M_{11} \quad M_{12}] \begin{bmatrix} e \\ v \end{bmatrix}.$$

From

$$\begin{aligned} \begin{bmatrix} y_c \\ y_m \end{bmatrix} &= \begin{bmatrix} M_{23} \\ M_{33} \end{bmatrix} [0 \quad I] \begin{bmatrix} \xi \\ y \end{bmatrix} + \begin{bmatrix} M_{21} & M_{22} \\ M_{31} & M_{32} \end{bmatrix} \begin{bmatrix} e \\ v \end{bmatrix} \\ &= \begin{bmatrix} M_{23} \\ M_{33} \end{bmatrix} y + \begin{bmatrix} M_{21} & M_{22} \\ M_{31} & M_{32} \end{bmatrix} \begin{bmatrix} e \\ v \end{bmatrix} = \tilde{G} \begin{bmatrix} e \\ v \end{bmatrix}, \end{aligned}$$

it then follows that

$$\tilde{G} = \begin{bmatrix} M_{21} & M_{22} \\ M_{31} & M_{32} \end{bmatrix} + \begin{bmatrix} M_{23} \\ M_{33} \end{bmatrix} (I - GM_{13})^{-1} G \,[\, M_{11} \quad M_{12} \,].$$

Thus,

$$\tilde{G}_{22} = [0 \quad I\,]\,\tilde{G} \begin{bmatrix} 0 \\ I \end{bmatrix} = M_{32} + M_{33}(I - GM_{13})^{-1} GM_{12}.$$

Now in (2.24), G_{22} is the matrix in the second block row, second block column of the generalized plant transfer matrix which in this example is given by $\tilde{G}_{22}$.

(c) For the standard configuration, it is evident that

$$M_{13} = \left[\begin{array}{cc|c} 0 & 0 & 0 \\ 0 & 0 & I \\ \hline 0 & I & 0 \end{array}\right],$$

$$G = \operatorname{diag}\{G_1, G_2\} = \operatorname{diag}\{\tilde{G}, C\} = \operatorname{diag}\left\{\begin{bmatrix} \tilde{G}_{11} & \tilde{G}_{12} \\ \tilde{G}_{21} & \tilde{G}_{22} \end{bmatrix}, C\right\},$$

$$\det T = (\det T_1)(\det T_2) = (\det T_{\tilde{G}})(\det T_C).$$

Hence,

$$\begin{aligned}(\det T)\det(I - GM_{13}) &= (\det T_{\tilde{G}})\,(\det T_C)\det \begin{bmatrix} I & 0 & -\tilde{G}_{12} \\ 0 & I & -\tilde{G}_{22} \\ 0 & -C & I \end{bmatrix} \\ &= (\det T_{\tilde{G}})\,(\det T_C)\det \begin{bmatrix} I & -\tilde{G}_{22} \\ -C & I \end{bmatrix} \\ &= (\det T_{\tilde{G}})\,(\det T_C)\det \begin{bmatrix} I & -\tilde{G}_{22} \\ 0 & (I - C\tilde{G}_{22}) \end{bmatrix} = (\det T_{\tilde{G}})\,(\det T_C)\det(I - C\tilde{G}_{22}),\end{aligned}$$

which corresponds to (2.24).

Example 2.3 Suppose $G_{22} = -A^{-1}B = -B_1A_1^{-1}$, $AX + BY = I$, and $\tilde{X}_1A_1 + \tilde{Y}_1B_1 = I$. In accordance with Theorem 2.1 for the standard configuration, any stabilizing controller transfer matrix is given by $C = (\tilde{X}_1 - \tilde{K}B)^{-1}(\tilde{Y}_1 + \tilde{K}A)$, where $\tilde{K}$ is a stable real rational matrix. Suppose also with $M \not\equiv 0$ that

$$\begin{bmatrix} A & B \\ \tilde{Y}_1 & -\tilde{X}_1 \end{bmatrix} \begin{bmatrix} X & B_1 \\ Y & -A_1 \end{bmatrix} = \begin{bmatrix} I & 0 \\ -M & I \end{bmatrix}$$

and so a relationship in the form of the generalized Bezout identity (2.62) does not hold. (a) Show nevertheless that as in Lemma 2.2 it is still true that $C = (Y + A_1K)(X - B_1K)^{-1}$, where K is a stable real rational matrix. (b) Show that the pair $\tilde{X}_1, \tilde{Y}_1$ is not unique and that there always exists another choice for which a relationship in the form of (2.62) does hold.

Solution (a) Setting

$$C = (\tilde{X}_1 - \tilde{K}B)^{-1}(\tilde{Y}_1 + \tilde{K}A) = (Y + A_1K)(X - B_1K)^{-1}$$

leads to

$$\begin{aligned}
&(\tilde{Y}_1 + \tilde{K}A)(X - B_1K) = (\tilde{X}_1 - \tilde{K}B)(Y + A_1K) \\
&\tilde{Y}_1X - \tilde{Y}_1B_1K + \tilde{K}AX - \tilde{K}AB_1K = \tilde{X}_1Y + \tilde{X}_1A_1K - \tilde{K}BY - \tilde{K}BA_1K \\
&K = \tilde{K} - \tilde{X}_1Y + \tilde{Y}_1X = \tilde{K} - M.
\end{aligned}$$

So K is a stable real rational matrix since M is a real polynomial matrix and $\tilde{K}$ is a stable real rational matrix.

(b) Clearly,

$$(\tilde{X}_1 - \tilde{M}B)A_1 + (\tilde{Y}_1 + \tilde{M}A)B_1 = I.$$

Hence, $\tilde{X}_1, \tilde{Y}_1$ can be replaced with $(\tilde{X}_1 - \tilde{M}B), (\tilde{Y}_1 + \tilde{M}A)$, where $\tilde{M}$ is any polynomial matrix. Then as in the generalized Bezout identity (2.62), one gets

$$\begin{bmatrix} A & B \\ (\tilde{Y}_1 + \tilde{M}A) & -(\tilde{X}_1 - \tilde{M}B) \end{bmatrix} \begin{bmatrix} X & B_1 \\ Y & -A_1 \end{bmatrix} = \begin{bmatrix} I & 0 \\ (\tilde{M} - M) & I \end{bmatrix} = \begin{bmatrix} I & 0 \\ 0 & I \end{bmatrix}$$

when $\tilde{M} = M$ is chosen.

Example 2.4 For the two coprime polynomial matrix fraction descriptions of $-G_{22}$ given by (2.107) and (2.108) it follows from (2.114) that $\tilde{X}_1 = N_1^{-1}(X_1 - MB), \tilde{Y}_1 = N_1^{-1}(Y_1 + MA)$, where M is an arbitrary real polynomial matrix. (a) In similar fashion, establish when $AX + BY = I$ that $\tilde{X} = (X - B_1M_1)N^{-1}$, $\tilde{Y} = (Y + A_1M_1)N^{-1}$, where M_1 is an arbitrary real polynomial matrix. In addition, show when the generalized Bezout identity (2.62) is satisfied and $M_1 = M$ is chosen that

$$\begin{bmatrix} \tilde{A} & \tilde{B} \\ \tilde{Y}_1 & -\tilde{X}_1 \end{bmatrix} \begin{bmatrix} \tilde{X} & \tilde{B}_1 \\ \tilde{Y} & -\tilde{A}_1 \end{bmatrix} = \begin{bmatrix} \tilde{X} & \tilde{B}_1 \\ \tilde{Y} & -\tilde{A}_1 \end{bmatrix} \begin{bmatrix} \tilde{A} & \tilde{B} \\ \tilde{Y}_1 & -\tilde{X}_1 \end{bmatrix} = \begin{bmatrix} I & 0 \\ 0 & I \end{bmatrix}.$$

(b) It follows from (a) and from Lemma 2.2 by analogy that

$$\tilde{C} = (\tilde{X}_1 - \tilde{K}\tilde{B})^{-1}(\tilde{Y}_1 + \tilde{K}\tilde{A}) = (\tilde{Y} + \tilde{A}_1\tilde{K})(\tilde{X} - \tilde{B}_1\tilde{K})^{-1}$$

when $M_1 = M$. Confirm that the set of all $\tilde{C}$ as $\tilde{K}$ ranges over all stable real rational matrices is identical to the set of all $C = (X_1 - KB)^{-1}(Y_1 + KA) = (Y + A_1K)(X - B_1K)^{-1}$ as K varies over all stable real rational matrices.

Solution

(a) $\tilde{A}\tilde{X} + \tilde{B}\tilde{Y} = N(A\tilde{X} + B\tilde{Y}) = I = NN^{-1}$ gives $A\tilde{X}N + B\tilde{Y}N = I = AX + BY$ or $A(\tilde{X}N - X) + B(\tilde{Y}N - Y) = 0$. Thus, $(\tilde{X}N - X) = -A^{-1}B(\tilde{Y}N - Y) = -B_1A_1^{-1}(\tilde{Y}N - Y)$ and so

$$\begin{aligned} A_1^{-1}(\tilde{Y}N - Y) &= (X_1A_1 + Y_1B_1)A_1^{-1}(\tilde{Y}N - Y) \\ &= X_1(\tilde{Y}N - Y) + Y_1B_1A_1^{-1}(\tilde{Y}N - Y) \\ &= X_1(\tilde{Y}N - Y) - Y_1(\tilde{X}N - X) = M_1 \end{aligned}$$

a polynomial matrix. That is, $\tilde{Y}N - Y = A_1M_1$, $\tilde{X}N - X = -B_1M_1$ or $\tilde{Y} = (Y + A_1M_1)N^{-1}$, $\tilde{X} = (X - B_1M_1)N^{-1}$. It is now easy to confirm that $\tilde{A}\tilde{X} + \tilde{B}\tilde{Y} = I$. Hence, when the generalized Bezout identity (2.62) is satisfied and $M_1 = M$ is chosen,

$$\begin{aligned} \tilde{Y}_1\tilde{X} - \tilde{X}_1\tilde{Y} &= N_1^{-1}(Y_1 + MA)(X - B_1M_1)N^{-1} - N_1^{-1}(X_1 - MB)(Y + A_1M_1)N^{-1} \\ &= N_1^{-1}(Y_1X - Y_1B_1M_1 + MAX - MAB_1M_1 - X_1Y \\ &\quad - X_1A_1M_1 + MBY + MBA_1M_1)N^{-1} \\ &= N_1^{-1}(Y_1X - X_1Y - M_1 + M)N^{-1} = 0 \end{aligned}$$

and it follows that

$$\begin{bmatrix} \tilde{A} & \tilde{B} \\ \tilde{Y}_1 & -\tilde{X}_1 \end{bmatrix}\begin{bmatrix} \tilde{X} & \tilde{B}_1 \\ \tilde{Y} & -\tilde{A}_1 \end{bmatrix} = \begin{bmatrix} \tilde{X} & \tilde{B}_1 \\ \tilde{Y} & -\tilde{A}_1 \end{bmatrix}\begin{bmatrix} \tilde{A} & \tilde{B} \\ \tilde{Y}_1 & -\tilde{X}_1 \end{bmatrix} = \begin{bmatrix} I & 0 \\ 0 & I \end{bmatrix}.$$

(b) One only needs to show the following since (2.115) and (2.117) are already established. By straightforward substitution

$$\begin{aligned} \tilde{C} &= (\tilde{Y} + \tilde{A}_1\tilde{K})(\tilde{X} - \tilde{B}_1\tilde{K})^{-1} = (Y + A_1M_1 + A_1N_1\tilde{K}N)(X - B_1M_1 - B_1N_1\tilde{K}N)^{-1} \\ &= (Y + A_1K)(X - B_1K)^{-1} \text{ where } K = M_1 + N_1\tilde{K}N = M + N_1\tilde{K}N \end{aligned}$$

as in (2.118). Since $M_1 = M$ is polynomial and N and N_1 are unimodular, there is a one-to-one correspondence between every stable K and every stable $\tilde{K}$. So the set of all $\tilde{C}$ and the set of all C are identical.

Example 2.5 For the 2DOF standard configuration shown in Fig. 1.3, establish Definition 2.3 and Theorem 2.2 directly from the polynomial matrix description of the system instead of using the results for the standard configuration as was done in Sect. 2.4. Hint: One can always write $C = [C_1 \quad C_2] = A_C^{-1}B_C = A_C^{-1}[B_{C_y} \quad B_{C_u}]$, where A_C, B_C is a left coprime polynomial matrix pair. Then first establish that $C = (X_1 - KB)^{-1}[(Y_1 + KA) \quad H]$ is necessary, where K and H are stable real rational matrices.

Solution Equations (2.122) through (2.124) describe the generalized plant and are repeated here for convenience:

$$\begin{aligned} T_G\xi_G &= U_1 d + U_2 v + i_{T_G} \\ [y_s' \quad y_t']' &= V_1\xi_G + W_{11}d + W_{12}v + i_c \\ y &= V_2\xi_G + W_{21}d + W_{22}v + i_y. \end{aligned}$$

Next a polynomial matrix description for the controller is introduced:

$$\begin{aligned} T_C\xi_C &= U_{C_1}y + U_{C_2}u + i_{T_C} \\ v &= V_C\xi_C + W_{C_1}y + W_{C_2}u + i_v. \end{aligned}$$

Thus,

$$\begin{aligned} T_C\xi_C &= U_{C_1}V_2\xi_G + U_{C_1}W_{21}d + U_{C_1}W_{22}v + U_{C_1}i_y + U_{C_2}u + i_{T_C} \\ v &= V_C\xi_C + W_{C_1}V_2\xi_G + W_{C_1}W_{21}d + W_{C_1}W_{22}v + W_{C_1}i_y + W_{C_2}u + i_v \end{aligned}$$

and one gets

$$\begin{bmatrix} T_G & 0 & -U_2 \\ -U_{C_1}V_2 & T_C & -U_{C_1}W_{22} \\ -W_{C_1}V_2 & -V_C & (I - W_{C_1}W_{22}) \end{bmatrix} \begin{bmatrix} \xi_G \\ \xi_C \\ v \end{bmatrix} = \begin{bmatrix} U_1 & 0 \\ U_{C_1}W_{21} & U_{C_2} \\ W_{C_1}W_{21} & W_{C_2} \end{bmatrix} \begin{bmatrix} d \\ u \end{bmatrix} + \begin{bmatrix} i_{T_G} \\ U_{C_1}i_y + i_{T_C} \\ W_{C_1}i_y + i_v \end{bmatrix}.$$

Hence, the system is stable iff T^{-1} is stable, where

$$T = \begin{bmatrix} T_G & 0 & -U_2 \\ -U_{C_1}V_2 & T_C & -U_{C_1}W_{22} \\ -W_{C_1}V_2 & -V_C & (I - W_{C_1}W_{22}) \end{bmatrix}.$$

Now

$$
\begin{aligned}
\det T &= \det \begin{bmatrix} T_G & 0 & 0 \\ -U_{C_1}V_2 & T_C & -U_{C_1}(V_2T_G^{-1}U_2+W_{22}) \\ -W_{C_1}V_2 & -V_C & (I-W_{C_1}(V_2T_G^{-1}U_2+W_{22})) \end{bmatrix} \\
&= \det \begin{bmatrix} T_G & 0 & 0 \\ -U_{C_1}V_2 & T_C & -U_{C_1}G_{22} \\ -W_{C_1}V_2 & -V_C & (I-W_{C_1}G_{22}) \end{bmatrix} \\
&= \det \begin{bmatrix} T_G & 0 & 0 \\ -U_{C_1}V_2 & T_C & 0 \\ -W_{C_1}V_2 & -V_C & (I-(V_CT_C^{-1}U_{C_1}+W_{C_1})\,G_{22}) \end{bmatrix} = (\det T_G)(\det T_C)\det(I-C_1G_{22})
\end{aligned}
$$

since in accordance with (1.4) and (1.5)

$$
C = [C_1 \quad C_2] = [(V_CT_C^{-1}U_{C_1}+W_{C_1}) \quad (V_CT_C^{-1}U_{C_2}+W_{C_2})] = [-C_y \quad C_u].
$$

Writing

$$
C = A_C^{-1}B_C = A_C^{-1}[B_{C_y} \quad B_{C_u}],\, G_{22} = -B_1A_1^{-1} = -A^{-1}B,
$$

where A_C, B_C and A, B are left coprime pairs and B_1, A_1 is a right coprime pair, one then gets

$$
\det T = \left(\frac{\det T_G}{\det A_1}\right)\left(\frac{\det T_C}{\det A_C}\right)\det\,(A_CA_1+B_{C_y}B_1)
$$

or

$$
\begin{aligned}
\Delta = \eta \det T &= \eta\left(\frac{\Delta_G}{\psi_{G_{22}}}\right)\left(\frac{\Delta_C}{\psi_C}\right)\det\,(A_CA_1+B_{C_y}B_1) \\
&= \eta h_G h_C \left(\frac{\psi_G}{\psi_{G_{22}}}\right)\det\,(A_CA_1+B_{C_y}B_1).
\end{aligned}
$$

Since G_{22} is a submatrix of G, $\psi_G/\psi_{G_{22}}$ is a polynomial. Hence, a necessary condition for the existence of a stabilizing controller is that $h_G = \Delta_G/\psi_G$ and $\psi_G/\psi_{G_{22}}$ be stable polynomials or, according to Definition 2.3, that the generalized plant be admissible. Given an admissible plant, one must then pick C so that h_C is stable (i.e., the controller has no unstable hidden poles) and $A_CA_1 + B_{C_y}B_1 = L$, where L is a real polynomial matrix whose determinant is stable. Employing the same methodology that yielded (2.48) and (2.49) one then gets that A_C and B_{C_y} are given by $A_C = LX_1 - MB$, $B_{C_y} = LY_1 + MA$. It then follows that

$$\begin{aligned} C &= (LX_1 - MB)^{-1}\left[\,(LY_1 + MA) \quad B_{C_u}\right] \\ &= (X_1 - KB)^{-1}\left[\,(Y_1 + KA) \quad L^{-1}B_{C_u}\right] = (X_1 - KB)^{-1}\left[\,(Y_1 + KA) \quad H\right], \end{aligned}$$

where $K = L^{-1}M$ and $H = L^{-1}B_{C_u}$ are stable matrices.

It remains to show that any such C is a stabilizing controller. First, given a stable K one can always write $K = L^{-1}M$, where L, M is a left coprime polynomial matrix pair and $\det L$ is stable. Then

$$\begin{aligned} C &= (X_1 - L^{-1}MB)^{-1}\left[\,(Y_1 + L^{-1}MA) \quad H\right] = (LX_1 - MB)^{-1}\left[\,(LY_1 + MA) \quad LH\right] \\ &= \left[\,C_1 \quad C_2\,\right]. \end{aligned}$$

Using the same argument as with (2.51) through (2.56) yields $LX_1 - MB$, $LY_1 + MA$ is a left coprime polynomial pair. Hence, the characteristic denominator for C_1 is

$$\psi_{C_1} = \eta_{C_1} \det\,(LX_1 - MB).$$

It now follows from

$$\det T = (\det T_G)(\det T_C)\det\,(I - C_1 G_{22})$$

that

$$\begin{aligned} \Delta = \eta \det T &= \eta \left(\frac{\Delta_G}{\psi_{G_{22}}}\right)\left(\frac{\Delta_C}{\psi_{C_1}}\right)\det\left[(LX_1 - MB)A_1 + (LY_1 + MA)B_1\right] \\ &= \eta h_G h_C \left(\frac{\psi_G}{\psi_{G_{22}}}\right)\left(\frac{\psi_C}{\psi_{C_1}}\right)\det L. \end{aligned}$$

When h_G, h_C, and $\psi_G/\psi_{G_{22}}$ are stable polynomials, it follows that Δ is a stable polynomial provided ψ_C/ψ_{C_1} is a stable polynomial. This is indeed the case for the following reasons. When p_o is an unstable finite pole of C, one has that $\delta(C;p_o) \leq \delta((LX_1 - MB)^{-1};p_o)$ because $\left[\,(LY_1 + MA) \quad LH\,\right]$ is a stable matrix. On the other hand, $\delta(C,p_o) \geq \delta((LX_1 - MB)^{-1};p_o)$ since $C_1 = (LX_1 - MB)^{-1}$ $(LY_1 + MA)$ is a submatrix of C. Thus, $\delta(C;p_o) = \delta((LX_1 - MB)^{-1};p_o) = m_o$, where m_o is the multiplicity of p_o as a zero of $\det\,(LX_1 - MB)$ or, equivalently, ψ_{C_1}. But m_o is also the multiplicity of p_o as a zero of ψ_C. So every such unstable zero of ψ_C is canceled exactly by the corresponding zero of ψ_{C_1} and ψ_C/ψ_{C_1} is a stable polynomial: that ψ_{C_1} divides ψ_C without remainder is a consequence of the fact that ψ_{C_1} is the characteristic denominator of a submatrix of C.

Example 2.6 Suppose AFP_d, L, P_a, and PA_1 are stable. Suppose in addition that F is square and F and F^{-1} are stable. Use Lemma 2.7 for the 3DOF system to establish that ψ_G/ψ_{FP} is a stable polynomial.

Solution Obviously, all one has to show is that the conditions of Lemma 2.7 are satisfied. Clearly, this is the case if P_aA_1, $P_aA_1Y_1FP_d$, and $(I - PA_1Y_1F)P_d$ are stable. That P_aA_1 is stable follows immediately from P_a stable and A_1 polynomial. It is also apparent that $P_aA_1Y_1FP_d = P_aYAFP_d$ is stable. Finally, on account of the assumed properties of F, one has that $(I - PA_1Y_1F)P_d$ is stable because

$$F(I - PA_1Y_1F)P_d = (I - FPA_1Y_1)FP_d = (I - B_1Y_1)FP_d = XAFP_d$$

is stable.

Example 2.7 Suppose for (2.216) in the case of the 3DOF system one has

$$P_o = \begin{bmatrix} 0 & P_a \\ P_d & P \end{bmatrix} = \begin{bmatrix} 0 & P_a \\ P_pP_b & P_pP_a \end{bmatrix},$$

where P_a and P_b are square stable matrices with stable inverses. Show that $\delta(P_o;p_o) = \delta(P;p_o)$ for any finite unstable pole p_o of P_o and, therefore, that ψ_{P_o}/ψ_P is a stable polynomial. Hence, this condition in Lemma 2.8 is always satisfied in these cases.

Solution From

$$P_o = \begin{bmatrix} 0 & P_a \\ 0 & 0 \end{bmatrix} + \begin{bmatrix} 0 & 0 \\ P_pP_b & P_pP_a \end{bmatrix}$$

and the stability of P_a and P_b one gets that $\delta(P_o;p_o) = \delta(P_p[P_b \;\; P_a];p_o) \le \delta(P_p;p_o)$. Since P_a and P_a^{-1} are stable, $\delta(P;p_o) = \delta(P_pP_a;p_o) = \delta(P_p;p_o)$. Hence, $\delta(P_o;p_o) \le \delta(P;p_o)$. But P is a submatrix of P_o. Thus, $\delta(P_o;p_o) \ge \delta(P;p_o)$ and ψ_{P_o}/ψ_P is a polynomial. It then follows that $\delta(P_o;p_o) = \delta(P;p_o)$ and ψ_{P_o}/ψ_P is a stable polynomial.

Example 2.8 From Lemma 2.8, a necessary condition for the 3DOF system to be admissible is that $\psi_F\psi_P/\psi_{FP}$ must be a stable polynomial. Show that this condition in turn requires that PA_1 and AF be stable where $A^{-1}B$ and $B_1A_1^{-1}$ are left coprime and right coprime polynomial matrix fraction descriptions for FP, respectively. Hint: Introduce the coprime matrix fraction descriptions $F = A_f^{-1}B_f$ and $P = B_{p_1}A_{p_1}^{-1}$. Let the polynomial matrix M be the greatest common right divisor of $B_fB_{p_1}$ and A_{p_1}. Let the polynomial matrix N be the greatest common left divisor of A_f and $B_fB_{p_1}$. Then establish first that det M and det N must be stable polynomials. Next show that $FPA_1 = B_1$ and $AFP = B$ lead to PA_1 and AF stable, respectively.

Solution With M and N as defined, $B_f B_{p_1} = B_{m_1} M$, $A_{p_1} = A_{m_1} M$, where B_{m_1}, A_{m_1} is a right coprime polynomial matrix pair and $A_f = NA_n$, $B_f B_{p_1} = NB_n$, where A_n, B_n is a left coprime polynomial matrix pair. That A_{m_1}, B_{m_1} is right coprime is easily established. For if M_1 is a common right divisor of B_{m_1}, A_{m_1} then $M_1 M$ is also a common right divisor of $B_f B_{p_1}, A_{p_1}$ and so must divide M, the greatest common right divisor, on the right. That is, $M(M_1 M)^{-1} = M_1^{-1}$ must be a polynomial matrix. Hence, $\det(M_1^{-1}) = 1/\det M_1$ must be a polynomial or $\det M_1$ must be a constant. Thus, M_1 must be unimodular. That A_n, B_n is left coprime is also easily established. For if N_1 is a common left divisor of A_n, B_n a dual argument yields that N_1 must be unimodular. Clearly,

$$FP = A^{-1}B = B_1 A_1^{-1} = A_f^{-1} B_f B_{p_1} A_{p_1}^{-1} = A_f^{-1} B_{m_1} A_{m_1}^{-1} = A_n^{-1} B_n A_{p_1}^{-1}$$

and it follows that at any finite point p_o

$$\delta(FP; p_o) = \delta(A^{-1}; p_o) \leq \delta(A_f^{-1}; p_o) + \delta(A_{m_1}^{-1}; p_o)$$

and

$$\delta(FP; p_o) = \delta(A^{-1}; p_o) \leq \delta(A_n^{-1}; p_o) + \delta(A_{p_1}^{-1}; p_o).$$

Hence, $\det A$ is a zero-remainder divisor of both $(\det A_f)(\det A_{m_1})$ and $(\det A_n)(\det A_{p_1})$. So from

$$\begin{aligned}\frac{\psi_F \psi_P}{\psi_{FP}} &= \eta \frac{(\det A_f)(\det A_{p_1})}{\det A} = \eta \frac{(\det A_f)(\det A_{m_1})(\det M)}{\det A} \\ &= \eta \frac{(\det N)(\det A_n)(\det A_{p_1})}{\det A}\end{aligned}$$

one gets that $\psi_F \psi_P / \psi_{FP}$ is a stable polynomial only if $\det M$ and $\det N$ are stable polynomials or, equivalently, M^{-1} and N^{-1} are stable matrices. One also has from

$$FPA_1 = A_f^{-1} B_f B_{p_1} A_{p_1}^{-1} A_1 = A_f^{-1} B_{m_1} A_{m_1}^{-1} A_1 = B_1$$

that $B_{m_1} A_{m_1}^{-1} A_1 = A_f B_1$ is stable. Since B_{m_1}, A_{m_1} is right coprime, it follows that $A_{m_1}^{-1} A_1$ is stable. Hence, $M^{-1} A_{m_1}^{-1} A_1 = A_{p_1}^{-1} A_1$ and, therefore, $PA_1 = B_{p_1} A_{p_1}^{-1} A_1$ must be stable. Similarly, from

$$AFP = AA_f^{-1} B_f B_{p_1} A_{p_1}^{-1} = AA_n^{-1} B_n A_{p_1}^{-1} = B$$

one gets that $AA_n^{-1}B_n = BA_{p_1}$ is stable. Since A_n, B_n is a left coprime pair, it follows that AA_n^{-1} must be stable. Hence, $AA_n^{-1}N^{-1} = AA_f^{-1}$ and, therefore, $AF = AA_f^{-1}B_f$ must be stable.

Example 2.9 Let $G_g = A_g^{-1}B_g = B_{g_1}A_{g_1}^{-1}$, where the real polynomial matrix pairs A_g, B_g and B_{g_1}, A_{g_1} satisfy

$$\begin{bmatrix} A_g & B_g \\ Y_{g_1} & -X_{g_1} \end{bmatrix}\begin{bmatrix} X_g & B_{g_1} \\ Y_g & -A_{g_1} \end{bmatrix} = \begin{bmatrix} X_g & B_{g_1} \\ Y_g & -A_{g_1} \end{bmatrix}\begin{bmatrix} A_g & B_g \\ Y_{g_1} & -X_{g_1} \end{bmatrix} = \begin{bmatrix} I & 0 \\ 0 & I \end{bmatrix}.$$

Establish that all stable real rational matrices K_a, K_b that satisfy $K_aG_g + K_b = \Gamma$ when $\tilde{\Gamma} = \Gamma A_{g_1}$ is a stable matrix are given by

$$K_a = \tilde{\Gamma}Y_{g_1} + K_cA_g, K_b = \tilde{\Gamma}X_{g_1} - K_cB_g,$$

where K_c is a stable real rational matrix.

Solution It is evident that $X_{g_1}A_{g_1} + Y_{g_1}B_{g_1} = I$. Since $\tilde{\Gamma}$ is a stable matrix, it then follows that a particular solution pair for the equation $K_aG_g + K_b = \Gamma$ or, equivalently, $K_aB_{g_1} + K_bA_{g_1} = \tilde{\Gamma}$ is $K_a = \tilde{\Gamma}Y_{g_1}$, $K_b = \tilde{\Gamma}X_{g_1}$. Hence, all potential solution pairs can be written as $K_a = \tilde{\Gamma}Y_{g_1} + K_\alpha$, $K_b = \tilde{\Gamma}X_{g_1} + K_\beta$, where K_α and K_β are stable real rational matrices satisfying $K_\alpha B_{g_1} + K_\beta A_{g_1} = 0$ or $K_\beta = -K_\alpha B_{g_1}A_{g_1}^{-1} = -K_\alpha A_g^{-1}B_g$. Now A_g, B_g is a left coprime polynomial matrix pair. Thus, with K_α stable, K_β is stable iff $K_\alpha A_g^{-1} = K_c$ or $K_\alpha = K_cA_g$, where K_c is a stable real rational matrix. It immediately follows that $K_\beta = -K_\alpha A_g^{-1}B_g = -K_cB_g$. That is, all solution pairs are given by

$$K_a = \tilde{\Gamma}Y_{g_1} + K_cA_g$$
$$K_b = \tilde{\Gamma}X_{g_1} - K_cB_g,$$

where K_c is a stable real rational matrix.

Example 2.10 Under the assumptions in Theorem 2.6, (2.297) is the parameterization, in terms of stable matrices K_1 and K_g, of all the allowable K_1 and H_2 in (2.275) for which v_d and ε_d are stable. (a) Show by direct substitution of (2.297) into (2.275) that the parameterization of H directly in terms of K_1 and K_g is given by $H = (Y_1 + K_1A)FP_d + K_gB_gA_d$. (b) Show that this parameterization of H directly in terms of K_1 and K_g agrees with the parameterization one obtains starting instead from the solution of (2.284) for H; specifically, from

$$H = A_1^{-1}(Y_p\tilde{P}_d + A_{p_1}K_m)A_d = A_1^{-1}Y_pA_pP_d + \tilde{K}_mA_d$$

with $\tilde{K}_m$ given by (2.295). (c) Confirm that v_d and ε_d are indeed stable when H is parameterized as in part (a).

Solution (a) Using (2.275), (2.291), and (2.297), one gets

$$\begin{aligned} H &= (Y_1 + K_1A)FP_d + H_2L = (Y_1 + K_1A)FP_d + K_gA_gA_g^{-1}B_gA_d \\ &= (Y_1 + K_1A)FP_d + K_gB_gA_d. \end{aligned}$$

(b) Using (2.292) and (2.295), one gets

$$\begin{aligned} H &= A_1^{-1}Y_pA_pP_d + \tilde{K}_mA_d = A_1^{-1}Y_pA_pP_d + (K_gB_g + K_1Q_d - \Gamma_d)A_d \\ &= A_1^{-1}Y_pA_pP_d - \Gamma_dA_d + K_gB_gA_d + K_1AFP_d. \end{aligned}$$

One also has from (2.288) that

$$\begin{aligned} \Gamma_dA_d &= A_1^{-1}(Y_p\tilde{P}_dA_d - YAFP_d) = A_1^{-1}(Y_pA_pP_d - A_1Y_1FP_d) \\ &= A_1^{-1}Y_pA_pP_d - Y_1FP_d. \end{aligned}$$

Hence, $A_1^{-1}Y_pA_pP_d - \Gamma_dA_d = Y_1FP_d$ and it follows that

$$\begin{aligned} H &= A_1^{-1}Y_pA_pP_d - \Gamma_dA_d + K_gB_gA_d + K_1AFP_d \\ &= Y_1FP_d + K_gB_gA_d + K_1AFP_d = (Y_1 + K_1A)FP_d + K_gB_gA_d \end{aligned}$$

as in part (a).

(c) From (2.273),

$$\begin{aligned} v_d &= -A_1Hd = -A_1[(Y_1 + K_1A)FP_d + K_gB_gA_d]A_d^{-1}b_d \\ &= -[(Y + A_1K_1)AFP_d + A_1K_gB_gA_d]A_d^{-1}b_d = -[(Y + A_1K_1)Q_d + A_1K_gB_g]b_d \end{aligned}$$

is stable because K_1, K_g, and Q_d are stable. From (2.274),

$$\begin{aligned} \varepsilon_d &= -(P_d - PA_1H)d = -P_dd - Pv_d \\ &= -[P_dA_d^{-1} - P(Y + A_1K_1)Q_d - PA_1K_gB_g]b_d \\ &= -[P_dA_d^{-1} - PYQ_d - PA_1(K_1Q_d + K_gB_g)]b_d. \end{aligned}$$

Now

$$\begin{aligned} P_dA_d^{-1} - PYQ_d &= P_dA_d^{-1} - PYAFP_dA_d^{-1} = A_p^{-1}\tilde{P}_d - PYAFA_p^{-1}\tilde{P}_d \\ &= [A_p^{-1} - PYAF(X_p + PY_p)]\tilde{P}_d = [A_p^{-1} - PYBY_p - PA_1Y_1FX_p]\tilde{P}_d \\ &= [A_p^{-1} - P(I - A_1X_1)Y_p - PA_1Y_1FX_p]\tilde{P}_d \\ &= [X_p + PA_1(X_1Y_p - Y_1FX_p)]\tilde{P}_d. \end{aligned}$$

Hence,

$$\varepsilon_d = -[X_p + PA_1(X_1Y_p - Y_1FX_p)]\tilde{P}_d b_d - PA_1(K_1Q_d + K_gB_g)b_d$$

is stable because K_1, K_g, Q_d, PA_1, $\tilde{P}_d$, and F are stable.

Example 2.11 Some of the comments made in Remark 2.13 are confirmed in this problem. (a) Suppose A_r is given and $r = A_r^{-1}b_r$, where the elements of A_r and b_r are real polynomials. Establish for an admissible 3DOF system that r is trackable for all b_r when T_d and F are stable iff $A_pT_dA_r^{-1}$ is stable. Moreover, all stabilizing controllers for which r is trackable for all b_r are given by (2.208) with

$$H_1 = A_1^{-1}A_{p_1}(Y_{p_1}T_d + K_rA_r),$$

where K_r is a stable real rational matrix. (b) Suppose A_d is given and $d = A_d^{-1}b_d$, where the elements of A_d and b_d are real polynomials. Establish for an admissible 3DOF system that d is rejectable for all b_d when F is stable and the real polynomial matrix pair A_g, B_g satisfies (2.291) and (2.292) iff $A_pP_dA_d^{-1}$ is stable. Moreover, all stabilizing controllers for which d is rejectable for all b_d are given by (2.208) with $[\,K_1 \quad H_2\,] = [\,K_1 \quad K_gA_g\,]$, where K_1 and K_g are stable real rational matrices belonging to the nonempty set of such matrices for which $\det(X_1 - K_1B) \not\equiv 0$.

Solution (a) From (2.234) through (2.238),

$$\begin{aligned} v_r &= A_1H_1r = A_1H_1A_r^{-1}b_r = \Psi b_r \\ \varepsilon_r &= (T_d - PA_1H_1)r = (T_d - PA_1H_1)A_r^{-1}b_r = \Phi b_r \end{aligned}$$

must be stable for all b_r with real polynomial elements. Hence, they must be stable when all elements of the column vector b_r are zero except for unity in the j-row. Thus, the j-column of Ψ and the j-column of Φ must be stable. Clearly, this holds no matter what the choice of j and so the real rational matrices Ψ and Φ must be stable. Conversely, if this is the case, v_r and ε_r are stable for all b_r. That is, v_r and ε_r are stable iff Ψ and Φ in (2.237) and (2.238) are stable. One again gets (2.242) which has stable solutions Φ and Ψ iff $\tilde{T}_d = A_pT_dA_r^{-1}$ is stable and these solutions are given by (2.249) and (2.250). Any such solution pair is realized iff the H_1 given by (2.253) is chosen. But a stabilizing controller is required and so H_1 must be stable. The fact that K_r, F, and T_d are stable guarantees this is the case when Lemma 2.10 is invoked. (b) From (2.273), (2.274), (2.279), and (2.280)

$$\begin{aligned} v_d &= -A_1Hd = -A_1HA_d^{-1}b_d = \Psi b_d \\ \varepsilon_d &= -(P_d - PA_1H)d = -(P_d - PA_1H)A_d^{-1}b_d = -\Phi b_d. \end{aligned}$$

Applying the same reasoning as in part (a), one concludes that v_d and ε_d are stable for all b_d iff the real rational matrices Ψ and Φ are stable. From (2.280), one gets

$$A_p\Phi + B_p\Psi = A_pP_dA_d^{-1} = \tilde{P}_d$$

which has stable solutions iff $\tilde{P}_d$ is stable and these solutions are given by (2.282) and (2.283). So proceeding exactly as was done in the steps leading to Theorem 2.6 in which the stability of F is invoked one finds that all stabilizing controllers for which d is rejectable for all b_d are the ones described in Theorem 2.6.

Example* **2.12** The tracking and disturbance rejection capabilities of the 2DOF standard configuration shown in Fig. 1.3 are investigated in this problem. It is assumed that the system is admissible and the controller is a stabilizing one as described in Theorem 2.2 with transfer matrix given by (2.161). The reference noise vector n is typically stochastic in nature and, therefore, devoid of any deterministic or shape-deterministic components. The elements of the disturbance input vector d and the reference input vector r, however, often contain deterministic or shape-deterministic components that are transforms of such signals as steps, ramps, and exponentially growing functions of time. In this case, d and r possess unstable poles. The objective here is to identify the class of inputs d and r of this type for which the tracking error ε and the controller output v have stable transforms and to parameterize the subset of stabilizing controllers that achieve this objective. Since a stabilizing controller is assumed, the contributions to ε and v that result from initial conditions in the analog case and initial data strings in the digital case are automatically stable and can therefore be ignored. That is, one can work exclusively with transfer matrices. Specifically, for Fig. 1.3, $[y_s' \;\; y_t' \;\; y']' = G[d' \;\; v']'$, where

$$G = \left[\begin{array}{c|c} G_{sd} & G_{sv} \\ G_{td} & G_{tv} \\ \hline G_{21} & G_{22} \end{array}\right] = \left[\begin{array}{c|c} G_{11} & G_{12} \\ \hline G_{21} & G_{22} \end{array}\right]$$

and the G_{ij} for $i,j = 1,2$ are given by (2.134) and (2.137) through (2.139).
(a) Establish that

$$\begin{aligned} v &= A_1Hr + A_1(Y_1 + KA)G_{21}d + A_1Hn \\ \varepsilon &= (T_d - G_{tv}A_1H)r - [G_{td} + G_{tv}A_1(Y_1 + KA)G_{21}]d - G_{tv}A_1Hn. \end{aligned}$$

(b) Introduce the coprime polynomial matrix fraction descriptions $G_{tv} = A_p^{-1}B_p = B_{p_1}A_{p_1}^{-1}$ that satisfy (2.241). Then establish when (1) $r = A_r^{-1}b_r$ is a left coprime real polynomial matrix fraction description and (2) T_d and $A_1^{-1}A_{p_1}$ are stable that r is trackable iff $A_p T_d A_r^{-1}$ is stable or, equivalently, $A_p T_d r = A_p T_d A_r^{-1} b_r$ is stable. Moreover, all H for which r is trackable are given by $H = A_1^{-1} A_{p_1} (Y_{p_1} T_d + K_r A_r)$, where K_r is a stable real rational matrix.

(c) Let $d = A_d^{-1} b_d$ be a left coprime real polynomial matrix fraction description. Introduce the coprime polynomial matrix fraction descriptions

$$G_{tv} = A_p^{-1} B_p = B_{p_1} A_{p_1}^{-1}, AG_{21} A_d^{-1} = A_g^{-1} B_g$$

satisfying (2.241) and (2.292), respectively. Then establish that d is rejectable iff $\tilde{G}_{td} = A_p G_{td} A_d^{-1}$ is stable when $A_1^{-1} A_{p_1}$ and $\Gamma_d = -(A_1^{-1} Y_p \tilde{G}_{td} + Y_1 G_{21} A_d^{-1})$ are stable (Note that the comments in Remark 2.11 apply here as well when P_d is replaced by G_{td}). Moreover, any K for which d is rejectable is given by $K = K_g A_g$, where K_g is a stable real rational matrix.

(d) Show that the matrix Γ_d in (c) is stable when $[G_{21} \quad G_{22}] = M[G_{td} \quad G_{tv}]$ and M is a stable matrix. This is the case for the system shown in Fig. 1.1 when $L \equiv 0$ (a 2DOF system), the feedback sensor transfer matrix F is stable, and one is concerned only about the contribution to ε and v caused by the disturbance d

(e) This part concerns the case when Γ_d is not stable. Now one needs to introduce the coprime polynomial matrix fraction descriptions $AG_{21} A_d^{-1} = A_g^{-1} B_g = B_{g_1} A_{g_1}^{-1}$ satisfying

$$\begin{bmatrix} A_g & B_g \\ Y_{g_1} & -X_{g_1} \end{bmatrix} \begin{bmatrix} X_g & B_{g_1} \\ Y_g & -A_{g_1} \end{bmatrix} = \begin{bmatrix} X_g & B_{g_1} \\ Y_g & -A_{g_1} \end{bmatrix} \begin{bmatrix} A_g & B_g \\ Y_{g_1} & -X_{g_1} \end{bmatrix} = \begin{bmatrix} I & 0 \\ 0 & I \end{bmatrix}.$$

Establish that d is rejectable when $A_1^{-1} A_{p_1}$ is stable iff $\tilde{G}_{td}$ and $\tilde{\Gamma}_d = \Gamma_d A_{g_1}$ are stable and there exists a real rational matrix K_d for which $\det [X_1 - (\tilde{\Gamma}_d Y_{g_1} + K_d A_g) B] \not\equiv 0$.

Solution (a) From $y = G_{21} d + G_{22} v$, one gets

$$v = -C_y y + C_u u = -C_y G_{21} d - C_y G_{22} v + C_u r + C_u n$$

or

$$v = (I + C_y G_{22})^{-1} [-C_y G_{21} d + C_u r + C_u n].$$

Now from (2.161)

$$\begin{aligned} R_y &= -(I - C_y G_{22})^{-1} C_y = [I + (X_1 - KB)^{-1} (Y_1 + KA) B_1 A_1^{-1}]^{-1} (X_1 - KB)^{-1} (Y_1 + KA) \\ &= A_1 [(X_1 - KB) A_1 + (Y_1 + KA) B_1]^{-1} (Y_1 + KA) = A_1 (Y_1 + KA) \end{aligned}$$

and

$$R_u = (I + C_y G_{22})^{-1} C_u = [I + (X_1 - KB)^{-1}(Y_1 + KA)B_1 A_1^{-1}]^{-1}(X_1 - KB)^{-1}H$$
$$= A_1[(X_1 - KB)A_1 + (Y_1 + KA)B_1]^{-1}H = A_1 H.$$

Hence,

$$v = R_u r + R_y G_{21} d + R_u n = A_1 H r + A_1(Y_1 + KA)G_{21} d + A_1 H n.$$

Then

$$\begin{aligned} \varepsilon &= T_d r - y_t = T_d r - (G_{td} d + G_{tv} v) \\ &= T_d r - G_{td} d - G_{tv}[A_1 H r + A_1(Y_1 + KA)G_{21} d + A_1 H n] \\ &= (T_d - G_{tv} A_1 H) r - [G_{td} + G_{tv} A_1 (Y_1 + KA) G_{21}] d - G_{tv} A_1 H n. \end{aligned}$$

(b) The reference input r is trackable iff

$$\begin{aligned} v_r &= A_1 H r = A_1 H A_r^{-1} b_r \\ \varepsilon_r &= (T_d - G_{tv} A_1 H) r = (T_d - G_{tv} A_1 H) A_r^{-1} b_r \end{aligned}$$

are stable. Since an admissible 2DOF standard configuration is assumed, it follows from Lemma 2.5 that

$$G_{12} A_1 = \begin{bmatrix} G_{sv} \\ G_{tv} \end{bmatrix} A_1 = \begin{bmatrix} G_{sv} A_1 \\ G_{tv} A_1 \end{bmatrix}$$

is stable. Hence, $G_{tv} A_1$ is stable. Also, since a stabilizing controller is assumed, H is stable. Thus, with T_d stable one has that $A_1 H$ and $(T_d - G_{tv} A_1 H)$ are stable. It then follows from the fact that A_r, b_r is a left coprime polynomial matrix pair that r is trackable iff $\Psi = A_1 H A_r^{-1}$ and

$$\Phi = (T_d - G_{tv} A_1 H) A_r^{-1} = T_d A_r^{-1} - G_{tv} \Psi$$

are stable. Introducing $G_{tv} = A_p^{-1} B_p$ yields

$$A_p \Phi + B_p \Psi = A_p T_d A_r^{-1} = \tilde{T}_d.$$

Hence, r is trackable iff there exist stable real rational matrices Φ, Ψ satisfying this equation for which $H = A_1^{-1}\Psi A_r$ is a stable real rational matrix. Now by direct analogy with (2.249) and (2.250) in the 3DOF case, stable solution pairs Φ, Ψ exist iff $\tilde{T}_d$ is stable and all such pairs are given by

$$\Phi = X_p\tilde{T}_d - B_{p_1}K_r, \Psi = Y_p\tilde{T}_d + A_{p_1}K_r,$$

where K_r is a stable real rational matrix. Moreover, when $A_1^{-1}A_{p_1}$ is stable so is the real rational matrix

$$H = A_1^{-1}\Psi A_r = A_1^{-1}(Y_p\tilde{T}_d + A_{p_1}K_r)A_r = A_1^{-1}A_{p_1}(Y_{p_1}T_d + K_rA_r)$$

which gives all H for which r is trackable.

(c) The disturbance input d is rejectable iff

$$\begin{aligned} v_d &= A_1(Y_1 + KA)G_{21}A_d^{-1}b_d \\ \varepsilon_d &= -\left[G_{td} + G_{tv}A_1(Y_1 + KA)G_{21}\right]A_d^{-1}b_d \end{aligned}$$

are stable. Since an admissible 2DOF standard configuration is assumed, it follows from Lemma 2.5 that AG_{21} is stable. Hence, $A_1(Y_1 + KA)G_{21} = (Y + A_1K)AG_{21}$ is stable since K is stable for a stabilizing controller. Also from Lemma 2.5,

$$G_{11} + G_{12}A_1Y_1G_{21} = \begin{bmatrix} G_{sd} \\ G_{td} \end{bmatrix} + \begin{bmatrix} G_{sv} \\ G_{tv} \end{bmatrix} A_1Y_1G_{21} = \begin{bmatrix} G_{sd} + G_{sv}A_1Y_1G_{21} \\ G_{td} + G_{tv}A_1Y_1G_{21} \end{bmatrix}$$

is stable. Hence,

$$G_{td} + G_{tv}A_1(Y_1 + KA)G_{21} = G_{td} + G_{tv}A_1Y_1G_{21} + G_{tv}A_1KA\,G_{21}$$

is stable. It then follows from the fact that A_d, b_d is a left coprime polynomial matrix pair that d is rejectable iff $\Psi = A_1H_\psi A_d^{-1}$ and

$$\Phi = (G_{td} + G_{tv}A_1H_\psi)A_d^{-1} = G_{td}A_d^{-1} + G_{tv}\Psi$$

are stable, where $H_\psi = (Y_1 + KA)G_{21}$. Introducing $G_{tv} = A_p^{-1}B_p$ in the expression for Φ leads to

$$A_p\Phi - B_p\Psi = A_pG_{td}A_d^{-1} = \tilde{G}_{td}.$$

Hence, d is rejectable iff there exist stable real rational matrices Φ, Ψ satisfying this equation for which $H_\psi = (Y_1 + KA)G_{21} = A_1^{-1}\Psi A_d$ has a stable solution for K satisfying $\det(X_1 - KB) \not\equiv 0$. Now by the same arguments that yielded (2.282) and (2.283) in the 3DOF case, one can conclude that stable solution pairs Φ, Ψ exist iff $\tilde{G}_{td}$ is stable and all such pairs are given by

$$\Phi = X_p\tilde{G}_{td} + B_{p_1}K_m, \Psi = -Y_p\tilde{G}_{td} + A_{p_1}K_m,$$

where K_m is a stable real rational matrix. Using these results in the expression for H_ψ then gives

$$H_\psi = (Y_1 + KA)G_{21} = A_1^{-1}(-Y_p\tilde{G}_{td} + A_{p_1}K_m)A_d$$

or

$$KAG_{21}A_d^{-1} - A_1^{-1}A_{p_1}K_m = -\left(A_1^{-1}Y_p\tilde{G}_{td} + Y_1G_{21}A_d^{-1}\right) = \Gamma_d.$$

Thus, d is rejectable iff there exist stable real rational matrices K, K_m satisfying this equation for which $\det(X_1 - KB) \not\equiv 0$. Now $A_1^{-1}A_{p_1}$ stable is given and admissibility requires $G_{tv}A_1 = B_{p_1}A_{p_1}^{-1}A_1$ stable. Hence, $A_{p_1}^{-1}A_1 = (A_1^{-1}A_{p_1})^{-1}$ is also stable. So one can introduce $\tilde{K}_m = A_1^{-1}A_{p_1}K_m$, which is stable iff K_m is stable, and $AG_{21}A_d^{-1} = A_g^{-1}B_g$ to get that d is rejectable iff there exist stable real rational matrices $K, \tilde{K}_m$ satisfying

$$KA_g^{-1}B_g = \Gamma_d + \tilde{K}_m$$

for which $\det(X_1 - KB) \not\equiv 0$. The first observation to make is that the left-hand side of this equation must be stable because the right-hand side, $\Gamma_d + \tilde{K}_m$, is stable. For stable K it immediately follows that this is the case iff $KA_g^{-1} = K_g$, a stable matrix. Hence, all solution pairs are given by

$$K = K_gA_g, \tilde{K}_m = -\Gamma_d + K_gB_g,$$

where K_g is a stable real rational matrix. The second observation to make is that for $K_g = 0$ the solution pair is $K = 0, \tilde{K}_m = -\Gamma_d$ for which $\det(X_1 - KB) = \det X_1 \not\equiv 0$ since one can always choose X_1 so that this is the case. That is, d is rejectable and all stabilizing controllers for which d is rejectable are given by (2.161) with $K = K_gA_g$, where K_g is a stable real rational matrix.

(d) $$\begin{aligned}\Gamma_d &= -A_1^{-1}Y_p\tilde{G}_{td} - Y_1G_{21}A_d^{-1} = -A_1^{-1}Y_p\tilde{G}_{td} - Y_1MA_p^{-1}A_pG_{td}A_d^{-1} \\ &= -\left[(X_1 - Y_1G_{22})Y_p + Y_1M(X_p + G_{tv}Y_p)\right]\tilde{G}_{td} \\ &= -\left(X_1Y_p - Y_1MG_{tv}Y_p + Y_1MX_p + Y_1MG_{tv}Y_p\right)\tilde{G}_{td} = -\left(X_1Y_p + Y_1MX_p\right)\tilde{G}_{td}\end{aligned}$$

which is stable when $\tilde{G}_{td}$ and M are stable. (e) The initial steps in part (c) up to the equation $KA_g^{-1}B_g = \Gamma_d + \tilde{K}_m$ apply here also. However, since Γ_d is not stable now, one cannot proceed as before. Instead, the substitution $A_g^{-1}B_g = B_{g_1}A_{g_1}^{-1}$ is made to get $KB_{g_1} - \tilde{K}_m A_{g_1} = \Gamma_d A_{g_1} = \tilde{\Gamma}_d$. Clearly, $\tilde{\Gamma}_d$ stable for stable solutions $K, \tilde{K}_m$ is necessary. It is also sufficient since in this case $K = \tilde{\Gamma}_d Y_{g_1}$, $\tilde{K}_m = -\tilde{\Gamma}_d X_{g_1}$ are particular solutions. Using the results of Example 2.9, it is clear that all solutions are given by

$$K = \tilde{\Gamma}_d Y_{g_1} + K_d A_g, \tilde{K}_m = -\tilde{\Gamma}_d X_{g_1} + K_d B_g,$$

where K_d is a stable real rational matrix. Hence, d is rejectable under the assumption $A_1^{-1}A_{p_1}$ is stable iff $\tilde{G}_{td}$ and $\tilde{\Gamma}_d$ are stable and there exists a stable real rational matrix K_d for which

$$\det(X_1 - KB) = \det[X_1 - (\tilde{\Gamma}_d Y_{g_1} + K_d A_g)B] \not\equiv 0.$$

Example* 2.13 Consider an admissible 3DOF system with stabilizing controller given by (2.208). This controller reduces to a 1DOF controller when $H_1 = H_2 = Y_1 + K_1 A$. For in this case

$$C = (X_1 - K_1 B)^{-1}(Y_1 + K_1 A)[-I \quad -I \quad I] = C_1[-I \quad -I \quad I]$$

and the 1DOF design freedom available resides in the choice of the stable real rational matrix K_1. The objective in this example is to investigate the tracking capabilities of this 1DOF system with respect to the reference input r. In this case, r is the only exogenous input which is not identically zero and it follows from Fig. 1.1 and Eq. (1.1) that $v = C_1 q$, where

$$q = -w - z + u = -Fy + r$$

since $n = 0$, $z = 0$, and $n_m = 0$. That is, the issue of r being trackable is equivalent to r being trackable for the system shown in Fig. 2.4. It is assumed throughout this example that F and T_d are stable.

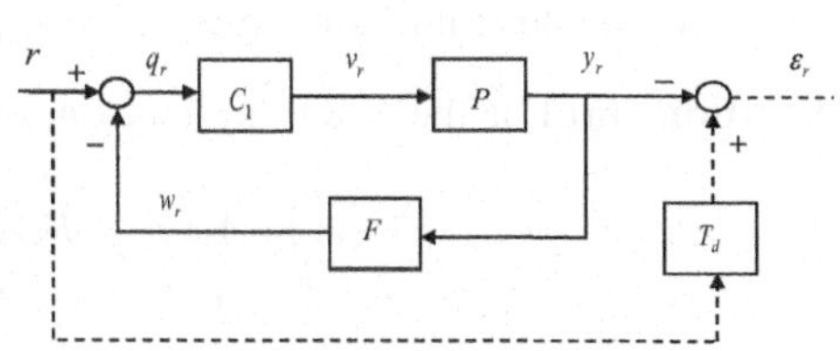

Fig. 2.4 1DOF tracking model for reference input

(a) Let $r = A_r^{-1} b_r$ be a left coprime real polynomial matrix fraction description for the reference input r. Introduce the coprime polynomial matrix fraction descriptions

$$AA_r^{-1} = B_{h_1} A_{h_1}^{-1} = A_h^{-1} B_h$$

that satisfy

$$\begin{bmatrix} A_h & B_h \\ Y_{h_1} & -X_{h_1} \end{bmatrix} \begin{bmatrix} X_h & B_{h_1} \\ Y_h & -A_{h_1} \end{bmatrix} = \begin{bmatrix} X_h & B_{h_1} \\ Y_h & -A_{h_1} \end{bmatrix} \begin{bmatrix} A_h & B_h \\ Y_{h_1} & -X_{h_1} \end{bmatrix} = \begin{bmatrix} I & 0 \\ 0 & I \end{bmatrix}$$

and let $\Gamma_h = A_1^{-1} Y_p \tilde{T}_d - Y_1 A_r^{-1}$, where $\tilde{T}_d = A_p T_d A_r^{-1}$. Then establish that r is trackable iff $\tilde{\Gamma}_h = \Gamma_h A_{h_1}$ and $\tilde{T}_d$ are stable and there exists a stable matrix K_h for which $\det(X_1 - K_1 B) \not\equiv 0$ when the choice $K_1 = \Gamma_h A_{h_1} Y_{h_1} + K_h A_h$ is made. (b) When $\tilde{T}_d$ is stable show that $\tilde{\Gamma}_h$ is stable iff $(FT_d - I)A_r^{-1} A_{h_1}$ is stable. That is, r is trackable iff $\tilde{T}_d$ and $(FT_d - I)A_r^{-1} A_{h_1}$ are stable and there exists a stable matrix K_h for which $\det(X_1 - K_1 B) \not\equiv 0$ when the choice $K_1 = \Gamma_h A_{h_1} Y_{h_1} + K_h A_h$ is made. Hint: Establish and use the fact that $\tilde{\Gamma}_h$ is stable iff $B_1 \tilde{\Gamma}_h$ is stable. (c) From Fig. 2.4,

$$q_r = r - Fy_r = r - F(T_d r - \varepsilon_r) = (I - FT_d)r + F\varepsilon_r.$$

With a trackable r, $F\varepsilon_r$ is stable. Hence, q_r is stable iff $(I - FT_d)r$ or, equivalently, $(FT_d - I)A_r^{-1}$ is stable. This is not assured by the stability of $(FT_d - I)A_r^{-1} A_{h_1}$ and it is possible for r to be trackable with q_r unstable. Show that this is indeed the case for an analog 1DOF system when $F = 1$, $P = 1/s$, $T_d = 1 - s$, and $r = 1/s(1 - s)$. Also, demonstrate for

$$K_h = -s + \frac{1}{s+1} + \frac{1}{(s+1)^2} = -s + \frac{s+2}{(s+1)^2}$$

that q_r, v_r, and ε_r are strictly proper and C_1 is proper. (d) Suppose that $\tilde{T}_d$ and $(FT_d - I)A_r^{-1}$ are stable. Establish that AA_r^{-1} is stable. Also confirm that r is trackable and q_r is stable with any stabilizing controller. (e) Show that Γ_h is stable when T_d^{-1}, $(T_d F - I)P$, and $\tilde{T}_d$ are stable. Then establish in this case that r is trackable and $K_1 = K_s A_h$, where K_s is a stable real rational matrix. (f) Under the assumptions for (e), confirm directly that v_r and ε_r are stable for $K_1 = K_s A_h$, where K_s is a stable real rational matrix. Also prove that q_r is stable iff XA_h^{-1} is stable.

Solution (a) For the 1DOF system one gets from (2.234) and (2.235) that

$$v_r = A_1(Y_1 + K_1 A)A_r^{-1} b_r$$
$$\varepsilon_r = [T_d - PA_1(Y_1 + K_1 A)]A_r^{-1} b_r.$$

Now K_1 is stable for a stabilizing controller, PA_1 is stable for an admissible system from Lemma 2.7, and T_d is stable by assumption. Since A_r, b_r is a left coprime pair, it follows that v_r and ε_r are stable iff

$$\Psi = A_1(Y_1 + K_1 A)A_r^{-1}$$
$$\Phi = [T_d - PA_1(Y_1 + K_1 A)]A_r^{-1} = T_d A_r^{-1} - P\Psi$$

are stable matrices. So r is trackable iff there exist stable real rational matrices Φ and Ψ satisfying $\Phi + P\Psi = T_d A_r^{-1}$ for which $K_1 = \left(A_1^{-1}\Psi A_r - Y_1\right)A^{-1}$ is stable and $\det(X_1 - K_1 B) \not\equiv 0$. Introduce the coprime polynomial matrix fraction descriptions $P = A_p^{-1}B_p = B_{p_1}A_{p_1}^{-1}$ satisfying (2.241). Then Φ and Ψ are solutions of $A_p\Phi + B_p\Psi = A_p T_d A_r^{-1} = \tilde{T}_d$. It then follows from the same reasoning that led to (2.249) and (2.250) that these solutions are stable iff $\tilde{T}_d$ is stable. In this case, the solutions are given by

$$\Phi = X_p\tilde{T}_d - B_{p_1}K_2, \Psi = Y_p\tilde{T}_d + A_{p_1}K_2,$$

where K_2 is a stable real rational matrix. Substituting the solution for Ψ into the expression for K_1 yields $K_1 = \left[A_1^{-1}\left(Y_p\tilde{T}_d + A_{p_1}K_2\right)A_r - Y_1\right]A^{-1}$ or

$$K_1 AA_r^{-1} - A_1^{-1}A_{p_1}K_2 = A_1^{-1}Y_p\tilde{T}_d - Y_1 A_r^{-1} = \Gamma_h.$$

Now $\tilde{K}_2 = A_1^{-1}A_{p_1}K_2$ is stable iff K_2 is stable since $A_1^{-1}A_{p_1}$ is stable by Lemma 2.10 and admissibility requires $PA_1 = B_{p_1}A_{p_1}^{-1}A_1$ stable or, equivalently, $A_{p_1}^{-1}A_1$ stable. Thus, the tracking problem has been reduced to finding stable matrices K_1 and $\tilde{K}_2$ satisfying $K_1 AA_r^{-1} - \tilde{K}_2 = \Gamma_h$ or

$$K_1 B_{h_1} - \tilde{K}_2 A_{h_1} = \tilde{\Gamma}_h$$

for which $\det(X_1 - K_1 B) \not\equiv 0$. It follows from the same reasoning as used in Example 2.8 that stable solutions $K_1, \tilde{K}_2$ exist iff $\tilde{\Gamma}_h$ is stable and all such solutions are given by

$$K_1 = \tilde{\Gamma}_h Y_{h_1} + K_h A_h, \tilde{K}_2 = -\tilde{\Gamma}_h X_{h_1} + K_h B_h.$$

Hence, r is trackable under the assumptions for this problem iff $\tilde{T}_d$ and $\tilde{\Gamma}_h$ are stable and there exists a stable real rational matrix K_h for which $\det(X_1 - K_1 B) \not\equiv 0$.

(b) It follows from

$$\begin{aligned}\tilde{\Gamma}_h &= (A_1^{-1}Y_p\tilde{T}_d - Y_1A_r^{-1})A_{h_1} = A_1^{-1}(Y_p\tilde{T}_dA_{h_1} - A_1Y_1A_r^{-1}A_{h_1})\\ &= A_1^{-1}(Y_p\tilde{T}_dA_{h_1} - YAA_r^{-1}A_{h_1}) = A_1^{-1}(Y_p\tilde{T}_dA_{h_1} - YB_{h_1})\end{aligned}$$

that $\tilde{\Gamma}_h$ is stable iff $B_1\tilde{\Gamma}_h$ is stable because $\left(Y_p\tilde{T}_dA_{h_1} - YB_{h_1}\right)$ is stable. Clearly,

$$\begin{aligned}B_1\tilde{\Gamma}_h &= B_1A_1^{-1}(Y_p\tilde{T}_dA_{h_1} - YB_{h_1}) = FP(Y_pA_pT_dA_r^{-1}A_{h_1} - YAA_r^{-1}A_{h_1})\\ &= (FPY_pA_pT_d - FPA_1Y_1)A_r^{-1}A_{h_1} = \left[F\left(A_p^{-1} - X_p\right)A_pT_d - B_1Y_1\right]A_r^{-1}A_{h_1}\\ &= FT_d - FX_pA_pT_d - I + XA)A_r^{-1}A_{h_1} = (FT_d - I)A_r^{-1}A_{h_1} - (FX_pA_pT_d - XA)A_r^{-1}A_{h_1}\\ &= (FT_d - I)A_r^{-1}A_{h_1} - FX_p\tilde{T}_dA_{h_1} + XB_{h_1}\end{aligned}$$

is stable iff $(FT_d - I)A_r^{-1}A_h$ is stable because $FX_p\tilde{T}_dA_{h_1}$ and XB_{h_1} are stable.
(c) From the given data:

$$FP = P = \frac{1}{s} \Rightarrow \begin{cases} A = A_1 = A_p = A_{p_1} = s, \ B = B_1 = B_p = B_{p_1} = 1, \\ \quad X = X_1 = X_p = X_{p_1}, \ Y = Y_1 = Y_p = Y_{p_1}. \end{cases}$$

$$AX + BY = 1 \Rightarrow sX + Y = 1 \Rightarrow X = 1, \ Y = 1 - s$$

$$r = \frac{1}{s(1-s)} \Rightarrow A_r = s(s-1), \ b_r = -1$$

$$AA_r^{-1} = \frac{1}{s-1} \Rightarrow A_h = A_{h_1} = s - 1, \ B_h = B_{h_1} = 1, \ X_h = X_{h_1}, \ Y_h = Y_{h_1}.$$

$$A_hX_h + B_hY_h = 1 \Rightarrow (s-1)X_h + Y_h = 1 \Rightarrow X_h = -1, \ Y_h = s.$$

Clearly, $(FT_d - I)A_r^{-1}A_{h_1} = -1$ is stable, but $(FT_d - I)A_r^{-1} = -A_{h_1}^{-1} = -(s-1)^{-1}$ is unstable. In addition, $\tilde{T}_d = A_pT_dA_r^{-1} = -1$ and

$$\Gamma_h = A_p^{-1}Y_p\tilde{T}_d - Y_1A_r^{-1} = \frac{s-1}{s} + \frac{1}{s} = 1.$$

Thus,

$$K_1 = \Gamma_hA_{h_1}Y_{h_1} + K_hA_h = s(s-1) + K_h(s-1) = (s + K_h)(s-1)$$

and

$$X_1 - K_1B = 1 - K_1 = -(s^2 - s - 1) - K_h(s-1) \not\equiv 0$$

for any real rational stable K_h. So r is trackable and q_r is unstable. In particular,

$$C_1 = (X_1 - K_1B)^{-1}(Y_1 + K_1A) = -\frac{(s-1)(s^2 + K_h s - 1)}{s^2 + (K_h - 1)s - (K_h + 1)}$$

and it easily follows from $q_r = r - FPC_1q_r$ that

$$q_r = (I + FPC_1)^{-1}r = s - \frac{1}{s-1} + K_h$$

which is unstable for any stable K_h. When

$$K_h = -s + \frac{1}{s+1} + \frac{1}{(s+1)^2} = -s + \frac{s+2}{(s+1)^2}$$

a little algebra yields

$$C_1 = -\frac{s-1}{s+3}, \quad q_r = -\frac{s+3}{(s-1)(s+1)^2}, \quad v_r = C_1q_r = \frac{1}{(s+1)^2},$$

and

$$\varepsilon_r = T_d r - Pv_r = \frac{1}{s} - \frac{1}{s(s+1)^2} = \frac{s+2}{(s+1)^2}.$$

Clearly, C_1 is proper and q_r, v_r, and ε_r are strictly proper. Moreover, v_r and ε_r are stable; hence, the analog performance measure in (1.7) with $y_c = [\varepsilon_r' \;\; v_r']'$ is finite in this case.

(d) It is given that $M_1 = \tilde{T}_d = A_pT_dA_r^{-1}$ and $M_2 = (FT_d - I)A_r^{-1}$ are stable. Hence,

$$A_r^{-1} = FT_dA_r^{-1} - M_2 = FA_p^{-1}M_1 - M_2$$

and

$$AA_r^{-1} = AFA_p^{-1}M_1 - AM_2.$$

Since AF is stable, AFA_p^{-1} is stable iff $AFA_p^{-1}B_p = AFP = B$ is stable. This is indeed the case and it follows that $AA_r^{-1} = A_h^{-1}B_h$ is stable. Therefore, A_h^{-1} is stable. It is also obvious that $M_2A_{h_1} = (FT_d - I)A_r^{-1}A_{h_1}$ is stable. Moreover, for any

stabilizing controller, K_1 is stable and $\det(X_1 - K_1 B) \not\equiv 0$. One then gets from (b) that r is trackable if the solution of $K_1 = \tilde{\Gamma}_h Y_{h_1} + K_h A_h$ for K_h is stable. Clearly, $K_h = (K_1 - \tilde{\Gamma}_h Y_{h_1}) A_h^{-1}$ is indeed stable since K_1, $\tilde{\Gamma}_h$, and A_h^{-1} are stable. Hence, from (c), q_r is stable because r is trackable and $M_2 = (FT_d - I)A_r^{-1}$ is stable.
(e) Since $\tilde{T}_d$ is stable,

$$\Gamma_h = A_1^{-1} Y_p \tilde{T}_d - Y_1 A_r^{-1} = (A_1^{-1} Y_p - Y_1 T_d^{-1} A_p^{-1}) \tilde{T}_d$$

is stable if

$$\begin{aligned}(A_1^{-1} Y_p - Y_1 T_d^{-1} A_p^{-1}) &= (X_1 + Y_1 FP) Y_p - Y_1 T_d^{-1} (X_p + P Y_p) \\ &= X_1 Y_p - Y_1 T_d^{-1} X_p + Y_1 T_d^{-1} (T_d F - I) P Y_p\end{aligned}$$

is stable. This is indeed the case when T_d^{-1} and $(T_d F - I)P$ are stable. Hence, $\tilde{\Gamma}_h = \Gamma_h A_{h_1}$ is stable. So from part (a) r is trackable if one can choose a stable K_h so that with $K_1 = \tilde{\Gamma}_h Y_{h_1} + K_h A_h = \Gamma_h A_{h_1} Y_{h_1} + K_h A_h$ one gets $\det(X_1 - K_1 B) \not\equiv 0$. Clearly, $K_h = -\Gamma_h Y_h$ is stable and gives $\det(X_1 - K_1 B) = \det X_1 \not\equiv 0$ since $K_1 = 0$ in this case and X_1 can always be selected so that $\det X_1 \not\equiv 0$. Moreover, for any stable K_h, $K_1 = \Gamma_h Y_h A_h + K_h A_h = K_s A_h$, where $K_s = \Gamma_h Y_h + K_h$ is stable since Γ_h is stable with the assumptions for this part of the example.
(f) The response v_r is stable iff $\Psi = A_1(Y_1 + K_1 A) A_r^{-1}$ is stable. With $K_1 = K_s A_h$,

$$\begin{aligned}\Psi &= A_1(Y_1 + K_s A_h A) A_r^{-1} = (Y + A_1 K_s A_h) A A_r^{-1} \\ &= Y A A_r^{-1} + A_1 K_s B_h = A_1 Y_1 A_r^{-1} + A_1 K_s B_h = Y_p \tilde{T}_d - A_1 \Gamma_h + A_1 K_s B_h\end{aligned}$$

is indeed stable since $\tilde{T}_d$, Γ_h, and K_s are stable. The response ε_r is stable iff $\Phi = [T_d - P A_1(Y_1 + K_1 A)] A_r^{-1}$ is stable. With $K_1 = K_s A_h$,

$$\begin{aligned}\Phi &= T_d A_r^{-1} - P A_1 Y_1 A_r^{-1} - P A_1 K_s A_h A A_r^{-1} = T_d A_r^{-1} - P(Y_p \tilde{T}_d - A_1 \Gamma_h) - P A_1 K_s B_h \\ &= A_p^{-1}(I - B_p Y_p)\tilde{T}_d + P A_1(\Gamma_h - K_s B_h) = X_p \tilde{T}_d + P A_1 (\Gamma_h - K_s B_h)\end{aligned}$$

is indeed stable since $\tilde{T}_d$, PA_1, Γ_h, and K_s are stable. Clearly,

$$\begin{aligned}q_r &= r - F y_r = r - F P v_r = r - F P A_1 (Y_1 + K_1 A) r = [I - B_1(Y_1 + K_1 A)] r \\ &= (I - B_1 Y_1 - B_1 K_1 A) A_r^{-1} b_r = (X - B_1 K_1) A A_r^{-1} b_r\end{aligned}$$

is stable iff

$$Q_r = (X - B_1K_1)AA_r^{-1} = (X - B_1K_sA_h)A_h^{-1}B_h$$

or, equivalently,

$$\tilde{Q}_r = (X - B_1K_sA_h)A_h^{-1} = XA_h^{-1} - B_1K_s$$

is stable because A_r, b_r and A_h, B_h are coprime pairs. Thus, q_r is stable iff XA_h^{-1} is stable.

Example 2.14 (a) Show that Theorem 2.4 is unaffected by the choice of coprime polynomial matrix fraction descriptions used for r, P, and FP. (b) Show that Theorem 2.6 is unaffected by the choice of coprime polynomial matrix fraction descriptions used for d, P, FP, and LA_d^{-1}.

Solution (a) Let $\tilde{A}_r^{-1}\tilde{b}_r$, $P = \tilde{A}_p^{-1}\tilde{B}_p = \tilde{B}_{p_1}\tilde{A}_{p_1}^{-1}$, and $FP = \tilde{A}^{-1}\tilde{B} = \tilde{B}_1\tilde{A}_1^{-1}$ be alternative coprime descriptions for Theorem 2.4. Then $\tilde{A}_r = N_rA_r$, $\tilde{A}_p = N_pA_p$, $\tilde{A}_{p_1} = A_{p_1}N_{p_1}$, $\tilde{A}_1 = A_1N_1$, and $N_{p_1}\tilde{Y}_{p_1} - Y_{p_1} = M_pA_p$, where N_r, N_p, N_{p_1}, and N_1 are unimodular polynomial matrices and M_p is an arbitrary real polynomial matrix (see the material leading to Lemma 2.4 and Example 2.4 and its solution). So $\tilde{A}_1^{-1}\tilde{A}_{p_1} = N_1^{-1}A_1^{-1}A_{p_1}N_{p_1}$ and $\tilde{A}_pT_d\tilde{A}_r^{-1} = N_pA_pT_dA_r^{-1}N_r^{-1}$ are stable iff $A_1^{-1}A_{p_1}$ and $A_pT_dA_r^{-1}$ are stable, respectively. Hence, in light of (2.234) and 2.235), it only remains to show that the set of all

$$\tilde{A}_1\tilde{H}_1 = \tilde{A}_{p_1}(\tilde{Y}_{p_1}T_d + \tilde{K}_r\tilde{A}_r) = A_{p_1}N_{p_1}(\tilde{Y}_{p_1}T_d + \tilde{K}_rN_rA_r)$$

as $\tilde{K}_r$ ranges over all stable real rational matrices is the same as the set of all A_1H_1 given by (2.257) as K_r ranges over all stable real rational matrices. Now,

$$\begin{aligned}\tilde{A}_1\tilde{H}_1 &= A_{p_1}(N_{p_1}\tilde{Y}_{p_1}T_d + N_{p_1}\tilde{K}_rN_rA_r)\\ &= A_{p_1}(Y_{p_1}T_d + M_pA_pT_d + N_{p_1}\tilde{K}_rN_rA_r) = A_{p_1}(Y_{p_1}T_d + K_rA_r) = A_1H_1\end{aligned}$$

when

$$K_r = M_pA_pT_dA_r^{-1} + N_{p_1}\tilde{K}_rN_r = M_p\tilde{T}_d + N_{p_1}\tilde{K}_rN_r.$$

Since, N_{p_1} and N_r are unimodular and $\tilde{T}_d$ is stable, K_r is stable iff $\tilde{K}_r$ is stable. So there is a one-to-one correspondence between every realizable A_1H_1 and $\tilde{A}_1\tilde{H}_1$. (b) Let $\tilde{A}_d^{-1}\tilde{b}_d$ be an alternative coprime description for d. Then $\tilde{A}_d = N_dA_d$, N_d a unimodular polynomial matrix. Hence, $\tilde{A}_pP_d\tilde{A}_d^{-1} = N_pA_pP_dA_d^{-1}N_d^{-1}$ is stable iff

$A_p P_d A_d^{-1}$ is stable. So, in light of (2.273), (2.274), and (2.276), it only remains to show that the set of all

$$\tilde{A}_1 \tilde{H}_1 = (\tilde{Y} + \tilde{A}_1 \tilde{K}_1)\tilde{A}FP_d + \tilde{A}_1 \tilde{H}_2 L = (\tilde{Y} + \tilde{A}_1 \tilde{K}_1)\tilde{A}FP_d + \tilde{A}_1 \tilde{K}_g \tilde{A}_g L$$

as $\tilde{K}_1$ and $\tilde{K}_g$ range over all stable real rational matrices is the same as the set of all

$$A_1 H_1 = (Y + A_1 K_1)AFP_d + A_1 H_2 L = (Y + A_1 K_1)AFP_d + A_1 K_g A_g L$$

as K_1 and K_g range over all stable real rational matrices. Now one has

$$L\tilde{A}_d^{-1} = LA_d^{-1}N_d^{-1} = A_g^{-1}B_g N_d^{-1} = \tilde{A}_g^{-1}\tilde{B}_g.$$

Moreover,

$$[A_g \quad B_g N_d^{-1}] = [A_g \quad B_g]\begin{bmatrix} I & 0 \\ 0 & N_d^{-1} \end{bmatrix}$$

has full row rank everywhere since A_g, B_g is a left coprime pair and N_d is unimodular. Hence, $A_g, B_g N_d^{-1}$ is another left coprime polynomial matrix pair in addition to $\tilde{A}_g, \tilde{B}_g$ which can be used in the matrix fraction description for $L\tilde{A}_d^{-1}$. That is, $\tilde{A}_g = N_g A_g, \tilde{B}_g = N_g B_g N_d^{-1}$, where N_g is unimodular. So one gets

$$\begin{aligned}\tilde{A}_1 \tilde{H}_1 &= [(Y + A_1 M)N^{-1} + A_1 N_1 \tilde{K}_1]NAFP_d + A_1 N_1 \tilde{K}_g N_g A_g L \\ &= [Y + A_1(M + N_1 \tilde{K}_1 N)]AFP_d + A_1(N_1 \tilde{K}_g N_g)A_g L,\end{aligned}$$

where M is an arbitrary real polynomial matrix. Clearly, $K_1 = M + N_1 \tilde{K}_1 N$ and $K_g = N_1 \tilde{K}_g N_g$ are stable iff $\tilde{K}_1$ and $\tilde{K}_g$ are stable and there is a one-to-one correspondence between every realizable $A_1 H_1$ and $\tilde{A}_1 \tilde{H}_1$.

Example* 2.15 Confirm the validity of (2.373).

Solution First note that $\tilde{A}\tilde{B}_1 = \Omega^{-1}AB_1\Lambda^{-1} = \Omega^{-1}BA_1\Lambda^{-1} = \tilde{B}\tilde{A}_1$. Then it suffices to show that $\tilde{A}\tilde{X} + \tilde{B}\tilde{Y} = I$, $\tilde{X}_1\tilde{A}_1 + \tilde{Y}_1\tilde{B}_1 = I$, and $\tilde{Y}_1 X = X_1 Y$. Clearly,

$$\tilde{A}\tilde{X}+\tilde{B}\tilde{Y} = \tilde{A}(\tilde{A}^{-1} - \tilde{B}_1\{\tilde{A}_1^{-1}Y\Omega\}_{sp}) + \tilde{B}\tilde{A}_1\{\tilde{A}_1^{-1}Y\Omega\}_{sp} = I$$
$$\tilde{X}_1\tilde{A}_1 + \tilde{Y}_1\tilde{B}_1 = (\tilde{A}_1^{-1} - \{\tilde{A}_1^{-1}Y\Omega\}_{sp}\tilde{B})\tilde{A}_1 + \{\tilde{A}_1^{-1}Y\Omega\}_{sp}\tilde{A}\tilde{B}_1 = I$$
$$\tilde{X}_1\tilde{Y} = (\tilde{A}_1^{-1} - \{\tilde{A}_1^{-1}Y\Omega\}_{sp}\tilde{B})\tilde{A}_1\{\tilde{A}_1^{-1}Y\Omega\}_{sp} = (I - \{\tilde{A}_1^{-1}Y\Omega\}_{sp}\tilde{B}\tilde{A}_1)\{\tilde{A}_1^{-1}Y\Omega\}_{sp}$$
$$\tilde{Y}_1\tilde{X} = \{\tilde{A}_1^{-1}Y\Omega\}_{sp}\tilde{A}(\tilde{A}^{-1} - \tilde{B}_1\{\tilde{A}_1^{-1}Y\Omega\}_{sp}) = \{\tilde{A}_1^{-1}Y\Omega\}_{sp}(I - \tilde{A}\tilde{B}_1\{\tilde{A}_1^{-1}Y\Omega\}_{sp})$$
$$= (I - \{\tilde{A}_1^{-1}Y\Omega\}_{sp}\tilde{B}\tilde{A}_1)\{\tilde{A}_1^{-1}Y\Omega\}_{sp} = \tilde{X}_1\tilde{Y}.$$

Example 2.16 The purpose of this problem is to specialize Theorem 2.8 to the case of the 3DOF system shown in Fig. 1.1 when $FP = A^{-1}B = B_1A_1^{-1}$ and the generalized Bezout identity (2.62) is satisfied. In particular, when the 3DOF system is admissible, FP is strictly proper, and the controller is free of unstable hidden poles, establish that all proper stabilizing controller transfer matrices are given by (2.208) with

$$K_1 = \Lambda^{-1}(Z_w - \{\Lambda A_1^{-1}Y\Omega\}_\infty)\Omega^{-1}, H_1 = \Lambda^{-1}Z_u, H_2 = \Lambda^{-1}Z_z,$$

where Z_u, Z_w, and Z_z are proper stable real rational matrices and the real rational matrices Λ and Ω are stable together with their inverses and selected so that $A_1\Lambda^{-1}$ and $\Omega^{-1}A$ are biproper.

Solution The approach that is taken here is to specialize the G_{22} for the standard configuration to that of the 3DOF system. This is easily accomplished in a two-step process from the results already available by first specializing to the 2DOF standard configuration and then specializing from the 2DOF standard configuration to the 3DOF system. In particular, from (2.141) it is clear that $G_{22} \to [G_{22}' \quad 0]'$ on the first step. On the second and final step, it follows from (2.193) that $[G_{22}' \quad 0]' \to [(FP)' \quad 0 \quad 0]'$. So G_{22} in the standard configuration in the particular case of the 3DOF system is given by the strictly proper matrix

$$G_{22} = \begin{bmatrix} FP \\ 0 \\ 0 \end{bmatrix} = \begin{bmatrix} A^{-1}B \\ 0 \\ 0 \end{bmatrix} = \begin{bmatrix} B_1A_1^{-1} \\ 0 \\ 0 \end{bmatrix},$$

where the first block row has the same number of rows as w, the second block row has the same number of rows as z, and the third block row has the same number of rows as r. Hence,

$$G_{22} = -\begin{bmatrix} A & 0 & 0 \\ 0 & I & 0 \\ 0 & 0 & I \end{bmatrix}^{-1}\begin{bmatrix} -B \\ 0 \\ 0 \end{bmatrix} = -\begin{bmatrix} -B_1 \\ 0 \\ 0 \end{bmatrix}A_1^{-1},$$

$$X_1A_1+\begin{bmatrix}-Y_1 & 0 & 0\end{bmatrix}\begin{bmatrix}-B_1\\0\\0\end{bmatrix}=I, X_1\begin{bmatrix}-Y & 0 & 0\end{bmatrix}=\begin{bmatrix}-Y_1 & 0 & 0\end{bmatrix}\begin{bmatrix}X & 0 & 0\\0 & I & 0\\0 & 0 & I\end{bmatrix},$$

and

$$\begin{bmatrix}A & 0 & 0\\0 & I & 0\\0 & 0 & I\end{bmatrix}\begin{bmatrix}X & 0 & 0\\0 & I & 0\\0 & 0 & I\end{bmatrix}+\begin{bmatrix}-B\\0\\0\end{bmatrix}\begin{bmatrix}-Y & 0 & 0\end{bmatrix}=\begin{bmatrix}I & 0 & 0\\0 & I & 0\\0 & 0 & I\end{bmatrix}.$$

Notice that the matrices X_1 and A_1 are unaffected. Also, in conformity with (2.157) and (2.205), the matrix K is replaced with $\begin{bmatrix}-K_1 & -H_2 & H_1\end{bmatrix}$. In addition, it should be clear from the equations above that for the Ω-matrix in Theorem 2.8 one can choose the block diagonal matrix diag $\{\Omega, I, I\}$. Then in place of (2.365) one gets

$$\begin{aligned}C&=\left(X_1-\begin{bmatrix}-K_1 & -H_2 & H_1\end{bmatrix}\begin{bmatrix}-B\\0\\0\end{bmatrix}\right)^{-1}\\&\quad\left(\begin{bmatrix}-Y_1 & 0 & 0\end{bmatrix}+\begin{bmatrix}-K_1 & -H_2 & H_1\end{bmatrix}\begin{bmatrix}A & 0 & 0\\0 & I & 0\\0 & 0 & I\end{bmatrix}\right)\\&=(X_1-K_1B)^{-1}\begin{bmatrix}-(Y_1+K_1A) & -H_2 & H_1\end{bmatrix}=\begin{bmatrix}-C_w & -C_z & C_u\end{bmatrix},\end{aligned}$$

which is in agreement with (2.208). In place of (2.366), one gets

$$\begin{aligned}\begin{bmatrix}-K_1 & -H_2 & H_1\end{bmatrix}&=\Lambda^{-1}\left(\begin{bmatrix}-Z_w & -Z_z & Z_u\end{bmatrix}-\left\{\Lambda A_1^{-1}\begin{bmatrix}-Y & 0 & 0\end{bmatrix}\begin{bmatrix}\Omega & 0 & 0\\0 & I & 0\\0 & 0 & I\end{bmatrix}\right\}_{\infty}\right)\\&\quad\begin{bmatrix}\Omega^{-1} & 0 & 0\\0 & I & 0\\0 & 0 & I\end{bmatrix}\\&=\Lambda^{-1}(\begin{bmatrix}-Z_w & -Z_z & Z_u\end{bmatrix}+\{\begin{bmatrix}\Lambda A_1^{-1}Y\Omega & 0 & 0\end{bmatrix}\}_{\infty})\begin{bmatrix}\Omega^{-1} & 0 & 0\\0 & I & 0\\0 & 0 & I\end{bmatrix}\\&=\begin{bmatrix}-\Lambda^{-1}(Z_w-\{\Lambda A_1^{-1}Y\Omega\}_{\infty})\Omega^{-1} & -\Lambda^{-1}Z_z & \Lambda^{-1}Z_u\end{bmatrix}.\end{aligned}$$

Chapter 3
$\mathcal{H}_2$ Design of Multivariable Control Systems

3.1 Overview

The design of analog and digital linear multivariable control systems for which performance is measured by a quadratic or $\mathcal{H}_2$ cost functional of the kind described in Chap. 1 and Appendix A is treated in this chapter. First, the parameterization of the tracking and disturbance rejection controllers for the 3DOF system presented in Chap. 2 is extended to the standard configuration of Fig. 1.2. Next, for this subset of stabilizing controllers, the ones for which the cost functional is finite are parameterized and the one for which the cost functional is minimum is established. Specialized results for 1DOF and 3DOF systems are then presented.

3.2 Preliminary Considerations

It is important to keep separate the modeling of an interconnected physical system from the modeling of the performance measures and the input signals. Otherwise, important issues with regard to system stability can be obscured and/or confused. That is, it is best to employ the standard configuration of Fig. 1.2 to represent all possible interconnections of the physical components in any feedback configuration, but it is best to keep separate the modeling of the exogenous inputs e and the desired responses y_d as in Fig. 3.1 where $\widetilde{G}$ and C represent the physical components while $\widetilde{G}_e$ and $\widetilde{G}_d$ account for the mathematical models of the exogenous signals e and the desired system responses y_d, respectively. In particular, it is important to recognize that $\widetilde{G}_e$ and $\widetilde{G}_d$ do not represent physical components and have no impact on the stability of the system as determined by the response to initial conditions or data strings within the blocks $\widetilde{G}$ and C. The modeling of e and y_d is accomplished by setting $q = [\,\sigma' \quad \mu'\,]'$ where σ is a vector whose elements are

J. J. Bongiorno Jr. and K. Park, *Design of Linear Multivariable Feedback Control Systems*, https://doi.org/10.1007/978-3-030-44356-6_3

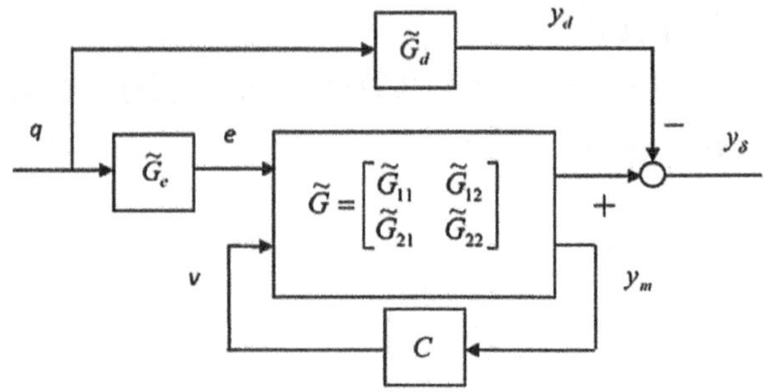

Fig. 3.1 Persistent input standard configuration (© 2009 Taylor & Francis Ltd, http://www.tandfonline.com. Reprinted, with permission, from Park and Bongiorno (2009))

real random variables with zero mean and covariance matrix $<\sigma\sigma'> = I$ and μ is a stochastic vector whose power spectral density matrix is given by $\Phi_{\mu\mu} = I$. It is assumed that σ and μ are independent processes. Clearly,

$$e = \tilde{G}_e q = \begin{bmatrix} \tilde{G}_{e1} & \tilde{G}_{e2} \end{bmatrix} \begin{bmatrix} \sigma \\ \mu \end{bmatrix} = \tilde{G}_{e1}\sigma + \tilde{G}_{e2}\mu = e_1 + e_2 \tag{3.1}$$

and

$$y_d = \tilde{G}_d q = \begin{bmatrix} \tilde{G}_{d1} & \tilde{G}_{d2} \end{bmatrix} \begin{bmatrix} \sigma \\ \mu \end{bmatrix} = \tilde{G}_{d1}\sigma + \tilde{G}_{d2}\mu = y_{d1} + y_{d2} \tag{3.2}$$

where e_1 and y_{d1} account for the shape-deterministic components of e and y_d, respectively, and e_2 and y_{d2} account for the stochastic components of e and y_d, respectively. With persistent inputs such as steps, ramps, and growing exponentials the matrix $\tilde{G}_{e1}$ is unstable. However, there is no loss in generality in assuming that $\tilde{G}_{e2}$ is stable when the power spectral density matrix for e_2 is good and nonsingular on the finite $j\omega$-axis in the analog case or is good and nonsingular on $|\lambda| = 1$ in the digital case since, in accordance with Corollaries C.3 and C.4 of Appendix C, one can choose for $\tilde{G}_{e2}$ a stable left Wiener–Hopf factor of the power spectral density matrix for e_2. Similar reasoning applies to $\tilde{G}_{d2}$ and so it also is assumed to be stable.

The elements of y_c are the physical plant outputs (e.g., y_t and y_s in Fig. 2.1) that are required to track reference inputs as well as other signals likely to cause saturation whose deviations from a required response must be kept small. That is, a fundamental objective is that y_c track y_d. As a first step toward satisfying this requirement, C is chosen from among the stabilizing controllers so that $y_{\delta 1}$, the shape-deterministic component of y_δ, is stable. Then the inverse transform of each element of $y_{\delta 1}$ approaches zero asymptotically. Of course, no progress can be made without the assumption that $\tilde{G}$ is admissible. Otherwise, no stabilizing controllers even exist. An additional consideration in the choice of the controller is the need in response to persistent exogenous inputs to prevent unbounded inputs to the actual plant (e.g., the inputs to P_p in Fig. 2.1) in order to avoid saturation. Hence, the shape-deterministic component of the actual plant input must be stable except possibly for simple poles on the stability boundary. Typically, the controller output v drives any precompensation of the actual plant and for robustness any needed simple poles of the actual plant input on the stability boundary are provided by the precompensation. (For example, in Fig. 2.1 any precompensation is included in P_a.) Then all permissible actual plant inputs can be generated with stable v along the lines described in the last few paragraphs of Sect. 1.3. So attention is restricted here to optimizing system performance over the subset of controllers for which both v_1, the shape-deterministic component of v, and $y_{\delta 1}$ are stable. The first step in the process involves establishing the conditions under which such a subset of controllers exists and to parameterize this subset.

Remark 3.1 It is possible to include the elements of v among the elements of y_c and to pick the corresponding rows of $\widetilde{G}_d$ to contain only zeros so that the elements of v are also included among the elements of y_δ. Then the stability of v_1 is assured by that of $y_{\delta 1}$. However, since v is an explicit system variable, it is simpler and more insightful to take into account directly the requirement that v_1 be stable whether or not v is included in y_c.

3.3 Tracking and Disturbance Rejection with the Standard Configuration[1]

Clearly,

$$y_\delta = (\widetilde{G}_{11}\widetilde{G}_e - \widetilde{G}_d)q + \widetilde{G}_{12}v = G_{11}\widetilde{G}_w q + G_{12}v \tag{3.3}$$

and

$$y_m = \widetilde{G}_{21}\widetilde{G}_e q + \widetilde{G}_{22}v = G_{21}\widetilde{G}_w q + G_{22}v \tag{3.4}$$

where

$$G_{11} = \begin{bmatrix} \widetilde{G}_{11} & -I \end{bmatrix}, G_{12} = \widetilde{G}_{12}, G_{21} = \begin{bmatrix} \widetilde{G}_{21} & 0 \end{bmatrix}, G_{22} = \widetilde{G}_{22}, \tag{3.5}$$

and

$$\widetilde{G}_w = \begin{bmatrix} \widetilde{G}_e \\ \widetilde{G}_d \end{bmatrix} = \begin{bmatrix} \widetilde{G}_{e1} & \widetilde{G}_{e2} \\ \widetilde{G}_{d1} & \widetilde{G}_{d2} \end{bmatrix} = \begin{bmatrix} \widetilde{G}_{w1} & \widetilde{G}_{w2} \end{bmatrix}. \tag{3.6}$$

[1]This section contains, with permission, some material equivalent to that in Park and Bongiorno (2009),

Using (3.4) in $v = Cy_m$, solving for v, and substituting the result into (3.3) then yields

$$v = (I - CG_{22})^{-1}CG_{21}\widetilde{G}_w q \tag{3.7}$$

and

$$y_\delta = [G_{11} + G_{12}(I - CG_{22})^{-1}CG_{21}]\widetilde{G}_w q. \tag{3.8}$$

Now in accordance with Lemma B.5, one can introduce coprime polynomial matrix fraction descriptions

$$G_{22} = \widetilde{G}_{22} = -A^{-1}B = -B_1A_1^{-1} \tag{3.9}$$

satisfying

$$\begin{bmatrix} A & B \\ Y_1 & -X_1 \end{bmatrix}\begin{bmatrix} X & B_1 \\ Y & -A_1 \end{bmatrix} = \begin{bmatrix} X & B_1 \\ Y & -A_1 \end{bmatrix}\begin{bmatrix} A & B \\ Y_1 & -X_1 \end{bmatrix} = \begin{bmatrix} I & 0 \\ 0 & I \end{bmatrix}. \tag{3.10}$$

(The coprime polynomial matrix fraction description $A^{-1}B = B_1A_1^{-1}$ is used to represent G, $-G_{22}$, or $-P_{22}$ in different sections of this monograph. It is always made clear which is intended so that no confusion arises.) It then follows from Theorem 2.1 and Lemma 2.2 that all stabilizing controller transfer matrices are given by

$$C = (X_1 - KB)^{-1}(Y_1 + KA) = (Y + A_1K)(X - B_1K)^{-1} \tag{3.11}$$

where K is a stable rational matrix such that det $(X_1 - KB) \not\equiv 0$. Substituting the first expression for C in (3.11) into (3.7) and (3.8) gives after straightforward calculations

$$v = G_v\widetilde{G}_w q = G_v\widetilde{G}_{w1}\sigma + G_v\widetilde{G}_{w2}\mu = v_1 + v_2 \tag{3.12}$$

and

$$y_\delta = G_\delta\widetilde{G}_w q = G_\delta\widetilde{G}_{w1}\sigma + G_\delta\widetilde{G}_{w2}\mu = y_{\delta 1} + y_{\delta 2} \tag{3.13}$$

where

$$G_v = A_1(Y_1 + KA)G_{21} = (Y + A_1K)AG_{21} \tag{3.14}$$

and

$$G_\delta = G_{11} + G_{12}A_1(Y_1 + KA)G_{21} = G_{11} + G_{12}G_v. \tag{3.15}$$

The following lemma plays an important role in the sequel and provides some useful results associated with the admissibility of the generalized plant $\widetilde{G}$.

Lemma 3.1 *The matrices* $G_{12}A_1$, AG_{21}, *and* $G_{11} + G_{12}A_1Y_1G_{21}$ *are all stable when* $\widetilde{G}$ *is admissible. Moreover, when* $\widetilde{G}$ *is admissible and* K *is stable, then* G_v *and* G_δ *are stable.*

Proof On account of Lemma 2.3, when $\widetilde{G}$ is admissible it follows that $\widetilde{G}_{12}A_1$, $A\widetilde{G}_{21}$, and $\widetilde{G}_{11} + \widetilde{G}_{12}A_1Y_1\widetilde{G}_{21}$ are stable. Hence,

$$G_{12}A_1 = \widetilde{G}_{12}A_1, AG_{21} = A\begin{bmatrix}\widetilde{G}_{21} & 0\end{bmatrix} = \begin{bmatrix}A\widetilde{G}_{21} & 0\end{bmatrix}, \tag{3.16}$$

and

$$\begin{aligned} G_{11} + G_{12}A_1Y_1G_{21} &= \begin{bmatrix}\widetilde{G}_{11} & -I\end{bmatrix} + \widetilde{G}_{12}A_1Y_1\begin{bmatrix}\widetilde{G}_{21} & 0\end{bmatrix} \\ &= \begin{bmatrix}(\widetilde{G}_{11} + \widetilde{G}_{12}A_1Y_1\widetilde{G}_{21}) & -I\end{bmatrix} \end{aligned} \tag{3.17}$$

are stable. It now readily follows from (3.14) and (3.15) that G_v and G_δ are stable since K is stable.

Remark 3.2 One can instead of as in (3.5) choose $G_{11} = \begin{bmatrix}\widetilde{G}_{11} & -I\end{bmatrix}\widetilde{G}_w$ and $G_{21} = \begin{bmatrix}\widetilde{G}_{21} & 0\end{bmatrix}\widetilde{G}_w$ so that the system in Fig. 3.1 is absorbed into a standard configuration with

$$\begin{bmatrix} y_\delta \\ y_m \end{bmatrix} = \begin{bmatrix} G_{11} & G_{12} \\ G_{21} & G_{22} \end{bmatrix}\begin{bmatrix} q \\ v \end{bmatrix} \tag{3.18}$$

and impose on this structure the admissibility constraints AG_{21}, $G_{12}A_1$, $G_{11} + G_{12}A_1Y_1G_{21}$ stable. But it is unwise to do so. For it then follows from $y_m = G_{21}q + G_{22}v = G_{21}q + G_{22}Cy_m$ or, equivalently, $y_m = (I - G_{22}C)^{-1}G_{21}q$ that the shape-deterministic component of y_m,

$$y_{m1} = [I + A^{-1}B(Y + A_1K)(X - B_1K)^{-1}]^{-1}G_{21}\begin{bmatrix}\sigma \\ 0\end{bmatrix} = (X - B_1K)AG_{21}\begin{bmatrix} I \\ 0\end{bmatrix}\sigma, \tag{3.19}$$

must be stable. Now for the 3DOF system shown in Fig. 1.1 and the 2DOF standard configuration shown in Fig. 1.3, the elements of the controller input $u = r + n$ are included in y_m. Since the noise n is typically purely stochastic, the shape-deterministic components of u are those of r which then have to be stable.

That is, systems of the type shown in Figs. 1.1 and 1.3 with persistent shape-deterministic reference inputs are excluded from consideration. In contrast, the choices made in (3.5) lead to no such restriction. Specifically, (3.3) and (3.4) correspond to the generalized persistent input standard configuration shown in Fig. 3.2 for which G, on account of Lemma 3.1, does satisfy $G_{12}A_1$, AG_{21}, $G_{11} + G_{12}A_1Y_1G_{21}$ stable without imposing the undesirable restrictions on 2DOF and 3DOF systems cited above. It is also important to point out for Fig. 3.2 that the transfer matrices from w to v and from w to y_δ are, respectively, G_v and G_δ. In addition, Fig. 3.2 is a more general representation in that it includes Fig. 3.1 when the choices (3.5) are made, but can represent other configurations in which $\widetilde{G}$ might be embedded (e.g., see Remark 3.6 in the sequel). Attention is of course restricted to those cases in which $\widetilde{G}$ is admissible, C belongs to the set of all stabilizing controllers for $\widetilde{G}$, and $G_{22} = \widetilde{G}_{22}$. In this regard, it is useful to introduce

Definition 3.1 The subsystem G in Fig. 3.2 is said to be *compatible* when the physical system $\widetilde{G}$ within it is admissible; $G_{22} = \widetilde{G}_{22}$; and $G_{12}A_1$, AG_{21}, and $G_{11} + G_{12}A_1Y_1G_{21}$ are stable.

It is assumed that the set of all possible outcomes for σ is sufficiently broad so that $v_1 = G_v\widetilde{G}_{w1}\sigma$ and $y_{\delta 1} = G_\delta\widetilde{G}_{w1}\sigma$ are both stable for all outcomes σ iff $G_v\widetilde{G}_{w1}$ and $G_\delta\widetilde{G}_{w1}$ are both stable. For example, this is the case when the set of all possible outcomes includes all the possibilities in which only one element of σ is nonzero. Since G_v, G_δ, and $\widetilde{G}_{w2} = \begin{bmatrix} \widetilde{G}'_{e2} & \widetilde{G}'_{d2} \end{bmatrix}'$ are stable, it follows that $G_v\widetilde{G}_{w1}$ and $G_\delta\widetilde{G}_{w1}$ are both stable iff $G_v\widetilde{G}_w$ and $G_\delta\widetilde{G}_w$ are both stable. In the optimization problem treated in the sequel one has to identify the subset of all stabilizing controllers for which G_vG_{wo} and $G_\delta G_{wo}$ are both stable where with α_1 a positive constant and α_2 a nonnegative constant, $G_{wo} = \begin{bmatrix} \sqrt{\alpha_1}\widetilde{G}_{w1} & \sqrt{\alpha_2}\widetilde{G}_{w2} \end{bmatrix}$. Clearly, when $\widetilde{G}_{w2}$ is stable, the subset of all stabilizing controllers for which $G_v\widetilde{G}_{w1}$ and $G_\delta\widetilde{G}_{w1}$ are both stable and the subset of all stabilizing controllers for which G_vG_{wo} and $G_\delta G_{wo}$ are both stable is the same. This observation is used to advantage in Sects. 3.7 and 3.8 and provides an alternative approach to the one taken in Sects. 3.5 and 3.6 (see Example 3.9). Clearly, conditions on an arbitrary matrix G_w under which one can choose a stable K so that G_vG_w and $G_\delta G_w$ are both stable are of interest. These conditions are now derived and a parameterization of all such K in

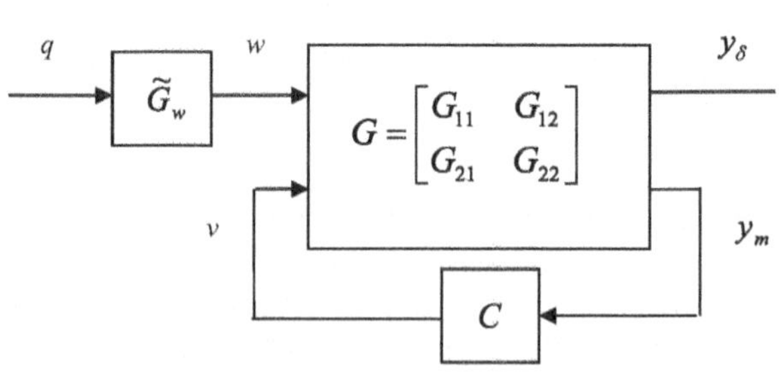

Fig. 3.2 Generalized persistent input standard configuration (© 2009 Taylor & Francis Ltd, http://www.tandfonline.com. Reprinted, with permission, from Park and Bongiorno (2009))

terms of an arbitrary stable rational matrix is given. For this purpose, it is useful to introduce

Definition 3.2 A matrix G_w is said to be *acceptable* for a compatible G when there exists a stable matrix K such that $G_v G_w$ and $G_\delta G_w$ are both stable when G_v and G_δ are given by (3.14) and (3.15), respectively.

It is sometimes appropriate and possible with persistent inputs to choose a stable K so that in addition to v_1 and $y_{\delta 1}$ being stable, $y_{m1} = G_m \widetilde{G}_{w1}\sigma$ is also stable, where G_m is the transfer matrix between w and y_m in Fig. 3.2: $y_m = G_m w$ and $G_m = (G_{21} + G_{22}G_v)$. With regard to this case, it is convenient to introduce

Definition 3.3 A matrix G_w is said to be *strictly acceptable* for a compatible G when there exists a stable matrix K such that $G_v G_w$, $G_\delta G_w$ and $G_m G_w$ are stable.

Treatment of this concept is relegated to Examples 3.21 and 3.22 for the standard configuration in Fig. 3.2 and justified in connection with 1DOF systems in Sect. 3.7.

The first step in establishing the conditions under which G_w is acceptable for a compatible G is to introduce the left-coprime polynomial matrix description $G_w = A_w^{-1}B_w$. Since G_v and G_δ are stable, it follows from Lemma B.8 that $G_v G_w$ and $G_\delta G_w$ are stable iff

$$\Psi = G_v A_w^{-1} = (Y + A_1 K)AG_{21}A_w^{-1} \tag{3.20}$$

and

$$\Phi = G_\delta A_w^{-1} = (G_{11} + G_{12}G_v)A_w^{-1} = G_{11}A_w^{-1} + G_{12}\Psi \tag{3.21}$$

are stable. Of course this is the case when A_w^{-1} is stable. When persistent shape-deterministic inputs are present, however, $\widetilde{G}_{w1}$ and, therefore, A_w^{-1} are unstable. In this case, the choice of stable K in the controller must be further constrained so that both Φ and Ψ are stable.

Clearly, there is no choice of stable K for which both Φ and Ψ are stable unless (3.21) or, equivalently,

$$\Phi - G_{12}\Psi = G_{11}A_w^{-1} \tag{3.22}$$

admits stable solution pairs Φ, Ψ. For the purpose of addressing this issue, it is convenient to introduce the coprime polynomial matrix fraction descriptions

$$G_{12} = -A_o^{-1}B_o = -B_{o1}A_{o1}^{-1} \tag{3.23}$$

where

$$\begin{bmatrix} A_o & B_o \\ Y_{o1} & -X_{o1} \end{bmatrix} \begin{bmatrix} X_o & B_{o1} \\ Y_o & -A_{o1} \end{bmatrix} = \begin{bmatrix} X_o & B_{o1} \\ Y_o & -A_{o1} \end{bmatrix} \begin{bmatrix} A_o & B_o \\ Y_{o1} & -X_{o1} \end{bmatrix} = \begin{bmatrix} I & 0 \\ 0 & I \end{bmatrix}. \tag{3.24}$$

Then (3.22) is equivalent to

$$A_o\Phi + B_o\Psi = A_oG_{11}A_w^{-1} \stackrel{\Delta}{=} \Gamma_w. \tag{3.25}$$

Since Φ and Ψ must be stable, Γ_w must be a stable matrix; this is the first necessary condition on the data. Moreover, Φ and Ψ must be stable solutions of (3.25). Equation (3.25) has the same form as (2.242) with A_p, B_p, and $\widetilde{T}_d = A_pT_dA_r^{-1}$ replaced by A_o, B_o, and Γ_w, respectively. Hence, by the same reasoning that yielded (2.249) and (2.250) one gets when Γ_w is stable that all possible stable solution pairs Φ, Ψ of (3.25) are given by

$$\Phi = X_o\Gamma_w - B_{o1}K_w, \Psi = Y_o\Gamma_w + A_{o1}K_w \tag{3.26}$$

where K_w is a stable rational matrix. These solution pairs assure that (3.21) is satisfied. So all that remains is the satisfaction of (3.20) when Ψ is given by (3.26). In particular, one has to identify the stable K, K_w pairs for which

$$\Psi A_w = (Y_o\Gamma_w + A_{o1}K_w)A_w = (Y + A_1K)AG_{21} \tag{3.27}$$

or, equivalently,

$$(Y + A_1K)AG_{21}A_w^{-1} = (Y_o\Gamma_w + A_{o1}K_w). \tag{3.28}$$

In order to complete this final step, one more coprime polynomial matrix fraction description is needed. Specifically,

$$AG_{21}A_w^{-1} = A_g^{-1}B_g = B_{g1}A_{g1}^{-1} \tag{3.29}$$

where

$$\begin{bmatrix} A_g & B_g \\ Y_{g1} & -X_{g1} \end{bmatrix} \begin{bmatrix} X_g & B_{g1} \\ Y_g & -A_{g1} \end{bmatrix} = \begin{bmatrix} X_g & B_{g1} \\ Y_g & -A_{g1} \end{bmatrix} \begin{bmatrix} A_g & B_g \\ Y_{g1} & -X_{g1} \end{bmatrix} = \begin{bmatrix} I & 0 \\ 0 & I \end{bmatrix}. \tag{3.30}$$

It readily follows from (3.28) and (3.29) that

$$(Y + A_1K)B_{g1} = (Y_o\Gamma_w + A_{o1}K_w)A_{g1} \tag{3.31}$$

or

$$K_oB_{g1} - K_wA_{g1} = A_{o1}^{-1}(Y_o\Gamma_wA_{g1} - YB_{g1}) \triangleq \Gamma_o \tag{3.32}$$

where

$$K_o = A_{o1}^{-1}A_1K. \tag{3.33}$$

When G is compatible, $G_{12}A_1 = -B_{o1}A_{o1}^{-1}A_1$ is stable; this is the case according to Lemma B.8 iff $A_{o1}^{-1}A_1$ is stable. So K_o must be stable and it follows from (3.32) that Γ_o must be stable. This is the second necessary condition on the data. It also follows from the solution of Example 2.9 that when Γ_o is stable all stable solution pairs K_oK_w of (3.32) are given by

$$K_o = \Gamma_oY_{g1} + K_mA_g, K_w = -\Gamma_oX_{g1} + K_mB_g \tag{3.34}$$

where K_m is a stable matrix. So from (3.33), a stable solution pair K,K_w exists iff

$$K = A_1^{-1}A_{o1}K_o = A_1^{-1}A_{o1}(\Gamma_oY_{g1} + K_mA_g) \tag{3.35}$$

is stable for some stable K_m. When $A_1^{-1}A_{o1}$ is stable this is the case for every stable K_m. Moreover, $K_g = A_1^{-1}A_{o1}K_m$ is stable iff K_m is stable because as previously pointed out $A_{o1}^{-1}A_1$ stable follows from $G_{12}A_1$ stable. So when $A_1^{-1}A_{o1}$ is stable, (3.35) is equivalent to $K = A_1^{-1}A_{o1}\Gamma_oY_{g1} + K_gA_g$ where K_g is a stable matrix. Thus, without the need of further proof one has

Theorem 3.1 *A matrix* $G_w = A_w^{-1}B_w$ *is acceptable for a compatible G iff* $\Gamma_w = A_oG_{11}A_w^{-1}$ *and* $\Gamma_o = A_{o1}^{-1}(Y_o\Gamma_wA_{g1} - YB_{g1})$ *are both stable and one can choose a stable matrix* K_m *in* (3.35) *for which K is stable. Moreover, when* $A_1^{-1}A_{o1}$ *is stable one gets for every stable matrix* K_m *a stable matrix K in* (3.35) *and all such K are given by*

$$K = A_1^{-1}A_{o1}(\Gamma_oY_{g1} + K_mA_g) = A_1^{-1}A_{o1}\Gamma_oY_{g1} + K_gA_g \tag{3.36}$$

where K_g is a stable matrix.

Remark 3.3 There is no guarantee under Theorem 3.1 that det $(X_1 - KB) \not\equiv 0$ for any of the K given by (3.36) as required by (3.11). However, practical design objectives for analog systems usually demand that

$$S = (I - CG_{22})^{-1} = A_1(X_1 - KB) \to I, \quad s \to \infty. \tag{3.37}$$

Hence, for any K meeting this requirement one gets

$$\det S = (\det A_1)\det (X_1 - KB) \to 1, \quad s \to \infty \tag{3.38}$$

and $\det (X_1 - KB) \not\equiv 0$. For digital systems it is usually the case in practice that $\det A_1(\lambda) \neq 0$, $B_1(\lambda) = 0$, and $B(\lambda) = 0$ at $\lambda = 0$. Then, since a stable $K(\lambda)$ is analytic in $|\lambda| \leq 1$,

$$X_1(0) - K(0)B(0) = X_1(0). \tag{3.39}$$

In addition,

$$X_1(0)A_1(0) + Y_1(0)B_1(0) = X_1(0)A_1(0) = I; \tag{3.40}$$

hence, $\det X_1(0) = 1/\det A_1(0) \neq 0$. That is,

$$\det [X_1(0) - K(0)B(0)] = \det X_1(0) = 1/\det A_1(0) \neq 0 \tag{3.41}$$

and so $\det (X_1 - KB) \not\equiv 0$ for any stable K. For additional insights regarding the realization of C when (3.36) applies see Example 3.19. Also, see Example 3.3 for alternative expressions one can use to check whether or not Γ_w and Γ_o are stable.

Remark 3.4 When, irrespective of whether or not $A_1^{-1}A_{o1}$ is stable, there exists a stable matrix K_{mo} for which

$$K_s = A_1^{-1}A_{o1}(\Gamma_o Y_{g1} + K_{mo}A_g) \tag{3.42}$$

is stable and one chooses

$$K_m = K_{mo} + A_{o1}^{-1}A_1 K_g, \tag{3.43}$$

then K_m is stable for all stable K_g and (3.35) yields

$$K = K_s + K_g A_g \tag{3.44}$$

which is stable for any stable matrix K_g. That is, when there is one stable K_m for which the right-hand side of (3.35) is stable, then there is a family of such matrices given by (3.43) as K_g ranges over all stable matrices. However, this family may not be the complete set of stable matrices K_m for which the right-hand side of (3.35) is stable when $A_1^{-1}A_{o1}$ is unstable: in this case there may be unstable K_g for which both $A_{o1}^{-1}A_1K_g$ and K_gA_g are stable.

Remark 3.5 When $AG_{21}A_w^{-1}$ is stable, it follows from (3.29) that A_g^{-1} is stable. Hence, in this case, (3.35) has a stable solution for K_m given by $K_m = (A_{o1}^{-1}A_1K - \Gamma_o Y_{g1})A_g^{-1}$ for every stable matrix K. That is, when G is compatible and when $AG_{21}A_w^{-1}$, Γ_w, and Γ_o are stable then $G_v G_w$ and $G_\delta G_w$ are stable for any stable K (See Example 3.2).

Remark 3.6 It is sometimes the case, as with the 3DOF system tracking and disturbance rejection problem addressed in Sect. 3.4, that $\mu = 0$ and $\widetilde{G}_{d1} = \widetilde{T}_d \widetilde{G}_{e1}$ where $\widetilde{T}_d$ is a stable matrix. Then the final forms of (3.3) and (3.4) still hold with $\widetilde{G}_{e1}$ in place of $\widetilde{G}_w$, σ in place of q, and

$$G_{11} = \widetilde{G}_{11} - \widetilde{T}_d, \; G_{12} = \widetilde{G}_{12}, \; G_{21} = \widetilde{G}_{21}, \; G_{22} = \widetilde{G}_{22} \tag{3.45}$$

in place of (3.5). Now

$$G_{11} + G_{12}A_1Y_1G_{21} = \widetilde{G}_{11} + \widetilde{G}_{12}A_1Y_1\widetilde{G}_{21} - \widetilde{T}_d \tag{3.46}$$

is stable when $\widetilde{G}$ is admissible since $\widetilde{T}_d$ is stable. So when $\widetilde{G}$ is admissible and $\widetilde{T}_d$ is stable it again readily follows from (3.45) and (3.46) that G is compatible because $G_{12}A_1$, AG_{21}, and $G_{11} + G_{12}A_1Y_1G_{21}$ are all stable and $G_{22} = \widetilde{G}_{22}$. Hence, when $G_w = \widetilde{G}_{e1}$, the G_{ij} are given by (3.45), and $\widetilde{T}_d$ is stable, Theorem 3.1 gives the necessary and sufficient conditions for $\widetilde{G}_{e1}$ to be acceptable (See Example 3.7).

3.4 Tracking and Disturbance Rejection in the 3DOF System Revisited

It is instructive and informative to reconsider the tracking and disturbance rejection problem treated in Sect. 2.6 for the 3DOF system using Theorem 3.1 and Remark 3.6 instead. Specifically, the system shown in Fig. 1.1 is considered with r and d solely deterministic and the only nonzero inputs. That is,

$$e = e_1 = \begin{bmatrix} d \\ r \end{bmatrix} = \begin{bmatrix} A_d^{-1}B_d\sigma_d \\ A_r^{-1}B_r\sigma_r \end{bmatrix} = A_e^{-1}B_e \begin{bmatrix} \sigma_d \\ \sigma_r \end{bmatrix} \tag{3.47}$$

where

$$A_e = \begin{bmatrix} A_d & 0 \\ 0 & A_r \end{bmatrix} \; B_e = \begin{bmatrix} B_d & 0 \\ 0 & B_r \end{bmatrix}. \tag{3.48}$$

When A_d, B_d and A_r, B_r are both left-coprime polynomial matrix pairs, it follows that

$$x'[A_e \quad B_e] = [x_1' \quad x_2'] \begin{bmatrix} A_d & 0 & B_d & 0 \\ 0 & A_r & 0 & B_r \end{bmatrix} = 0 \tag{3.49}$$

iff

$$x_1'[A_d \quad B_d] = 0, \; x_2'[A_r \quad B_r] = 0 \tag{3.50}$$

or $x_1' = 0$ and $x_2' = 0$ since $[A_d \quad B_d]$ and $[A_r \quad B_r]$ have full row rank everywhere (Lemma B.4). Hence, $x = 0$ is the only solution possible in (3.49) and the pair A_e, B_e is left-coprime. Then one can set $\widetilde{G}_{e1} = A_w^{-1}B_w$ with $A_w = A_e$, $B_w = B_e$.

The controlled output y_c is taken to be y and obviously $y_m = [w' \quad z' \quad u']'$. The noise inputs n, n_l, and n_m are usually purely stochastic and, therefore, do not contribute to the shape-deterministic components of y_c and y_m. This justifies the restriction of concern here to the case in which only r and d are nonzero inputs. Clearly,

$$y_c = y = Pv + P_d d = [P_d \quad 0]e + Pv \tag{3.51}$$

and

$$y_m = \begin{bmatrix} w \\ z \\ u \end{bmatrix} = \begin{bmatrix} FPv + FP_d d \\ Ld \\ r \end{bmatrix} = \begin{bmatrix} FP_d & 0 \\ L & 0 \\ 0 & I \end{bmatrix} e + \begin{bmatrix} FP \\ 0 \\ 0 \end{bmatrix} v. \tag{3.52}$$

It readily follows that

$$\widetilde{G}_{11} = [P_d \quad 0], G_{12} = \widetilde{G}_{12} = P, \tag{3.53}$$

and

$$G_{21} = \widetilde{G}_{21} = \begin{bmatrix} FP_d & 0 \\ L & 0 \\ 0 & I \end{bmatrix}, G_{22} = \widetilde{G}_{22} = \begin{bmatrix} FP \\ 0 \\ 0 \end{bmatrix} \tag{3.54}$$

when (3.45) is invoked. Moreover, y_δ is connected to the error ε by

$$y_\delta = y_c - y_d = y - T_d r = -\varepsilon \tag{3.55}$$

and

$$y_d = T_d r = \begin{bmatrix} 0 & T_d \end{bmatrix} e = \widetilde{T}_d e. \tag{3.56}$$

Thus,

$$G_{11} = \widetilde{G}_{11} - \widetilde{T}_d = \begin{bmatrix} P_d & -T_d \end{bmatrix}. \tag{3.57}$$

Attention is restricted to the case in which T_d is stable. Clearly (see Remark 3.6), G is then compatible when $\widetilde{G}$ is admissible and Theorem 3.1 can be invoked.

In order to proceed, the coprime polynomial matrix fraction descriptions defined in (3.9), (3.23), and (3.29) have to be identified. For this purpose, the following coprime polynomial matrix fraction descriptions are introduced:

$$FP = A^{-1}B = B_1 A_1^{-1} \tag{3.58}$$

with

$$\begin{bmatrix} A & B \\ Y_1 & -X_1 \end{bmatrix}\begin{bmatrix} X & B_1 \\ Y & -A_1 \end{bmatrix} = \begin{bmatrix} X & B_1 \\ Y & -A_1 \end{bmatrix}\begin{bmatrix} A & B \\ Y_1 & -X_1 \end{bmatrix} = \begin{bmatrix} I & 0 \\ 0 & I \end{bmatrix} \tag{3.59}$$

and

$$P = A_p^{-1} B_p = B_{p1} A_{p1}^{-1} \tag{3.60}$$

with

$$\begin{bmatrix} A_p & B_p \\ Y_{p1} & -X_{p1} \end{bmatrix}\begin{bmatrix} X_p & B_{p1} \\ Y_p & -A_{p1} \end{bmatrix} = \begin{bmatrix} X_p & B_{p1} \\ Y_p & -A_{p1} \end{bmatrix}\begin{bmatrix} A_p & B_p \\ Y_{p1} & -X_{p1} \end{bmatrix} = \begin{bmatrix} I & 0 \\ 0 & I \end{bmatrix}. \tag{3.61}$$

Clearly,

$$G_{22} = \begin{bmatrix} FP \\ 0 \\ 0 \end{bmatrix} = -\widetilde{A}^{-1}\widetilde{B} = -\widetilde{B}_1 \widetilde{A}_1^{-1} \tag{3.62}$$

where

$$\widetilde{A} = \begin{bmatrix} A & 0 & 0 \\ 0 & I & 0 \\ 0 & 0 & I \end{bmatrix}, \quad \widetilde{B} = \begin{bmatrix} -B \\ 0 \\ 0 \end{bmatrix} \tag{3.63}$$

and

$$\widetilde{A}_1 = A_1, \quad \widetilde{B}_1 = \begin{bmatrix} -B_1 \\ 0 \\ 0 \end{bmatrix}. \tag{3.64}$$

Moreover, one can introduce

$$\widetilde{X} = \begin{bmatrix} X & 0 & 0 \\ 0 & I & 0 \\ 0 & 0 & I \end{bmatrix}, \quad \widetilde{Y} = \begin{bmatrix} -Y & 0 & 0 \end{bmatrix} \tag{3.65}$$

and

$$\widetilde{X}_1 = X_1, \quad \widetilde{Y}_1 = \begin{bmatrix} -Y_1 & 0 & 0 \end{bmatrix}. \tag{3.66}$$

That is, for the 3DOF system, $\widetilde{A}$, $\widetilde{B}$, $\widetilde{A}_1$, $\widetilde{B}_1$, $\widetilde{X}$, $\widetilde{Y}$, $\widetilde{X}_1$, and $\widetilde{Y}_1$ are the replacements for A, B, A_1, B_1, X, Y, X_1, and Y_1, respectively, in (3.9) and (3.10). It should also be obvious from (3.23) and (3.53) that

$$A_o = A_p, \quad B_o = -B_p, \quad A_{o1} = A_{p1}, \quad B_{o1} = -B_{p1} \tag{3.67}$$

and

$$X_o = X_p, \quad Y_o = -Y_p, \quad X_{o1} = X_{p1}, \quad Y_{o1} = -Y_{p1}. \tag{3.68}$$

The next step involves the determination of the coprime polynomial matrix fraction descriptions in (3.29). Clearly, they are determined by

$$\widetilde{A}G_{21}A_w^{-1} = \begin{bmatrix} A & 0 & 0 \\ 0 & I & 0 \\ 0 & 0 & I \end{bmatrix} \begin{bmatrix} FP_d & 0 \\ L & 0 \\ 0 & I \end{bmatrix} \begin{bmatrix} A_d & 0 \\ 0 & A_r \end{bmatrix}^{-1} = \begin{bmatrix} AFP_dA_d^{-1} & 0 \\ LA_d^{-1} & 0 \\ 0 & A_r^{-1} \end{bmatrix} \tag{3.69}$$

for the 3DOF system. Further progress is possible with the introduction of the coprime polynomial matrix fraction descriptions

$$\begin{bmatrix} AFP_dA_d^{-1} \\ LA_d^{-1} \end{bmatrix} = A_q^{-1}B_q = B_{q1}A_{q1}^{-1} \tag{3.70}$$

with

$$\begin{bmatrix} A_q & B_q \\ Y_{q1} & -X_{q1} \end{bmatrix} \begin{bmatrix} X_q & B_{q1} \\ Y_q & -A_{q1} \end{bmatrix} = \begin{bmatrix} X_q & B_{q1} \\ Y_q & -A_{q1} \end{bmatrix} \begin{bmatrix} A_q & B_q \\ Y_{q1} & -X_{q1} \end{bmatrix} = \begin{bmatrix} I & 0 \\ 0 & I \end{bmatrix}. \tag{3.71}$$

Then (3.69) yields

$$\widetilde{A}G_{21}A_w^{-1} = \begin{bmatrix} A_q & 0 \\ 0 & A_r \end{bmatrix}^{-1} \begin{bmatrix} B_q & 0 \\ 0 & I \end{bmatrix} = \begin{bmatrix} B_{q1} & 0 \\ 0 & I \end{bmatrix} \begin{bmatrix} A_{q1} & 0 \\ 0 & A_r \end{bmatrix}^{-1}. \tag{3.72}$$

It is now easy to confirm that

$$A_g = \begin{bmatrix} A_q & 0 \\ 0 & A_r \end{bmatrix}, \; B_g = \begin{bmatrix} B_q & 0 \\ 0 & I \end{bmatrix}, \; A_{g1} = \begin{bmatrix} A_{q1} & 0 \\ 0 & A_r \end{bmatrix}, \; B_{g1} = \begin{bmatrix} B_{q1} & 0 \\ 0 & I \end{bmatrix} \tag{3.73}$$

and

$$X_g = \begin{bmatrix} X_q & 0 \\ 0 & 0 \end{bmatrix}, \; Y_g = \begin{bmatrix} Y_q & 0 \\ 0 & I \end{bmatrix}, \; X_{g1} = \begin{bmatrix} X_{q1} & 0 \\ 0 & 0 \end{bmatrix}, \; Y_{g1} = \begin{bmatrix} Y_{q1} & 0 \\ 0 & I \end{bmatrix} \tag{3.74}$$

satisfy (3.30). Hence, coprime polynomial matrix fraction descriptions that can be associated with (3.29) are provided by (3.73).

All the information needed for the evaluation of Γ_w and Γ_o in Theorem 3.1 is now available. One easily gets

$$\Gamma_w = A_o G_{11} A_w^{-1} = A_p [P_d \quad -T_d] \begin{bmatrix} A_d & 0 \\ 0 & A_r \end{bmatrix}^{-1} = [A_p P_d A_d^{-1} \quad -A_p T_d A_r^{-1}]. \tag{3.75}$$

Also,

$$\Gamma_o = A_{o1}^{-1}(Y_o \Gamma_w A_{g1} - \widetilde{Y} B_{g1}) = -A_{p1}^{-1} Y_p \Gamma_w \begin{bmatrix} A_{q1} & 0 \\ 0 & A_r \end{bmatrix} - A_{p1}^{-1}[-Y \quad 0 \mid 0] \begin{bmatrix} B_{q1} & 0 \\ \hline 0 & I \end{bmatrix} \tag{3.76}$$

or $\Gamma_o = [\Gamma_{o1} \quad \Gamma_{o2}]$ where

$$\Gamma_{o1} = -A_{p1}^{-1} Y_p A_p P_d A_d^{-1} A_{q1} + A_{p1}^{-1}[Y \quad 0] B_{q1} = -Y_{p1} P_d A_d^{-1} A_{q1} + A_{p1}^{-1}[Y \quad 0] B_{q1} \tag{3.77}$$

and

$$\Gamma_{o2} = A_{p1}^{-1} Y_p A_p T_d = Y_{p1} T_d. \tag{3.78}$$

Clearly, Γ_{o2} is stable when T_d is stable. When F and Γ_w are also stable, it can be shown as follows that Γ_{o1} is stable. Solving (3.70) for B_{q1} and using the result in (3.77) yields

$$\Gamma_{o1} = -Y_{p1} P_d A_d^{-1} A_{q1} + A_{p1}^{-1} YAFP_d A_d^{-1} A_{q1} = A_{p1}^{-1}(-Y_p + YAFA_p^{-1}) A_p P_d A_d^{-1} A_{q1}. \tag{3.79}$$

Now

$$
\begin{aligned}
-Y_p + YAFA_p^{-1} &= YAF(X_p + PY_p) - Y_p = YAFX_p + (YB - I)Y_p \\
&= A_1(Y_1FX_p - X_1Y_p)
\end{aligned} \tag{3.80}
$$

and one gets from (3.79)

$$
\Gamma_{o1} = A_{p1}^{-1}A_1(Y_1FX_p - X_1Y_p)A_pP_dA_d^{-1}A_{q1}. \tag{3.81}
$$

Since G is compatible, one has that

$$
G_{12}\widetilde{A}_1 = \widetilde{G}_{12}\widetilde{A}_1 = PA_1 = B_{p1}A_{p1}^{-1}A_1 \tag{3.82}
$$

is stable; hence, $A_{p1}^{-1}A_1$ is stable. It is now easy to see when F and T_d are stable that Γ_w and Γ_o are both stable iff $A_pP_dA_d^{-1}$ and $A_pT_dA_r^{-1}$ are both stable. Necessity is obvious from (3.75) and sufficiency follows immediately from (3.75), (3.78), and (3.81). That is, when F and T_d are stable and the 3DOF system is admissible then $G_w = G_{e1} = A_e^{-1}B_e$ is acceptable iff $A_pP_dA_d^{-1}$ and $A_pT_dA_r^{-1}$ are both stable. Clearly, this result is consistent with Theorems 2.4 and 2.6. Moreover,

$$
\widetilde{A}_1^{-1}A_{o1} = A_1^{-1}A_{p1} = (X_1 + Y_1FP)A_{p1} = X_1A_{p1} + Y_1FB_{p1} \tag{3.83}
$$

is stable and (3.36) applies.

It is also of interest to examine the formula (3.36) which for the 3DOF system becomes $K = A_1^{-1}A_{p1}\Gamma_oY_{g1} + K_gA_g$. After the appropriate substitutions one gets

$$
K = \begin{bmatrix} \Gamma_dA_{q1}Y_{q1} & A_1^{-1}A_{p1}Y_{p1}T_d \end{bmatrix} + \begin{bmatrix} K_{g1} & K_{g2} \end{bmatrix}\begin{bmatrix} A_q & 0 \\ 0 & A_r \end{bmatrix} = \begin{bmatrix} [-K_1 & -H_2] & H_1 \end{bmatrix} \tag{3.84}
$$

where

$$
\Gamma_d = (Y_1FX_p - X_1Y_p)A_pP_dA_d^{-1}. \tag{3.85}
$$

The final expression in (3.84) evolves from (2.157) and (2.205). One immediately gets

$$
H_1 = A_1^{-1}A_{p1}Y_{p1}T_d + K_{g2}A_r \tag{3.86}
$$

where K_{g2} is an arbitrary stable matrix. Since $A_{p1}^{-1}A_1$ and $A_1^{-1}A_{p1}$ are both stable, there is no loss in generality to set $K_{g2} = A_1^{-1}A_{p1}K_r$ in (3.86) to get in agreement with (2.257)

$$H_1 = A_1^{-1}A_{p1}(Y_{p1}T_d + K_rA_r) \tag{3.87}$$

where K_r is an arbitrary stable matrix. Equation (3.84) also gives, since $A_{q1}Y_{q1} = Y_qA_q$,

$$-\begin{bmatrix} K_1 & H_2 \end{bmatrix} = (\Gamma_d Y_q + K_{g1})A_q = K_{go}A_q \tag{3.88}$$

where K_{go} is an arbitrary stable matrix: one can always choose $K_{g1} = K_{go} - \Gamma_d Y_q$ because Γ_d is stable.

The final expression in (3.88) is not yet in agreement with (2.297). Further progress requires exploiting the fact that the left-hand side of (3.70) can be expressed as

$$\begin{bmatrix} M \\ N \end{bmatrix} = \begin{bmatrix} AFP_dA_d^{-1} \\ LA_d^{-1} \end{bmatrix} = \begin{bmatrix} A_m^{-1}B_m \\ A_n^{-1}B_n \end{bmatrix} = \begin{bmatrix} A_m & 0 \\ 0 & A_n \end{bmatrix}^{-1} \begin{bmatrix} B_m \\ B_n \end{bmatrix} \tag{3.89}$$

where A_m, B_m and A_n, B_n are left-coprime polynomial matrix pairs and recognizing that

$$M = AFP_dA_d^{-1} = (AFA_p^{-1})(A_pP_dA_d^{-1}) \tag{3.90}$$

is stable when $A_pP_dA_d^{-1}$ and AFA_p^{-1} are stable. The former follows from (3.75) and the latter is the case because

$$AFA_p^{-1}B_p = AFP = B \tag{3.91}$$

is stable and this is true when F is stable iff AFA_p^{-1} is stable. So A_m^{-1} is a stable matrix. When, moreover,

$$A_q = \begin{bmatrix} A_m & 0 \\ 0 & A_n \end{bmatrix}, \quad B_q = \begin{bmatrix} B_m \\ B_n \end{bmatrix} \tag{3.92}$$

constitutes a left-coprime polynomial matrix pair, then it follows from (3.70) and (3.89) that with $K_{go} = -\begin{bmatrix} K_mA_m^{-1} & K_n \end{bmatrix}$ in (3.88) one gets

$$\begin{bmatrix} K_1 & H_2 \end{bmatrix} = \begin{bmatrix} K_mA_m^{-1} & K_n \end{bmatrix} \begin{bmatrix} A_m & 0 \\ 0 & A_n \end{bmatrix} = \begin{bmatrix} K_m & K_nA_n \end{bmatrix} \tag{3.93}$$

where K_m and K_n are arbitrary stable matrices. Equation (3.93) is in agreement with (2.297) when notational differences are taken into account.

The fact that A_m, B_m and A_n, B_n are left-coprime polynomial matrix pairs, however, is not sufficient to assure that the pair A_q, B_q given by (3.92) is left-coprime although this is true when $\det A_m$ and $\det A_n$ have no common zeros which is typically the case (see Example 3.6). Nevertheless, one can still arrive at (3.93) by directly specializing (3.28) to the 3DOF system beginning with

$$(\widetilde{Y} + \widetilde{A}_1 K)\widetilde{A} G_{21} A_w^{-1} = (Y_o \Gamma_w + A_{o1} K_w). \tag{3.94}$$

From (3.69) and (3.89)

$$\widetilde{A} G_{21} A_w^{-1} = \begin{bmatrix} M & 0 \\ N & 0 \\ 0 & A_r^{-1} \end{bmatrix}. \tag{3.95}$$

One then gets from (3.94) that

$$([-Y \quad 0 \quad 0] + A_1[-K_1 \quad -H_2 \quad H_1]) \begin{bmatrix} M \\ N \\ 0 \end{bmatrix} = -Y_p A_p P_d A_d^{-1} + A_{p1} K_{w1} \tag{3.96}$$

where K_{w1} is the appropriate first block column of K_w. Hence,

$$(Y + A_1 K_1)M + A_1 H_2 N = Y_p A_p P_d A_d^{-1} - A_{p1} K_{w1} \tag{3.97}$$

or

$$H_2 N = A_1^{-1}(Y_p A_p P_d A_d^{-1} - YM) - (A_1^{-1} A_{p1} K_{w1} + K_1 M). \tag{3.98}$$

Now $(A_1^{-1} A_{p1} K_{w1} + K_1 M)$ is stable and using $A_1^{-1} = X_1 + Y_1 FP$ and $A_p^{-1} = X_p + PY_p$ one gets

$$\begin{aligned} A_1^{-1}(Y_p A_p P_d A_d^{-1} - YM) &= A_1^{-1} Y_p A_p P_d A_d^{-1} - A_1^{-1} YAFP_d A_d^{-1} \\ &= (A_1^{-1} Y_p - Y_1 F A_p^{-1}) A_p P_d A_d^{-1} = (X_1 Y_p - Y_1 F X_p) A_p P_d A_d^{-1} \end{aligned} \tag{3.99}$$

which is also stable when F and $A_p P_d A_d^{-1}$ are stable. So $H_2 N = H_2 A_n^{-1} B_n$ is stable and this is true for stable H_2 iff $H_2 = K_s A_n$ where K_s is an arbitrary stable matrix. In particular, for any stable K_1 and stable K_s one can always pick a stable K_{w1} so that (3.98) and, therefore, (3.96) is satisfied. Hence, consistent with (2.297) when notational differences are taken into account,

$$[K_1 \quad H_2] = [K_1 \quad K_s A_n] \tag{3.100}$$

where K_1 and K_s are arbitrary stable matrices.

3.5 $\mathcal{H}_2$ Design of Digital Multivariable Control Systems

The case in which the generalized persistent input standard configuration of Fig. 3.2 represents a digital system is considered in this section. In this instance the $\mathcal{H}_2$ performance measure used for transient responses is

$$\begin{aligned} J_{tr} &= \frac{1}{2\pi j} \oint_{|\lambda|=1} \text{Tr}\ [\langle y_{\delta 1}(\lambda) y_{\delta 1*}(\lambda)\rangle] \frac{d\lambda}{\lambda} \\ &= \frac{1}{2\pi j} \oint_{|\lambda|=1} \text{Tr}\ \left[G_\delta \widetilde{G}_{w1} \langle \sigma\sigma' \rangle \widetilde{G}_{w1*} G_{\delta *}\right] \frac{d\lambda}{\lambda} \\ &= \frac{1}{2\pi j} \oint_{|\lambda|=1} \text{Tr}\ \left[G_\delta \widetilde{G}_{w1} \widetilde{G}_{w1*} G_{\delta *}\right] \frac{d\lambda}{\lambda} \end{aligned} \tag{3.101}$$

and

$$\begin{aligned} J_{ss} &= \frac{1}{2\pi j} \oint_{|\lambda|=1} \text{Tr}\ [\Phi_{y_{\delta 2} y_{\delta 2}}(\lambda)] \frac{d\lambda}{\lambda} \\ &= \frac{1}{2\pi j} \oint_{|\lambda|=1} \text{Tr}\ \left[G_\delta \widetilde{G}_{w2} \Phi_{\mu\mu} \widetilde{G}_{w2*} G_{\delta *}\right] \frac{d\lambda}{\lambda} \\ &= \frac{1}{2\pi j} \oint_{|\lambda|=1} \text{Tr}\ \left[G_\delta \widetilde{G}_{w2} \widetilde{G}_{w2*} G_{\delta *}\right] \frac{d\lambda}{\lambda} \end{aligned} \tag{3.102}$$

is used to measure steady-state stochastic performance. These two measures are incorporated into the single one

$$J = \frac{1}{2\pi j} \oint_{|\lambda|=1} \text{Tr}\ [Q G_\delta G_w G_{w*} G_{\delta *}] \frac{d\lambda}{\lambda} \tag{3.103}$$

where flexibility for weighting the relative importance of the rows of $G_\delta G_w$ or, equivalently, the elements of y_δ is accomplished with the addition of a parahermitian-positive weighting matrix Q that is good. In addition,

$$G_w = \left[\sqrt{\alpha_1}\widetilde{G}_{w1} \quad \sqrt{\alpha_2}\widetilde{G}_{w2}\right] \tag{3.104}$$

where α_1 is a positive constant and α_2 is a nonnegative constant which allow for weighting the relative importance of the transient and steady-state contributions to system performance. Invoking Theorem C.2, one can always write $Q = W_* W$ where W is a stable matrix because Q is good. Then

$$J = \frac{1}{2\pi j} \oint_{|\lambda|=1} \text{Tr}\ [(WG_\delta G_w)(WG_\delta G_w)_*] \frac{d\lambda}{\lambda}. \tag{3.105}$$

The design objective is to find the stabilizing controller C or, equivalently, find the stable matrix K for which $G_v G_w$ and $G_\delta G_w$ are stable and the cost functional J is minimum when G is compatible and G_w is acceptable for G. Attention is restricted to the case in which $A_1^{-1} A_{o1}$ is stable so that in accordance with Theorem 3.1 all permissible K for which $G_v G_w$ and $G_\delta G_w$ are stable are given by either one of the expressions in (3.36) with K_m or K_g an arbitrary stable matrix. It is preferable to work with K_m as the arbitrary stable matrix parameter when W is a polynomial matrix: in this case, the matrices that must be factored to obtain the solution of the optimization problem being treated are polynomial. This is the approach taken here. For all such K_m, $WG_\delta G_w$ is stable and the cost function J has practical significance as pointed out in Chap. 1; hence, it does make sense to seek that K_m for which J is a minimum.

It easily follows from (3.21), (3.26), and (3.34) that

$$WG_\delta G_w = W\Phi B_w = W(X_o \Gamma_w + B_{o1} \Gamma_o X_{g1} - B_{o1} K_m B_g) B_w = \nabla_a - \nabla_b K_m \nabla_c \tag{3.106}$$

where

$$\nabla_a = W(X_o \Gamma_w + B_{o1} \Gamma_o X_{g1}) B_w, \quad \nabla_b = WB_{o1}, \quad \nabla_c = B_g B_w \tag{3.107}$$

are all stable and determined by the given data. So the task remaining is to find the stable K_m for which J is minimized. In this regard it is convenient to introduce

$$\begin{aligned} \rho &= \text{Tr}\ [(\nabla_a - \nabla_b K_m \nabla_c)(\nabla_a - \nabla_b K_m \nabla_c)_*] \\ &= \text{Tr}\ [(\nabla_a - \nabla_b K_m \nabla_c)(\nabla_{a*} - \nabla_{c*} K_{m*} \nabla_{b*})] \end{aligned} \tag{3.108}$$

which is $2\pi j\lambda$ times the integrand of J. Using Tr $(\text{M}+\text{N}) =$ Tr $(\text{M}) +$ Tr (N) and Tr $(\text{MN}) =$ Tr (NM) one easily gets

$$\rho = \text{Tr}\ (\nabla_a \nabla_{a*}) + \text{Tr}\ (-K_m \nabla_c \nabla_{a*} \nabla_b - \nabla_{b*} \nabla_a \nabla_{c*} K_{m*} + K_m \nabla_c \nabla_{c*} K_{m*} \nabla_{b*} \nabla_b). \tag{3.109}$$

It is important to recognize that the two terms on the right-hand side of (3.109) cannot in general be combined into one because neither ∇_a nor ∇_{a*} need have the same number of rows as K_m. It follows from (3.109) that

$$\rho = \rho_a + \text{Tr } (-K_m\Gamma_* - \Gamma K_{m*} + K_m\Omega\Omega_* K_{m*}\Lambda_*\Lambda) \tag{3.110}$$

where

$$\rho_a = \text{Tr } (\nabla_a\nabla_{a*}), \Gamma = \nabla_{b*}\nabla_a\nabla_{c*}, \tag{3.111}$$

and Ω and Λ are the Wiener–Hopf spectral factors satisfying

$$G_b = \nabla_{b*}\nabla_b = \Lambda_*\Lambda, G_c = \nabla_c\nabla_{c*} = \Omega\Omega_*. \tag{3.112}$$

Clearly, G_b and G_c are square and parhermitian-positive. Since ∇_b and ∇_c are good, so are G_b and G_c. Also, it is assumed here that both of the square matrices G_b and G_c are nonsingular on the unit circle. Then, in accordance with Theorem C.2 and Remark C.3, the matrices Ω and Λ can be chosen so that in addition to being square they are together with their inverses analytic in $|\lambda| \leq 1$. There is no need to distinguish between left and right inverses since Ω^{-1} is both a left and right inverse of Ω and similarly for Λ^{-1} when G_b and G_c have full rank on $|\lambda| = 1$ as assumed.

It easily follows from (3.110) utilizing the properties of the trace and then completing the square that

$$\rho = \rho_o + \text{Tr } [(\Lambda K_m\Omega - \widetilde{\Gamma})(\Lambda K_m\Omega - \widetilde{\Gamma})_*] \tag{3.113}$$

where

$$\widetilde{\Gamma} = \Lambda_*^{-1}\Gamma\Omega_*^{-1}, \quad \rho_o = \rho_a - \text{Tr } (\widetilde{\Gamma}\widetilde{\Gamma}_*) \tag{3.114}$$

are both good because ∇_a, ∇_b, ∇_c, Λ^{-1}, and Ω^{-1} are good. Clearly, consistent with the notation for partial fraction expansions introduced in Sect. 1.4,

$$\Lambda K_m\Omega - \widetilde{\Gamma} = Z - \{\widetilde{\Gamma}\}_- \tag{3.115}$$

where

$$Z = \Lambda K_m\Omega - \{\widetilde{\Gamma}\}_+ - \{\widetilde{\Gamma}\}_\infty \tag{3.116}$$

is stable for every stable K_m: the matrix $\{\widetilde{\Gamma}\}_+$ is stable because the matrix $\widetilde{\Gamma}$ is good. Conversely, for every stable Z

$$K_m = \Lambda^{-1}(Z + \{\widetilde{\Gamma}\}_+ + \{\widetilde{\Gamma}\}_\infty)\Omega^{-1} \tag{3.117}$$

is stable. So minimizing J over all stable K_m is equivalent to minimizing

$$J = \frac{1}{2\pi j}\oint\limits_{|\lambda|=1} \rho(\lambda)\frac{d\lambda}{\lambda} = \frac{1}{2\pi j}\oint\limits_{|\lambda|=1} \left\{\mathrm{Tr}\,[(Z - \{\widetilde{\Gamma}\}_-)(Z - \{\widetilde{\Gamma}\}_-)_*] + \rho_o\right\}\frac{d\lambda}{\lambda} \tag{3.118}$$

over all stable Z.

Now for any real rational matrix $M(\lambda)$ one gets

$$\frac{1}{2\pi j}\oint\limits_{|\lambda|=1} \mathrm{Tr}\,[M_*(\lambda)]\frac{d\lambda}{\lambda} = \frac{1}{2\pi j}\oint\limits_{|\lambda|=1} \mathrm{Tr}\left[M\left(\frac{1}{\lambda}\right)\right]\frac{d\lambda}{\lambda} = \frac{1}{2\pi}\int_{-\pi}^{\pi} \mathrm{Tr}\,[M(e^{j\theta})]d\theta \tag{3.119}$$

and

$$\frac{1}{2\pi j}\oint\limits_{|\lambda|=1} \mathrm{Tr}\,[M(\lambda)]\frac{d\lambda}{\lambda} = \frac{1}{2\pi}\int_{-\pi}^{\pi} \mathrm{Tr}\,[M(e^{-j\theta})]d\theta = \frac{1}{2\pi}\int_{-\pi}^{\pi} \mathrm{Tr}\,[M(e^{j\theta})]d\theta. \tag{3.120}$$

Thus, both integrals are equal to one another. Hence,

$$\frac{1}{2\pi j}\oint\limits_{|\lambda|=1} \mathrm{Tr}\,[Z\{\widetilde{\Gamma}\}_{-*}]\frac{d\lambda}{\lambda} = \frac{1}{2\pi j}\oint\limits_{|\lambda|=1} \mathrm{Tr}\,[\{\widetilde{\Gamma}\}_- Z_*]\}\frac{d\lambda}{\lambda} = 0 \tag{3.121}$$

where the last equality holds for the following reasons. First, Z and $\{\widetilde{\Gamma}\}_{-*}$ are analytic in $|\lambda| \leq 1$. Moreover, $\{\widetilde{\Gamma}\}_-$ is strictly proper and, therefore, every nonzero element of $\{\widetilde{\Gamma}\}_-$ has a zero at $\lambda = \infty$. Thus, every element of $\{\widetilde{\Gamma}\}_{-*}$ has a zero at $\lambda = 0$ and the integrand of the first integral in (3.121) is analytic in $|\lambda| \leq 1$; hence, the integral is zero as a consequence of the Cauchy Integral Theorem (Appendix A). It now readily follows from (3.118) that

$$J = \frac{1}{2\pi j}\oint\limits_{|\lambda|=1} \{\mathrm{Tr}\,[ZZ_*] + \mathrm{Tr}\,[\{\widetilde{\Gamma}\}_-\{\widetilde{\Gamma}\}_{-*}] + \rho_o\}\frac{d\lambda}{\lambda}. \tag{3.122}$$

When Z is an $n_r \times n_c$ matrix,

$$\frac{1}{2\pi j}\oint\limits_{|\lambda|=1} \text{Tr }[ZZ_*]\frac{d\lambda}{\lambda} = \frac{1}{2\pi j}\oint\limits_{|\lambda|=1} \text{Tr }[Z_*Z]\frac{d\lambda}{\lambda}$$

$$= \frac{1}{2\pi}\int\limits_{-\pi}^{\pi} \text{Tr }[Z'(e^{j\theta})Z(e^{-j\theta})]d\theta = \frac{1}{2\pi}\int\limits_{-\pi}^{\pi}\sum_{r=1}^{n_r}\sum_{c=1}^{n_c}\left|z_{rc}(e^{j\theta})\right|^2 d\theta \geq 0 \tag{3.123}$$

and the performance functional J is minimized when $Z \equiv 0$ is chosen. In this case one gets from (3.117) and (3.122) that K_m and J are given by, respectively,

$$K_{mo} = \Lambda^{-1}(\{\widetilde{\Gamma}\}_+ + \{\widetilde{\Gamma}\}_\infty)\Omega^{-1} \tag{3.124}$$

and

$$J_o = \frac{1}{2\pi j}\oint\limits_{|\lambda|=1} \{\text{Tr }[\{\widetilde{\Gamma}\}_-\{\widetilde{\Gamma}\}_{-*}] + \rho_o\}\frac{d\lambda}{\lambda}. \tag{3.125}$$

Using the same reasoning as in connection with (3.123) it should be clear from (3.105) that $J \geq 0$ for every choice of stable K_m including K_{mo}. Thus, $J_o \geq 0$.

It is now convenient and possible without the need of further proof to state the key results derived above for digital systems as a theorem.

Theorem 3.2 *Under the assumptions*

(1) ∇_a, ∇_b, and ∇_c *stable;*
(2) det $(\nabla_{b*}\nabla_b) \neq 0$, $|\lambda| = 1$; *and*
(3) det $(\nabla_c\nabla_{c*}) \neq 0$, $|\lambda| = 1$;

the performance functional

$$J = \frac{1}{2\pi j}\oint\limits_{|\lambda|=1} \rho\frac{d\lambda}{\lambda} = \frac{1}{2\pi j}\oint\limits_{|\lambda|=1} \text{Tr }[(\nabla_a - \nabla_b K_m \nabla_c)(\nabla_a - \nabla_b K_m \nabla_c)_*]\frac{d\lambda}{\lambda} \tag{3.126}$$

is finite for all stable K_m. Moreover, one can construct Wiener–Hopf spectral factors Λ and Ω that are square and together with their inverses are analytic in $|\lambda| \leq 1$ such that

$$\nabla_{b*}\nabla_b = \Lambda_*\Lambda, \quad \nabla_c\nabla_{c*} = \Omega\Omega_* \tag{3.127}$$

is satisfied. In addition, any stable K_m can be written as

$$K_m = \Lambda^{-1}(Z + \{\widetilde{\Gamma}\}_+ + \{\widetilde{\Gamma}\}_\infty)\Omega^{-1} \tag{3.128}$$

where Z is a stable matrix and where $\widetilde{\Gamma} = \Lambda_*^{-1}\Gamma\Omega_*^{-1} = \Lambda_*^{-1}\nabla_{b*}\nabla_a\nabla_{c*}\Omega_*^{-1}$. In this case,

$$J = \frac{1}{2\pi j}\oint_{|\lambda|=1} \{\text{Tr}\,[ZZ_*] + \text{Tr}\,(\nabla_a\nabla_{a*}) - \text{Tr}\,(\widetilde{\Gamma}\widetilde{\Gamma}_*) + \text{Tr}\,[\{\widetilde{\Gamma}\}_-\{\widetilde{\Gamma}\}_{-*}]\}\frac{d\lambda}{\lambda}, \tag{3.129}$$

which is equivalent to (see the solution for Example 3.25*)*

$$J = \frac{1}{2\pi j}\oint_{|\lambda|=1} \{\text{Tr}\,[ZZ_*] + \text{Tr}\,[\nabla_a\nabla_{a*}] - \text{Tr}\,[(\{\widetilde{\Gamma}\}_+ + \{\widetilde{\Gamma}\}_\infty)(\{\widetilde{\Gamma}\}_+ + \{\widetilde{\Gamma}\}_\infty)_*]\}\frac{d\lambda}{\lambda}, \tag{3.130}$$

and the choice $Z \equiv 0$ results in the minimum value for J being realized.

Remark 3.7 Theorem 3.2 still holds when Assumption (1) is replaced with ∇_a, ∇_b, and ∇_c good. However, in this case $\nabla_a - \nabla_b K_m \nabla_c$ need not be stable for stable K_m and the relevance of J as a measure of system performance no longer pertains: the inverse transform of $\nabla_a - \nabla_b K_m \nabla_c$ may contain elements whose amplitude can increase without bound. (See Example 3.10 for additional insights.)

Remark 3.8 The implications of Assumptions (2) and (3) of Theorem 3.2 on the identifications made in (3.107) is clearly of interest. The first observation to make is that Assumption (2) holds iff $\nabla_b = WB_{o1}$ has full column rank on $|\lambda| = 1$. Similarly, Assumption (3) holds when $\nabla_c = B_g B_w$ iff $B_g B_w$ has full row rank on $|\lambda| = 1$.

At this point there is no guarantee that using (3.128) for K_m in (3.36) yields a K such that

$$\det\,(X_1 - KB) \not\equiv 0 \tag{3.131}$$

when Z is admissible. However, this is typically the case on account of Remark 3.3. For easy reference and completeness, the results for digital systems with persistent inputs obtained in this section are now summarized without any additional proof needed in

Theorem 3.3 *When G is compatible and G_w is acceptable for G, $A_1^{-1}A_{o1}$ is stable, W is stable and on the unit circle WB_{o1} has full column rank and $B_g B_m$ has full row rank, then the controller transfer matrix C given by* (3.11) *for which $G_v G_w$ and $G_\delta G_w$ are stable and the performance functional*

$$J = \frac{1}{2\pi j} \oint_{|\lambda|=1} \text{Tr } [(WG_\delta G_w)(WG_\delta G_w)_*] \frac{d\lambda}{\lambda} \tag{3.132}$$

is finite is the one for which

$$K = A_1^{-1} A_{o1} (\Gamma_o Y_{g1} + K_m A_g) \tag{3.133}$$

is chosen with

$$K_m = \Lambda^{-1}(Z + \{\widetilde{\Gamma}\}_+ + \{\widetilde{\Gamma}\}_\infty)\Omega^{-1}, \tag{3.134}$$

provided det $(X_1 - KB) \not\equiv 0$ *which is typically the case on account of Remark* 3.3. *In the formula for K_m, Z is an arbitrary stable matrix,*

$$\widetilde{\Gamma} = \Lambda_*^{-1} \Gamma \Omega_*^{-1} = \Lambda_*^{-1} \nabla_{b*} \nabla_a \nabla_{c*} \Omega_*^{-1} \tag{3.135}$$

where

$$\nabla_a = W(X_o \Gamma_w + B_{o1} \Gamma_o X_{g1}) B_w, \;\; \nabla_b = W B_{o1}, \;\; \nabla_c = B_g B_w, \tag{3.136}$$

and Λ and Ω are square Wiener–Hopf spectral factors satisfying

$$\nabla_{b*} \nabla_b = \Lambda_* \Lambda, \;\; \nabla_c \nabla_{c*} = \Omega \Omega_*. \tag{3.137}$$

All other matrices needed are as defined in Sect. 3.3. *Finally, the minimum value for J is obtained when $Z \equiv 0$ is chosen and is given by*

$$J = \frac{1}{2\pi j} \oint_{|\lambda|=1} \{\text{Tr } (\nabla_a \nabla_{a*}) - \text{Tr } (\widetilde{\Gamma} \widetilde{\Gamma}_*) + \text{Tr } [\{\widetilde{\Gamma}\}_- \{\widetilde{\Gamma}\}_{-*}]\} \frac{d\lambda}{\lambda}, \tag{3.138}$$

which is equivalent to

$$J = \frac{1}{2\pi j} \oint_{|\lambda|=1} \{\text{Tr } [\nabla_a \nabla_{a*}] - \text{Tr } [(\{\widetilde{\Gamma}\}_+ + \{\widetilde{\Gamma}\}_\infty)(\{\widetilde{\Gamma}\}_+ + \{\widetilde{\Gamma}\}_\infty)_*]\} \frac{d\lambda}{\lambda}. \tag{3.139}$$

3.6 $\mathcal{H}_2$ Design of Analog Multivariable Control Systems

The case in which the generalized persistent input standard configuration of Fig. 3.2 represents an analog system is considered in this section. In this instance the $\mathcal{H}_2$ performance measure used for transient responses is

$$\begin{aligned} J_{tr} &= \frac{1}{2\pi j} \int_{-j\infty}^{j\infty} \operatorname{Tr}\left[\langle y_{\delta 1}(s) y_{\delta 1*}(s) \rangle\right] ds \\ &= \frac{1}{2\pi j} \int_{-j\infty}^{j\infty} \operatorname{Tr}\left[G_\delta \widetilde{G}_{w1} \langle \sigma\sigma' \rangle \widetilde{G}_{w1*} G_{\delta*}\right] ds \\ &= \frac{1}{2\pi j} \int_{-j\infty}^{j\infty} \operatorname{Tr}\left[G_\delta \widetilde{G}_{w1} \widetilde{G}_{w1*} G_{\delta*}\right] ds \end{aligned} \tag{3.140}$$

and

$$\begin{aligned} J_{ss} &= \frac{1}{2\pi j} \int_{-j\infty}^{j\infty} \operatorname{Tr}\left[\Phi_{y_{\delta 2} y_{\delta 2}}(s)\right] ds \\ &= \frac{1}{2\pi j} \int_{-j\infty}^{j\infty} \operatorname{Tr}\left[G_\delta \widetilde{G}_{w2} \Phi_{\mu\mu} \widetilde{G}_{w2*} G_{\delta*}\right] ds \\ &= \frac{1}{2\pi j} \int_{-j\infty}^{j\infty} \operatorname{Tr}\left[G_\delta \widetilde{G}_{w2} \widetilde{G}_{w2*} G_{\delta*}\right] ds \end{aligned} \tag{3.141}$$

is used to measure steady-state stochastic performance. These two measures are incorporated into the single one

$$J = \frac{1}{2\pi j} \int_{-j\infty}^{j\infty} \operatorname{Tr}\left[Q G_\delta G_w G_{w*} G_{\delta*}\right] ds \tag{3.142}$$

where flexibility for weighting the relative importance of the rows of $G_\delta G_w$ or, equivalently, the elements of y_δ is accomplished with the addition of a parahermitian-positive weighting matrix Q that is good. In addition,

$$G_w = \left[\sqrt{\alpha_1}\widetilde{G}_{w1} \quad \sqrt{\alpha_2}\widetilde{G}_{w2}\right] \tag{3.143}$$

where the positive constant α_1 and the nonnegative constant α_2 allow for weighting the relative importance of the transient and steady-state contributions to system performance. Invoking Theorem C.1, one can always write $Q = W_* W$ where W is a stable matrix because Q is good. Then

$$J = \frac{1}{2\pi j} \int_{-j\infty}^{j\infty} \text{Tr}\,[(WG_\delta G_w)(WG_\delta G_w)_*]ds. \tag{3.144}$$

The design objective is to find the stabilizing controller C or, equivalently, find the stable matrix K for which $G_v G_w$ and $G_\delta G_w$ are stable and the cost functional J is minimum when G is compatible and G_w is acceptable for G. Attention is restricted to the case in which $A_1^{-1}A_{o1}$ is stable so that in accordance with Theorem 3.1 all permissible K for which $G_v G_w$ and $G_\delta G_w$ are stable are given by either one of the expressions in (3.36) with K_m or K_g an arbitrary stable matrix. It is preferable to work with K_m as the arbitrary stable matrix parameter when W is a polynomial matrix: in this case, the matrices that must be factored to obtain the solution of the optimization problem being treated are polynomial. This is the approach taken here and the alternative approach with K_g an arbitrary stable matrix is treated in Example 3.20. For all stable K_m, $WG_\delta G_w$ is stable. In contrast with the treatment for digital systems, however, the cost functional J is finite in the case of analog systems only if $WG_\delta G_w$ is strictly proper as well. As a consequence, a modification of the approach taken for digital systems is needed in order to find that stable K_m, if one exists, for which J is a minimum.

From (3.14) and (3.15),

$$G_\delta = G_{11} + G_{12}(Y + A_1 K)AG_{21}. \tag{3.145}$$

Then using the first expression for K in (3.36) one gets

$$\nabla = WG_\delta G_w = WG_{11}G_w + WG_{12}(Y + A_{o1}\Gamma_o Y_{g1} + A_{o1}K_m A_g)AG_{21}G_w. \tag{3.146}$$

When (3.23) and (3.29) are recalled, it easily follows that

$$\nabla = WG_{11}G_w - WB_{o1}(\widetilde{\Gamma}_o + K_m)B_g B_w \tag{3.147}$$

where

$$\widetilde{\Gamma}_o = A_{o1}^{-1}(Y + A_{o1}\Gamma_o Y_{g1})A_g^{-1}. \tag{3.148}$$

The stability of ∇ is automatically guaranteed for all stable K_m by Theorem 3.1. The next step is the parameterization of the subset of such K_m for which ∇ is strictly proper as well.

It is again convenient to introduce the Wiener–Hopf spectral factors Λ and Ω satisfying

$$G_b = \nabla_{b*}\nabla_b = \Lambda_*\Lambda, \quad G_c = \nabla_c\nabla_{c*} = \Omega\Omega_* \tag{3.149}$$

where

$$\nabla_b = WB_{o1}, \ \nabla_c = B_g B_w. \tag{3.150}$$

Since ∇_b and ∇_c are stable, G_b and G_c are good. Also, attention is restricted here to the case in which both of the square matrices G_b and G_c are nonsingular on the finite $j\omega$-axis. Then, in accordance with Theorem C.1 and Remark 3.2, the matrices Ω and Λ can be chosen so that in addition to being square they are together with their inverses analytic in Re $s \geq 0$. There is no need to distinguish between left and right inverses since Ω^{-1} is both a left and right inverse of Ω and similarly for Λ^{-1} when G_b and G_c have full rank on the finite $j\omega$-axis as assumed. It is also important to recognize from

$$\Lambda_*^{-1} \nabla_{b*} \nabla_b \Lambda^{-1} = I, \ \ \Omega^{-1} \nabla_c \nabla_{c*} \Omega_*^{-1} = I \tag{3.151}$$

that

$$U_b = \nabla_b \Lambda^{-1} = WB_{o1} \Lambda^{-1} \tag{3.152}$$

and

$$U_c = \Omega^{-1} \nabla_c = \Omega^{-1} B_g B_w \tag{3.153}$$

are paraunitary matrices:

$$U_{b*} U_b = I, \ \ U_c U_{c*} = I. \tag{3.154}$$

Since attention is restricted to real rational matrices, it is easily inferred from Lemma C.1 that U_b, U_{b*}, U_c, and U_{c*} are proper matrices.

Using (3.152) and (3.153) in (3.147) yields

$$\nabla = WG_{11} G_w - U_b (\Lambda \widetilde{\Gamma}_o \Omega + \Lambda K_m \Omega) U_c. \tag{3.155}$$

A reasonable assumption consistent with practical applications is that

$$WG_{11} G_w \leq 0(s^{-1}). \tag{3.156}$$

Clearly, in this case, ∇ is strictly proper iff

$$\widetilde{\nabla} = U_b (\Lambda \widetilde{\Gamma}_o \Omega + \Lambda K_m \Omega) U_c \tag{3.157}$$

is strictly proper. Since

$$U_{b*}\widetilde{\nabla}U_{c*} = U_{b*}U_b(\Lambda\widetilde{\Gamma}_o\Omega + \Lambda K_m\Omega)U_cU_{c*} = (\Lambda\widetilde{\Gamma}_o\Omega + \Lambda K_m\Omega), \tag{3.158}$$

$\widetilde{\nabla}$ is strictly proper iff $(\Lambda\widetilde{\Gamma}_o\Omega + \Lambda K_m\Omega)$ is strictly proper. For the choice

$$K_m = \widetilde{K}_m - \Lambda^{-1}\{\Lambda\widetilde{\Gamma}_o\Omega\}_\infty\Omega^{-1}, \tag{3.159}$$

this is the case iff $\widetilde{K}_m$ is such that $\Lambda\widetilde{K}_m\Omega$ is strictly proper. Moreover, K_m is stable iff $\widetilde{K}_m$ is stable. Any $\widetilde{K}_m$ satisfying these restrictions is called admissible. Clearly, (3.159) parameterizes the subset of all stable K_m for which ∇ is strictly proper and stable when (3.156) holds in terms of the set of admissible $\widetilde{K}_m$ which includes $\widetilde{K}_m = 0$.

Using (3.159) in (3.155) and setting

$$\nabla_a = WG_{11}G_w - U_b(\{\Lambda\widetilde{\Gamma}_o\Omega\}_+ + \{\Lambda\widetilde{\Gamma}_o\Omega\}_-)U_c \tag{3.160}$$

one gets

$$\nabla = WG_\delta G_w = \nabla_a - U_b\Lambda\widetilde{K}_m\Omega U_c = \nabla_a - \nabla_b\widetilde{K}_m\nabla_c. \tag{3.161}$$

Since ∇ is strictly proper and stable for all admissible $\widetilde{K}_m$ including $\widetilde{K}_m = 0$, it is evident that ∇_a is strictly proper and stable (also see Example 3.8). In addition, it is clear from (3.150) that ∇_b and ∇_c are stable. That is, (3.161) is identical in form with (3.106) except that now ∇_a has the additional property of being strictly proper as required for the analog case.

Following the same steps that lead to (3.113) again yields

$$\rho = \text{Tr}\,(\nabla\nabla_*) = \rho_o + \text{Tr}\,[(\Lambda\widetilde{K}_m\Omega - \widetilde{\Gamma})(\Lambda\widetilde{K}_m\Omega - \widetilde{\Gamma})_*] \tag{3.162}$$

where

$$\widetilde{\Gamma} = \Lambda_*^{-1}\Gamma\Omega_*^{-1} = \Lambda_*^{-1}\nabla_{b*}\nabla_a\nabla_{c*}\Omega_*^{-1} = U_{b*}\nabla_aU_{c*} \tag{3.163}$$

and

$$\rho_o = \text{Tr}\,(\nabla_a\nabla_{a*}) - \text{Tr}\,(\widetilde{\Gamma}\widetilde{\Gamma}_*). \tag{3.164}$$

Clearly, $\widetilde{\Gamma}$ is strictly proper and good because ∇_a is. Thus, $\{\widetilde{\Gamma}\}_\infty = 0$ and $\{\widetilde{\Gamma}\}_+$ is stable. In addition, $\rho_o \le 0(s^{-2})$. Moreover, every stable matrix $\widetilde{K}_m$ can be expressed as

$$\widetilde{K}_m = \Lambda^{-1}(Z + \{\widetilde{\Gamma}\}_+)\Omega^{-1} \tag{3.165}$$

where Z is a stable matrix. Indeed, the matrix $\widetilde{K}_m$ is stable iff Z is stable and $\Lambda\widetilde{K}_m\Omega$ is strictly proper iff Z is strictly proper. That is, under the assumptions made here any stable K_m for which ∇ is stable and strictly proper is given by

$$K_m = \widetilde{K}_m - \Lambda^{-1}\{\Lambda\widetilde{\Gamma}_o\Omega\}_\infty\Omega^{-1} = \Lambda^{-1}(Z + \{\widetilde{\Gamma}\}_+ - \{\Lambda\widetilde{\Gamma}_o\Omega\}_\infty)\Omega^{-1} \tag{3.166}$$

where Z is admissible (i.e., is an arbitrary stable and strictly proper rational matrix). It now easily follows for any such K_m that the performance functional (3.144) is given by

$$\begin{aligned} J &= \frac{1}{2\pi j}\int_{-j\infty}^{j\infty} \rho ds = \frac{1}{2\pi j}\int_{-j\infty}^{j\infty} \mathrm{Tr}\,(\nabla\nabla_*)ds \\ &= \frac{1}{2\pi j}\int_{-j\infty}^{j\infty} \{\rho_o + \mathrm{Tr}\,[(Z - \{\widetilde{\Gamma}\}_-)(Z - \{\widetilde{\Gamma}\}_-)_*]\}ds \end{aligned} \tag{3.167}$$

and is finite for all admissible Z. Hence, the problem of minimizing J over all stable K_m is equivalent to minimizing J over all stable and strictly proper Z.

The solution to the optimization problem easily follows after some key observations are made. First (Appendix A), for any real rational matrix $M(s) \le 0(s^{-2})$ one has that

$$\int_{-j\infty}^{j\infty} \mathrm{Tr}\,[M(s)]ds = \int_{-j\infty}^{j\infty} \mathrm{Tr}\,[M_*(s)]ds = \int_C \mathrm{Tr}\,[M(s)]ds \tag{3.168}$$

where the contour C includes the $j\omega$-axis and is closed by a semicircular arc into either the left or right-half planes. Hence,

$$\int_{-j\infty}^{j\infty} \mathrm{Tr}\,[Z\{\widetilde{\Gamma}\}_{-*}]ds = \int_{-j\infty}^{j\infty} \mathrm{Tr}\,[\{\widetilde{\Gamma}\}_- Z_*]ds = \int_C \mathrm{Tr}\,[\{\widetilde{\Gamma}\}_- Z_*]ds = 0 \tag{3.169}$$

because $\{\widetilde{\Gamma}\}\_Z_*$ is analytic in Re $s \leq 0$, $\{\widetilde{\Gamma}\}\_Z_* \leq 0(s^{-2})$, and the contour C can be closed into the left-half plane. It follows from (3.167) that

$$J = \frac{1}{2\pi j} \int_{-j\infty}^{j\infty} \{\rho_o + \text{Tr}\,[ZZ_* + \{\widetilde{\Gamma}\}_-\{\widetilde{\Gamma}\}_{-*}]\}ds. \tag{3.170}$$

When Z is an $n_r \times n_c$ matrix,

$$\frac{1}{2\pi j} \int_{-j\infty}^{j\infty} \text{Tr}\,[ZZ_*]ds = \frac{1}{2\pi} \int_{-\infty}^{\infty} \text{Tr}\,[Z(j\omega)Z'(-j\omega)]d\omega = \frac{1}{2\pi} \int_{-\infty}^{\infty} \sum_{r=1}^{n_r} \sum_{c=1}^{n_c} |z_{rc}(j\omega)|^2 d\omega \geq 0 \tag{3.171}$$

and the performance functional J is minimized when $Z = 0$.

It is now convenient and possible without the need of further proof to state the key results derived above for analog systems as a theorem.

Theorem 3.4 *Under the assumptions*

(1) ∇_a, ∇_b, *and* ∇_c *stable;*
(2) det $(\nabla_{b*}\nabla_b) \neq 0$ *on the finite* $j\omega$*-axis;*
(3) det $(\nabla_c\nabla_{c*}) \neq 0$ *on the finite* $j\omega$*-axis; and*
(4) ∇_a *strictly proper;*

one can construct Wiener–Hopf spectral factors Λ *and* Ω *that are square and together with their inverses are analytic in* Re $s \geq 0$ *such that*

$$\nabla_{b*}\nabla_b = \Lambda_*\Lambda, \;\; \nabla_c\nabla_{c*} = \Omega\Omega_* \tag{3.172}$$

is satisfied. In addition, any stable $\widetilde{K}_m$ *can be written as*

$$\widetilde{K}_m = \Lambda^{-1}(Z + \{\widetilde{\Gamma}\}_+)\Omega^{-1} \tag{3.173}$$

where Z *is a stable matrix and where* $\widetilde{\Gamma} = \Lambda_*^{-1}\Gamma\Omega_*^{-1} = \Lambda_*^{-1}\nabla_{b*}\nabla_a\nabla_{c*}\Omega_*^{-1}$ *is strictly proper and good. In this case, the performance functional*

$$J = \frac{1}{2\pi j} \int_{-j\infty}^{j\infty} \rho ds = \frac{1}{2\pi j} \int_{-j\infty}^{j\infty} \text{Tr}\,[(\nabla_a - \nabla_b\widetilde{K}_m\nabla_c)(\nabla_a - \nabla_b\widetilde{K}_m\nabla_c)_*]ds \tag{3.174}$$

is finite iff Z is strictly proper and is given by

$$J = \frac{1}{2\pi j} \int_{-j\infty}^{j\infty} \{\text{Tr}\ [ZZ_*] + \text{Tr}\ (\nabla_a \nabla_{a*}) - \text{Tr}\ (\widetilde{\Gamma}\widetilde{\Gamma}_*) + \text{Tr}\ [\{\widetilde{\Gamma}\}_-\{\widetilde{\Gamma}\}_{-*}]\} ds, \tag{3.175}$$

which is equivalent to (see the solution for Example 3.25)

$$J = \frac{1}{2\pi j} \int_{-j\infty}^{j\infty} \{\text{Tr}\ [ZZ_*] + \text{Tr}\ [\nabla_a \nabla_{a*}] - \text{Tr}\ [\{\widetilde{\Gamma}\}_+\{\widetilde{\Gamma}\}_{+*}]\} ds, \tag{3.176}$$

and the choice $Z \equiv 0$ *results in the minimum value for J being realized.*

Remark 3.9 Theorem 3.4 still holds when Assumption (1) is replaced with ∇_a, ∇_b, and ∇_c good. However, in this case $\nabla_a - \nabla_b \widetilde{K}_m \nabla_c$ need not be stable for stable $\widetilde{K}_m$ and the relevance of J as a measure of system performance no longer pertains: the inverse transform of $\nabla_a - \nabla_b \widetilde{K}_m \nabla_c$ may contain elements whose amplitude can increase without bound. (See Example 3.10 for additional insights.)

Remark 3.10 The implications of Assumptions (2) and (3) of Theorem 3.4 on the identifications made in (3.150) is clearly of interest. The first observation to make is that Assumption (2) holds iff $\nabla_b = WB_{o1}$ has full column rank on the finite $j\omega$-axis. Similarly, Assumption (3) holds when $\nabla_c = B_g B_w$ iff $B_g B_w$ has full row rank on the finite $j\omega$-axis.

At this point there is no guarantee that using (3.166) for K_m in (3.36) yields a K such that

$$\det\ (X_1 - KB) \not\equiv 0 \tag{3.177}$$

when Z is admissible. However, this can be assured as described in the sequel with the introduction of order restrictions on the data. In this regard it is useful to recognize from (3.166) that

$$K_m = \Lambda^{-1}\Gamma_a\Omega^{-1} - \widetilde{\Gamma}_o \tag{3.178}$$

where

$$\Gamma_a = Z + \{\widetilde{\Gamma}\}_+ + \{\Lambda\widetilde{\Gamma}_o\Omega\}_+ + \{\Lambda\widetilde{\Gamma}_o\Omega\}_- \tag{3.179}$$

is strictly proper when Z is admissible. Using (3.178) in (3.36) and recalling (3.148) then yields

$$K = A_1^{-1}(A_{o1}\Lambda^{-1}\Gamma_a\Omega^{-1}A_g - Y) \tag{3.180}$$

and it follows that

$$A_1(X_1 - KB) = A_1X_1 + YB - A_{o1}\Lambda^{-1}\Gamma_a\Omega^{-1}A_gB = I - A_{o1}\Lambda^{-1}\Gamma_a\Omega^{-1}A_gB. \tag{3.181}$$

Clearly, $A_1(X_1 - KB) \to I,\ s \to \infty$ when the second term in the final expression of (3.181) is strictly proper and in this case det $(X_1 - KB) \not\equiv 0$. Conditions that assure this behavior are easily obtained with the help of the following lemma.

Lemma 3.2 *A necessary and sufficient condition for* $M_*M \leq 0(s^{2k})$ *or* $MM_* \leq 0(s^{2k})$ *is* $M \leq 0(s^k)$.

Proof For any rational matrix $M(s)$ one can always write

$$M(s) = \frac{N(s)}{d(s)} = \frac{1}{d(s)}\sum_{i=1}^{q} N_i s^i \tag{3.182}$$

where $d(s)$ is the *lcd* of $M(s)$ and $N(s)$ is a polynomial matrix with $N_q \neq 0$. Clearly, $N_q^*N_q \neq 0$ because with $N_q \neq 0$ there is at least one diagonal entry of $N_q^*N_q$ which is not zero: each diagonal element of $N_q^*N_q$ is the norm of a different column of N_q and these cannot all be zero. Hence,

$$M_*M = \frac{\sum_{i=1}^{q}\sum_{j=1}^{q}(-1)^i N_i^*N_j s^{i+j}}{dd_*} \leq 0(s^{2k}) \tag{3.183}$$

iff $q - \delta(d) \leq k$ or $M \leq 0(s^k)$. The modification of this proof to accommodate the case MM_* is obvious.

It follows from (3.149) and (3.150) that

$$(\nabla_{b*}\nabla_b)^{-1} = (B_{o1*}W_*WB_{o1})^{-1} = \Lambda^{-1}\Lambda_*^{-1} \tag{3.184}$$

or

$$(A_{o1}\Lambda^{-1})(A_{o1}\Lambda^{-1})_* = (A_{o1*}^{-1}B_{o1*}W_*WB_{o1}A_{o1}^{-1})^{-1} = (G_{12*}W_*WG_{12})^{-1}. \tag{3.185}$$

Hence, by Lemma 3.2,

$$(G_{12*}W_*WG_{12})^{-1} \leq 0(s^{2k_1}) \Leftrightarrow A_{o1}\Lambda^{-1} \leq 0(s^{k_1}). \tag{3.186}$$

In addition,

$$(\nabla_c\nabla_{c*})^{-1} = (B_gB_wB_{w*}B_{g*})^{-1} = [(A_gAG_{21}G_w)(A_gAG_{21}G_w)_*]^{-1} = \Omega_*^{-1}\Omega^{-1} \tag{3.187}$$

or

$$(G_{21}G_wG_{w*}G_{21*})^{-1} = (\Omega^{-1}A_gA)_*(\Omega^{-1}A_gA). \tag{3.188}$$

Hence, by Lemma 3.2,

$$(G_{21}G_wG_{w*}G_{21*})^{-1} \leq 0(s^{2k_2}) \Leftrightarrow \Omega^{-1}A_gA \leq 0(s^{k_2}). \tag{3.189}$$

Moreover, from

$$\Omega^{-1}A_gAG_{22} = -\Omega^{-1}A_gB \tag{3.190}$$

one gets

$$\Omega^{-1}A_gB \leq 0(s^{k_2+k_3}) \tag{3.191}$$

when

$$G_{22} \leq 0(s^{k_3}). \tag{3.192}$$

Finally, when the order relationships (3.186), (3.189), and (3.192) hold and Z is admissible so that Γ_a is strictly proper, then

$$A_{o1}\Lambda^{-1}\Gamma_a\Omega^{-1}A_gB \leq 0(s^{k_1+k_2+k_3-1}). \tag{3.193}$$

Hence, from (3.181), det $(X_1 - KB) \not\equiv 0$ when $k_1 + k_2 + k_3 \leq 0$ and Z is admissible.

It is also of interest to investigate the implication of these order conditions on the controller obtained with the K given by (3.180). With the aid of (3.181) it is clear that

$$C = (X_1 - KB)^{-1}A_1^{-1}A_1(Y_1 + KA) = (I - A_{o1}\Lambda^{-1}\Gamma_a\Omega^{-1}A_gB)^{-1}(A_1Y_1 + A_1KA). \tag{3.194}$$

Since

$$A_1KA = A_{o1}\Lambda^{-1}\Gamma_a\Omega^{-1}A_gA - YA \tag{3.195}$$

and $YA = A_1Y_1$, one easily gets

$$C = (I - A_{o1}\Lambda^{-1}\Gamma_a\Omega^{-1}A_gB)^{-1}A_{o1}\Lambda^{-1}\Gamma_a\Omega^{-1}A_gA. \tag{3.196}$$

When $k_1 + k_2 + k_3 \leq 0$ and Z is admissible, it immediately follows that

$$C \leq 0(s^{k_1+k_2-1}) \tag{3.197}$$

because $A_{o1}\Lambda^{-1}\Gamma_a\Omega^{-1}A_gB$ is strictly proper in this case. It is very typical in practical applications that (3.186) and (3.189) are satisfied with $k_1 = k_2 = 0$; it then follows from (3.197) that $C(s)$ is strictly proper.

Some simplification in the expressions for $\tilde{\Gamma}_o$ and Γ_a are possible. Substituting into (3.148) the expression for Γ_o defined in (3.32) yields

$$\begin{aligned}\tilde{\Gamma}_o &= A_{o1}^{-1}(Y + Y_o\Gamma_wA_{g1}Y_{g1} - YB_{g1}Y_{g1})A_g^{-1}\\ &= A_{o1}^{-1}[Y + Y_o\Gamma_wY_gA_g - Y(I - X_gA_g)]A_g^{-1} = A_{o1}^{-1}Y_a\end{aligned} \tag{3.198}$$

where

$$Y_a = Y_o\Gamma_wY_g + YX_g. \tag{3.199}$$

Using (3.160) in (3.163) and recalling (3.154) gives

$$\tilde{\Gamma} = U_{b*}WG_{11}G_wU_{c*} - \{\Lambda\tilde{\Gamma}_o\Omega\}_+ - \{\Lambda\tilde{\Gamma}_o\Omega\}_-. \tag{3.200}$$

It then follows from (3.179) that

$$\Gamma_a = Z + \{U_{b*}WG_{11}G_wU_{c*}\}_+ + \{\Lambda\tilde{\Gamma}_o\Omega\}_- \tag{3.201}$$

or

$$\Gamma_a = Z + \{\Lambda_*^{-1}\nabla_{b*}WG_{11}G_w\nabla_{c*}\Omega_*^{-1}\}_+ + \{\Lambda A_{o1}^{-1}Y_a\Omega\}_-. \tag{3.202}$$

For easy reference and completeness, the results for analog systems obtained in this section are now summarized without the need for any additional proof in

Theorem 3.5 *When G is compatible and G_w is acceptable for G; when $A_1^{-1}A_{o1}$ and W are stable; when WB_{o1} has full column rank and B_gB_w has full row rank on the finite $j\omega$-axis; and when the order relationships*

$$
\begin{aligned}
&WG_{11}G_w \leq 0(s^{-1}), (G_{12*}W_*WG_{12})^{-1} \leq 0(s^{2k_1}),\\
&(G_{21}G_wG_{w*}G_{21*})^{-1} \leq 0(s^{2k_2}), G_{22} \leq 0(s^{k_3})
\end{aligned}
\tag{3.203}
$$

are satisfied with $k_1 + k_2 + k_3 \leq 0$; *then all controller transfer matrices* C *given by* (3.11) *for which* G_vG_w *and* $G_\delta G_w$ *are stable and the performance functional*

$$
J = \frac{1}{2\pi j}\int_{-j\infty}^{j\infty} \mathrm{Tr}\ [(WG_\delta G_w)(WG_\delta G_w)_*]ds \tag{3.204}
$$

is finite are the ones for which

$$
K = A_1^{-1}(A_{o1}\Lambda^{-1}\Gamma_a\Omega^{-1}A_g - Y) \tag{3.205}
$$

is chosen where

$$
\Gamma_a = Z + \{\Lambda_*^{-1}\nabla_{b*}WG_{11}G_w\nabla_{c*}\Omega_*^{-1}\}_+ + \{\Lambda A_{o1}^{-1}Y_a\Omega\}_- \tag{3.206}
$$

with Z *a strictly proper stable matrix and*

$$
Y_a = Y_o\Gamma_wY_g + YX_g. \tag{3.207}
$$

In the formulas for K *and* Γ_a,

$$
\nabla_b = WB_{o1},\ \nabla_c = B_gB_w \tag{3.208}
$$

and Λ *and* Ω *are the square stable Wiener–Hopf spectral factors with stable inverses satisfying*

$$
\nabla_{b*}\nabla_b = \Lambda_*\Lambda,\ \ \nabla_c\nabla_{c*} = \Omega\Omega_*. \tag{3.209}
$$

In this case, the controller transfer matrix is given by

$$
C = (I - A_{o1}\Lambda^{-1}\Gamma_a\Omega^{-1}A_gB)^{-1}A_{o1}\Lambda^{-1}\Gamma_a\Omega^{-1}A_gA \tag{3.210}
$$

and satisfies

$$
C \leq 0(s^{k_1+k_2-1}). \tag{3.211}
$$

Finally, the minimum value for J *is obtained when* $Z \equiv 0$ *is chosen and is given by*

$$J = \frac{1}{2\pi j} \int_{-j\infty}^{j\infty} \{\mathrm{Tr}\,(\nabla_a \nabla_{a*}) - \mathrm{Tr}\,(\widetilde{\Gamma}\widetilde{\Gamma}_*) + \mathrm{Tr}\,[\{\widetilde{\Gamma}\}_-\{\widetilde{\Gamma}\}_{-*}]\} ds \tag{3.212}$$

which is equivalent to

$$J = \frac{1}{2\pi j} \int_{-j\infty}^{j\infty} \{\mathrm{Tr}\,[\nabla_a \nabla_{a*}] - \mathrm{Tr}\,[\{\widetilde{\Gamma}\}_+\{\widetilde{\Gamma}\}_{+*}]\} ds \tag{3.213}$$

where

$$\nabla_a = WG_{11}G_w - \nabla_b \Lambda^{-1}(\{\Lambda A_{o1}^{-1} Y_a \Omega\}_+ + \{\Lambda A_{o1}^{-1} Y_a \Omega\}_-)\Omega^{-1}\nabla_c \tag{3.214}$$

easily follows from (3.152), (3.153), (3.160), *and* (3.198) *and*

$$\widetilde{\Gamma} = \Lambda_*^{-1} \nabla_{b*} WG_{11}G_w \nabla_{c*} \Omega_*^{-1} - \{\Lambda A_{o1}^{-1} Y_a \Omega\}_+ - \{\Lambda A_{o1}^{-1} Y_a \Omega\}_- \tag{3.215}$$

easily follows from (3.200).

3.7 $\mathcal{H}_2$ Design of 1DOF Systems

It is instructive and insightful and a useful illustration of the theory to consider the 1DOF system shown in Fig. 3.3. Without loss in generality, $n = n_m + n_l - n_r$ incorporates any stochastic noises n_r, n_l, and n_m contaminating respectively r, l, and m. Also, L must be stable because the feedforward transducer is a physical block

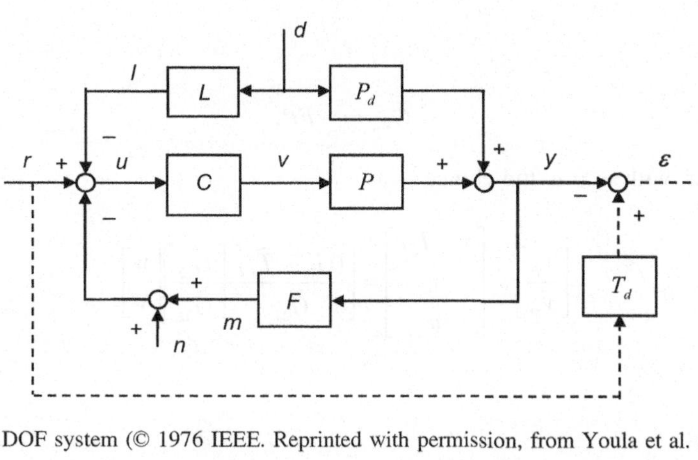

Fig. 3.3 1DOF system (© 1976 IEEE. Reprinted with permission, from Youla et al. (1976b))

outside the feedback loop. The first step is to establish the conditions for and the implications of admissibility of the system. In this regard it is convenient to set

$$P_o = FP_d + L \tag{3.216a}$$

and to make the identifications

$$\left.\begin{aligned} y_m &= u \\ y_c &= [y' \quad v']' \\ e &= [d' \quad r' \quad n']' \end{aligned}\right\}. \tag{3.216b}$$

It then follows from

$$\begin{bmatrix} y \\ v \\ \hline u \end{bmatrix} = \begin{bmatrix} P_d d + Pv \\ v \\ \hline r - n - P_o d - FPv \end{bmatrix} = \left[\begin{array}{ccc|c} P_d & 0 & 0 & P \\ 0 & 0 & 0 & I \\ \hline -P_o & I & -I & -FP \end{array}\right] \begin{bmatrix} d \\ r \\ n \\ \hline v \end{bmatrix} \tag{3.217}$$

that

$$\begin{bmatrix} y_c \\ y_m \end{bmatrix} = \widetilde{G} \begin{bmatrix} e \\ v \end{bmatrix} = \begin{bmatrix} \widetilde{G}_{11} & \widetilde{G}_{12} \\ \widetilde{G}_{21} & \widetilde{G}_{22} \end{bmatrix} \begin{bmatrix} e \\ v \end{bmatrix} \tag{3.218}$$

where

$$\widetilde{G}_{11} = \begin{bmatrix} P_d & 0 & 0 \\ 0 & 0 & 0 \end{bmatrix}, \tag{3.219}$$

$$\widetilde{G}_{12} = \begin{bmatrix} P \\ I \end{bmatrix}, \tag{3.220}$$

$$\widetilde{G}_{21} = [-P_o \quad I \quad -I], \tag{3.221}$$

and

$$\widetilde{G}_{22} = -FP. \tag{3.222}$$

Moreover, with $w = e$, the choice

$$\begin{bmatrix} y_\delta \\ \hline y_m \end{bmatrix} = \begin{bmatrix} y - T_d r \\ v \\ \hline u \end{bmatrix} = \left[\begin{array}{c|c} (\widetilde{G}_{11} - \widetilde{T}_d) & \widetilde{G}_{12} \\ \hline \widetilde{G}_{21} & \widetilde{G}_{22} \end{array}\right] \begin{bmatrix} w \\ \hline v \end{bmatrix} \tag{3.223}$$

where

$$\widetilde{T}_d = \begin{bmatrix} 0 & T_d & 0 \\ 0 & 0 & 0 \end{bmatrix} \tag{3.224}$$

is obviously of interest with regard to the final objective of keeping $\hat{v}(t)$ and $\widehat{\varepsilon}(t)$ small. Hence, it follows that the 1DOF system is a particular case of the one shown in Fig. 3.2 with

$$G_{12} = \widetilde{G}_{12},\ G_{21} = \widetilde{G}_{21},\ G_{22} = \widetilde{G}_{22}, \tag{3.225}$$

and

$$G_{11} = \begin{bmatrix} P_d & -T_d & 0 \\ 0 & 0 & 0 \end{bmatrix} = \widetilde{G}_{11} - \widetilde{T}_d. \tag{3.226}$$

The usual coprime polynomial matrix fraction descriptions for FP and P given by (3.58) through (3.61) are again introduced. Then from

$$G_{12} = \begin{bmatrix} P \\ I \end{bmatrix} = \begin{bmatrix} B_{p1} \\ A_{p1} \end{bmatrix} A_{p1}^{-1} = \begin{bmatrix} A_p & 0 \\ 0 & I \end{bmatrix}^{-1} \begin{bmatrix} B_p \\ I \end{bmatrix}, \tag{3.227}$$

$$\begin{bmatrix} A_p & 0 \\ 0 & I \end{bmatrix} \begin{bmatrix} X_p & 0 \\ -Y_p & I \end{bmatrix} + \begin{bmatrix} -B_p \\ -I \end{bmatrix} \begin{bmatrix} -Y_p & 0 \end{bmatrix} = \begin{bmatrix} I & 0 \\ 0 & I \end{bmatrix}, \tag{3.228}$$

and

$$X_{p1}A_{p1} + \begin{bmatrix} -Y_{p1} & 0 \end{bmatrix} \begin{bmatrix} -B_{p1} \\ -A_{p1} \end{bmatrix} = I \tag{3.229}$$

it is clear that (3.23) and (3.24) are satisfied with

$$A_o = \begin{bmatrix} A_p & 0 \\ 0 & I \end{bmatrix},\ X_o = \begin{bmatrix} X_p & 0 \\ -Y_p & I \end{bmatrix} \tag{3.230}$$

$$A_{o1} = A_{p1},\ X_{o1} = X_{p1} \tag{3.231}$$

$$B_o = -\begin{bmatrix} B_p \\ I \end{bmatrix},\ B_{o1} = -\begin{bmatrix} B_{p1} \\ A_{p1} \end{bmatrix} \tag{3.232}$$

$$Y_o = \begin{bmatrix} -Y_p & 0 \end{bmatrix},\ Y_{o1} = \begin{bmatrix} -Y_{p1} & 0 \end{bmatrix}. \tag{3.233}$$

On account of Definition 2.2 and Lemma 2.3, the 1DOF system is admissible only if

$$\widetilde{G}_{12}A_1 = G_{12}A_1 = \begin{bmatrix} PA_1 \\ A_1 \end{bmatrix}, \tag{3.234}$$

$$A\widetilde{G}_{21} = AG_{21} = \begin{bmatrix} -AP_o & A & -A \end{bmatrix}, \tag{3.235}$$

and

$$\widetilde{G}_{11} + \widetilde{G}_{12}A_1Y_1\widetilde{G}_{21} = G_{11} + G_{12}A_1Y_1G_{21} + \widetilde{T}_d \tag{3.236}$$

are all stable. Hence, PA_1 and $AFP_d = AP_o - AL$ must be stable. It then follows from

$$\widetilde{G}_{11} + \widetilde{G}_{12}A_1Y_1\widetilde{G}_{21} = \begin{bmatrix} (P_d - PA_1Y_1P_o) & PA_1Y_1 & -PA_1Y \\ -YAP_o & A_1Y_1 & -A_1Y_1 \end{bmatrix} \tag{3.237}$$

that

$$P_d - PA_1Y_1P_o = (I - PA_1Y_1F)P_d - PA_1Y_1L \tag{3.238}$$

must be stable. Hence, $(I - PA_1Y_1F)P_d$ must be stable. In fact, it is not difficult to confirm that

Lemma 3.3 *The matrices $\widetilde{G}_{12}A_1$, $A\widetilde{G}_{21}$, and $\widetilde{G}_{11} + \widetilde{G}_{12}A_1Y_1\widetilde{G}_{21}$ are stable when L is stable iff PA_1, AFP_d, and $(I - PA_1Y_1F)P_d$ are stable.*

Since $G_{11} + G_{12}A_1Y_1G_{21}$ stable is also needed for Theorem 3.1, it follows from (3.236) that $\widetilde{T}_d$ stable or, equivalently, T_d stable should be assumed as well.

One also has from *(ii)* of Definition 2.2 that the 1DOF system is admissible iff $\Delta_{\widetilde{G}} / \psi_{\widetilde{G}_{22}}$ is a stable polynomial. It readily follows from $\widetilde{G}_{22} = -FP$ that $\psi_{\widetilde{G}_{22}} = \psi_{FP}$. The determination of $\Delta_{\widetilde{G}}$ can be accomplished using the procedure of Example 2.1. In particular, with the identifications

$$\left.\begin{array}{c} G_1 = \begin{bmatrix} P_d & P \end{bmatrix},\ G_2 = F,\ G_3 = L,\ e = \begin{bmatrix} d' & r' & n' \end{bmatrix}' \\ u_1 = \begin{bmatrix} d' & v' \end{bmatrix}',\ u_2 = y = y_1,\ u_3 = d = \begin{bmatrix} I & 0 \end{bmatrix} u_1,\ y_2 = m,\ y_3 = l \end{array}\right\} \tag{3.239}$$

one easily gets

$$\left.\begin{bmatrix} u_1 \\ u_2 \\ u_3 \end{bmatrix}\right|_{\substack{e=0 \\ v=0}} = M_{13}\begin{bmatrix} y_1 \\ y_2 \\ y_3 \end{bmatrix} = \begin{bmatrix} 0 & 0 & 0 \\ I & 0 & 0 \\ 0 & 0 & 0 \end{bmatrix}\begin{bmatrix} y_1 \\ y_2 \\ y_3 \end{bmatrix} \tag{3.240}$$

and

$$GM_{13} = \begin{bmatrix} G_1 & 0 & 0 \\ 0 & G_2 & 0 \\ 0 & 0 & G_3 \end{bmatrix} \begin{bmatrix} 0 & 0 & 0 \\ I & 0 & 0 \\ 0 & 0 & 0 \end{bmatrix} = \begin{bmatrix} 0 & 0 & 0 \\ F & 0 & 0 \\ 0 & 0 & 0 \end{bmatrix}. \tag{3.241}$$

Hence,

$$\begin{aligned} \Delta_{\tilde{G}} &= \eta(\det T)\det(I - GM_{13}) = \eta(\det T_1)(\det T_2)(\det T_3) = \Delta_{G_1}\Delta_{G_2}\Delta_{G_3} \\ &= h_o\psi_{G_1}\psi_{G_2}\psi_{G_3} \end{aligned} \tag{3.242}$$

where h_o is a polynomial accounting for any hidden poles. Now because P is a submatrix of G_1, $\psi_{G_1} = h_1\psi_P$ where h_1 is a polynomial. Moreover, $\psi_{G_2} = \psi_F$ and $\psi_{G_3} = \psi_L$. Thus,

$$\frac{\Delta_{\tilde{G}}}{\psi_{\tilde{G}_{22}}} = \frac{h_o h_1 \psi_F \psi_P \psi_L}{\psi_{FP}} \tag{3.243}$$

and the 1DOF system is admissible iff h_o, h_1, ψ_L, and $\psi_F\psi_P/\psi_{FP}$ are all stable polynomials which is assumed to be the case. It would then also be necessary to assume that F is stable in order to invoke (3.36) of Theorem 3.1: it is only under this circumstance from Lemma 2.10 that

$$A_1^{-1}A_{p1} = A_1^{-1}A_{o1} \tag{3.244}$$

would be stable. Moreover, in some circumstances (see Example 3.12) it is necessary that $(T_dF - I)P$ be stable. It then follows from Remark 2.5 that when the 1DOF system is admissible and both T_d and T_d^{-1} are stable that F again must be stable. So attention is restricted here to admissible 1DOF systems with F stable. It is important to note as well that ψ_L is a stable polynomial iff L is stable and the necessity for this assumption is again clear.

Next, the modeling of the exogenous input w is considered. It is assumed that d and r are modeled as

$$d = A_d^{-1}B_d\sigma_d + \Omega_d\mu_d, \; r = A_r^{-1}B_r\sigma_r + \Omega_r\mu_r \tag{3.245}$$

where Ω_d and Ω_r are stable left Wiener–Hopf factors associated with Φ_{dd} and Φ_{rr}, the power spectral density matrices for the stochastic components of the signals d and r, respectively; where the elements of the vectors μ_d and μ_r are zero-mean independent white-noise processes with unit variance; where A_d, B_d and A_r, B_r are left-coprime polynomial matrix pairs; and where σ_d and σ_r are vectors whose elements are random variables satisfying

$$\left\langle \begin{bmatrix} \sigma_d \\ \sigma_r \end{bmatrix} [\sigma'_d \quad \sigma'_r] \right\rangle = I. \tag{3.246}$$

The input n is modeled as purely stochastic with a power spectral density matrix $\Phi_{nn} = \Omega_n \Omega_{n*}$ with Ω_n a stable left Wiener–Hopf factor. Hence, one can write

$$n = \Omega_n \mu_n \tag{3.247}$$

where the elements of μ_n are zero-mean independent white-noise processes with unit variance. Moreover, $\Phi_{\mu\mu} = I$ where

$$\mu = [\mu'_d \quad \mu'_r \quad \mu'_n]'. \tag{3.248}$$

Thus,

$$w = \begin{bmatrix} d \\ r \\ n \end{bmatrix} = \left[\begin{array}{cc|ccc} A_d^{-1}B_d & 0 & \Omega_d & 0 & 0 \\ 0 & A_r^{-1}B_r & 0 & \Omega_r & 0 \\ 0 & 0 & 0 & 0 & \Omega_n \end{array}\right] \left[\begin{array}{c} \sigma_d \\ \sigma_r \\ \hline \mu_d \\ \mu_r \\ \mu_n \end{array}\right] = [\widetilde{G}_{w1} \mid \widetilde{G}_{w2}] \left[\begin{array}{c} \sigma \\ \hline \mu \end{array}\right] = \widetilde{G}_w q \,. \tag{3.249}$$

Hence, for $\widetilde{G}_{w1} = A_w^{-1}B_w$ one can take

$$A_w = \text{diag}\,\{A_d, A_r, I\} \tag{3.250}$$

$$B_w = \begin{bmatrix} B_d & 0 \\ 0 & B_r \\ 0 & 0 \end{bmatrix} \tag{3.251}$$

for the left-coprime polynomial matrix pair A_w, B_w. Since $\widetilde{G}_{w2}$ is stable, it follows for any constants $\alpha_1 > 0$ and $\alpha_2 \geq 0$ that $G_w = \left[\sqrt{\alpha_1}\widetilde{G}_{w1} \quad \sqrt{\alpha_2}\widetilde{G}_{w2} \right]$ is acceptable iff $\widetilde{G}_{w1}$ is acceptable.

In so far as optimization of the performance of the 1DOF system is concerned, it should be clear that the most design freedom is available when no restriction on the controller is imposed other than that it be a stabilizing one. Moreover, this is the case in Youla et al. (1976b) and results obtained here can then be compared with those obtained there. So in accordance with Remark 3.5 and with $\widetilde{G}_{w1} = A_w^{-1}B_w$ in place of G_w in Theorem 3.1, it is natural to consider the case in which

$$\Gamma_g = AG_{21}A_w^{-1} = \begin{bmatrix} -AP_oA_d^{-1} & AA_r^{-1} & -A \end{bmatrix} \tag{3.252}$$

is a stable matrix. Clearly, Γ_g is stable iff

$$\widetilde{\Gamma}_g = \begin{bmatrix} -AP_oA_d^{-1} & AA_r^{-1} \end{bmatrix} \tag{3.253}$$

is stable. (In Example 2.13, some situations are considered where this is not the case.) Under these circumstances, it follows from Theorem 3.1 that $\widetilde{G}_{w1}$ is acceptable for all stabilizing controllers iff Γ_w and Γ_o are also stable. Now

$$\Gamma_w = A_oG_{11}A_w^{-1} = \begin{bmatrix} A_pP_dA_d^{-1} & -A_pT_dA_r^{-1} & 0 \\ 0 & 0 & 0 \end{bmatrix} \tag{3.254}$$

is stable iff

$$\widetilde{\Gamma}_w = \begin{bmatrix} A_pP_dA_d^{-1} & -A_pT_dA_r^{-1} \end{bmatrix} \tag{3.255}$$

is stable.

A number of important necessary conditions follow from the imposition of the requirement that $\widetilde{\Gamma}_g$ and $\widetilde{\Gamma}_w$ be stable. Obviously, $\widetilde{\Gamma}_w$ is stable iff $A_pP_dA_d^{-1}$ and $A_pT_dA_r^{-1}$ are stable and $\widetilde{\Gamma}_g$ is stable iff AA_r^{-1} and

$$AP_oA_d^{-1} = AFP_dA_d^{-1} + ALA_d^{-1} \tag{3.256}$$

are stable. Clearly, since $AFA_p^{-1}B_p = B$ is a stable matrix, one gets when F is stable that AFA_p^{-1} is stable. Hence,

$$AFP_dA_d^{-1} = (AFA_p^{-1})(A_pP_dA_d^{-1}) \tag{3.257}$$

is stable and it follows from (3.256) that $AP_oA_d^{-1}$ is stable iff ALA_d^{-1} is stable. In fact, it is established in the sequel in the proof of Lemma 3.4 that LA_d^{-1} must be stable for Γ_o to be stable.

There is one more key necessary condition to establish. One gets from Fig. 3.3 that the shape-deterministic component of u is given by

$$u_1 = r_1 - (L + FP_d)d_1 - FPCu_1 \tag{3.258}$$

or

$$u_1 = (I + FPC)^{-1}[r_1 - (L + FP_d)d_1] \tag{3.259}$$

where $r_1 = A_r^{-1}B_r\sigma_r$ and $d_1 = A_d^{-1}B_d\sigma_d$. Setting $C = (Y + A_1K)(X - B_1K)^{-1}$ which represents all stabilizing controller transfer matrices it readily follows that

$$u_1 = r_1 - Fy_1 = r_1 - F(\varepsilon_1 - T_d r_1) = (I - FT_d)A_r^{-1}B_r\sigma_r + F\varepsilon_1. \tag{3.260}$$

When F, T_d, and ε_1 are stable, it then follows that u_1 is stable for all σ_r iff $(I - FT_d)A_r^{-1}$ is stable. Indeed,

Lemma 3.4 *When the 1DOF system is admissible, then (1) $G = [G_{ij}]$ is compatible iff T_d is stable and (2) G_w is acceptable for all stabilizing controllers, provided F, T_d, and $\widetilde{G}_{w2}$ are stable, iff $A_pP_dA_d^{-1}$, LA_d^{-1}, $A_pT_dA_r^{-1}$, and $(I - FT_d)A_r^{-1}$ are stable.*

Proof That (1) is true is an immediate consequence of Eqs. (3.224) and (3.234) through (3.236). Except for LA_d^{-1}, the necessity for the conditions given in (2) has already been established. Clearly, the conditions are sufficient for Γ_w stable and for $AP_oA_d^{-1}$ stable. With $M_1 = A_pT_dA_r^{-1}$ and $M_2 = (I - FT_d)A_r^{-1}$ stable matrices it follows that

$$AA_r^{-1} = A(M_2 + FT_dA_r^{-1}) = AM_2 + AFA_p^{-1}M_1 \tag{3.261}$$

is stable. Hence, Γ_g is stable. It only remains to show that Γ_o is stable and that for this to be the case it is necessary that LA_d^{-1} be stable.

The matrix Γ_o is defined in (3.32) and with the aid of (3.29) and (3.252) can be written as

$$\Gamma_o = A_{o1}^{-1}(Y_o\Gamma_w - Y\Gamma_g)A_{g1}. \tag{3.262}$$

It readily follows from (3.231) and (3.233) when $Y_pA_p = A_{p1}Y_{p1}$ is recalled that

$$\Gamma_o = A_{p1}^{-1}([-Y_p \quad 0]\Gamma_w - Y\Gamma_g)A_{g1} = \begin{bmatrix}\Gamma_a & \Gamma_b & A_{p1}^{-1}YA\end{bmatrix}A_{g1} \tag{3.263}$$

where

$$\Gamma_a = -Y_{p1}P_dA_d^{-1} + A_{p1}^{-1}YAP_oA_d^{-1} \tag{3.264}$$

and

$$\Gamma_b = Y_{p1}T_dA_r^{-1} - A_{p1}^{-1}YAA_r^{-1}. \tag{3.265}$$

Now

$$A_{p1}^{-1}YA = A_{p1}^{-1}A_1Y_1 \tag{3.266}$$

is stable because admissibility requires $PA_1 = B_{p1}A_{p1}^{-1}A_1$ stable and this is the case iff $A_{p1}^{-1}A_1$ is stable. Since $\Gamma_g = B_{g1}A_{g1}^{-1}$ is stable iff A_{g1}^{-1} is stable, it is clear that when the 1DOF system is admissible and Γ_g is stable then Γ_o is stable iff Γ_a and Γ_b are stable. Now

$$\begin{aligned}\Gamma_a &= (-Y_{p1}A_p^{-1} + A_{p1}^{-1}YAFA_p^{-1})(A_pP_dA_d^{-1}) + A_{p1}^{-1}YALA_d^{-1}\\ &= \widetilde{\Gamma}_aA_pP_dA_d^{-1} + A_{p1}^{-1}YALA_d^{-1}\end{aligned} \tag{3.267}$$

where

$$\begin{aligned}\widetilde{\Gamma}_a &= -Y_{p1}A_p^{-1} + A_{p1}^{-1}YAFA_p^{-1}\\ &= A_{p1}^{-1}[-Y_p + YAF(X_p + PY_p)]\\ &= A_{p1}^{-1}[A_1Y_1FX_p + (YB - I)Y_p]\\ &= A_{p1}^{-1}A_1(Y_1FX_p - X_1Y_p]\end{aligned} \tag{3.268a}$$

is stable because F is stable and PA_1 stable implies $A_{p1}^{-1}A_1$ stable. Since $A_pP_dA_d^{-1}$ is stable, it follows that Γ_a is stable iff $A_{p1}^{-1}A_1Y_1LA_d^{-1}$ is stable. This is the case iff $Y_1LA_d^{-1}$ is stable because $A_1^{-1}A_{p1}$ is stable by Lemma 2.10. Hence, it is necessary that

$$B_1Y_1LA_d^{-1} = (I - XA)LA_d^{-1} = LA_d^{-1} - XALA_d^{-1} \tag{3.268b}$$

be stable. Clearly, $XALA_d^{-1}$ is stable because ALA_d^{-1} stable follows from (3.256). So it is necessary that LA_d^{-1} be stable and when this is the case one gets from (3.267) that Γ_a is stable.

Next it is established that

$$\begin{aligned}\Gamma_b &= (Y_{p1}T_dA^{-1} - A_{p1}^{-1}Y)AA_r^{-1} = Y_{p1}T_dA_r^{-1} - A_{p1}^{-1}YAA_r^{-1}\\ &= Y_{p1}T_dA_r^{-1} - (X_{p1} + Y_{p1}P)YAA_r^{-1} = Y_{p1}(T_dA_r^{-1} - PYAA_r^{-1}) - X_{p1}YAA_r^{-1}\\ &= Y_{p1}(A_p^{-1}M_1 - PYAM_2 - PYAFA_p^{-1}M_1) - X_{p1}YAA_r^{-1}\\ &= Y_{p1}(I - PA_1Y_1F)A_p^{-1}M_1 - Y_{p1}PA_1Y_1M_2 - X_{p1}YAA_r^{-1}\end{aligned} \tag{3.269}$$

is stable. Since $(I - PA_1Y_1F)$ is stable, $(I - PA_1Y_1F)A_p^{-1}$ is stable iff

$$(I - PA_1Y_1F)A_p^{-1}B_p = (I - PYAF)P = P(I - YAFP) = P(I - YB) = PA_1X_1 \tag{3.270}$$

is stable which is the case. It immediately follows from (3.269) that Γ_b and, therefore, Γ_o are stable. This completes the proof of the lemma.

Clearly, under the conditions of Lemma 3.4, it follows from Theorem 3.1 and Remark 3.5 for the A_w given by (3.250) that $\widetilde{G}_{w1} = A_w^{-1}B_w$, and, therefore, G_w is acceptable for the 1DOF system with any stabilizing controller. It is important to note in addition that when $S_r = AA_r^{-1}$ and $S_d = A_pP_dA_d^{-1}$ are stable, then $r = A_r^{-1}B_r\sigma_r = A^{-1}\widetilde{S}_r$ where $\widetilde{S}_r = S_rB_r\sigma_r$ is a stable vector and $P_dd = P_dA_d^{-1}B_d\sigma_d = A_p^{-1}\widetilde{S}_d$ where $\widetilde{S}_d = S_dB_d\sigma_d$ is a stable vector. Hence, any unstable poles of r must be unstable poles of A^{-1} or, equivalently, unstable poles of FP which when F is stable can only be unstable poles of P. Also, any unstable poles of P_dd must be poles of A_p^{-1} or, equivalently, poles of P.

Typically, $L = L_1L_2$ where L_2 is the transfer matrix of the disturbance sensors and L_1 represents any associated cascade compensation that is introduced. Since L_1 and L_2 are physical blocks outside the feedback loop both must be stable. The assumption that LA_d^{-1} is stable imposes a constraint on L_1. In particular, with $L_2A_d^{-1} = \mathcal{A}^{-1}\mathcal{B}$ a left coprime polynomial matrix fraction description, LA_d^{-1} is stable iff $L_1\mathcal{A}^{-1} = \widetilde{L}_1$ is stable. That is, one must choose L_1 so that $\widetilde{L}_1$ is a stable matrix.

Attention is now turned to the $\mathcal{H}_2$ design of a digital 1DOF system with F and T_d stable. It is assumed that the system is admissible, $\widetilde{G}_{w2}$ is stable, and for all stabilizing controllers, G_w is acceptable for G. That is,

$$\nabla = W[G_{11} + G_{12}(Y + A_1K)AG_{21}]G_w = WG_\delta G_w \tag{3.271}$$

is stable for all stable K when W is stable. Setting

$$-\nabla_b = WG_{12}A_1 = W\begin{bmatrix} P \\ I \end{bmatrix}A_1, \tag{3.272}$$

$$\nabla_c = A\ddot{G}_{21}\ddot{G}_w = \left[\sqrt{\alpha_1}\Gamma_gB_w \quad \sqrt{\alpha_2}AG_{21}\ddot{G}_{w2}\right] = A[-P_o \quad I \quad -I]G_w, \tag{3.273}$$

and

$$\nabla_a = WG_{11}G_w - \nabla_bA_1^{-1}Y\nabla_c \tag{3.274}$$

yields

$$\nabla = \nabla_a - \nabla_b K \nabla_c. \tag{3.275}$$

Since ∇ is stable for all stable K which includes $K \equiv 0$, it follows that ∇_a is stable. Clearly, ∇_b and ∇_c are stable because W, PA_1, Γ_g, AG_{21}, and $\widetilde{G}_{w2}$ are stable for the assumptions made here.

It is of interest to consider the choice

$$W = \begin{bmatrix} I & 0 \\ 0 & W_q \end{bmatrix} \tag{3.276a}$$

for which one gets from (3.223) that

$$Wy_\delta = \begin{bmatrix} I & 0 \\ 0 & W_q \end{bmatrix} \begin{bmatrix} y - T_d r \\ v \end{bmatrix} = \begin{bmatrix} -\varepsilon \\ W_q v \end{bmatrix}. \tag{3.276b}$$

Hence, the matrix W_q weights the relative importance of v with respect to the error ε. For example, the choice $W_q = 0$ eliminates v from having any impact on the performance functional and the emphasis is solely on keeping $\widehat{\varepsilon}(k)$ small. On the other hand, other choices for W_q can be considered to tradeoff error against the size of the control signal $\hat{v}(k)$.

It is also useful for comparison with what is obtained for analog 1DOF systems in Youla et al. (1976b) to consider the choice (3.276a) with

$$W_{q*} W_q = kQ \tag{3.277}$$

where k is a positive constant. Clearly, Q is good because W_q is stable and also Q is a parahermitian-positive matrix. Then

$$\nabla_{b*} \nabla_b = A_{1*}(P_* P + kQ) A_1. \tag{3.278}$$

It is also convenient to introduce

$$G_t = P_o G_d P_{o*} + G_r + G_n \tag{3.279}$$

where

$$\left. \begin{aligned} G_d &= \alpha_1 A_d^{-1} B_d B_{d*} A_{d*}^{-1} + \alpha_2 \Omega_d \Omega_{d*} \\ G_r &= \alpha_1 A_r^{-1} B_r B_{r*} A_{r*}^{-1} + \alpha_2 \Omega_r \Omega_{r*} \\ G_n &= \alpha_2 \Omega_n \Omega_{n*} \end{aligned} \right\}. \tag{3.280}$$

Then

$$\nabla_c \nabla_{c*} = AG_t A_*. \tag{3.281}$$

Hence, when

$$\det\left[A_{1*}(P_*P + kQ)A_1\right] \neq 0, \; |\lambda| = 1 \tag{3.282}$$

and

$$\det\left[AG_t A\right] \neq 0, \; |\lambda| = 1, \tag{3.283}$$

Theorem 3.2 can be invoked with K in place of K_m and with Wiener–Hopf spectral factors Λ and Ω that satisfy

$$A_{1*}(P_*P + kQ)A_1 = \Lambda_*\Lambda, \quad AG_t A_* = \Omega\Omega_*. \tag{3.284}$$

In particular, the performance functional J given by (3.126) is finite for all stable K and all such K are given by

$$K = \Lambda^{-1}(Z + \{\widetilde{\Gamma}\}_+ + \{\widetilde{\Gamma}\}_\infty)\Omega^{-1} \tag{3.285}$$

where Z is a stable matrix and where

$$\widetilde{\Gamma} = \Lambda_*^{-1}\nabla_{b*}\nabla_a\nabla_{c*}\Omega_*^{-1}. \tag{3.286}$$

Moreover, the minimum J is realized with the choice $Z \equiv 0$.

Substituting (3.274) for ∇_a into (3.286) and introducing

$$\Gamma_i = \nabla_{b*} W G_{11} G_w \nabla_{c*} \tag{3.287}$$

one gets

$$\widetilde{\Gamma} = \Lambda_*^{-1}\Gamma_i\Omega_*^{-1} - \Lambda A_1^{-1} Y \Omega \tag{3.288}$$

where

$$\begin{aligned}\Gamma_i &= -A_{1*}\begin{bmatrix} P_* & W_{q*} \end{bmatrix}\begin{bmatrix} I & 0 \\ 0 & W_q \end{bmatrix}\begin{bmatrix} P_d & -T_d & 0 \\ 0 & 0 & 0 \end{bmatrix}\begin{bmatrix} G_d & 0 & 0 \\ 0 & G_r & 0 \\ 0 & 0 & G_n \end{bmatrix}\begin{bmatrix} -P_{o*} \\ I \\ -I \end{bmatrix} A_* \\ &= A_{1*}P_*(T_d G_r + P_d G_d P_{o*})A_*. \end{aligned} \tag{3.289}$$

Using (3.288) in (3.285) and recognizing that

$$\{\Lambda A_1^{-1} Y\Omega\}_+ + \{\Lambda A_1^{-1} Y\Omega\}_\infty = \Lambda A_1^{-1} Y\Omega - \{\Lambda A_1^{-1} Y\Omega\}_- \tag{3.290}$$

gives finally

$$K = \Lambda^{-1}(Z + \{\Lambda_*^{-1}\Gamma_i\Omega_*^{-1}\}_+ + \{\Lambda_*^{-1}\Gamma_i\Omega_*^{-1}\}_\infty + \{\Lambda A_1^{-1} Y\Omega\}_-)\Omega^{-1} - A_1^{-1} Y. \tag{3.291}$$

One can expect typically that det $(X_1 - KB) \not\equiv 0$ in accordance with Remark 3.3 so that the required controller with transfer matrix $C = (X_1 - KB)^{-1}(Y_1 + KA)$ is realizable. Also, except for the $\{\Lambda_*^{-1}\Gamma_i\Omega_*^{-1}\}_\infty$ term, the expression for K in (3.291) is identical in form with the one given in Youla et al. (1976b) for the analog case. For insights regarding the computation of the controller transfer matrix C see Example 3.18.

When the 1DOF system is analog instead of digital, additional assumptions are needed. In particular, attention is turned to the analog case under the practical assumptions that F, L, W, T_d, and P_d are proper and that $A_r^{-1}B_r$, $A_d^{-1}B_d$, Ω_r, and Ω_d are strictly proper. It then follows from (3.226) and (3.249) that $WG_{11}G_w \leq 0(s^{-1})$ because

$$WG_{11}\widetilde{G}_w = W\begin{bmatrix} P_d A_d^{-1} B_d & -T_d A_r^{-1} B_r & P_d\Omega_d & -T_d\Omega_r & 0 \\ 0 & 0 & 0 & 0 & 0 \end{bmatrix} \leq 0(s^{-1}). \tag{3.292}$$

Making the same choices for W, ∇_b, and ∇_c as in the digital case one gets from (3.271) that

$$\nabla = WG_{11}G_w - \nabla_b(A_1^{-1}Y + K)\nabla_c \tag{3.293}$$

where W, ∇_b, and ∇_c are again stable for the assumptions made here. Hence, when for finite ω

$$\det(\nabla_{b*}\nabla_b) = \det(\Lambda_*\Lambda) = \det(A_{1*}(P_*P + kQ)A_1) \neq 0, \; s = j\omega \tag{3.294}$$

and

$$\det(\nabla_c\nabla_{c*}) = \det(\Omega\Omega_*) = \det(AG_tA_*) \neq 0, \; s = j\omega \tag{3.295}$$

are assumed, the Wiener–Hopf factors Λ and Ω together with their inverses are stable. Moreover, one has that $\nabla_b\Lambda^{-1}$ and $\Omega^{-1}\nabla_c$ are paraunitary matrices. Writing (3.293) as

$$\nabla = WG_{11}G_w - \nabla_b\Lambda^{-1}(\Lambda A_1^{-1}Y\Omega + \Lambda K\Omega)\Omega^{-1}\nabla_c \tag{3.296}$$

and recognizing that any stable matrix K can be written as

$$K = \widetilde{K}_m - \Lambda^{-1}\{\Lambda A_1^{-1}Y\Omega\}_\infty\Omega^{-1} \tag{3.297}$$

where $\widetilde{K}_m$ is a stable matrix then yields

$$\nabla = \nabla_a - \nabla_b\widetilde{K}_m\nabla_c \tag{3.298}$$

where

$$\nabla_a = WG_{11}G_w - \nabla_b\Lambda^{-1}(\{\Lambda A_1^{-1}Y\Omega\}_+ + \{\Lambda A_1^{-1}Y\Omega\}_-)\Omega^{-1}\nabla_c. \tag{3.299}$$

Again ∇ is stable for all stable K and in particular for $\widetilde{K}_m \equiv 0$. Hence, ∇_a is stable. In addition, it is evident from (3.292) and (3.299) that ∇_a is strictly proper. Theorem 3.4 then yields that the performance functional J is finite iff $\widetilde{K}_m$ is given by (3.173) or, equivalently,

$$K = \Lambda^{-1}(Z + \{\widetilde{\Gamma}\}_+ - \{\Lambda A_1^{-1}Y\Omega\}_\infty)\Omega^{-1} \tag{3.300}$$

where Z is stable and strictly proper and now

$$\widetilde{\Gamma} = \Lambda_*^{-1}\nabla_{b*}\nabla_a\nabla_{c*}\Omega_*^{-1} = \Lambda_*^{-1}\Gamma_i\Omega_*^{-1} - \{\Lambda A_1^{-1}Y\Omega\}_+ - \{\Lambda A_1^{-1}Y\Omega\}_- \tag{3.301}$$

with Γ_i again given by the formula (3.289). Substituting (3.301) into (3.300), recognizing that

$$\{\{\Lambda A_1^{-1}Y\Omega\}_+ + \{\Lambda A_1^{-1}Y\Omega\}_-\}_+ = \{\Lambda A_1^{-1}Y\Omega\}_+, \tag{3.302}$$

and recalling (3.290) gives

$$K = \Lambda^{-1}(Z + \{\Lambda_*^{-1}\Gamma_i\Omega_*^{-1}\}_+ + \{\Lambda A_1^{-1}Y\Omega\}_-)\Omega^{-1} - A_1^{-1}Y \tag{3.303}$$

which except for the $\{\Lambda_*^{-1}\Gamma_i\Omega_*^{-1}\}_\infty$ term in (3.291) is identical in form with the result obtained for the digital 1DOF system. The expression for K is also identical in form with the one obtained in Youla et al. (1976b) for a 1DOF analog system. For insights regarding the computation of the controller transfer matrix $C = (X_1 - KB)^{-1}(Y_1 + KA)$ see Example 3.18.

A controller transfer matrix $C = (X_1 - KB)^{-1}(Y_1 + KA)$ associated with the matrix K given by (3.303) exists iff det $(X_1 - KB) \not\equiv 0$. Similar reasoning to that used in connection with (3.181) through (3.197) can be employed here (see

Example 3.13) to confirm that this is the case when the order conditions in (3.203) are met with $(k_1 + k_2 + k_3) \leq 0$. For the 1DOF system,

$$G_{12*}W_*WG_{12} = [P_* \quad W_{q*}]\begin{bmatrix} P \\ W_q \end{bmatrix} = P_*P + kQ. \tag{3.304}$$

Typically, $P \leq 0(s^{-1})$ and Q is biproper. Thus, with $k > 0$

$$(G_{12*}W_*WG_{12})^{-1} = (P_*P + kQ)^{-1} \tag{3.305}$$

is proper and $k_1 = 0$. Also,

$$(G_{21}G_wG_{w*}G_{21*})^{-1} = G_t^{-1} = (P_oG_dP_{o*} + G_r + G_n)^{-1} \tag{3.306}$$

is typically proper because G_n is biproper and with F and L proper the other terms in G_t are strictly proper. Hence, $k_2 = 0$. Finally, $G_{22} = -FP \leq 0(s^{-1})$ is typical and one gets $k_3 = -1$. Thus, $(k_1 + k_2 + k_3) = -1 \leq 0$ and C is realizable for the K given by (3.303). Moreover, C is strictly proper on account of (3.197).

A summary of the key assumptions made and their implications is now presented for easy reference and clarification. The key assumptions are

Assumption 3.1 The 1DOF system is admissible.

Assumption 3.2 The matrices F, T_d, LA_d^{-1}, $A_pP_dA_d^{-1}$, $A_pT_dA_r^{-1}$ and $(I - FT_d)A_r^{-1}$ are stable.

Assumption 3.3 The matrices Ω_d, Ω_r, Ω_n, and $W = \text{diag}\{I, W_q\}$ are stable.

Assumption 3.4 On $|\lambda| = 1$ for the digital case and on the finite $s = j\omega$ axis for the analog case, $\det[A_{1*}(P_*P + kQ)A_1] \neq 0$ and $\det[AG_tA_*] \neq 0$ where $kQ = W_{q*}W_q$.

Assumption 3.5 In the analog case, F, L, W_q, T_d, and P_d are proper and $A_r^{-1}B_r$, $A_d^{-1}B_d$, Ω_r, and Ω_d are strictly proper. In addition, the order conditions in (3.203) or, equivalently, $(P_*P + kQ)^{-1} \leq 0(s^{2k_1})$, $(P_oG_dP_{o*} + G_r + G_n)^{-1} \leq 0(s^{2k_2})$, and $FP \leq 0(s^{k_3})$ are met with $k_1 + k_2 + k_3 \leq 0$.

Assumption 3.1 assures that the 1DOF system can be stabilized and that all stabilizing controller transfer matrices are given by (3.11) with K a stable matrix. It holds iff the polynomial (3.243) is stable; hence, ψ_L^{-1} and $\psi_{FP}(\psi_F\psi_P)^{-1}$ must be stable. It follows that L must be stable and for F stable $A_1^{-1}A_{p1}$ is stable because of Lemma 2.10. It also follows from Lemma 2.3 that the matrices $\widetilde{G}_{12}A_1$, $A\widetilde{G}_{21}$, and $\widetilde{G}_{11} + \widetilde{G}_{12}A_1Y_1\widetilde{G}_{21}$ must be stable; hence, PA_1, AFP_d, and $(I - PA_1Y_1F)P_d$ must be stable by Lemma 3.3.

Assumption 3.2 assures under Lemma 3.4 that the $G = [G_{ij}]$ associated with the 1DOF system is compatible. It also assures that $G_w = \left[\sqrt{\alpha_1}\widetilde{G}_{w1} \quad \sqrt{\alpha_2}\widetilde{G}_{w2} \right]$, where $\widetilde{G}_{w1}$ and $\widetilde{G}_{w2}$ are defined by (3.249) and $\widetilde{G}_{w2}$ is stable because of Assumption 3.3, is acceptable for every stable K.

Collectively, Assumptions 3.1, 3.2, and 3.3 lead to $\nabla = WG_\delta G_w = \nabla_a - \nabla_b K \nabla_c$ stable for all stable K with ∇_a, ∇_b, and ∇_c stable. Then for the digital case, the additional Assumption 3.4 together with the choice (3.276a) and (3.277) for W leads by Theorem 3.2 to the expression (3.291) for K which for $Z \equiv 0$ yields the minimum value of the J given by (3.105). For the analog case, ∇_a is strictly proper with the additional Assumption 3.5 which together with the choice (3.276a) and (3.277) for W leads by Theorem 3.4 to the expression (3.303) for K which for $Z \equiv 0$ yields the minimum value of the J given by (3.144). Finally, the order restrictions on $(P_*P + kQ)^{-1}$, $(P_oG_dP_{o*} + G_r + G_n)^{-1}$, and FP assure that the controller is realizable.

The design of an analog 1DOF optimal $\mathcal{H}_2$ multivariable system for the control of submarine pitch and depth is presented in Example 3.23 to illustrate the application of the theory developed in this section. The model used for the submarine motion dynamics is the same as the one given in Sect. 10.4.4 of Grimble (2006) which provides an $\mathcal{H}_\infty$ solution.

It is important to recognize that significant simplification in the calculation of the controller transfer matrix C for the K given by (3.291) or (3.303) occurs when any unstable poles of FP lie only on $|\lambda| = 1$ in the digital case or on the finite $j\omega$-axis in the analog case. For details in this regard see Example 3.18(b) and (c).

It is also important to note that imposition of the requirement that G_w be acceptable for all 1DOF stabilizing controllers led to $y_{m1} = u_1$ in (3.259) being stable. That is, G_w is actually strictly acceptable for this requirement.

Finally, when the transfer matrix F for the feedback sensors is the identity matrix, a clever modification of the standard spectral separation process in connection with partial fraction expansions and the use of nonstandard matrix spectral factorizations in Stefanovski (2007) yield a significant simplification of the formula for the optimal controller in almost all cases.

3.8 $\mathcal{H}_2$ Design of 3DOF Systems

The 3DOF system shown in Fig. 1.1 is considered and it is assumed that the system is admissible as defined in Sect. 2.5. The exogenous inputs are modeled as

$$e = \begin{bmatrix} d \\ r \\ n_r \\ n_m \\ n_l \end{bmatrix} = \begin{bmatrix} d_1 \\ r_1 \\ 0 \\ 0 \\ 0 \end{bmatrix} + \begin{bmatrix} d_2 \\ r_2 \\ n_r \\ n_m \\ n_l \end{bmatrix} = e_1 + e_2 \tag{3.307}$$

where d_1 and r_1 are the shape-deterministic components of d and r and where d_2, r_2, n_r, n_m, and n_l are purely stochastic. In order to be more descriptive and to be consistent with the notation in Sect. 3.7, n_r is used here for the noise n in Fig. 1.1. Specifically,

$$d = A_d^{-1} B_d \sigma_d + \Omega_d \mu_d, \; r = A_r^{-1} B_r \sigma_r + \Omega_r \mu_r \tag{3.308}$$

where Ω_d and Ω_r are stable left Wiener–Hopf factors associated with Φ_{dd} and Φ_{rr}, the power spectral density matrices for the stochastic components of the signals d and r, respectively; where the elements of the vectors μ_d and μ_r are zero-mean independent white-noise processes with unit variance; where A_d, B_d and A_r, B_r are left-coprime polynomial matrix pairs; and where $\sigma = \begin{bmatrix} \sigma_d' & \sigma_r' \end{bmatrix}'$ is a vector whose elements are random variables satisfying

$$\langle \sigma\sigma' \rangle = \left\langle \begin{bmatrix} \sigma_d \\ \sigma_r \end{bmatrix} \begin{bmatrix} \sigma_d' & \sigma_r' \end{bmatrix} \right\rangle = I. \tag{3.309}$$

It is also assumed that the power spectral density matrices $\Phi_{n_r n_r}$, $\Phi_{n_m n_m}$, and $\Phi_{n_l n_l}$ for n_r, n_m, and n_l, respectively, are good. Then in accordance with Theorem C.1 and Theorem C.2 one can write

$$\Phi_{n_r n_r} = \Omega_{n_r} \Omega_{n_r *}, \;\; \Phi_{n_m n_m} = \Omega_{n_m} \Omega_{n_m *}, \;\; \Phi_{n_l n_l} = \Omega_{n_l} \Omega_{n_l *} \tag{3.310}$$

where Ω_{n_r}, Ω_{n_m}, and Ω_{n_l} are stable left Wiener–Hopf factors. Equivalently,

$$n_r = \Omega_{n_r} \mu_{n_r}, \;\; n_m = \Omega_{n_m} \mu_{n_m}, \;\; n_l = \Omega_{n_l} \mu_{n_l} \tag{3.311}$$

where the elements of $\widehat{\mu}_{n_r}$, $\widehat{\mu}_{n_m}$, and $\widehat{\mu}_{n_l}$ are zero-mean independent white-noise processes with unit variance. Setting

$$\mu = \begin{bmatrix} \mu_d' & \mu_r' & \mu_{n_r}' & \mu_{n_m}' & \mu_{n_l}' \end{bmatrix}', \tag{3.312}$$

$$q = \begin{bmatrix} \sigma' & \mu' \end{bmatrix}', \tag{3.313}$$

$$\Omega_s = \text{diag}\{\Omega_d, \Omega_r, \Omega_{n_r}, \Omega_{n_m}, \Omega_{n_l}\}, \tag{3.314}$$

and

$$A_e = \begin{bmatrix} A_d & 0 \\ 0 & A_r \end{bmatrix}, \;\; B_e = \begin{bmatrix} B_d & 0 \\ 0 & B_r \end{bmatrix} \tag{3.315}$$

one can write

$$e = \begin{bmatrix} \widetilde{G}_{e1} & \widetilde{G}_{e2} \end{bmatrix} \begin{bmatrix} \sigma \\ \mu \end{bmatrix} = \widetilde{G}_e q \tag{3.316}$$

where

$$\widetilde{G}_{e1} = \begin{bmatrix} I & 0 \\ 0 & I \\ 0 & 0 \\ 0 & 0 \\ 0 & 0 \end{bmatrix} A_e^{-1} B_e, \quad \widetilde{G}_{e2} = \Omega_s. \tag{3.317}$$

Clearly, A_e, B_e is a left-coprime polynomial matrix pair.

The design objective is to keep $\hat{v}(k)$ and $\widehat{\varepsilon}(k)$ small for the digital case and $\hat{v}(t)$ and $\widehat{\varepsilon}(t)$ small for the analog case. Thus,

$$\begin{bmatrix} y_\delta \\ \hline y_m \end{bmatrix} = \begin{bmatrix} y - T_d r \\ v \\ \hline w \\ z \\ u \end{bmatrix} = \left[\begin{array}{c|c} G_{11} & G_{12} \\ \hline G_{21} & G_{22} \end{array}\right] \begin{bmatrix} e \\ \hline v \end{bmatrix} \tag{3.318}$$

is of interest and the 3DOF system corresponds to the generalized persistent input standard configuration of Fig. 3.2 with $w = e$,

$$\widetilde{G}_w = \begin{bmatrix} \widetilde{G}_{w1} & \widetilde{G}_{w2} \end{bmatrix} = \begin{bmatrix} \widetilde{G}_{e1} & \widetilde{G}_{e2} \end{bmatrix} = \widetilde{G}_e, \tag{3.319}$$

$$G_{11} = \begin{bmatrix} P_d & -T_d & 0 & 0 & 0 \\ 0 & 0 & 0 & 0 & 0 \end{bmatrix}, \tag{3.320}$$

$$G_{12} = \begin{bmatrix} P \\ I \end{bmatrix}, \tag{3.321}$$

$$G_{21} = \begin{bmatrix} FP_d & 0 & 0 & I & 0 \\ L & 0 & 0 & 0 & I \\ 0 & I & I & 0 & 0 \end{bmatrix}, \tag{3.322}$$

and

$$G_{22} = \begin{bmatrix} FP \\ 0 \\ 0 \end{bmatrix}. \tag{3.323}$$

The G_{ij} are easily confirmed by (also see Example 1.1)

$$\left.\begin{aligned} y - T_d r &= [I \quad 0](G_{11}e + G_{12}v) = P_d d - T_d r + Pv \\ v &= [0 \quad I](G_{11}e + G_{12}v) = 0 + v \\ w &= [I \quad 0 \quad 0](G_{21}e + G_{22}v) = FP_d d + n_m + FPv \\ z &= [0 \quad I \quad 0](G_{21}e + G_{22}v) = Ld + n_l \\ u &= [0 \quad 0 \quad I](G_{21}e + G_{22}v) = r + n_r \end{aligned}\right\}. \tag{3.324}$$

The matrix G_{22} is the same as the one in (3.54) of Sect. 3.4 and so (3.58) through (3.66) are again applicable. It is then easy to verify that $G = [G_{ij}]$ is compatible when the 3DOF system is admissible and T_d is stable. In particular,

$$G_{12}\widetilde{A}_1 = \begin{bmatrix} P \\ I \end{bmatrix} A_1 = \begin{bmatrix} PA_1 \\ A_1 \end{bmatrix}, \tag{3.325}$$

$$\widetilde{A}G_{21} = \begin{bmatrix} A & 0 & 0 \\ 0 & I & 0 \\ 0 & 0 & I \end{bmatrix} \begin{bmatrix} FP_d & 0 & 0 & I & 0 \\ L & 0 & 0 & 0 & I \\ 0 & I & I & 0 & 0 \end{bmatrix} = \begin{bmatrix} AFP_d & 0 & 0 & A & 0 \\ L & 0 & 0 & 0 & I \\ 0 & I & I & 0 & 0 \end{bmatrix}, \tag{3.326}$$

and

$$\begin{aligned} G_{11} + G_{12}\widetilde{A}_1\widetilde{Y}_1 G_{21} &= \begin{bmatrix} P_d & -T_d & 0 & 0 & 0 \\ 0 & 0 & 0 & 0 & 0 \end{bmatrix} + \begin{bmatrix} PA_1 \\ A_1 \end{bmatrix} [-Y_1 \quad 0 \quad 0] \begin{bmatrix} FP_d & 0 & 0 & I & 0 \\ L & 0 & 0 & 0 & I \\ 0 & I & I & 0 & 0 \end{bmatrix} \\ &= \begin{bmatrix} (I - PA_1Y_1F)P_d & -T_d & 0 & -PA_1Y_1 & 0 \\ -YAFP_d & 0 & 0 & -A_1Y_1 & 0 \end{bmatrix} \end{aligned} \tag{3.327}$$

are all stable on account of Lemma 2.7.

Since $\widetilde{G}_{w2} = \widetilde{G}_{e2} = \Omega_s$ is a stable matrix, it is clear that attention must be restricted to the $\widetilde{G}_{w1} = \widetilde{G}_{e1}$ defined in (3.317) that are acceptable or, equivalently, to those for which

$$G_\delta \widetilde{G}_{w1} = G_{11}\widetilde{G}_{w1} + G_{12}\widetilde{A}_1(\widetilde{Y}_1 + K\widetilde{A})G_{21}\widetilde{G}_{w1} \tag{3.328}$$

and

$$G_v \widetilde{G}_{w1} = \widetilde{A}_1(\widetilde{Y}_1 + K\widetilde{A})G_{21}\widetilde{G}_{w1} \tag{3.329}$$

are stable. Now in light of (3.320) and (3.321),

$$[0 \quad I]G_\delta \widetilde{G}_{w1} = \widetilde{A}_1(\widetilde{Y}_1 + K\widetilde{A})G_{21}\widetilde{G}_{w1} = G_v\widetilde{G}_{w1}. \tag{3.330}$$

Hence, $G_\delta \widetilde{G}_{w1}$ and $G_v \widetilde{G}_{w1}$ are stable iff $G_\delta \widetilde{G}_{w1}$ is stable. That is, iff

$$G_\delta \widetilde{G}_{w1} = G_\delta \widetilde{G}_{e1} = \left\{ \begin{bmatrix} P_d & -T_d \\ 0 & 0 \end{bmatrix} + \begin{bmatrix} PA_1 \\ A_1 \end{bmatrix} (\widetilde{Y}_1 + K\widetilde{A}) \begin{bmatrix} FP_d & 0 \\ L & 0 \\ 0 & I \end{bmatrix} \right\} A_e^{-1} B_e \tag{3.331}$$

or, equivalently,

$$G_\delta \widetilde{G}_{w1} = \begin{bmatrix} \left\{ [P_d \quad -T_d] + PA_1(\widetilde{Y}_1 + K\widetilde{A}) \begin{bmatrix} FP_d & 0 \\ L & 0 \\ 0 & I \end{bmatrix} \right\} A_e^{-1} B_e \\ A_1(\widetilde{Y}_1 + K\widetilde{A}) \begin{bmatrix} FP_d & 0 \\ L & 0 \\ 0 & I \end{bmatrix} A_e^{-1} B_e \end{bmatrix} \tag{3.332}$$

is stable.

The expressions in the first and second block rows of (3.332) are the same as the ones that are obtained by replacing $\widetilde{G}_{w1}$ with $A_e^{-1}B_e$ and using the G_{ij} in (3.53), (3.54), and (3.57) of Sect. 3.4 in (3.328) and (3.329). Hence, the parameterization of all stable K for which (3.332) is stable is the same as the one derived in Sect. 3.4. Specifically, for an admissible 3DOF system with F, T_d, $A_pP_dA_d^{-1}$, and $A_pT_dA_r^{-1}$ stable, the set of all stable K for which (3.332) is stable is given by (see (3.84), (3.86), and (3.93))

$$\begin{aligned} K &= [-K_1 \quad -H_2 \quad H_1] = \left[-K_m \quad -K_nA_n \quad (A_1^{-1}A_{p1}Y_{p1}T_d + K_{g2}A_r)\right] \\ &= [-K_m \quad -K_n \quad K_{g2}]\widetilde{A}_t + \left[0 \quad 0 \quad A_1^{-1}A_{p1}Y_{p1}T_d\right] \end{aligned} \tag{3.333}$$

where K_m, K_n, and K_{g2} are arbitrary stable matrices; where A_n is defined by (3.89); and where

$$\widetilde{A}_t = \begin{bmatrix} I & 0 & 0 \\ 0 & A_n & 0 \\ 0 & 0 & A_r \end{bmatrix}. \tag{3.334}$$

Moreover, for every K in this set one has with W stable and

$$G_w = \left[\sqrt{\alpha_1}\widetilde{G}_{w1} \quad \sqrt{\alpha_2}\widetilde{G}_{w2}\right] = \left[\sqrt{\alpha_1}\widetilde{G}_{e1} \quad \sqrt{\alpha_2}\widetilde{G}_{e2}\right] \tag{3.335}$$

that

$$\nabla = WG_\delta G_w = W[G_{11} + G_{12}\widetilde{A}_1(\widetilde{Y}_1 + K\widetilde{A})G_{21}]G_w \tag{3.336}$$

is stable.

It is important to recognize when persistent inputs are present that $\widetilde{A}_t^{-1}$ is not stable because A_n^{-1} and/or A_r^{-1} are not stable. Hence, the subset of all stable matrices K defined by (3.333) is not the entire set of all stable matrices. In particular, the solution of (3.333) for

$$K_t = [\,K_m \quad K_n \quad -K_{g2}\,] \tag{3.337}$$

is not a stable matrix for every stable K. Thus, in contrast with the 1DOF system of Sect. 3.7, ∇ is not stable for all stable K; it is only stable for those K given by (3.333) with K_t stable.

Using (3.333) in (3.336), recalling that $\widetilde{A}_1 = A_1$, and letting

$$A_t = \widetilde{A}_t\widetilde{A} = \begin{bmatrix} I & 0 & 0 \\ 0 & A_n & 0 \\ 0 & 0 & A_r \end{bmatrix} \begin{bmatrix} A & 0 & 0 \\ 0 & I & 0 \\ 0 & 0 & I \end{bmatrix} = \begin{bmatrix} A & 0 & 0 \\ 0 & A_n & 0 \\ 0 & 0 & A_r \end{bmatrix} \tag{3.338}$$

yields

$$\nabla = \nabla_a - \nabla_b K_t \nabla_c \tag{3.339}$$

where

$$\nabla_b = WG_{12}A_1, \quad \nabla_c = A_t G_{21} G_w, \tag{3.340}$$

and

$$\begin{aligned} \nabla_a &= WG_{11}G_w + WG_{12}A_1(\widetilde{Y}_1 + [0 \quad 0 \quad A_1^{-1}A_{p1}Y_{p1}T_d]\widetilde{A})G_{21}G_w \\ &= WG_{11}G_w + \nabla_b[-Y_1 \quad 0 \quad A_1^{-1}A_{p1}Y_{p1}T_d]A_t^{-1}\nabla_c. \end{aligned} \tag{3.341}$$

Since ∇ is stable for all K_t, it is stable for $K_t = 0$. Hence, ∇_a is stable. Since the admissible 3DOF system is compatible, $G_{12}A_1$ is stable. Hence, ∇_b is stable because W is stable. Finally,

$$\nabla_c = \widetilde{A}_t \left[\sqrt{\alpha_1} \widetilde{A} G_{21} \widetilde{G}_{w1} \quad \sqrt{\alpha_2} \widetilde{A} G_{21} \widetilde{G}_{w2} \right] \tag{3.342}$$

is considered. Clearly, $\widetilde{A} G_{21} \widetilde{G}_{w2}$ is stable because $\widetilde{G}_{w2} = \widetilde{G}_{e2} = \Omega_s$ is stable and because $\widetilde{A} G_{21}$ is stable when the 3DOF system is compatible. Hence, $\widetilde{A}_t \widetilde{A} G_{21} \widetilde{G}_{w2}$ is stable. It also follows from

$$A_t G_{21} \widetilde{G}_{w1} = A_t G_{21} \widetilde{G}_{e1} = \begin{bmatrix} AFP_d & 0 \\ A_n L & 0 \\ 0 & A_r \end{bmatrix} A_e^{-1} B_e = \begin{bmatrix} AFP_d A_d^{-1} & 0 \\ B_n & 0 \\ 0 & I \end{bmatrix} \begin{bmatrix} B_d & 0 \\ 0 & B_r \end{bmatrix} \tag{3.343}$$

that $\widetilde{A}_t \widetilde{A} G_{21} \widetilde{G}_{w1} = A_t G_{21} \widetilde{G}_{w1}$ is stable because $B_n = A_n L A_d^{-1}$ is a polynomial matrix (see (3.89)) and because $AFP_d A_d^{-1}$ is stable on account of (3.257). Thus, ∇_c is stable and Theorem 3.2 for the digital case can now be applied assuming that $\det(\nabla_{b*} \nabla_b) \neq 0$ and $\det(\nabla_c \nabla_{c*}) \neq 0$ on $|\lambda| = 1$.

With W again chosen as in (3.276a) and (3.277) for the 1DOF system one gets here also that

$$\nabla_{b*} \nabla_b = A_{1*} (P_* P + kQ) A_1 = \Lambda_* \Lambda \tag{3.344}$$

where Q is good and parahermitian-positive and where Λ is a square matrix which together with its inverse is stable. Using (3.342) one also has

$$\nabla_c \nabla_{c*} = A_t G_{21} G_s G_{21*} A_{t*} \tag{3.345}$$

where

$$G_s = G_w G_{w*} = (\alpha_1 \widetilde{G}_{w1} \widetilde{G}_{w1*} + \alpha_2 \widetilde{G}_{w2} \widetilde{G}_{w2*}) = \alpha_1 \widetilde{G}_{e1} \widetilde{G}_{e1*} + \alpha_2 \Omega_s \Omega_{s*}. \tag{3.346}$$

Setting

$$\left. \begin{aligned} G_d &= \alpha_1 A_d^{-1} B_d B_{d*} A_{d*}^{-1} + \alpha_2 \Omega_d \Omega_{d*} \\ G_r &= \alpha_1 A_r^{-1} B_r B_{r*} A_{r*}^{-1} + \alpha_2 \Omega_r \Omega_{r*} \\ G_{n_r} &= \alpha_2 \Omega_{n_r} \Omega_{n_r*}, \; G_{n_m} = \alpha_2 \Omega_{n_m} \Omega_{n_m*}, \; G_{n_l} = \alpha_2 \Omega_{n_l} \Omega_{n_l*} \end{aligned} \right\} \tag{3.347}$$

it follows from (3.314), (3.315), and (3.317) that

$$G_s = \mathrm{diag}\{G_d, G_r, G_{n_r}, G_{n_m}, G_{n_l}\}. \tag{3.348}$$

Using (3.322), (3.338), and (3.348) in (3.345) leads to

$$\nabla_c \nabla_{c*} = \begin{bmatrix} G_t & 0 \\ 0 & G_o \end{bmatrix} = \begin{bmatrix} \Omega_t \Omega_{t*} & 0 \\ 0 & \Omega_o \Omega_{o*} \end{bmatrix} \tag{3.349}$$

where

$$G_t = \Omega_t \Omega_{t*} = \begin{bmatrix} A & 0 \\ 0 & A_n \end{bmatrix} \begin{bmatrix} (FP_d G_d P_{d*} F_* + G_{n_m}) & FP_d G_d L_* \\ LG_d P_{d*} F_* & LG_d L_* + G_{n_l} \end{bmatrix} \begin{bmatrix} A_* & 0 \\ 0 & A_{n*} \end{bmatrix} \tag{3.350}$$

and

$$G_o = \Omega_o \Omega_{o*} = A_r (G_r + G_{n_r}) A_{r*} \tag{3.351}$$

with Ω_t and Ω_o square stable Wiener–Hopf factors with stable inverses whose existence is assured by Theorem C.2 and Remark C.3. Clearly, $\nabla_c \nabla_{c*} = \Omega\Omega_*$ where

$$\Omega = \text{diag}\{\Omega_t, \Omega_o\} \tag{3.352}$$

is a square stable Wiener–Hopf factor with stable inverse. It immediately follows from Theorem 3.2 with K_t in place of K_m that the performance functional J is finite iff K_t is given by

$$K_t = \Lambda^{-1}(Z + \{\widetilde{\Gamma}\}_+ + \{\widetilde{\Gamma}\}_\infty)\Omega^{-1} \tag{3.353}$$

where Z is a stable matrix and where

$$\widetilde{\Gamma} = \Lambda_*^{-1} \nabla_{b*} \nabla_a \nabla_{c*} \Omega_*^{-1}. \tag{3.354}$$

Moreover, the minimum J is realized with the choice $Z \equiv 0$. Clearly, $\widetilde{\Gamma}$ is good because each of the factors in (3.286) are good; hence, K_t is stable as it should be.

It is clear from (3.341) and (3.354) that

$$\widetilde{\Gamma} = \widetilde{\Gamma}_1 + \widetilde{\Gamma}_2 \tag{3.355}$$

where

$$\begin{aligned} \widetilde{\Gamma}_1 &= \Lambda_*^{-1} \nabla_{b*} W G_{11} G_w \nabla_{c*} \Omega_*^{-1} = \Lambda_*^{-1} A_{1*} G_{12*} W_* W G_{11} G_w G_{w*} G_{21*} A_{t*} \Omega_*^{-1} \\ &= \Lambda_*^{-1} A_{1*} [P_* \quad I] \begin{bmatrix} I & 0 \\ 0 & kQ \end{bmatrix} G_{11} G_s G_{21*} A_{t*} \Omega_*^{-1} \\ &= \Lambda_*^{-1} A_{1*} P_* [P_d G_d P_{d*} F_* A_* \quad P_d G_d L_* A_{n*} \mid -T_d G_r A_{r*}] \left[\begin{array}{cc|c} \multicolumn{2}{c|}{\Omega_{t*}^{-1}} & \begin{matrix} 0 \\ 0 \end{matrix} \\ \hline 0 & 0 & \Omega_{o*}^{-1} \end{array}\right] \end{aligned} \tag{3.356}$$

and

$$\begin{aligned}\widetilde{\Gamma}_2 &= \Lambda_*^{-1}\nabla_{b*}\nabla_b[-Y_1A^{-1} \quad 0 \quad A_1^{-1}A_{p1}Y_{p1}T_dA_r^{-1}]\nabla_c\nabla_{c*}\Omega_*^{-1} \\ &= \Lambda A_1^{-1}[-Y \quad 0 \mid A_{p1}Y_{p1}T_dA_r^{-1}]\left[\begin{array}{cc|c} \Omega_t & & \begin{matrix}0\\0\end{matrix} \\ \hline 0 & 0 & \Omega_o\end{array}\right].\end{aligned} \tag{3.357}$$

It also readily follows from (3.333) and (3.337) that

$$H_1 = K_{g2}A_r + A_1^{-1}A_{p1}Y_{p1}T_d = -K_t\begin{bmatrix}0\\0\\I\end{bmatrix}A_r + A_1^{-1}A_{p1}Y_{p1}T_d \tag{3.358}$$

and

$$[K_1 \quad H_2] = [K_m \quad K_nA_n] = K_t\begin{bmatrix}I & 0\\0 & A_n\\0 & 0\end{bmatrix}. \tag{3.359}$$

Using (3.353) in the expression for H_1, writing

$$Z = [Z_m \quad Z_n \quad -Z_{g2}], \tag{3.360}$$

and recognizing that

$$\Omega^{-1}\begin{bmatrix}0\\0\\I\end{bmatrix} = \left[\begin{array}{cc|c} \Omega_t^{-1} & & \begin{matrix}0\\0\end{matrix} \\ \hline 0 & 0 & \Omega_o^{-1}\end{array}\right]\begin{bmatrix}0\\0\\I\end{bmatrix} = \begin{bmatrix}0\\0\\\Omega_o^{-1}\end{bmatrix} = \begin{bmatrix}0\\0\\I\end{bmatrix}\Omega_o^{-1} \tag{3.361}$$

yields

$$H_1 = \Lambda^{-1}(Z_{g2} - \{\widetilde{\Gamma}_o\}_+ - \{\widetilde{\Gamma}_o\}_\infty)\Omega_o^{-1}A_r + A_1^{-1}A_{p1}Y_{p1}T_d \tag{3.362}$$

where

$$\begin{aligned}\widetilde{\Gamma}_o = \widetilde{\Gamma}\begin{bmatrix}0\\0\\I\end{bmatrix} &= (\widetilde{\Gamma}_1 + \widetilde{\Gamma}_2)\begin{bmatrix}0\\0\\I\end{bmatrix} \\ &= -\Lambda_*^{-1}A_{1*}P_*T_dG_rA_{r*}\Omega_{o*}^{-1} + \Lambda A_1^{-1}A_{p1}Y_{p1}T_dA_r^{-1}\Omega_o.\end{aligned} \tag{3.363}$$

Clearly,

$$H_1 = \Lambda^{-1}(Z_{g2} - \{\widetilde{\Gamma}_o\}_+ - \{\widetilde{\Gamma}_o\}_\infty + \Lambda A_1^{-1}A_{p1}Y_{p1}T_dA_r^{-1}\Omega_o)\Omega_o^{-1}A_r \tag{3.364}$$

and one easily gets

$$H_1 = \Lambda^{-1}(Z_{g2} + \{\widetilde{\Gamma}_r\}_+ + \{\widetilde{\Gamma}_r\}_\infty + \{\Lambda A_1^{-1} A_{p1} Y_{p1} T_d A_r^{-1} \Omega_o\}_-)\Omega_o^{-1} A_r \quad (3.365)$$

where

$$\widetilde{\Gamma}_r = \Lambda_*^{-1} A_{1*} P_* T_d G_r A_{r*} \Omega_{o*}^{-1}. \quad (3.366)$$

Using (3.353) in (3.359) leads to

$$[K_1 \quad H_2] = \Lambda^{-1}(Z + \{\widetilde{\Gamma}\}_+ + \{\widetilde{\Gamma}\}_\infty) \begin{bmatrix} I & 0 \\ 0 & I \\ 0 & 0 \end{bmatrix} \Omega_t^{-1} \begin{bmatrix} I & 0 \\ 0 & A_n \end{bmatrix}. \quad (3.367)$$

It is now useful to define

$$Z_{mn} = [Z_m \quad Z_n] = Z \begin{bmatrix} I & 0 \\ 0 & I \\ 0 & 0 \end{bmatrix}, \quad (3.368)$$

$$[\widetilde{\Gamma}_w \quad \widetilde{\Gamma}_z] = \widetilde{\Gamma}_1 \begin{bmatrix} I & 0 \\ 0 & I \\ 0 & 0 \end{bmatrix} = \Lambda_*^{-1} A_{1*} P_* [P_d G_d P_{d*} F_* A_* \quad P_d G_d L_* A_{n*}] \Omega_{t*}^{-1}, \quad (3.369)$$

and

$$[\widetilde{\Gamma}_a \quad \widetilde{\Gamma}_b] = -\widetilde{\Gamma}_2 \begin{bmatrix} I & 0 \\ 0 & I \\ 0 & 0 \end{bmatrix} = \Lambda A_1^{-1} [Y \quad 0] \Omega_t. \quad (3.370)$$

Then

$$[K_1 \quad H_2] = \Lambda^{-1}(Z_{mn} + \{\widetilde{\Gamma}_{mn}\}_+ + \{\widetilde{\Gamma}_{mn}\}_\infty)\Omega_t^{-1} \begin{bmatrix} I & 0 \\ 0 & A_n \end{bmatrix} \quad (3.371)$$

where

$$\widetilde{\Gamma}_{mn} = [\widetilde{\Gamma}_w \quad \widetilde{\Gamma}_z] - [\widetilde{\Gamma}_a \quad \widetilde{\Gamma}_b]. \quad (3.372)$$

The final objective is the determination of the controller transfer matrix C which follows from (2.208) using the formulas (3.365) and (3.371). Typically, the considerations in Remark 3.3 also apply here and det $(X_1 - K_1 B) \not\equiv 0$ so that the controller is realizable.

Also of interest are the closed-loop transfer matrices defined in (2.228) and given by (2.233) when $K = [-K_1 \quad -H_2 \quad H_1]$ is chosen in accordance with the formulas (3.365) and (3.371). Obviously,

$$R_u = A_1 H_1 = A_1 \Lambda^{-1}(Z_{g2} + \{\widetilde{\Gamma}_r\}_+ + \{\widetilde{\Gamma}_r\}_\infty + \{\Lambda A_1^{-1} A_{p1} Y_{p1} T_d A_r^{-1} \Omega_o\}_-)\Omega_o^{-1} A_r. \tag{3.373}$$

It also easily follows from (3.371) and

$$[R_w \quad R_z] = [A_1(Y_1 + K_1 A) \quad A_1 H_2] = [A_1 Y_1 \quad 0] + A_1[K_1 \quad H_2]\begin{bmatrix} A & 0 \\ 0 & I \end{bmatrix} \tag{3.374}$$

that

$$\begin{aligned}[R_w \quad R_z] = [A_1 Y_1 \quad 0] + A_1 \Lambda^{-1}(Z_{mn} + \{[\widetilde{\Gamma}_w \quad \widetilde{\Gamma}_z]\}_+ + \{[\widetilde{\Gamma}_w \quad \widetilde{\Gamma}_z]\}_\infty \\ - \Lambda A_1^{-1}[Y \quad 0]\Omega_t + \{\Lambda A_1^{-1}[Y \quad 0]\Omega_t\}_-)\Omega_t^{-1}\begin{bmatrix} A & 0 \\ 0 & A_n \end{bmatrix}\end{aligned} \tag{3.375}$$

which reduces to

$$[R_w \quad R_z] = A_1 \Lambda^{-1}(Z_{mn} + \{[\widetilde{\Gamma}_w \quad \widetilde{\Gamma}_z]\}_+ + \{[\widetilde{\Gamma}_w \quad \widetilde{\Gamma}_z]\}_\infty + \{[\widetilde{\Gamma}_a \quad \widetilde{\Gamma}_b]\}_-)\Omega_t^{-1}\begin{bmatrix} A & 0 \\ 0 & A_n \end{bmatrix} \tag{3.376}$$

when $A_1 Y_1 = YA$ is recalled. Clearly,

$$\begin{aligned}&[R_w \mid R_z] = \\ &A_1 \Lambda^{-1}[(Z_m + \{\widetilde{\Gamma}_w\}_+ + \{\widetilde{\Gamma}_w\}_\infty + \{\widetilde{\Gamma}_a\}_-) \mid (Z_n + \{\widetilde{\Gamma}_z\}_+ + \{\widetilde{\Gamma}_z\}_\infty + \{\widetilde{\Gamma}_b\}_-)]\Omega_t^{-1}\left[\begin{array}{c|c} A & 0 \\ \hline 0 & A_n \end{array}\right]\end{aligned} \tag{3.377}$$

When all of the unstable poles of the shape-deterministic inputs r_1 and d_1 are confined to the unit circle $|\lambda| = 1$, some simplification in the expressions for the closed-loop transfer matrices can be achieved. In this case, A_r^{-1} and A_d^{-1} are analytic in $|\lambda| < 1$. It also follows from L stable and $LA_d^{-1} = A_n^{-1} B_n$ that A_n^{-1} is analytic in $|\lambda| < 1$. In addition, one has from (3.349) and the assumption det $(\nabla_c \nabla_{c*}) \neq 0$ on $|\lambda| = 1$ that det $(G_t) \neq 0$ and det $(G_o) \neq 0$ on $|\lambda| = 1$. Hence, one gets from (3.350) and (3.351) that det $(\widetilde{G}_t) \neq 0$ and det $(\widetilde{G}_o) \neq 0$ on $|\lambda| = 1$ where

$$\widetilde{G}_o = G_r + G_{n_r} = \widetilde{\Omega}_o \widetilde{\Omega}_{o*}; \tag{3.378}$$

$$\widetilde{G}_t = \begin{bmatrix} A & 0 \\ 0 & I \end{bmatrix} \begin{bmatrix} (FP_dG_dP_{d*}F_* + G_{n_m}) & FP_dG_dL_* \\ LG_dP_{d*}F_* & LG_dL_* + G_{n_l} \end{bmatrix} \begin{bmatrix} A_* & 0 \\ 0 & I \end{bmatrix} = \widetilde{\Omega}_t\widetilde{\Omega}_{t*}; \tag{3.379}$$

and $\widetilde{\Omega}_o$ and $\widetilde{\Omega}_t$ are square stable Wiener–Hopf factors which together with their inverses are analytic in $|\lambda| < 1$. Clearly,

$$\Omega_o = A_r\widetilde{\Omega}_o \tag{3.380}$$

and

$$\Omega_t = \begin{bmatrix} I & 0 \\ 0 & A_n \end{bmatrix}\widetilde{\Omega}_t = \widetilde{A}_n\widetilde{\Omega}_t. \tag{3.381}$$

are analytic in $|\lambda| < 1$. Both are also analytic on $|\lambda| = 1$. Otherwise (see Example 3.15), $G_o = (A_r\widetilde{\Omega}_o)(A_r\widetilde{\Omega}_o)_*$ and/or $G_t = (\widetilde{A}_n\widetilde{\Omega}_t)(\widetilde{A}_n\widetilde{\Omega}_t)_*$ have poles on $|\lambda| = 1$ in contradiction of (3.349) since ∇_c is stable. Moreover, when det $(G_o) \neq 0$ and det $(G_t) \neq 0$ on $|\lambda| = 1$, det $(A_r\widetilde{\Omega}_o) \neq 0$ and det $(\widetilde{A}_n\widetilde{\Omega}_t) \neq 0$ on $|\lambda| = 1$. Since $A_r\widetilde{\Omega}_o$ and $\widetilde{A}_n\widetilde{\Omega}_t$ are analytic on $|\lambda| = 1$, it immediately follows that $(A_r\widetilde{\Omega}_o)^{-1}$ and $(\widetilde{A}_n\widetilde{\Omega}_t)^{-1}$ which are analytic in $|\lambda| < 1$ are also analytic on $|\lambda| = 1$. That is, (3.380) and (3.381) are acceptable choices for the Wiener–Hopf factors Ω_o and Ω_t which together with their inverses are stable.

It is also true when $A_1^{-1}A_{p1}$ and T_d are stable and A_r^{-1} has no poles in $|\lambda| < 1$ that

$$\{\Lambda A_1^{-1}A_{p1}Y_{p1}T_dA_r^{-1}\Omega_o\}_- = 0. \tag{3.382}$$

It immediately follows from (3.366) and (3.373) that

$$R_u = A_1\Lambda^{-1}(Z_{g2} + \{\widetilde{\Gamma}_r\}_+ + \{\widetilde{\Gamma}_r\}_\infty)\widetilde{\Omega}_o^{-1} \tag{3.383}$$

where

$$\widetilde{\Gamma}_r = \Lambda_*^{-1}A_{1*}P_*T_dG_r\widetilde{\Omega}_{o*}^{-1}. \tag{3.384}$$

Equations (3.369), (3.370), and (3.377) also reduce to

$$[R_w \mid R_z] = A_1\Lambda^{-1}[(Z_m + \{\widetilde{\Gamma}_w\}_+ + \{\widetilde{\Gamma}_w\}_\infty + \{\widetilde{\Gamma}_a\}_-) \mid (Z_n + \{\widetilde{\Gamma}_z\}_+ + \{\widetilde{\Gamma}_z\}_\infty + \{\widetilde{\Gamma}_b\}_-)]\widetilde{\Omega}_t^{-1}\left[\begin{array}{c|c} A & 0 \\ \hline 0 & I \end{array}\right] \tag{3.385}$$

where

$$[\,\widetilde{\Gamma}_w \quad \widetilde{\Gamma}_z\,] = \Lambda_*^{-1} A_{1*} P_* [\,P_d G_d P_{d*} F_* A_* \quad P_d G_d L_*\,] \widetilde{\Omega}_{t*}^{-1} \tag{3.386}$$

and

$$[\,\widetilde{\Gamma}_a \quad \widetilde{\Gamma}_b\,] = \Lambda A_1^{-1} [\,Y \quad 0\,] \begin{bmatrix} I & 0 \\ 0 & A_n \end{bmatrix} \widetilde{\Omega}_t = \Lambda A_1^{-1} [\,Y \quad 0\,] \widetilde{\Omega}_t. \tag{3.387}$$

Attention is next turned to consideration of an analog 3DOF system with the same assumptions as in the digital case except that now $\det(\nabla_{b*}\nabla_b) \neq 0$ and $\det(\nabla_c \nabla_{c*}) \neq 0$ on $s = j\omega$, ω finite and in addition

$$WG_{11}G_w = [\,\sqrt{\alpha_1} WG_{11} \widetilde{G}_{w1} \quad \sqrt{\alpha_2} WG_{11} \widetilde{G}_{w2}\,] \leq 0(s^{-1}). \tag{3.388}$$

The condition (3.388) is usually satisfied because WG_{11} is proper when W, P_d, and T_d are proper and because $\widetilde{G}_{w1} = \widetilde{G}_{e1}$ and $\widetilde{G}_{w2} = \widetilde{G}_{e2}$ are strictly proper when $A_d^{-1} B_d$, $A_r^{-1} B_r$, and Ω_s are strictly proper. The expression for $\nabla = WG_\delta G_w$ in the analog case is the same as the one obtained from (3.339) through (3.341) for the digital case with s in place of λ. It is more convenient in the analog case, however, to write

$$\nabla = WG_{11}G_w - \nabla_b (\Gamma_y + K_t) \nabla_c \tag{3.389}$$

where

$$\Gamma_y = \left[\, Y_1 A^{-1} \quad 0 \quad -A_1^{-1} A_{p1} Y_{p1} T_d A_r^{-1} \,\right]. \tag{3.390}$$

Again with $\nabla_{b*}\nabla_b = \Lambda_* \Lambda$ and $\nabla_c \nabla_{c*} = \Omega \Omega_*$ the usual Wiener–Hopf spectral factorizations (Theorem C.1 and Remark C.1) one gets

$$\nabla = WG_{11}G_w - \nabla_b \Lambda^{-1} (\Lambda \Gamma_y \Omega + \Lambda K_t \Omega) \Omega^{-1} \nabla_c. \tag{3.391}$$

Moreover, every stable K_t can be written as

$$K_t = \widetilde{K}_t - \Lambda^{-1} \{\Lambda \Gamma_y \Omega\}_\infty \Omega^{-1} \tag{3.392}$$

where $\widetilde{K}_t$ is a stable matrix. Hence,

$$\nabla = \nabla_a - \nabla_b \widetilde{K}_t \nabla_c \tag{3.393}$$

where now

$$\nabla_a = WG_{11}G_w - \nabla_b\Lambda^{-1}(\{\Lambda\Gamma_y\Omega\}_+ + \{\Lambda\Gamma_y\Omega\}_-)\Omega^{-1}\nabla_c \tag{3.394}$$

is strictly proper because $\nabla_b\Lambda^{-1}$ and $\Omega^{-1}\nabla_c$ are paraunitary matrices (see Lemma C.1). In addition, ∇_a is stable because when G_w is acceptable ∇ is stable for all stable K_t and, therefore, for $\widetilde{K}_t \equiv 0$. Thus, assumptions (1) through (4) of Theorem 3.4 hold and (3.173) yields

$$\widetilde{K}_t = \Lambda^{-1}(Z + \{\widetilde{\Gamma}\}_+)\Omega^{-1} \tag{3.395}$$

where

$$\begin{aligned}\widetilde{\Gamma} &= \Lambda_*^{-1}\nabla_{b*}\nabla_a\nabla_{c*}\Omega_*^{-1} \\ &= \Lambda_*^{-1}\nabla_{b*}WG_{11}G_w\nabla_{c*}\Omega_*^{-1} - (\{\Lambda\Gamma_y\Omega\}_+ + \{\Lambda\Gamma_y\Omega\}_-) \\ &= \widetilde{\Gamma}_1 - \{\Lambda\Gamma_y\Omega\}_+ - \{\Lambda\Gamma_y\Omega\}_-.\end{aligned} \tag{3.396}$$

Using the fact that

$$\{\{\Lambda\Gamma_y\Omega\}_-\}_+ = 0, \quad \{\{\Lambda\Gamma_y\Omega\}_+\}_+ = \{\Lambda\Gamma_y\Omega\}_+ \tag{3.397}$$

one easily gets from (3.392), (3.395), and (3.396) that

$$K_t = \Lambda^{-1}(Z + \{\widetilde{\Gamma}_1\}_+ - \{\Lambda\Gamma_y\Omega\}_+ - \{\Lambda\Gamma_y\Omega\}_\infty)\Omega^{-1}. \tag{3.398}$$

Again the stable matrices K_t and Z can be partitioned as in (3.337) and (3.360). Then (3.358) and (3.361) can be invoked to get

$$H_1 = \Lambda^{-1}(Z_{g2} + \{\widetilde{\Gamma}_r\}_+ + \{\Lambda\widetilde{\Gamma}_y\}_+ + \{\Lambda\widetilde{\Gamma}_y\}_\infty)\Omega_o^{-1}A_r + A_1^{-1}A_{p1}Y_{p1}T_d \tag{3.399}$$

where

$$\widetilde{\Gamma}_r = -\widetilde{\Gamma}_1\begin{bmatrix}0\\0\\I\end{bmatrix} = \Lambda_*^{-1}A_{1*}P_*T_dG_rA_{r*}\Omega_{o*}^{-1} \tag{3.400}$$

and

$$\widetilde{\Gamma}_y = \Gamma_y\begin{bmatrix}0\\0\\\Omega_o\end{bmatrix} = -A_1^{-1}A_{p1}Y_{p1}T_dA_r^{-1}\Omega_o. \tag{3.401}$$

It then readily follows from

$$\{\Lambda\widetilde{\Gamma}_y\}_+ + \{\Lambda\widetilde{\Gamma}_y\}_\infty = \Lambda\widetilde{\Gamma}_y - \{\Lambda\widetilde{\Gamma}_y\}_- \tag{3.402}$$

that

$$H_1 = \Lambda^{-1}(Z_{g2} + \{\widetilde{\Gamma}_r\}_+ + \{\Lambda A_1^{-1}A_{p1}Y_{p1}T_dA_r^{-1}\Omega_o\}_-)\Omega_o^{-1}A_r \tag{3.403}$$

and

$$R_u = A_1H_1 = A_1\Lambda^{-1}(Z_{g2} + \{\widetilde{\Gamma}_r\}_+ + \{\Lambda A_1^{-1}A_{p1}Y_{p1}T_dA_r^{-1}\Omega_o\}_-)\Omega_o^{-1}A_r. \tag{3.404}$$

One also gets from (3.359), (3.398), and

$$\Omega^{-1}\begin{bmatrix} I & 0 \\ 0 & A_n \\ 0 & 0 \end{bmatrix} = \begin{bmatrix} I & 0 \\ 0 & I \\ 0 & 0 \end{bmatrix}\Omega_t^{-1}\begin{bmatrix} I & 0 \\ 0 & A_n \end{bmatrix} \tag{3.405}$$

that

$$\begin{aligned}[K_1 \quad H_2] = \Lambda^{-1}([Z_m \quad Z_n] + \{[\widetilde{\Gamma}_w \quad \widetilde{\Gamma}_z]\}_+ - \{[\widetilde{\Gamma}_a \quad \widetilde{\Gamma}_b]\}_+ \\ - \{[\widetilde{\Gamma}_a \quad \widetilde{\Gamma}_b]\}_\infty)\Omega_t^{-1}\begin{bmatrix} I & 0 \\ 0 & A_n \end{bmatrix}\end{aligned} \tag{3.406}$$

where just as in (3.369)

$$[\widetilde{\Gamma}_w \quad \widetilde{\Gamma}_z] = \widetilde{\Gamma}_1\begin{bmatrix} I & 0 \\ 0 & I \\ 0 & 0 \end{bmatrix} = \Lambda_*^{-1}A_{1*}P_*[P_dG_dP_{d*}F_*A_* \quad P_dG_dL_*A_{n*}]\Omega_{t*}^{-1} \tag{3.407}$$

and where again

$$[\widetilde{\Gamma}_a \quad \widetilde{\Gamma}_b] = \Lambda\Gamma_y\Omega\begin{bmatrix} I & 0 \\ 0 & I \\ 0 & 0 \end{bmatrix} = \Lambda[A_1^{-1}Y \quad 0]\Omega_t = \Lambda[Y_1A^{-1} \quad 0]\Omega_t \tag{3.408}$$

follows now from (3.390).

It is also useful to recognize that (3.406) yields

$$\begin{aligned}[K_1 \quad H_2] = \Lambda^{-1}([Z_m \quad Z_n] + \{[\widetilde{\Gamma}_w \quad \widetilde{\Gamma}_z]\}_+ + \{[\widetilde{\Gamma}_a \quad \widetilde{\Gamma}_b]\}_-)\Omega_t^{-1}\begin{bmatrix} I & 0 \\ 0 & A_n \end{bmatrix} \\ - [Y_1A^{-1} \quad 0]\end{aligned} \tag{3.409}$$

when

$$\{[\widetilde{\Gamma}_a \quad \widetilde{\Gamma}_b]\}_+ + \{[\widetilde{\Gamma}_a \quad \widetilde{\Gamma}_b]\}_\infty = \Lambda[Y_1A^{-1} \quad 0]\Omega_t - \{[\widetilde{\Gamma}_a \quad \widetilde{\Gamma}_b]\}_- \tag{3.410}$$

is introduced. It then readily follows from (3.374) that

$$[R_w \quad R_z] = A_1\Lambda^{-1}([Z_m \quad Z_n] + \{[\widetilde{\Gamma}_w \quad \widetilde{\Gamma}_z]\}_+ + \{[\widetilde{\Gamma}_a \quad \widetilde{\Gamma}_b]\}_-)\Omega_t^{-1}\begin{bmatrix} A & 0 \\ 0 & A_n \end{bmatrix} \tag{3.411}$$

or

$$[R_w \mid R_z] = A_1\Lambda^{-1}[(Z_m + \{\widetilde{\Gamma}_w\}_+ + \{\widetilde{\Gamma}_a\}_-) \mid (Z_n + \{\widetilde{\Gamma}_z\}_+ + \{\widetilde{\Gamma}_b\}_-)]\Omega_t^{-1}\left[\begin{array}{c|c} A & 0 \\ \hline 0 & A_n \end{array}\right]. \tag{3.412}$$

Finally, it follows from (2.208) or alternatively from $C = (\widetilde{X}_1 - K\widetilde{B})^{-1}(\widetilde{Y}_1 + K\widetilde{A})$ and (3.58) through (3.66) that

$$C = [-C_w \quad -C_z \quad C_u] = (X_1 - K_1B)^{-1}[-(Y_1 + K_1A) \quad -H_2 \quad H_1] \tag{3.413}$$

provided det $(X_1 - K_1B) \not\equiv 0$. This can be assured as follows. With

$$Z_\gamma = [Z_m \quad Z_n] + \{[\widetilde{\Gamma}_w \quad \widetilde{\Gamma}_z]\}_+ + \{[\widetilde{\Gamma}_a \quad \widetilde{\Gamma}_b]\}_- \tag{3.414}$$

one has from (3.409) that

$$[K_1 \quad H_2] = \Lambda^{-1}Z_\gamma\Omega_t^{-1}\begin{bmatrix} I & 0 \\ 0 & A_n \end{bmatrix} - [Y_1A^{-1} \quad 0]. \tag{3.415}$$

Thus,

$$K_1 = \Lambda^{-1}Z_\gamma\Omega_t^{-1}\begin{bmatrix} I \\ 0 \end{bmatrix} - Y_1A^{-1} \tag{3.416}$$

and

$$\begin{aligned} A_1(X_1 - K_1B) &= A_1X_1 - A_1\Lambda^{-1}Z_\gamma\Omega_t^{-1}\begin{bmatrix} I \\ 0 \end{bmatrix}B + YB \\ &= I - A_1\Lambda^{-1}Z_\gamma\Omega_t^{-1}\begin{bmatrix} A & 0 \\ 0 & A_n \end{bmatrix}\begin{bmatrix} I \\ 0 \end{bmatrix}FP. \end{aligned} \tag{3.417}$$

So conditions for which the final expression in (3.417) approaches I as $s \to \infty$ are sufficient to assure that det $(X_1 - K_1B) \not\equiv 0$. When

$$(P_*P+kQ)^{-1} \leq 0(s^{2k_1}), \tag{3.418}$$

$$A_1\Lambda^{-1}\Lambda_*^{-1}A_{1*} = (P_*P+kQ)^{-1} \leq 0(s^{2k_1}) \tag{3.419}$$

and

$$A_1\Lambda^{-1} \leq 0(s^{k_1}). \tag{3.420}$$

With

$$\Phi = \begin{bmatrix} (FP_dG_dP_{d*}F_* + G_{n_m}) & FP_dG_dL_* \\ LG_dP_{d*}F_* & LG_dL_* + G_{n_l} \end{bmatrix} \tag{3.421}$$

it follows from (3.350) when $\Phi^{-1} \leq 0(s^{2k_2})$ that

$$\begin{bmatrix} A_* & 0 \\ 0 & A_{n*} \end{bmatrix} \Omega_{t*}^{-1}\Omega_t^{-1} \begin{bmatrix} A & 0 \\ 0 & A_n \end{bmatrix} = \Phi^{-1} \leq 0(s^{2k_2}) \tag{3.422}$$

and

$$\Omega_t^{-1} \begin{bmatrix} A & 0 \\ 0 & A_n \end{bmatrix} \leq 0(s^{k_2}). \tag{3.423}$$

With $FP \leq 0(s^{k_3})$ and $k_1 + k_2 + k_3 \leq 0$ it then follows from (3.417) and the fact that $Z_\gamma \leq 0(s^{-1})$ when $Z \leq 0(s^{-1})$ that

$$A_1(X_1 - K_1B) \rightarrow I, \ s \rightarrow \infty \tag{3.424}$$

and det $(X_1 - K_1B) \not\equiv 0$.

From (3.411) it is clear when $Z_\Gamma \leq 0(s^{-1})$ that

$$[R_w \quad R_z] \leq 0(s^{k_1+k_2-1}). \tag{3.425}$$

Then

$$R_wFP \leq 0(s^{k_1+k_2+k_3-1}) \tag{3.426}$$

is strictly proper when $k_1 + k_2 + k_3 \leq 0$. Hence, $(I - R_wFP)^{-1}$ is proper. Moreover, when $(G_r + G_{n_r})^{-1} \leq 0(s^{2k_o})$ one gets from (3.351) that

$$A_{r*}\Omega_{o*}^{-1}\Omega_o^{-1}A_r = (G_r + G_{n_r})^{-1} \leq 0(s^{2k_o}) \tag{3.427}$$

and it follows that

$$\Omega_o^{-1}A_r \leq 0(s^{k_o}). \tag{3.428}$$

Hence, from (3.404),

$$R_u \leq 0(s^{k_o+k_1-1}). \tag{3.429}$$

When $k_1+k_2+k_3 \leq 0$, Eq. (2.228) then yields

$$C_u = (I - R_w FP)^{-1} R_u \leq 0(s^{k_o+k_1-1}) \tag{3.430}$$

and

$$[C_w \quad C_z] = (I - R_w FP)^{-1}[R_w \quad R_z] \leq 0(s^{k_1+k_2-1}). \tag{3.431}$$

Clearly, the controller is proper (strictly proper) in this case when $k_o+k_1 \leq 1$ and $k_1+k_2 \leq 1$ (when $k_o+k_1 \leq 0$ and $k_1+k_2 \leq 0$).

When all the finite unstable poles of the shape-deterministic inputs r_1 and d_1 are confined to the $j\omega$-axis, one can introduce the analog equivalents to (3.378) and (3.379) and again work with (3.380) and (3.381). In this case,

$$R_u = A_1 \Lambda^{-1}(Z_{g2} + \{\tilde{\Gamma}_r\}_+)\tilde{\Omega}_o^{-1} \tag{3.432}$$

where

$$\tilde{\Gamma}_r = \Lambda_*^{-1} A_{1*} P_* T_d G_r \tilde{\Omega}_{o*}^{-1}. \tag{3.433}$$

Moreover,

$$[R_w \mid R_z] = A_1\Lambda^{-1}[(Z_m + \{\tilde{\Gamma}_w\}_+ + \{\tilde{\Gamma}_a\}_-) \mid (Z_n + \{\tilde{\Gamma}_z\}_+ + \{\tilde{\Gamma}_b\}_-)]\tilde{\Omega}_t^{-1}\left[\begin{array}{c|c} A & 0 \\ \hline 0 & I \end{array}\right] \tag{3.434}$$

where

$$[\tilde{\Gamma}_w \quad \tilde{\Gamma}_z] = \Lambda_*^{-1} A_{1*} P_* [P_d G_d P_{d*} F_* A_* \quad P_d G_d L_*]\tilde{\Omega}_{t*}^{-1} \tag{3.435}$$

and

$$[\tilde{\Gamma}_a \quad \tilde{\Gamma}_b] = \Lambda A_1^{-1}[Y \quad 0]\begin{bmatrix} I & 0 \\ 0 & A_n \end{bmatrix}\tilde{\Omega}_t = \Lambda A_1^{-1}[Y \quad 0]\tilde{\Omega}_t. \tag{3.436}$$

It is important to clarify the results presented here for R_u and $[R_w \mid R_z]$ in the analog case to the ones presented in Park and Bongiorno (1990). When the notational differences are recognized, one can see that the reduced Eqs. (3.432) and (3.434) agree with the formulas given in Theorems 2 and 3 of Park and Bongiorno (1990) for the case when plant uncertainty is not taken into account (i.e., when their

$\mu = 0$). However, (3.404) and (3.412) are more general in the following regard. In Park and Bongiorno (1990), the restriction (3.333) needed on K when A_r^{-1} and/or A_n^{-1} are unstable is not imposed. So whenever A_r^{-1} and/or A_n^{-1} have unstable poles in Re $s > 0$ one can expect the formulas in Park and Bongiorno (1990) to fail. When the finite unstable poles of A_r^{-1} and/or A_n^{-1} are confined to the $j\omega$-axis, it is true never-the-less that the formulas in Park and Bongiorno (1990) are valid for the assumptions there assure that the matrix K will be stable even though the restriction (3.333) is not explicitly imposed.

A summary of the key assumptions made and their implications is now presented for easy reference and clarification.

Assumption 3.6 The 3DOF system is admissible.

Assumption 3.7 The matrices F, T_d, $A_p T_d A_r^{-1}$, and $A_p P_d A_d^{-1}$ are stable.

Assumption 3.8 The matrices Ω_d, Ω_r, Ω_{n_r}, Ω_{n_m}, Ω_{n_l} and $W = \text{diag}\{I, W_q\}$ are stable.

Assumption 3.9 On $|\lambda| = 1$ for the digital case and on the finite $s = j\omega$ axis for the analog case, $\det\left[A_{1*}(P_* P + kQ)A_1\right] \neq 0$ and $\det\left[A_t G_{21} G_s G_{21*} A_{t*}\right] \neq 0$ where $kQ = W_{q*} W_q$.

Assumption 3.10 In the analog case, W_q, T_d, and P_d are proper and $A_r^{-1}B_r$, $A_d^{-1}B_d$, and Ω_s are strictly proper. In addition, $(P_* P + kQ)^{-1} \leq 0(s^{2k_1})$, $\Phi^{-1} \leq 0(s^{2k_2})$, and $FP \leq 0(s^{k_3})$ with $k_1 + k_2 + k_3 \leq 0$.

Assumption 3.6 assures that the 3DOF system can be stabilized and that all stabilizing controller transfer matrices are given by (2.208) with $K = \left[-K_1 \quad -H_2 \quad H_1\right]$ a stable matrix. It also follows from Lemma 2.8 that ψ_L^{-1} and $\psi_{FP}(\psi_F \psi_P)^{-1}$ must be stable; hence, L must be stable and with F stable, $A_1^{-1}A_{p1}$ is stable because of Lemma 2.10. Moreover, PA_1, AFP_d, and $(I - PA_1 Y_1 F)P_d$ must be stable from Lemma 2.7.

Assumption 3.7 then assures that the $G = [G_{ij}]$ associated with the 3DOF system is compatible. It also assures that $G_w = \left[\sqrt{\alpha_1}\widetilde{G}_{w1} \quad \sqrt{\alpha_2}\widetilde{G}_{w2}\right]$, where $\widetilde{G}_{w1}$ and $\widetilde{G}_{w2}$ are defined by (3.317) and (3.319) and where $\widetilde{G}_{w2}$ is stable because of Assumption 3.8, is acceptable for every stable K given by (3.333) iff $K_t = \left[K_m \quad K_n \quad -K_{g2}\right]$ a stable matrix.

Collectively, Assumptions 3.6, 3.7, and 3.8 lead to $\nabla = WG_\delta G_w = \nabla_a - \nabla_b K_t \nabla_c$ stable for all stable K_t in the digital case and $\nabla = WG_\delta G_w = \nabla_a - \nabla_b \widetilde{K}_t \nabla_c$ stable for all stable $\widetilde{K}_t$ in the analog case with ∇_a, ∇_b, and ∇_c stable. Then the additional Assumption 3.9 together with the choice (3.276a) and (3.277) for W leads by Theorem 3.2 to the expressions (3.364) and (3.371) for the submatrices of $K = \left[-K_1 \quad -H_2 \quad H_1\right]$ which in the digital case guarantees that the performance functional J defined in (3.105) is finite. With $Z = \left[Z_m \quad Z_n \quad -Z_{g2}\right] \equiv 0$ in these submatrices, the minimum value is obtained for the performance functional

J. For the analog case, the additional Assumption 3.10 assures that $WG_{11}G_w$ and, therefore, ∇_a are strictly proper. With the choice (3.276a) and (3.277) for W, Theorem 3.4 then leads to the expressions (3.3) and (3.9) for the submatrices of $K = [-K_1 \quad -H_2 \quad H_1]$ which guarantee that the performance functional J defined in (3.144) is finite. With $Z = [Z_m \quad Z_n \quad -Z_{g2}] \equiv 0$ in these submatrices, the minimum value is obtained for the performance functional J. Finally, the order restrictions on $(P_*P + kQ)^{-1}$, Φ^{-1}, and FP in Assumption 3.10 assure that the controller is realizable.

It is also important to remember that when the unstable poles of r_1 and d_1 lie only on $|\lambda| = 1$ for the digital case and on $s = j\omega$ for the analog case, then the required factorizations of G_t and G_o can be carried out in accordance with (3.378) through (3.381) leading to the reduced formulas (3.383) through (3.387) for the digital case and (3.432) through (3.436) for the analog case.

It should also be pointed out that when $(T_dF - I)P$ is stable and the other assumptions in Theorems 2.5 and 2.7 are satisfied then instead of beginning with (3.333) for the subset of suitable K one could take instead

$$K = [-K_1 \quad -H_2 \quad H_1] = [-K_m \quad -K_nA_n \quad (Y_1 + K_rA_r)] \tag{3.437}$$

where $LA_d^{-1} = A_n^{-1}B_n$ and K_m, K_n, and K_r are stable matrices. Then the need to find a coprime polynomial matrix fraction description for P can be avoided. For brevity this approach is not considered here since the one presented is more general.

The design of an analog 2DOF optimal $\mathcal{H}_2$ multivariable system for the control of the Rosenbrock process which exhibits a severe coupling phenomenon (Åström et al. 2002) is presented in Example 3.24 to illustrate the application of the theory developed in this section.

3.9 Trade-off of Optimal Performance for Reduced Sensitivity to Plant Uncertainty

3.9.1 *Overview*

The performance functionals employed in Sects. 3.5 and 3.6 of this chapter take into account tracking error and saturation for nominal plant models. The stabilizing controllers for which the performance functional is finite are parameterized in terms of a real rational stable matrix parameter Z. The performance functional is minimized when $Z \equiv 0$. When non-null choices are made for the matrix parameter, the value of the performance function increases. It follows from Theorem 3.3 and (3.129) and from Theorem 3.5 and (3.175), respectively, that the increase is given simply by

$$\Delta J = J - J_o = \frac{1}{2\pi j} \oint_{|\lambda|=1} \mathrm{Tr}(ZZ_*) d\lambda/\lambda \tag{3.438}$$

in the digital case and by

$$\Delta J = J - J_o = \frac{1}{2\pi j} \int_{-j\infty}^{j\infty} \mathrm{Tr}(ZZ_*) ds \tag{3.439}$$

in the analog case. Clearly, the possibility exists to trade off optimal nominal system performance for improvement in stability margin and reduced sensitivity to plant parameter variations. That is, the possibility exists for choosing a $Z \not\equiv 0$ for which $J \leq N$ and for which the non-optimal nominal design obtained is more robust with respect to uncertainty in the nominal plant model than the optimal design. Attention is restricted to the more difficult analog case: One can easily apply the same methodology for the digital case.

A straightforward Wiener–Hopf methodology for achieving this trade-off is described in the sequel. The methodology begins with the determination of two functionals: one that penalizes designs that are too sensitive to small plant parameter changes and one that penalizes designs for stability margins that are too small. The focus on the increase in the value of the quadratic performance functional here contrasts with the one taken in the loop transfer recovery (LTR) approach (Maciejowski 1989) and/or the modification of weighting functions approach (Grimble 2006). In particular, the matrix parameter Z is used to establish a more transparent methodology for trading-off optimality for more robustness; especially in multivariable cases where modification of weighting functions is not as clear as in the scalar (single variable) case.

3.9.2 Sensitivity to Small Plant Parameter Changes

It easily follows from Fig. 3.2 that

$$v = CG_{21}w + CG_{22}v \Leftrightarrow v = (I - CG_{22})^{-1}CG_{21}w = RG_{21}w \tag{3.440}$$

and

$$y_\delta = G_{11}w + G_{12}v = G_{11}w + G_{12}(I - CG_{22})^{-1}CG_{21}w = (G_{11} + G_{12}RG_{21})w \tag{3.441}$$

where

$$R = SC, \quad S = (I - CG_{22})^{-1}. \tag{3.442}$$

These equations characterize the nominal system. The interest here is in characterizing the changes Δy_δ and Δv in y_δ and v, respectively, when $G = [G_{ij}]$ is actually $G + \Delta G = [G_{ij} + \Delta G_{ij}]$ while C and w remain unchanged. Clearly, $\Delta R = (\Delta S)C$. Moreover, when only first order changes are retained, one gets

$$\Delta(SS^{-1}) = (\Delta S)S^{-1} + S[\Delta(S^{-1})] = \Delta(I) = 0 \Leftrightarrow \Delta S = -S[\Delta(S^{-1})]S. \tag{3.443}$$

Hence,

$$\Delta R = (\Delta S)\mathrm{C} = -S[\Delta(I - CG_{22})]SC = SC(\Delta G_{22})SC = R(\Delta G_{22})R. \tag{3.444}$$

It now follows from (3.440) and (3.441) that

$$\begin{aligned}\Delta v &= [(\Delta R)G_{21} + R(\Delta G_{21})]w = [R(\Delta G_{22})RG_{21} + R(\Delta G_{21})]w \\ &= R(\Delta G_{22})v + R(\Delta G_{21})w = R[\,\Delta G_{21} \quad \Delta G_{22}\,][\,w' \quad v'\,]'\end{aligned} \tag{3.445}$$

Similarly,

$$\begin{aligned}\Delta y_\delta &= (\Delta G_{11})w + (\Delta G_{12})v + G_{12}(\Delta v) = \{[\,\Delta G_{11} \quad \Delta G_{12}\,] + G_{12}R[\,\Delta G_{21} \quad \Delta G_{22}\,]\}\begin{bmatrix} w \\ v \end{bmatrix} \\ &= [\,I \quad G_{12}R\,]\begin{bmatrix} \Delta G_{11} & \Delta G_{12} \\ \Delta G_{21} & \Delta G_{22} \end{bmatrix}\begin{bmatrix} w \\ v \end{bmatrix} = [\,I \quad G_{12}R\,]\Delta G\begin{bmatrix} w \\ v \end{bmatrix}.\end{aligned} \tag{3.446}$$

In (3.446), v is the nominal controller output; hence, the weighted matrix multiplier

$$\mathcal{S} = \begin{bmatrix} W_{11} & 0 \\ 0 & W_{22} \end{bmatrix}\begin{bmatrix} I & G_{12}R \\ 0 & R \end{bmatrix}(\Delta G) = W_1\mathcal{R}(\Delta G), \tag{3.447}$$

where the weighting matrices W_{11} and W_{22} can be chosen based on the relative importance of plant uncertainty on deviations of the controlled variables v and y_δ from their nominal values. (The weighting matrices W_{11} and W_{22} are introduced to allow for more generality; however, $W_{11} = \lambda_1 I$ and $W_{22} = \lambda_2 I$ should be the primary choice with λ_1 and λ_2 nonnegative constants.) It is then consistent with the Wiener–Hopf approach to use this indicator to measure sensitivity with the functional

$$J_{s1} = \frac{1}{2\pi j}\int_{-j\infty}^{j\infty} \langle Tr(\mathcal{S}\mathcal{S}_*)\rangle ds = \frac{1}{2\pi j}\int_{-j\infty}^{j\infty} \langle Tr(W_1 \mathcal{R} G_s \mathcal{R}_* W_{1*})\rangle ds \tag{3.448}$$

where

$$G_s = \langle (\Delta G)(\Delta G)_* \rangle \tag{3.449}$$

is a plant-uncertainty spectral density matrix. (J_{s2} is used in the sequel for the stability margin functional.) Clearly, G_s is a parahermetian-positive matrix and it follows from Appendix C (Theorem C.1) that there exists a left Wiener–Hopf factorization $G_s = \Omega_s \Omega_{s*}$ where Ω_s and its left inverse are analytic in Re > 0. Here it is assumed that G_s and W_1 satisfy

Assumption 3.11 The plant uncertainty matrix G_s is analytic and nonsingular on the finite $j\omega$-axis and it is strictly proper. The weighting matrix W_1 is stable and proper.

The properties imposed on G_s in Assumption 3.11 insure that Ω_s is square, strictly proper, and it and its inverse are analytic in $0 \leq \text{Re}\, s < \infty$. The assumption that G_s is analytic on the $j\omega$-axis is consistent with $\|\Delta G(j\omega)\|_\infty$ being bounded in $\mathcal{H}_\infty$ approaches (Maciejowski 1989; Grimble 2006).

Attention is also restricted here to those R that are given by Theorem 3.5. Specifically, one gets for the analog case that

$$R = A_1(Y_1 + KA) = A_{o1}\Lambda^{-1} Z \Omega^{-1} A_g A + R_o \tag{3.450}$$

where Z is stable and strictly proper and where

$$R_o = A_{o1}\Lambda^{-1}[\{\Lambda_*^{-1}\nabla_{b*} W G_{11} G_w \nabla_{c*} \Omega_*^{-1}\}_+ + \{\Lambda A_{o1}^{-1} Y_a \Omega\}_-]\Omega^{-1} A_g A \tag{3.451}$$

is the expression for R in the optimal nominal case: i.e., $Z \equiv 0$. Since G is compatible, $G_{12}A_1$ is stable; hence, so is $G_{12}R$ and it follows that

$$\nabla_{s1} = W_1 \begin{bmatrix} I & G_{12}R \\ 0 & R \end{bmatrix} \Omega_s = W_1 \begin{bmatrix} I & G_{12}R \\ 0 & R \end{bmatrix} \begin{bmatrix} \Omega_{s1} \\ \Omega_{s2} \end{bmatrix} = \nabla_{sa1} - \nabla_{sb1} Z \nabla_{sc1} \tag{3.452}$$

where

$$\left.\begin{aligned} \nabla_{sa1} &= \begin{bmatrix} W_{11}(\Omega_{s1} + G_{12}R_o\Omega_{s2}) \\ W_{22}R_o\Omega_{s2} \end{bmatrix} \\ \nabla_{sb1} &= -\begin{bmatrix} W_{11}G_{12}A_{o1}\Lambda^{-1} \\ W_{22}A_{o1}\Lambda^{-1} \end{bmatrix} \\ \nabla_{sc1} &= \Omega^{-1}A_g A \Omega_{s2} \end{aligned}\right\} \tag{3.453}$$

are all stable. The matrix ∇_{s1} is also strictly proper when $G_{12}R$ is proper. Recalling (3.186), (3.189), and Theorem 3.5, one easily gets from (3.450) and (3.451) that R is strictly proper when $k_1 + k_2 \leq 0$ and this inequality is usually satisfied. Hence, when G is proper, $G_{12}R$ is usually strictly proper and it is not unreasonable to impose.

Assumption 3.12 R and $G_{12}R$ are strictly proper.

Under Assumptions 3.11 and 3.12, ∇_{s1} is stable and strictly proper which assures that J_{s1} is finite. Moreover, ∇_{sa1} is strictly proper.

Clearly, R is linear in Z and one can consider choosing a $Z \not\equiv 0$ for which J_{s1} is minimized subject to the constraint $\Delta J \leq N$. Problems of this type have been addressed by Newton et al. (1957) and Chang (1961) for the scalar case. The same methodology was used for the multivariable case in Bongiorno et al. (1997). In particular, a Lagrange multiplier α^2 is introduced and the functional

$$2\pi j(J_{s1} + \alpha^2 \Delta J) = \int_{-j\infty}^{j\infty} \mathrm{Tr}\ (\nabla_{s1}\nabla_{s1*})ds + \alpha^2 \int_{-j\infty}^{j\infty} \mathrm{Tr}\ (ZZ_*)ds \qquad (3.454)$$

is minimized with respect to Z as a function of α^2; then α^2 is chosen so that $2\pi j\Delta J = N$. However, it would be premature to do so without first including an additional term in (3.454) to account for stability margin.

3.9.3 *Stability Margin*

Also important with regard to plant uncertainty is the question of stability. Specifically, it is important that a margin of safety exits so that the system remains stable for the expected range of plant parameter values and for other plant modelling errors. When C in Fig. 3.2 is a stabilizing controller for the nominal $G_{22} = \widetilde{G}_{22} = -A^{-1}B = -B_1A_1^{-1}$, it follows from (2.33) and(2.60) that

$$\phi = \det\ (A_cA_1 + B_cB_1) = \det\ (LX_1A_1 - MBA_1 + LY_1B_1 + MAB_1) = \det L \qquad (3.455)$$

is a stable polynomial. When G_{22} is actually given by the coprime polynomial matrix fraction description

$$G_{22} + \Delta G_{22} = -(B_1 + \Delta B_1)(A_1 + \Delta A_1)^{-1}, \qquad (3.456)$$

where ΔA_1 and ΔB_1 are polynomial matrices whose coefficient matrices are the changes in the coefficient matrices of A_1 and B_1 from their nominal values, respectively, then ϕ becomes

$$\phi_a = \det\left(A_cA_1 + B_cB_1 + [A_c \quad B_c]\begin{bmatrix}\Delta A_1 \\ \Delta B_1\end{bmatrix}\right) = \det\left(L + [(LX_1 - MB) \quad (LY_1 + \mathrm{MA})]\begin{bmatrix}\Delta A_1 \\ \Delta B_1\end{bmatrix}\right)$$
$$= (\det L)\det\left(I + [(X_1 - KB) \quad (Y_1 + KA)]\begin{bmatrix}\Delta A_1 \\ \Delta B_1\end{bmatrix}\right) = (\det L)\ \phi_\Delta = \phi\phi_\Delta. \tag{3.457}$$

Thus, the system remains stable iff it remains admissible and ϕ_Δ is free of zeros in Re $s \geq 0$.

It is also clear from (3.11), (3.37), and (3.442) that

$$\phi_\Delta = \det\left(I + A_1^{-1}[S \quad R]\begin{bmatrix}\Delta A_1 \\ \Delta B_1\end{bmatrix}\right) \tag{3.458}$$

and

$$S = I + RG_{22}. \tag{3.459}$$

One can always write

$$\begin{bmatrix}\Delta A_1 \\ \Delta B_1\end{bmatrix} = C_\Delta S_\Delta \tag{3.460}$$

where C_Δ contains the elements of the coefficient matrices in ΔA_1 and ΔB_1 and the elements of S_Δ are all powers of s with highest power equal to the largest degree of the elements in $B_1(s)$ and $A_1(s)$. It now follows from (3.458) that

$$\phi_\Delta = \det\,(I + S_\Delta A_1^{-1}[S \quad R]C_\Delta) = \det\,(I + \mathscr{M}C_\Delta) \tag{3.461}$$

where

$$\mathscr{M} = S_\Delta A_1^{-1}[S \quad R] = S_\Delta[(X_1 - KB) \quad (Y_1 + KA)]. \tag{3.462}$$

Clearly, $\mathscr{M}$ is stable. Moreover, when a weighting matrix W_2 is introduced and when

Assumption 3.13 G_{22} is proper; R and $S_\Delta A_1^{-1}$ are strictly proper; and W_2 is proper and stable;

holds, it easily follows that $\nabla_{s2} = W_2 \mathscr{M} \Omega_\Delta$ is strictly proper and stable where

$$G_\Delta = \langle C_\Delta C_{\Delta *} \rangle = \Omega_\Delta \Omega_{\Delta *} \tag{3.463}$$

is the parameter-uncertainty covariance matrix for C_Δ whose elements are modeled as zero-mean random variables. Hence,

$$J_{s2} = \frac{1}{2\pi j} \int_{-j\infty}^{j\infty} \langle \mathrm{Tr}(\nabla_{s2} \nabla_{s2*}) \rangle ds = \frac{1}{2\pi j} \int_{-j\infty}^{j\infty} \mathrm{Tr}(W_2 \mathscr{M} G_\Delta \mathscr{M}_* W_{2*}) ds \tag{3.464}$$

is finite. In addition, J_{s2} is a quadratic functional on Z because it follows from (3.450), (3.451), and (3.459) that

$$\nabla_{s2} = W_2 \mathscr{M} \Omega_\Delta = \nabla_{sa2} - \nabla_{sb2} Z \nabla_{sc2} \tag{3.465}$$

where

$$\nabla_{sa2} = W_2 S_\Delta A_1^{-1} [\,(I + R_o G_{22}) \quad R_o\,] \Omega_\Delta, \tag{3.466a}$$

$$\nabla_{sb2} = -W_2 S_\Delta A_1^{-1} A_{o1} \Lambda^{-1}, \tag{3.466b}$$

and

$$\nabla_{sc2} = \Omega^{-1} A_g A [\,G_{22} \quad I\,] \Omega_\Delta. \tag{3.466c}$$

Moreover, it follows that $\nabla_{sa2}, \nabla_{sb2}$, and ∇_{sc2} are stable and ∇_{sa2} is strictly proper since $A_1^{-1} A_{o1}$ stable is assumed in Theorem 3.5 and $S_\Delta A_1^{-1}$ is strictly proper. The assumption that $S_\Delta A_1^{-1}$ is strictly proper is not overly restrictive. Indeed, when G_{22} is strictly proper and A_1 is column reduced it is not difficult to establish (Bongiorno et al. 1997) that practical cases exist in which $S_\Delta A_1^{-1}$ is strictly proper.

With $\mathscr{M} = A_m^{-1} B_m$ a left coprime matrix fraction description, it follows from (3.461) that

$$\phi_\Delta = \det(I + \mathscr{M} C_\Delta) = \det(A_m + B_m C_\Delta)/(\det A_m) = n(s)/d(s). \tag{3.467}$$

Since $\mathscr{M} C_\Delta$ is stable, $d(s) = \det A_m(s)$ is a stable polynomial. Since $\mathscr{M} C_\Delta$ is strictly proper, ϕ_Δ has no zeros at $s = \infty$ and its zeros are those of the polynomial

$$n(s) = \det(A_m + B_m C_\Delta) \tag{3.468}$$

Clearly, when $C_\Delta = 0$ the zeros of $n(s)$ are those of $d(s)$ and all lie in Re $s < 0$. Moreover, the zeros of $n(s)$ are continuous functions of the elements of C_Δ. So if the elements of C_Δ are confined to a hypersphere of sufficiently small radius in the Euclidean space whose axes correspond to the elements of C_Δ then the zeros of $n(s)$ remain in Re $s < 0$. The only way zeros of $n(s)$ can occur in Re $s \geq 0$ is if the radius of the hypersphere is first increased enough for zeros of $n(s)$ to exist on the finite $j\omega$-axis. This is only possible from (3.463) if $\mathscr{M}(j\omega)C_\Delta$ has an eigenvalue equal to -1 for some $0 \leq \omega < \infty$. Any condition on C_Δ that assures this is not possible is then sufficient to ensure that no zeros of $\phi_\Delta(s)$ lie in Re $s \geq 0$ and the system remains stable. It is not difficult to show (see Example 3.26) that the eigenvalue of $\mathscr{M}(j\omega)C_\Delta$ with largest magnitude has a magnitude which is less than or equal to the largest singular value of $\mathscr{M}(j\omega)C_\Delta$ or, equivalently, is less than or equal to the positive square root of the largest eigenvalue of $[\mathscr{M}(j\omega)C_\Delta][\mathscr{M}(j\omega)C_\Delta]_*$. Hence, ϕ_Δ is free of zeros in Re $s \geq 0$ when

$$\|\mathscr{M}(j\omega)C_\Delta\|_\infty \leq \|\mathscr{M}(j\omega)\|_\infty \|C_\Delta\|_\infty < 1 \tag{3.469}$$

where $\|\bullet\|_\infty$ is the usual designation for the $\mathcal{H}_\infty$-norm of a real rational stable matrix; i.e., the least upper bound for all the singular values of the matrix on the imaginary axis.

Following the $\mathcal{H}_\infty$ approach, one might consider selecting Z so that the weighted norm

$$\|\nabla_{s2}\|_\infty = \|W_2 \mathscr{M} \Omega_\Delta\|_\infty = \|\nabla_{sa2} - \nabla_{sb2} Z \nabla_{sc2}\|_\infty \tag{3.470}$$

is minimized. In the $\mathcal{H}_\infty$ approach, however, a proper Z is allowed and no attention is paid to the value of ΔJ given by (3.439). Thus, the Z that minimizes $\|\nabla_{s2}\|_\infty$ might not be strictly proper and ΔJ could even be infinite. Unfortunately, minimizing the $\mathcal{H}_\infty$-norm subject to a constraint on the $\mathcal{H}_2$-norm increment is a difficult one for which to find an analytical solution. The difficulties are circumvented here by using $J_{s2} = \|\nabla_{s2}\|_2^2$ instead as an approximate measure of stability margin: The smaller J_{s2}, the better is the expected stability margin. This is in keeping with the use of an $\mathcal{H}_2$-norm as an approximate measure for plant-input saturation. Additional justification for the use of the $\mathcal{H}_2$-norm J_{s2} as an approximate measure of stability margin is that an optimal $\mathcal{H}_\infty$ problem can be viewed as a weighted Wiener–Hopf problem (Sideris 1990).

Thus, J_{s2} is a performance functional that takes into account stability margin and similarly to minimizing (3.454) one could minimize instead

$$2\pi j(J_{s2} + \alpha^2 \Delta J) = \int_{-j\infty}^{j\infty} \mathrm{Tr}(\nabla_{s2}\nabla_{s2*})ds + \alpha^2 \int_{-j\infty}^{j\infty} \mathrm{Tr}(ZZ_*)ds \tag{3.471}$$

with respect to Z and then choose α^2 so that $\Delta J = N$. In the next subsection this is done for sensitivity and stability margin combined.

3.9.4 The Analytical Solution for Sensitvity and Stability Margin Combined

Obviously, $J_s = J_{s1} + J_{s2}$ is a performance functional that takes sensitivity and stability margin into account. Hence, the general problem of interest is the minimization of

$$2\pi j(J_{s1} + J_{s2} + \alpha^2 \Delta J) = \int_{-j\infty}^{j\infty} \mathrm{Tr}(\nabla_{s1}\nabla_{s1*})ds + \int_{-j\infty}^{j\infty} \mathrm{Tr}(\nabla_{s2}\nabla_{s2*})ds + \alpha^2 \int_{-j\infty}^{j\infty} \mathrm{Tr}(ZZ_*)ds \tag{3.472}$$

with respect to Z and choosing α^2 so that $\Delta J = N$. More general problems of this type are addressed in Khargonekar and Rotea (1991) for the multivariable case; the methodology presented here yields the desired solution directly from Theorem 3.4.

The performance functional in (3.472) is a special case of the more general one

$$2\pi j\widetilde{J}_s = \sum_{k=1}^{q} \int_{-j\infty}^{j\infty} \mathrm{Tr}(\nabla_{sk}\nabla_{sk*})ds, \quad \nabla_{sk} = \nabla_{sak} - \nabla_{sbk}Z\nabla_{sck} \tag{3.473}$$

where $\nabla_{sak}, \nabla_{sbk},$ and ∇_{sck} are stable for $k = 1 \to q$ and ∇_{sak} is strictly proper for $k = 1 \to q$. As already noted in the next to last paragraph of Sect. 1.4, the Kronecker product $G \otimes R$ is the matrix whose ij-block is $g_{ij}R$ and $vec\, G = [g_2' \;\; g_2' \;\; \cdots \;\; g_2']'$ is the vector obtained by stacking the n columns $g_j, j = 1 \to n$ of G. Thus,

$$\mathrm{Tr}(\nabla_{sk}\nabla_{sk*}) = \mathrm{Tr}(\nabla_{sk*}\nabla_{sk}) = (\mathrm{vec}\ \nabla_{sk})_*(\mathrm{vec}\ \nabla_{sk}) = v_{sk*}v_{sk} \tag{3.474}$$

and

$$2\pi j\widetilde{J}_s = \int_{-j\infty}^{j\infty} \sum_{k=1}^{q} v_{sk*}v_{sk}ds = \int_{-j\infty}^{j\infty} v_{s*}v_s ds = \int_{-j\infty}^{j\infty} \mathrm{Tr}(v_s v_{s*})ds \tag{3.475}$$

where $v_{sk} = \mathrm{vec}\ \nabla_{sk}$ and

$$v_s' = [v_{s1}' \;\; v_{s2}' \;\; \cdots \;\; v_{sq}'] . \tag{3.476}$$

Clearly, one gets from (3.473) that

$$v_{sk} = \text{vec } \nabla_{sk} = \text{vec } \nabla_{sak} - \text{vec } (\nabla_{sbk} Z \nabla_{sck}). \tag{3.477}$$

Two useful identities which are not difficult to verify are

$$\left.\begin{aligned} \text{vec}(AVD) &= (D' \otimes A)\text{vec } V \\ (G_1 \otimes G_2)_* &= G_{1*} \otimes G_{2*} \end{aligned}\right\}. \tag{3.478}$$

It follows from the first identity and (3.477) that

$$v_{sk} = v_{sak} - \Phi_{sk}(\text{vec } Z) \tag{3.479}$$

where

$$\left.\begin{aligned} v_{sak} &= \text{vec } \nabla_{sak} \\ \Phi_{sk} &= \nabla'_{sck} \otimes \nabla_{sbk} \end{aligned}\right\}. \tag{3.480}$$

Introducing

$$v'_{sa} = \begin{bmatrix} v'_{sa1} & v'_{sa2} & \cdots & v'_{saq} \end{bmatrix} \tag{3.481}$$

and

$$\Phi'_s = \begin{bmatrix} \Phi'_{s1} & \Phi'_{s2} & \cdots & \Phi'_{sq} \end{bmatrix} \tag{3.482}$$

yields

$$2\pi j \widetilde{J}_s = \int_{-j\infty}^{j\infty} \text{Tr}[(v_s v_{s*}]ds = \int_{-j\infty}^{j\infty} \text{Tr}[(v_{sa} - \Phi_s \text{vec } Z)(v_{sa} - \Phi_s \text{vec } Z)_* ds. \tag{3.483}$$

Equation (3.483) has the same form as (3.174) when the identifications

$$\nabla_a = v_{sa},\ \nabla_b = \Phi_s,\ \nabla_c = 1,\ \widetilde{K}_m = \text{vec } Z \tag{3.484}$$

are made. Since $\nabla_{sak}, \nabla_{sbk}$, and ∇_{sck} are stable for $k = 1 \to q$ and ∇_{sak} is strictly proper for $k = 1 \to q$, it is not difficult to confirm from (3.480) that ∇_a and ∇_b are stable and it is obvious that $\nabla_c = 1$ is also. Clearly, ∇_a is strictly proper and $\det(\nabla_c \nabla_{c*}) = 1 \neq 0$ on the finite $j\omega$-axis. Moreover,

$$\nabla_{b*} \nabla_b = \Phi_{s*} \Phi_s = \sum_{k=1}^{q} \Phi_{sk*} \Phi_{sk} \tag{3.485}$$

is analytic on the finite $j\omega$-axis. When $\Phi_{s*}\Phi_s$ is also nonsingular on the finite $j\omega$-axis, it should be clear that the four assumptions in Theorem 3.4 are satisfied. In this case, one can construct Wiener–Hopf spectral factors $\widetilde{\Lambda}$ and $\widetilde{\Omega}$ that are square and together with their inverses are analytic in Re $s \geq 0$ such that

$$\nabla_{b*}\nabla_b = \Phi_{s*}\Phi_s = \widetilde{\Lambda}_*\widetilde{\Lambda} \tag{3.486}$$

and

$$\nabla_c\nabla_{c*} = \widetilde{\Omega}\widetilde{\Omega}_* = 1 \quad \Leftrightarrow \quad \widetilde{\Omega} = 1. \tag{3.487}$$

It then follows from (3.173) that the choice for vec Z which minimizes $\widetilde{J}_s$ is given by

$$\text{vec } Z_o = \widetilde{K}_{mo} = \widetilde{\Lambda}^{-1}\{\widetilde{\Gamma}\}_+ \tag{3.488}$$

where

$$\widetilde{\Gamma} = \widetilde{\Lambda}_*^{-1}\nabla_{b*}\nabla_a\nabla_{c*}\Omega_*^{-1} = \widetilde{\Lambda}_*^{-1}\Phi_{s*}v_{sa}. \tag{3.489}$$

This is so because the choice for $\widetilde{K}_m$ which minimizes J in Theorem 3.4 is obtained by setting the matrix parameter Z in (3.173) to zero.

That $\Phi_{s*}\Phi_s$ is nonsingular on the finite $j\omega$-axis is established next when in (3.473), as is the case in (3.472), $\nabla_{sq} = -\alpha Z$. Clearly, $\nabla_{sq} = -\alpha Z$ corresponds to

$$\nabla_{saq} = 0, \nabla_{sbq} = \alpha I, \nabla_{scq} = I. \tag{3.490}$$

Then from (3.480)

$$\Phi_{sq} = \nabla'_{scq} \otimes \nabla_{sbq} = I \otimes \alpha I = \alpha I \tag{3.491}$$

where the identity matrices are appropriately sized. It immediately follows that Φ_s has full column rank at all points in the finite s-plane; so, $\Phi_{s*}\Phi_s$ is indeed nonsingular on the finite $j\omega$-axis. Thus,

Theorem 3.6 *Under the assumptions*

(1) $\nabla_{sak}, \nabla_{sbk},$ and ∇_{sck} are stable for $k = 1 \rightarrow q$;
(2) ∇_{sak} is strictly proper for $k = 1 \rightarrow q$; *and*
(3) $\nabla_{sq} = -\alpha Z$;

the performance functional $\widetilde{J}_s$ *in* (3.473) *has the minimum value* $\widetilde{J}_{so}$ *when vec* Z *is given by* vec $Z_o = \widetilde{\Lambda}^{-1}\{\widetilde{\Gamma}\}_+$. *Specifically,* (3.483) *yields*

$$2\pi j \widetilde{J}_{so} = \int_{-j\infty}^{j\infty} \mathrm{Tr}[(v_{sa} - \Phi_s \widetilde{\Lambda}^{-1} \{\widetilde{\Gamma}\}_+)(v_{sa} - \Phi_s \widetilde{\Lambda}^{-1} \{\widetilde{\Gamma}\}_+)_*] ds \tag{3.492}$$

which simplifies to

$$2\pi j \widetilde{J}_{so} = \int_{-j\infty}^{j\infty} (v_{sa*} v_{sa} - \{\widetilde{\Gamma}\}_{+*} \{\widetilde{\Gamma}\}_+) ds. \tag{3.493}$$

Moreover, vec Z_o is strictly proper and stable assuring that ΔJ in (3.439) is finite for $Z = Z_o$.

Proof It only remains to show that vec Z_o is strictly proper and stable and that (3.492) simplifies to (3.493). Since Λ^{-1}, Φ_s, and v_{sa} are stable, it follows that $\widetilde{\Gamma}$ is good; hence, $\{\widetilde{\Gamma}\}_+$ is stable and it follows from (3.488) that so is vec Z_o. Then,

$$\nabla_{so} = v_{sa} - \Phi_s \mathrm{vec}\, Z_o = v_{sa} - \Phi_s \widetilde{\Lambda}^{-1} \{\widetilde{\Gamma}\}_+ \tag{3.494}$$

is also stable. Moreover, from

$$\widetilde{\Lambda}_*^{-1} \Phi_{s*} \Phi_s \widetilde{\Lambda}^{-1} = I \tag{3.495}$$

it follows that $\Phi_s \widetilde{\Lambda}^{-1}$ and $\widetilde{\Lambda}_*^{-1} \Phi_{s*}$ are proper. Hence, from $\Phi_{sq} \widetilde{\Lambda}^{-1} = \alpha \widetilde{\Lambda}^{-1}$ it is clear that $\widetilde{\Lambda}^{-1}$ is proper. So vec Z_o and ∇_{so} are not only stable, they are also strictly proper; this assures that both $\widetilde{J}_{so}$ and ΔJ with $Z = Z_o$ are finite. Since v_{sa} is strictly proper, it is also true that $\widetilde{\Gamma} = \widetilde{\Lambda}_*^{-1} \Phi_{s*} v_{sa}$ is strictly proper. Thus,

$$\widetilde{\Gamma} = \{\widetilde{\Gamma}\}_+ + \{\widetilde{\Gamma}\}_- \tag{3.496}$$

and

$$\{\widetilde{\Gamma}\}_{-*} \{\widetilde{\Gamma}\}_+ = [\{\widetilde{\Gamma}\}_{+*} \{\widetilde{\Gamma}\}_-]_* \lesssim 0(s^{-2}) \tag{3.497}$$

is analytic in Re $s \geq 0$. It then follows that

$$\int_{-j\infty}^{j\infty} (\{\widetilde{\Gamma}\}_{-*} \{\widetilde{\Gamma}\}_+ + \{\widetilde{\Gamma}\}_{+*} \{\widetilde{\Gamma}\}_-) ds = 0; \tag{3.498}$$

hence,

$$\left.\begin{aligned} 2\pi j\widetilde{J}_{so} &= \int_{-j\infty}^{j\infty} (v_{sa*} - \{\widetilde{\Gamma}\}_{+*}\widetilde{\Lambda}_{*}^{-1}\Phi_{s*})(v_{sa} - \Phi_s\widetilde{\Lambda}^{-1}\{\widetilde{\Gamma}\}_{+})ds \\ &= \int_{-j\infty}^{j\infty} (v_{sa*}v_{sa} - \widetilde{\Gamma}_{*}\{\widetilde{\Gamma}\}_{+} - \{\widetilde{\Gamma}\}_{+*}\widetilde{\Gamma} + \{\widetilde{\Gamma}\}_{+*}\{\widetilde{\Gamma}\}_{+})ds \\ &= \int_{-j\infty}^{j\infty} (v_{sa*}v_{sa} - \{\widetilde{\Gamma}\}_{+*}\{\widetilde{\Gamma}\}_{+})ds \end{aligned}\right\} \tag{3.499}$$

and the proof is complete.

Clearly, both ΔJ and the sum of the first $q-1$ terms on the right hand side of (3.473) given by

$$\widetilde{J}_{uo} = \widetilde{J}_{so} - \alpha^2 \Delta J \tag{3.500}$$

are functions of α^2 when $Z = Z_o$. Using the same arguments as in Appendix A of Bongiorno et al. (1997), it easily follows that when $\Delta J \leq N$ for all $\alpha^2 \geq \alpha_n^2$, the minimum J_{uo} subject to the constraint $\Delta J \leq N$ is realized with the choice $\alpha^2 = \alpha_n^2$. It is possible that $\alpha_n^2 = 0$; this is the case when $Z = Z_o$ minimizes J_u (Z) and $\Delta J(Z_o) \leq N$.

3.10 Historical Perspective and Commentary

The methodology presented in this chapter is based on Park and Bongiorno (2009) which begins with the identification of the acceptable persistent exogenous inputs for which there exists a parameterization of the subset of stabilizing controllers such that the error transform is stable and the actual plant inputs remain bounded with time. The optimal $\mathcal{H}_2$ solution is then obtained from within the subset of free parameters for which the cost functional is finite. A key element in this approach which distinguishes it from the earlier works is the explicit recognition and exploitation of the requirement that the controller output should be stable. Specifically, physical considerations are invoked to justify imposition of the requirement that the deterministic components of both the regulated variable y_δ and the controller output v be stable.

A necessary and sufficient condition for only the regulated variable y_δ to be stable is derived in Cheng and Pearson (1978, 1981) for the standard configuration and in Wolovich and Ferreira (1979) and Saeks and Murray (1981) for the conventional feedback configuration. So if one uses the results in these papers, there would be no guarantee that v is stable. To assure that this was the case, the elements of v would have to be included in y_δ. This would impose restrictions on the generalized plant that were not explicitly exploited in these works where, moreover, no attempt to find optimal $\mathcal{H}_2$ solutions was made. Here, the requirement that v be stable is dealt with directly instead.[2]

The analog 2DOF standard configuration with persistent inputs is treated in Corrêa and Da Silveira (1995), Xie et al. (2000), and Park et al. (2002). The first two papers allow persistent deterministic exogenous inputs with poles in Re $s \geq 0$ where the plant has none and insist on asymptotic tracking with zero steady-state error. In this case, the plant input must have poles in Re $s \geq 0$ and the $\mathcal{H}_2$-norm of the plant input is not finite. Corrêa and Da Silveira (1995) circumvent this issue by only incorporating the stable part of the plant input into the system performance measure. Xie et al. (2000) circumvent this issue by only incorporating the $\mathcal{H}_2$-norm of the transfer matrix between the reference input and the plant input into the system performance measure. As a consequence of these approaches, however, plant input saturation can become a problem. In Park et al. (2002) on the other hand, the only persistent deterministic exogenous inputs considered are ones with poles on the imaginary axis that the plant is capable of tracking with a stable plant input. In this case, the $\mathcal{H}_2$-norm of the plant input is finite and can be included in the system performance measure to avoid plant input saturation and, moreover, it is possible to make a connection between state-space and Wiener–Hopf design formulas.

In the papers by Doyle et al. (1989), Khargonekar and Rotea (1991), and Hunt et al. (1994), the standard configuration exogenous input is assumed to be a white noise or simply an indeterminate to indicate an input channel. In this setting, the dynamics of the persistent inputs are absorbed into the generalized plant block. However, in these papers certain stabilizability assumptions are made which lead to a distinct disadvantage: as explained in Remark 3.2, the results are not applicable to 2DOF and 3DOF systems with unstable reference inputs. In Park and Bongiorno (1989), this problem is avoided by keeping the model of the persistent inputs separate from the generalized plant; however, persistent inputs with unstable poles only on the stability boundary can be accommodated. Specifically, a parameterization of all stabilizing controllers that yield finite cost in these cases is established and the one for which the cost is minimized is identified. However, for shape-deterministic inputs with poles in Re $s > 0$, the regulated variables even for a stabilized system can have poles in Re $s > 0$. In this case, Parseval's formula does not hold; i.e., when $z(s)$ has poles in Re $s > 0$,

$$\int_{0}^{\infty} Tr[z(t)z^{*}(t)]dt \neq \frac{1}{2\pi j}\int_{-j\infty}^{j\infty} Tr[z(s)z_{*}(s)]ds. \tag{3.501}$$

[2]© 2009 Taylor & Francis Ltd, http://www.tandfonline.com. This paragraph is reprinted, with permission, from Park and Bongiorno (2009).

So the frequency-domain cost functional that was minimized in Park and Bongiorno (1989) has no meaningful physical significance in these cases.[3]

Additional insights regarding the material presented in this chapter are contained in the reply by Bongiorno and Park (2010) to comments by Aliev and Larin (2010).

3.11 Examples

Example 3.1 Confirm that the K given by (3.36) in Theorem 3.1 is stable for all stable K_g when Γ_w, Γ_o, and $A_1^{-1}A_{o1}$ are stable and G is compatible. Moreover, for all such K show that the matrices Ψ and Φ given by (3.20) and (3.21) are stable and, therefore, so are $G_v G_w = \Psi B_w$ and $G_\delta G_w = \Phi B_w$.

Solution Since $A_1^{-1}A_{o1}$ and Γ_o are stable and since Y_{g1} and A_g are polynomial matrices, the matrix K given by (3.36) is obviously stable for all stable K_g. Using (3.36) in (3.20) one gets

$$\begin{aligned}
\Psi &= (Y + A_{o1}\Gamma_o Y_{g1} + A_1 K_g A_g)A_g^{-1}B_g \\
&= (YB_{g1} + A_{o1}\Gamma_o Y_{g1}B_{g1})A_{g1}^{-1} + A_1 K_g B_g \\
&= [YB_{g1} + (Y_o\Gamma_w A_{g1} - YB_{g1})(I - X_{g1}A_{g1})]A_{g1}^{-1} + A_1 K_g B_g \\
&= [Y_o\Gamma_w - (Y_o\Gamma_w A_{g1} - YB_{g1})X_{g1}] + A_1 K_g B_g \\
&= Y_o\Gamma_w - A_{o1}\Gamma_o X_{g1} + A_1 K_g B_g
\end{aligned}$$

which is stable for all stable K_g when Γ_w and Γ_o are stable. Clearly, (3.21) then yields

$$\begin{aligned}
\Phi &= G_{11}A_w^{-1} + G_{12}\Psi = A_o^{-1}\Gamma_w - B_{o1}A_{o1}^{-1}(Y_o\Gamma_w - A_{o1}\Gamma_o X_{g1} + A_1 K_g B_g) \\
&= A_o^{-1}(I - B_o Y_o)\Gamma_w + B_{o1}(\Gamma_o X_{g1} - A_{o1}^{-1}A_1 K_g B_g) = X_o\Gamma_w + B_{o1}(\Gamma_o X_{g1} - A_{o1}^{-1}A_1 K_g B_g).
\end{aligned}$$

Since G is compatible, $G_{12}A_1 = -B_{o1}A_{o1}^{-1}A_1$ is stable or, equivalently, $A_{o1}^{-1}A_1$ is stable. Hence, Φ is stable when K_g, Γ_w, and Γ_o are stable. It is also of interest to note with $K_g = A_1^{-1}A_{o1}K_m$ that

$$\Phi = X_o\Gamma_w - B_{o1}(-\Gamma_o X_{g1} + K_m B_g) = X_o\Gamma_w - B_{o1}K_w$$

when (3.34) is recalled which is in agreement with (3.26).

Example* 3.2 Using (3.14) and (3.15), show directly that G_vG_w and $G_\delta G_w$ are stable for all stable K when G is compatible and Γ_w, Γ_o, and $AG_{21}A_w^{-1}$ are stable. That is, confirm Remark 3.5.

Solution Since G_v and G_δ are stable for all stable K when G is compatible, it suffices to show that $G_vA_w^{-1}$ and $G_\delta A_w^{-1}$ are stable. It is obvious from $G_vA_w^{-1} = (Y+A_1K)AG_{21}A_w^{-1}$ that $G_vA_w^{-1}$ is stable for all stable K when $AG_{21}A_w^{-1}$ is stable. Now the last term on the right-hand side of

$$G_\delta A_w^{-1} = G_{11}A_w^{-1} + G_{12}A_1Y_1G_{21}A_w^{-1} + G_{12}A_1KAG_{21}A_w^{-1}$$

is stable for all stable K when $AG_{21}A_w^{-1}$ is stable since $G_{12}A_1$ is stable when G is compatible. So with the help of (3.32) one gets that $G_\delta A_w^{-1}$ is stable provided

$$\begin{aligned} Q = G_{11}A_w^{-1} + G_{12}A_1Y_1G_{21}A_w^{-1} &= A_o^{-1}\Gamma_w + G_{12}YAG_{21}A_w^{-1} \\ = A_o^{-1}\Gamma_w + G_{12}YB_{g1}A_{g1}^{-1} &= A_o^{-1}\Gamma_w + G_{12}(Y_o\Gamma_wA_{g1} - A_{o1}\Gamma_o)A_{g1}^{-1} \\ = A_o^{-1}(I - B_oY_o)\Gamma_w + B_{o1}\Gamma_oA_{g1}^{-1} &= X_o\Gamma_w + B_{o1}\Gamma_oA_{g1}^{-1} \end{aligned}$$

is stable. This is indeed the case because $AG_{21}A_w^{-1} = B_{g1}A_{g1}^{-1}$ is stable iff A_{g1}^{-1} is stable.

Example* 3.3 Establish that $\Gamma_w = A_oG_{11}A_w^{-1}$ is stable when G is compatible iff $\Gamma_wB_w = A_oG_{11}G_w$ is stable. Show that Γ_o is stable when Γ_w is stable iff $\Gamma_z = (G_{11} + G_{12}A_1Y_1G_{21})A_w^{-1}A_{g1}$ is stable.

Solution Since G is compatible,

$$M = G_{11} + G_{12}A_1Y_1G_{21} = G_{11} + G_{12}YAG_{21} = G_{11} - A_o^{-1}B_oYAG_{21}$$

is stable. Hence, $A_oG_{11} = A_oM + B_oYAG_{21}$ is stable because AG_{21} is stable. Since A_w, B_w is a left-coprime polynomial matrix pair, it immediately follows that $\Gamma_w = A_oG_{11}A_w^{-1}$ is stable iff $A_oG_{11}A_w^{-1}B_w = A_oG_{11}G_w$ is stable. So one can check whether or not Γ_w is stable without the need to calculate A_w and B_w. When Γ_w is stable, $\Gamma_o = A_{o1}^{-1}\widetilde{\Gamma}_o$ where $\widetilde{\Gamma}_o = Y_o\Gamma_wA_{g1} - YB_{g1}$ is stable. Hence, Γ_o is stable iff

$$\begin{aligned} B_{o1}\Gamma_o &= -G_{12}(Y_o\Gamma_wA_{g1} - YB_{g1}) = -G_{12}(Y_o\Gamma_w - YAG_{21}A_w^{-1})A_{g1} \\ &= -G_{12}(Y_oA_oG_{11} - YAG_{21})A_w^{-1}A_{g1} = (A_o^{-1}B_oY_oA_oG_{11} + G_{12}YAG_{21})A_w^{-1}A_{g1} \\ &= [A_o^{-1}(I - A_oX_o)A_oG_{11} + G_{12}YAG_{21}]A_w^{-1}A_{g1} = (G_{11} + G_{12}A_1Y_1G_{21} - X_oA_oG_{11})A_w^{-1}A_{g1} \\ &= \Gamma_z - X_o\Gamma_wA_{g1} \end{aligned}$$

is stable. Clearly, this is the case iff Γ_z is stable.

Example 3.4 Confirm that the pair Φ, Ψ given by (3.26) satisfies (3.21).

Solution It suffices to show that the Φ given in (3.26) is obtained from (3.21) when the expression for Ψ given in (3.26) is used in (3.21). Clearly, this is the case from

$$\Phi = G_{12}\Psi + G_{11}A_w^{-1} = -B_{o1}A_{o1}^{-1}(Y_o\Gamma_w + A_{o1}K_w) + A_o^{-1}\Gamma_w$$
$$= -B_{o1}Y_{o1}A_o^{-1}\Gamma_w - B_{o1}K_w + A_o^{-1}\Gamma_w = (X_oA_o - I)A_o^{-1}\Gamma_w - B_{o1}K_w + A_o^{-1}\Gamma_w = X_o\Gamma_w - B_{o1}K_w.$$

Example 3.5 A matrix G_w is said to be matched to G if the ratio of characteristic denominators $\psi_{G_{22}}/\psi_{G_{11}G_w}$ is stable. Establish that a necessary condition for G_w to be acceptable for G when G is compatible is that G_w be matched to G. **Hint:** Begin with $G_\delta G_w = G_{11}G_w + G_{12}G_vG_w$ and use the fact that when G is compatible then $G_{12} = MA_1^{-1}$ where M is a stable matrix.

Solution When G is compatible and G_w is acceptable, there exists a stable K for which G_vG_w and $G_\delta G_w = G_{11}G_w + G_{12}G_vG_w$ are stable. Moreover, from

$$G_{11}G_w = G_\delta G_w - G_{12}G_vG_w = G_\delta G_w - MA_1^{-1}G_vG_w$$

it follows for any finite p_o in the unstable region of the complex plane that $\delta(G_{11}G_w; p_o) \leq \delta(A_1^{-1}; p_o)$ because $G_\delta G_w$, G_vG_w, and M are stable. That is, $\psi_{G_{22}}/\psi_{G_{11}G_w}$ is stable since to within a nonzero multiplicative constant $\psi_{G_{22}}$ is given by det A_1. So G_w is matched to G.

Example* 3.6 Establish that the polynomial matrix pair

$$A_q = \begin{bmatrix} A_m & 0 \\ 0 & A_n \end{bmatrix}, \quad B_q = \begin{bmatrix} B_m \\ B_n \end{bmatrix}$$

is left-coprime when A_m, B_m and A_n, B_n are left-coprime pairs and det A_m and det A_n have no common zeros.

Solution At all points where det $A_m \neq 0$ and det $A_n \neq 0$,

$$[A_q \quad B_q] = \left[\begin{array}{cc|c} A_m & 0 & B_m \\ 0 & A_n & B_n \end{array}\right]$$

has full row rank. When det $A_m \neq 0$ and det $A_n = 0$,

$$[x'_m \quad x'_n]\begin{bmatrix} A_m & 0 & B_m \\ 0 & A_n & B_n \end{bmatrix} = 0' \quad \Rightarrow x'_m A_m = 0' \quad \Rightarrow x'_m = 0'$$

and

$$x'_n[A_n \quad B_n] = 0' \quad \Rightarrow x'_n = 0'$$

because A_n, B_n is left-coprime. Hence, $[A_q \quad B_q]$ has full row rank at points where $\det A_m \neq 0$ and $\det A_n = 0$. Similarly, when $\det A_m = 0$ and $\det A_n \neq 0$, $[A_q \quad B_q]$ has full row rank because A_m, B_m is left-coprime. Since $\det A_m = 0$ and $\det A_n = 0$ simultaneously is ruled out, the pair A_q, B_q is left-coprime.

Typically, for the A_m and A_n defined by (3.89), $\det A_m$ and $\det A_n$ have no common zeros for the following reasons. First, A_d^{-1} usually has only unstable poles. But M in (3.89) is stable so none of these poles are poles of both A_m^{-1} and A_n^{-1}. Second, it would almost always be true that L and AFP_d have no stable poles in common. Thus, A_m^{-1} and A_n^{-1} would have no poles in common and, consequently, $\det A_m$ and $\det A_n$ would have no zeros in common.

Example* 3.7 Confirm Remark 3.6.

Solution With $\mu = 0$ and $\widetilde{G}_{d1} = \widetilde{T}_d \widetilde{G}_{e1}$ one gets from (3.3) and (3.4) that

$$y_\delta = (\widetilde{G}_{11} - \widetilde{T}_d)\widetilde{G}_{e1}\sigma + \widetilde{G}_{12}v = G_{11}\widetilde{G}_{e1}\sigma + G_{12}v$$
$$y_m = \widetilde{G}_{21}\widetilde{G}_{e1}\sigma + \widetilde{G}_{22}v = G_{21}\widetilde{G}_{e1}\sigma + G_{22}v$$

which is identical in form with the final expressions in (3.3) and (3.4) except that now $\widetilde{G}_{e1}\sigma$ is in place of $\widetilde{G}_w q$ and instead of (3.5) one has (3.45). It now follows in straightforward fashion that (3.12) and (3.13) become, respectively,

$$v = G_v\widetilde{G}_{e1}\sigma, \; y_\delta = G_\delta\widetilde{G}_{e1}\sigma$$

where G_v and G_δ are given by (3.14) and (3.15) with the G_{ij} given by (3.45). Moreover, as pointed out in Remark 3.6, when $\widetilde{G}$ is admissible and $\widetilde{T}_d$ is stable, then AG_{21}, $G_{12}A_1$, $G_{11} + G_{12}A_1Y_1G_{21}$ are all stable. So with $\widetilde{G}_{e1}$ in place of G_w the same arguments used to establish the circumstances under which G_vG_w and $G_\delta G_w$ are stable again apply. That is, when $\widetilde{G}_{e1} = G_w$, $\widetilde{T}_d$ is stable, and the G_{ij} are given by (3.45), Theorem 3.1 still holds.

Example* 3.8 Establish directly that the ∇_a given by (3.160) is stable when G is compatible and G_w is acceptable for G.

Solution From (3.160),

$$\nabla_a = WG_{11}G_w - U_b(\{\Lambda\widetilde{\Gamma}_o\Omega\}_+ + \{\Lambda\widetilde{\Gamma}_o\Omega\}_-)U_c$$

or, equivalently,

$$\nabla_a = WG_{11}G_w - U_b(\Lambda\widetilde{\Gamma}_o\Omega - \{\Lambda\widetilde{\Gamma}_o\Omega\}_\infty)U_c.$$

Since U_b, U_c, and $\{\Lambda\widetilde{\Gamma}_o\Omega\}_\infty$ are stable, it follows that ∇_a is stable iff

$$\widetilde{\nabla}_a = WG_{11}G_w - U_b\Lambda\widetilde{\Gamma}_o\Omega U_c$$

is stable. Using (3.148), (3.152), and (3.153) one gets

$$\begin{aligned}\widetilde{\nabla}_a &= WG_{11}G_w - WB_{o1}A_{o1}^{-1}(Y + A_{o1}\Gamma_o Y_{g1})A_g^{-1}B_gB_w \\ &= W(G_{11}A_w^{-1} - B_{o1}A_{o1}^{-1}YA_g^{-1}B_g - B_{o1}\Gamma_o Y_{g1}A_g^{-1}B_g)B_w.\end{aligned}$$

Recalling (3.25) then yields

$$\widetilde{\nabla}_a = W(A_o^{-1}\Gamma_w - B_{o1}A_{o1}^{-1}YB_{g1}A_{g1}^{-1} - B_{o1}\Gamma_o A_{g1}^{-1}Y_gB_g)B_w.$$

Introducing

$$A_o^{-1} = X_o + A_o^{-1}B_oY_o = X_o + B_{o1}A_{o1}^{-1}Y_o$$

one gets

$$\widetilde{\nabla}_a = W[X_o\Gamma_w + B_{o1}A_{o1}^{-1}(Y_o\Gamma_wA_{g1} - YB_{g1})A_{g1}^{-1} - B_{o1}\Gamma_oA_{g1}^{-1}Y_gB_g]B_w.$$

With the aid of (3.32) it follows that

$$\widetilde{\nabla}_a = W[X_o\Gamma_w + B_{o1}\Gamma_oA_{g1}^{-1}(I - Y_gB_g)]B_w = W(X_o\Gamma_w + B_{o1}\Gamma_oX_{g1})B_w.$$

Since Γ_w and Γ_o are stable when G is compatible and G_w is acceptable for G, it is clear that $\widetilde{\nabla}_a$ and, therefore, ∇_a are stable in this case.

Example* 3.9 In Sects. 3.5 and 3.6 on the $\mathcal{H}_2$ design of digital and analog multivariable control systems, it is assumed that G_w is acceptable and Theorem 3.1 is invoked with $G_w = A_w^{-1}B_w$. When $\widetilde{G}_{w2}$ is stable, however, G_w is acceptable iff $\widetilde{G}_{w1}$ is acceptable. Hence, one can use Theorem 3.1 with $\widetilde{G}_{w1} = A_w^{-1}B_w$ instead. (a) With $A_1^{-1}A_{o1}$ stable derive the alternative expressions to (3.107) for digital systems and to (3.150) and (3.160) for analog systems which one now gets for ∇_a, ∇_b, and ∇_c. Confirm that ∇_a, ∇_b, and ∇_c are stable and give conditions that assure ∇_a is strictly proper in the analog case. (b) When W is a polynomial matrix, so are the ∇_b and ∇_c given by (3.107) and (3.150). Is this still true for the ∇_b and ∇_c found in (a)? What other implications are there?

Solution (a) All of the results in Sect. 3.3 still hold, but with A_w, B_w now the left-coprime polynomial matrix pair associated with $\widetilde{G}_{w1}$. Again

$$\nabla = WG_\delta G_w = WG_\delta \big[\sqrt{\alpha_1}\widetilde{G}_{w1} \quad \sqrt{\alpha_2}\widetilde{G}_{w2}\big]$$

where

$$G_\delta = G_{11} + G_{12}A_1(Y_1 + KA)G_{21}.$$

Using (3.36) for K yields

$$G_\delta = G_a + G_{12}A_{o1}K_m A_g A G_{21}$$

where

$$G_a = G_{11} + G_{12}A_1 Y_1 G_{21} + G_{12}A_{o1}\Gamma_o Y_{g1} A G_{21}.$$

Thus,

$$\begin{aligned}\nabla = WG_\delta G_w &= WG_a G_w + WG_{12}A_{o1}K_m A_g A G_{21} G_w \\ &= WG_a G_w - WB_{o1}K_m \big[\sqrt{\alpha_1}A_g A G_{21}\widetilde{G}_{w1} \quad \sqrt{\alpha_2}A_g A G_{21}\widetilde{G}_{w2}\big].\end{aligned}$$

Now

$$A_g A G_{21}\widetilde{G}_{w1} = A_g(AG_{21}A_w^{-1})B_w = A_g(A_g^{-1}B_g)B_w = B_g B_w$$

and

$$A_g A G_{21}\widetilde{G}_{w2} = A_g(AG_{21}A_w^{-1})A_w\widetilde{G}_{w2} = A_g(A_g^{-1}B_g)A_w\widetilde{G}_{w2} = B_g A_w \widetilde{G}_{w2}.$$

Hence, for the digital case

$$\nabla = \nabla_a - \nabla_b K_m \nabla_c$$

where

$$\nabla_a = WG_a G_w, \ \nabla_b = WB_{o1}, \ \nabla_c = \big[\sqrt{\alpha_1}B_g B_w \quad \sqrt{\alpha_2}B_g A_w \widetilde{G}_{w2}\big].$$

Clearly, ∇_b and ∇_c are stable when W and $\widetilde{G}_{w2}$ are stable. Also, since ∇ is stable for $K_m \equiv 0$, it follows that ∇_a is stable and Theorem 3.2 can be applied.

For the analog case, one can again begin with (3.146) which leads to

$$\nabla = WG_{11}G_w - \nabla_b(\widetilde{\Gamma}_o + K_m)\nabla_c$$

where $\widetilde{\Gamma}_o$ is given by (3.148). Proceeding as previously done in Sect. 3.6 one now easily gets $\nabla = \nabla_a - \nabla_b \widetilde{K}_m \nabla_c$ where

$$\nabla_b = WB_{o1}, \ \nabla_c = \left[\sqrt{\alpha_1} B_g B_w \quad \sqrt{\alpha_2} B_g A_w \widetilde{G}_{w2} \right],$$

and ∇_a is now defined by (3.160). Clearly, ∇_b and ∇_c are stable when W and $\widetilde{G}_{w2}$ are stable. Moreover, ∇_a is strictly proper when $WG_{11}G_w \leq 0(s^{-1})$ and is stable because ∇ is stable when $\widetilde{K}_m \equiv 0$. Hence, Theorem 3.4 can be applied. (b) Obviously, $\nabla_b = WB_{o1}$ is still a polynomial matrix when W is. However, ∇_c may no longer be because $B_g A_w \widetilde{G}_{w2}$ may not be: $\widetilde{G}_{w2}$ is usually a proper rational matrix. Thus, the spectral factorization $\nabla_c \nabla_{c*} = \Omega\Omega_*$ now typically involves factoring a rational matrix instead of a polynomial matrix. Moreover, the expression for ∇_c which was $\nabla_c = B_g B_w$ in Sects. 3.5 and 3.6 is now the more complicated one $\nabla_c = \left[\sqrt{\alpha_1} B_g B_w \quad \sqrt{\alpha_2} B_g A_w \widetilde{G}_{w2} \right]$. On the other hand, there is a significant benefit from the fact that finding literal expressions for the coprime polynomial matrix A_w, B_w is possible as in Sects. 3.7 and 3.8.

Example[*] 3.10 The purpose of this example is to provide additional insights regarding $\mathcal{H}_2$ optimization for the analog case. The obvious modifications for the digital case are omitted. The performance functional considered is

$$J = \frac{1}{2\pi j} \int_{-j\infty}^{j\infty} \rho ds = \frac{1}{2\pi j} \int_{-j\infty}^{j\infty} \mathrm{Tr}[(\nabla_a - \nabla_b Q \nabla_c)(\nabla_a - \nabla_b Q \nabla_c)_*] ds$$

where ∇_a, ∇_b, and ∇_c are real rational matrices and no stability requirement on $\nabla = \nabla_a - \nabla_b Q \nabla_c$ is imposed initially. It is assumed that $G_b = \nabla_{b*}\nabla_b$ and $G_c = \nabla_c \nabla_{c*}$ are full rank matrices. (a) Determine the necessary and sufficient condition for which there exists a parameterization of all marginally stable (analytic in Re $s > 0$) Q for which J is finite or, equivalently, $\nabla = \nabla_a - \nabla_b Q \nabla_c$ is strictly proper and good and give the parameterization for Q. (b) Under the condition found in (a), which choice for Q minimizes J? (c) When G_b and G_c in addition to being of full rank are such that G_b^{-1} and G_c^{-1} are good and when ∇_a is strictly proper and good, establish that the necessary and sufficient condition found in (a) is satisfied and all the Q given by the parameterization are stable. (d) When the assumptions of Theorem 3.4 are satisfied, show that ∇ is strictly proper and stable and Q is stable as an alternative proof of Theorem 3.4 with Q in place of $\widetilde{K}_m$.

Solution (a) Since $G_b = \nabla_{b*}\nabla_b$ and $G_c = \nabla_c\nabla_{c*}$ are parahermitian-positive and have full rank, there exist by Theorem C.1 spectral factorizations $\nabla_{b*}\nabla_b = \Lambda_*\Lambda$ and $\nabla_c\nabla_{c*} = \Omega\Omega_*$ such that the square matrices Λ and Ω together with their inverses are marginally stable. Moreover, the paraunitary matrices $U_b = \nabla_b\Lambda^{-1}$ and $U_c = \Omega^{-1}\nabla_c$ are good and proper. Clearly, $Q = \Lambda^{-1}\widetilde{Q}\Omega^{-1}$ is marginally stable iff $\widetilde{Q}$ is and one can write

$$\nabla = \nabla_a - \nabla_b\Lambda^{-1}\widetilde{Q}\Omega^{-1}\nabla_c = \nabla_a - U_b\widetilde{Q}U_c.$$

Hence, with

$$\widetilde{\Gamma} = \Lambda_*^{-1}\nabla_{b*}\nabla_a\nabla_{c*}\Omega_*^{-1} = U_{b*}\nabla_a U_{c*}$$

it follows for finite J that

$$U_{b*}\nabla U_{c*} = U_{b*}\nabla_a U_{c*} - \widetilde{Q} = \widetilde{\Gamma} - \widetilde{Q}$$

must be strictly proper and good because ∇ must be strictly proper and good. Now for any marginally stable $\widetilde{Q}$ one can always write

$$\widetilde{Q} = Z + \{\widetilde{\Gamma}\}_\infty + \{\widetilde{\Gamma}\}_+$$

where Z is marginally stable. Clearly,

$$\widetilde{\Gamma} - \widetilde{Q} = \widetilde{\Gamma} - Z - \{\widetilde{\Gamma}\}_\infty - \{\widetilde{\Gamma}\}_+ = \{\widetilde{\Gamma}\}_- - Z$$

is strictly proper and good iff Z is. Since Z must be good in addition to being marginally stable, Z must be stable. Thus, for finite J it is necessary that Q be given by

$$Q = \Lambda^{-1}\widetilde{Q}\Omega^{-1} = \Lambda^{-1}(\{\widetilde{\Gamma}\}_+ + \{\widetilde{\Gamma}\}_\infty + Z)\Omega^{-1} = \Lambda^{-1}(\widetilde{\Gamma} - \{\widetilde{\Gamma}\}_- + Z)\Omega^{-1}$$

where Z is strictly proper and stable (i.e., admissible). With such a Q,

$$\nabla = \nabla_a - \nabla_b\Lambda^{-1}(\widetilde{\Gamma} - \{\widetilde{\Gamma}\}_- + Z)\Omega^{-1}\nabla_c = \nabla_o + \nabla_b\Lambda^{-1}(\{\widetilde{\Gamma}\}_- - Z)\Omega^{-1}\nabla_c$$

is strictly proper and good or, equivalently, J is finite iff

$$\nabla_o = \nabla_a - \nabla_b\Lambda^{-1}\widetilde{\Gamma}\Omega^{-1}\nabla_c$$

is strictly proper and good: the paraunitary matrices $U_b = \nabla_b \Lambda^{-1}$ and $U_c = \Omega^{-1}\nabla_c$ are proper and good and $\widetilde{Z} = \{\widetilde{\Gamma}\}_- - Z$ is strictly proper and good. That is a parameterization of all marginally stable Q exists for which J is finite iff ∇_o is strictly proper and good and the parameterization is given by $Q = \Lambda^{-1}(\{\widetilde{\Gamma}\}_+ + \{\widetilde{\Gamma}\}_\infty + Z)\Omega^{-1}$ where Z is a strictly proper and stable matrix.

(b) The choice of Q or, equivalently, Z for which J is minimized is obtained from the following considerations. Clearly,

$$\nabla = \nabla_o + \nabla_b \Lambda^{-1}\widetilde{Z}\Omega^{-1}\nabla_c.$$

Thus,

$$\begin{aligned}
\rho &= \mathrm{Tr}\,(\nabla\nabla_*) = \mathrm{Tr}\,\{[\nabla_o + \nabla_b\Lambda^{-1}\widetilde{Z}\Omega^{-1}\nabla_c][\nabla_{o*} + \nabla_{c*}\Omega_*^{-1}\widetilde{Z}_*\Lambda_*^{-1}\nabla_{b*}]\}\\
&= \mathrm{Tr}\,(\nabla_o\nabla_{o*} + \nabla_o\nabla_{c*}\Omega_*^{-1}\widetilde{Z}_*\Lambda_*^{-1}\nabla_{b*} + \nabla_b\Lambda^{-1}\widetilde{Z}\Omega^{-1}\nabla_c\nabla_{o*} + \nabla_b\Lambda^{-1}\widetilde{Z}\Omega^{-1}\nabla_c\nabla_{c*}\Omega_*^{-1}\widetilde{Z}_*\Lambda_*^{-1}\nabla_{b*})\\
&= \mathrm{Tr}\,(\nabla_o\nabla_{o*}) + \mathrm{Tr}\,(\Lambda_*^{-1}\nabla_{b*}\nabla_o\nabla_{c*}\Omega_*^{-1}\widetilde{Z}_*) + \mathrm{Tr}\,(\widetilde{Z}\Omega^{-1}\nabla_c\nabla_{o*}\nabla_b\Lambda^{-1}) + \mathrm{Tr}\,(\widetilde{Z}\Omega^{-1}\nabla_c\nabla_{c*}\Omega_*^{-1}\widetilde{Z}_*\Lambda_*^{-1}\nabla_{b*}\nabla_b\Lambda^{-1})\\
&= \mathrm{Tr}\,(\nabla_o\nabla_{o*}) + \mathrm{Tr}\,(\widetilde{Z}\widetilde{Z}_*)
\end{aligned}$$

because $\Omega^{-1}\nabla_c\nabla_{c*}\Omega_*^{-1} = I$, $\Lambda_*^{-1}\nabla_{b*}\nabla_b\Lambda^{-1} = I$, and

$$\begin{aligned}
\Lambda_*^{-1}\nabla_{b*}\nabla_o\nabla_{c*}\Omega_*^{-1} &= (\Omega^{-1}\nabla_c\nabla_{o*}\nabla_b\Lambda^{-1})_* = \Lambda_*^{-1}\nabla_{b*}\nabla_a\nabla_{c*}\Omega_*^{-1} - \widetilde{\Gamma} = \widetilde{\Gamma} - \widetilde{\Gamma}\\
&= 0.
\end{aligned}$$

Moreover,

$$\begin{aligned}
\rho_o &= \mathrm{Tr}\,(\nabla_o\nabla_{o*}) = \mathrm{Tr}[(\nabla_a - \nabla_b\Lambda^{-1}\widetilde{\Gamma}\Omega^{-1}\nabla_c)(\nabla_{a*} - \nabla_{c*}\Omega_*^{-1}\widetilde{\Gamma}_*\Lambda_*^{-1}\nabla_{b*})]\\
&= \mathrm{Tr}\,(\nabla_a\nabla_{a*} - \nabla_b\Lambda^{-1}\widetilde{\Gamma}\Omega^{-1}\nabla_c\nabla_{a*} - \nabla_a\nabla_{c*}\Omega_*^{-1}\widetilde{\Gamma}_*\Lambda_*^{-1}\nabla_{b*} + \nabla_b\Lambda^{-1}\widetilde{\Gamma}\Omega^{-1}\nabla_c\nabla_{c*}\Omega_*^{-1}\widetilde{\Gamma}_*\Lambda_*^{-1}\nabla_{b*})\\
&= \mathrm{Tr}\,(\nabla_a\nabla_{a*}) - \mathrm{Tr}\,(\widetilde{\Gamma}\Omega^{-1}\nabla_c\nabla_{a*}\nabla_b\Lambda^{-1}) - \mathrm{Tr}\,(\Lambda_*^{-1}\nabla_{b*}\nabla_a\nabla_{c*}\Omega_*^{-1}\widetilde{\Gamma}_*) + \mathrm{Tr}\,(\Lambda_*^{-1}\nabla_{b*}\nabla_b\Lambda^{-1}\widetilde{\Gamma}\Omega^{-1}\nabla_c\nabla_{c*}\Omega_*^{-1}\widetilde{\Gamma}_*)\\
&= \mathrm{Tr}\,(\nabla_a\nabla_{a*}) - \mathrm{Tr}\,(\widetilde{\Gamma}\widetilde{\Gamma}_*) - \mathrm{Tr}\,(\widetilde{\Gamma}\widetilde{\Gamma}_*) + \mathrm{Tr}\,(\widetilde{\Gamma}\widetilde{\Gamma}_*) = \mathrm{Tr}\,(\nabla_a\nabla_{a*}) - \mathrm{Tr}\,(\widetilde{\Gamma}\widetilde{\Gamma}_*).
\end{aligned}$$

In addition,

$$\begin{aligned}
\mathrm{Tr}\,(\widetilde{Z}\widetilde{Z}_*) &= \mathrm{Tr}\,[(\{\widetilde{\Gamma}\}_- - Z)(\{\widetilde{\Gamma}\}_- - Z)_*]\\
&= \mathrm{Tr}\,(\{\widetilde{\Gamma}\}_-\{\widetilde{\Gamma}\}_{-*}) - \mathrm{Tr}\,(\{\widetilde{\Gamma}\}_- Z_*) - \mathrm{Tr}\,(Z\{\widetilde{\Gamma}\}_{-*}) + \mathrm{Tr}\,(ZZ_*).
\end{aligned}$$

Also, since $\{\widetilde{\Gamma}\}_- Z_* \leq 0(s^{-2})$ and analytic in $\mathrm{Re}\ s \leq 0$ and since $Z\{\widetilde{\Gamma}\}_{-*} \leq 0(s^{-2})$ and analytic in $\mathrm{Re}\ s \geq 0$, it follows that

$$J = \frac{1}{2\pi j}\int_{-j\infty}^{j\infty} \mathrm{Tr}(\{\tilde{\Gamma}\}_{-} Z_*)ds = \frac{1}{2\pi j}\int_{-j\infty}^{j\infty} \mathrm{Tr}(Z\{\tilde{\Gamma}\}_{-*})ds = 0.$$

Hence,

$$J = \frac{1}{2\pi j}\int_{-j\infty}^{j\infty} \rho ds = \frac{1}{2\pi j}\int_{-j\infty}^{j\infty} \mathrm{Tr}(\nabla\nabla_*)ds = \frac{1}{2\pi j}\int_{-j\infty}^{j\infty} [\rho_o + \mathrm{Tr}(\tilde{Z}\tilde{Z}_*)]ds$$

$$= \frac{1}{2\pi j}\int_{-j\infty}^{j\infty} [\mathrm{Tr}(\nabla_a\nabla_{a*}) - \mathrm{Tr}(\tilde{\Gamma}\tilde{\Gamma}_*) + \mathrm{Tr}(\{\tilde{\Gamma}\}_{-}\{\tilde{\Gamma}\}_{-*}) + \mathrm{Tr}(ZZ_*)]ds.$$

Since $\mathrm{Tr}[Z(j\omega)Z_*(j\omega)] > 0$ for all $Z(j\omega) \not\equiv 0$, the minimum J is obtained iff $Z(s) \equiv 0$ and is given by

$$J_o = \frac{1}{2\pi j}\int_{-j\infty}^{j\infty} [\mathrm{Tr}(\nabla_a\nabla_{a*}) - \mathrm{Tr}(\tilde{\Gamma}\tilde{\Gamma}_*) + \mathrm{Tr}(\{\tilde{\Gamma}\}_{-}\{\tilde{\Gamma}\}_{-*})]ds$$

and the corresponding choice for Q is

$$Q_o = \Lambda^{-1}(\{\tilde{\Gamma}\}_+ + \{\tilde{\Gamma}\}_\infty)\Omega^{-1} = \Lambda^{-1}(\tilde{\Gamma} - \{\tilde{\Gamma}\}_-)\Omega^{-1}.$$

(c) When ∇_a is good and strictly proper, so are $\tilde{\Gamma}$ and ∇_o because the paraunitary matrices $U_b = \nabla_b\Lambda^{-1}$ and $U_c = \Omega^{-1}\nabla_c$ are proper and good. In addition, when G_b and G_c are of full rank and such that G_b^{-1} and G_c^{-1} are good then from $G_b^{-1} = \Lambda^{-1}\Lambda_*^{-1}$ and $G_c^{-1} = \Omega_*^{-1}\Omega^{-1}$ it follows that the square matrices Λ^{-1} and Ω^{-1} are good and, therefore, stable. Thus, one gets from $Q = \Lambda^{-1}(\{\tilde{\Gamma}\}_+ + \{\tilde{\Gamma}\}_\infty + Z)\Omega^{-1}$ that Q is stable since $\{\tilde{\Gamma}\}_+$ is stable when $\tilde{\Gamma}$ is good and since Z is stable. Moreover, $\{\tilde{\Gamma}\}_\infty = 0$ because $\tilde{\Gamma} \leq 0(s^{-1})$ and one gets the simplified expression $Q = \Lambda^{-1}(\{\tilde{\Gamma}\}_+ + Z)\Omega^{-1}$.

(d) Under the assumptions in Theorem 3.4, ∇_a is strictly proper and stable, ∇_b and ∇_c are stable, $\tilde{\Gamma}$ is strictly proper and good, and Λ^{-1} and Ω^{-1} are stable. Hence, when Z is admissible,

$$Q = \Lambda^{-1}(\{\widetilde{\Gamma}\}_+ + \{\widetilde{\Gamma}\}_\infty + Z)\Omega^{-1} = \Lambda^{-1}(\{\widetilde{\Gamma}\}_+ + Z)\Omega^{-1}$$

and

$$\nabla = \nabla_a - \nabla_b Q \nabla_c$$

are stable. Also,

$$\nabla = \nabla_a - \nabla_b \Lambda^{-1}(\{\widetilde{\Gamma}\}_+ + Z)\Omega^{-1}\nabla_c = \nabla_a - U_b(\{\widetilde{\Gamma}\}_+ + Z)U_c$$

is strictly proper because U_b and U_c are paraunitary matrices and ∇_a, $\{\widetilde{\Gamma}\}_+$, and Z are strictly proper.

Example 3.11 Establish for the 1DOF system of Fig. 3.3 using the notation of Sect. 3.7 that when

$$\nabla = WG_\delta G_w = \begin{bmatrix} I & 0 \\ 0 & W_q \end{bmatrix} G_\delta G_w$$

one gets in the integrand of the cost functional J

$$\begin{aligned}\operatorname{Tr}(\nabla\nabla_*) = \operatorname{Tr}[(T_d - PR)G_r(T_d - PR)_* + (P_d - PRP_o)G_d(P_d - PRP_o)_* \\ + (PR)G_n(PR)_*] + \operatorname{Tr}(kQRG_tR_*)\end{aligned}$$

where $R = (Y + A_1K)A = A_1(Y_1 + KA)$. Aside from notational differences, this expression is the same as the one used in Youla et al. (1976b).

Solution For the 1DOF system, Eqs. (3.225) and (3.226) lead to

$$\begin{aligned}G_\delta &= G_{11} + G_{12}A_1(Y_1 + KA)G_{21} = G_{11} + \widetilde{G}_{12}R\widetilde{G}_{21} \\ &= \begin{bmatrix} P_d & -T_d & 0 \\ 0 & 0 & 0 \end{bmatrix} + \begin{bmatrix} P \\ I \end{bmatrix} R[-P_o \quad I \quad -I]\end{aligned}$$

and

$$\begin{aligned}WG_\delta &= \begin{bmatrix} I & 0 \\ 0 & W_q \end{bmatrix} G_\delta = \begin{bmatrix} P_d & -T_d & 0 \\ 0 & 0 & 0 \end{bmatrix} + \begin{bmatrix} P \\ W_q \end{bmatrix} R[-P_o \quad I \quad -I] \\ &= \begin{bmatrix} (P_d - PRP_o) & (PR - T_d) & -PR \\ -W_qRP_o & W_qR & -W_qR \end{bmatrix}.\end{aligned}$$

Also, using (3.248), (3.280), and $G_w = \left[\sqrt{\alpha_1}\widetilde{G}_{w1} \quad \sqrt{\alpha_2}\widetilde{G}_{w2}\right]$ yields $G_w G_{w*} = \mathrm{diag}\{G_d, G_r, G_n\}$. Hence, $\mathrm{Tr}(\nabla\nabla_*) = \mathrm{Tr}[(WG_\delta)(G_w G_{w*})(WG_\delta)_*]$ leads to

$$\begin{aligned}\mathrm{Tr}(\nabla\nabla_*) &= \mathrm{Tr}\left\{\begin{bmatrix}(P_d - PRP_o)G_d & (PR - T_d)G_r & -PRG_n \\ -W_qRP_oG_d & W_qRG_r & -W_qRG_n\end{bmatrix}\begin{bmatrix}(P_d - PRP_o)_* & -(W_qRP_o)_* \\ (PR - T_d)_* & (W_qR)_* \\ -(PR)_* & -(W_qR)_*\end{bmatrix}\right\} \\ &= \mathrm{Tr}[(P_d - PRP_o)G_d(P_d - PRP_o)_* + (PR - T_d)G_r(PR - T_d)_* + PRG_n(PR)_*] \\ &\quad + \mathrm{Tr}[W_qR(P_oG_dP_{o*} + G_r + G_n)(W_qR)_*] \\ &= \mathrm{Tr}[(P_d - PRP_o)G_d(P_d - PRP_o)_* + (PR - T_d)G_r(PR - T_d)_* + PRG_n(PR)_*] + \mathrm{Tr}[kQRG_tR_*]\end{aligned}$$

when (3.279) is recalled.

Example[*] **3.12** (a) When the 1DOF system shown in Fig. 3.3 is admissible, a stabilizing controller is employed, and both d and n are zero, establish that

$$\varepsilon = (T_dA^{-1} - PY - PA_1K)Ar.$$

(b) When F and T_d are stable and when $r = A^{-1}\widetilde{S}_r$ where $\widetilde{S}_r$ is any stable vector, prove that ε is stable iff $(T_dF - I)P$ is stable. **Hint:** Establish the fact that MPY is stable when M is stable iff MP is stable using $YB = I - A_1X_1$. (c) For the reference inputs r of part (b), show that u and v are also stable for all stable K. (d) When $r = A_r^{-1}b_r$ is a left coprime polynomial matrix fraction description and AA_r^{-1} and T_d are stable, prove that ε is stable iff $(T_d - PA_1Y_1)A_r^{-1}$ is stable. (e) When F and T_d are stable, establish that the pair AA_r^{-1}, $(T_d - PA_1Y_1)A_r^{-1}$ is stable iff the pair $A_pT_dA_r^{-1}$, $(I - FT_d)A_r^{-1}$ is stable which is consistent with Lemma 3.4. Note: $M = (T_d - PA_1Y_1)A_r^{-1}$ is stable iff $T_d = MA_r + PA_1Y_1$ where M is a stable matrix and this provides a parameterization of all suitable stable T_d.

Solution (a) With the usual coprime polynomial matrix fraction descriptions for FP given by (3.58) and (3.59), any stabilizing controller is given by

$$C = (X_1 - KB)^{-1}(Y_1 + KA) = (Y + A_1K)(X - B_1K)^{-1}.$$

It then follows from

$$u = r - FPCu \Rightarrow u = (I + FPC)^{-1}r$$

that

$$\begin{aligned}u &= [I + A^{-1}B(Y + A_1K)(X - B_1K)^{-1}]^{-1}r \\ &= (X - B_1K)[A(X - B_1K) + B(Y + A_1K)]^{-1}Ar = (X - B_1K)Ar.\end{aligned}$$

Hence,

$$\begin{aligned} v &= Cu = (Y + A_1 K)Ar, \\ y &= PCu = P(Y + A_1 K)Ar, \end{aligned}$$

and it follows that

$$\varepsilon = T_d r - y = (T_d A^{-1} - PY - PA_1 K)Ar.$$

(b) Since PA_1 is stable because the system is admissible and K is stable because C is a stabilizing controller, it follows that ε is stable for all stable $Ar = \widetilde{S}_r$ iff

$$T_d A^{-1} - PY = T_d(X + FPY) - PY = T_d X + (T_d F - I)PY$$

is stable. Since $T_d X$ is stable, this is the case iff $(T_d F - I)PY$ is stable. Clearly, $(T_d F - I)P$ stable is sufficient. It is also necessary. For if $(T_d F - I)PY$ is stable then

$$(T_d F - I)PYB = (T_d F - I)P(I - A_1 X_1) = (T_d F - I)P - (T_d F - I)PA_1 X_1$$

must be stable. Thus, $(T_d F - I)P$ must be stable because T_d, F, and PA_1 are stable. (c) Clearly, $u = (X - B_1 K)\widetilde{S}_r$ and $v = (Y + A_1 K)\widetilde{S}_r$ are stable for all stable K. (d) The error ε is stable iff

$$\varepsilon = (T_d A^{-1} - PY)AA_r^{-1} b_r - PA_1 KAA_r^{-1} b_r = (T_d - PA_1 Y_1)A_r^{-1} b_r - PA_1 KAA_r^{-1} b_r$$

is stable. Since $PA_1 KAA_r^{-1} b_r$ is stable and $(T_d - PA_1 Y_1)$ is stable, ε is stable iff $(T_d - PA_1 Y_1)A_r^{-1}$ is stable. (e) When AA_r^{-1} and $(T_d - PA_1 Y_1)A_r^{-1}$ are stable, it easily follows from

$$A_p(T_d - PA_1 Y_1)A_r^{-1} = A_p T_d A_r^{-1} - B_p YAA_r^{-1}$$

that $A_p T_d A_r^{-1}$ must be stable. Also, from

$$\begin{aligned} F(T_d - PA_1 Y_1)A_r^{-1} &= (FT_d - B_1 Y_1)A_r^{-1} = (FT_d - I + XA)A_r^{-1} \\ &= (FT_d - I)A_r^{-1} + XAA_r^{-1} \end{aligned}$$

it follows that $(FT_d - I)A_r^{-1}$ must be stable. The converse that AA_r^{-1} and $(T_d - PA_1 Y_1)A_r^{-1}$ are stable when $M_1 = A_p T_d A_r^{-1}$ and $M_2 = (FT_d - I)A_r^{-1}$ are stable is established next. First, it immediately follows from (3.261) that AA_r^{-1} is stable. It also follows from (3.261) that

$$\begin{aligned}(T_d - PA_1Y_1)A_r^{-1} &= A_p^{-1}M_1 - PYAA_r^{-1} = A_p^{-1}M_1 - PYA(M_2 + FA_p^{-1}M_1) \\ &= (I - PYAF)A_p^{-1}M_1 - PA_1Y_1M_2 = (I - PA_1Y_1F)A_p^{-1}M_1 - PA_1Y_1M_2.\end{aligned}$$

Since $(I - PYAF) = (I - PA_1Y_1F)$ is stable and

$$(I - PYAF)A_p^{-1}B_p = (I - PYAF)P = P(I - YAFP) = P(I - YB) = PA_1X_1$$

is stable, it follows that $(I - PYAF)A_p^{-1}$ is stable and, therefore, so is $(T_d - PA_1Y_1)A_r^{-1}$.

Example* 3.13 (a) Establish for the analog 1DOF system treated in Sect. 3.7 that $\det(X_1 - KB) \not\equiv 0$ when K is given by (3.303) and when $(P_*P + kQ)^{-1} \leq 0(s^{2k_1})$, $G_t^{-1} \leq 0(s^{2k_2})$, and $FP \leq 0(s^{k_3})$ with $k_1 + k_2 + k_3 \leq 0$. (b) In this case, also establish that $C \leq 0(s^{k_1+k_2-1})$.

Solution (a) It suffices to establish that $A_1(X_1 - KB) \to I$ for $s \to \infty$. From (3.303),

$$K = \Lambda^{-1}\Gamma_a\Omega^{-1} - A_1^{-1}Y$$

where

$$\Gamma_a = Z + \{\Lambda_*^{-1}\Gamma_i\Omega_*^{-1}\}_+ + \{\Lambda A_1^{-1}Y\Omega\}_-$$

is strictly proper. Thus,

$$A_1(X_1 - KB) = A_1X_1 - A_1\Lambda^{-1}\Gamma_a\Omega^{-1}B + YB = I - A_1\Lambda^{-1}\Gamma_a\Omega^{-1}B.$$

Now (recall Lemma 3.2)

$$A_1\Lambda^{-1}\Lambda_*^{-1}A_{1*} = (P_*P + kQ)^{-1} \leq 0(s^{2k_1}) \Leftrightarrow A_1\Lambda^{-1} \leq 0(s^{k_1})$$

and

$$\begin{aligned}B_*\Omega_*^{-1}\Omega^{-1}B &= B_*A_*^{-1}A_*\Omega_*^{-1}\Omega^{-1}AA^{-1}B = (FP)_*G_t^{-1}(FP) \leq 0(s^{2k_2+2k_3}) \\ &\Leftrightarrow \Omega^{-1}B \leq 0(s^{k_2+k_3}).\end{aligned}$$

Hence,

$$A_1\Lambda^{-1}\Gamma_a\Omega^{-1}B \leq 0(s^{k_1+k_2+k_3-1})$$

is strictly proper when $k_1+k_2+k_3 \leq 0$ and in this case $A_1(X_1 - KB) \to I$ for $s \to \infty$. That is, $\det(X_1 - KB) \not\equiv 0$. (b) It easily follows from $C = (X_1 - KB)^{-1}A_1^{-1}A_1(Y_1 + KA)$ that

$$\begin{aligned} C &= (I - A_1\Lambda^{-1}\Gamma_a\Omega^{-1}B)^{-1}(A_1Y_1 + A_1\Lambda^{-1}\Gamma_a\Omega^{-1}A - YA) \\ &= (I - A_1\Lambda^{-1}\Gamma_a\Omega^{-1}B)^{-1}A_1\Lambda^{-1}\Gamma_a\Omega^{-1}A. \end{aligned}$$

Now

$$G_t^{-1} = A_*\Omega_*^{-1}\Omega^{-1}A \leq 0(s^{2k_2}) \Leftrightarrow \Omega^{-1}A \leq 0(s^{k_2})$$

and therefore $A_1\Lambda^{-1}\Gamma_a\Omega^{-1}A \leq 0(s^{k_1+k_2-1})$. Since $A_1\Lambda^{-1}\Gamma_a\Omega^{-1}B \leq 0(s^{k_1+k_2+k_3-1})$ is strictly proper when $k_1+k_2+k_3 \leq 0$ it follows from the expression for C that $C \leq 0(s^{k_1+k_2-1})$.

Example 3.14 Suppose for the 1DOF system treated in Sect. 3.7 that instead of choosing y_δ as in (3.223) one took

$$y_\delta = \begin{bmatrix} y - T_d r \\ P_s v \end{bmatrix} = \begin{bmatrix} P_d d + P v - T_d r \\ P_s v \end{bmatrix} = (\widetilde{G}_{11} - \widetilde{T}_d)w + \widetilde{P}_s\widetilde{G}_{12}v$$

where $\widetilde{P}_s = \mathrm{diag}\{I, P_s\}$. Then G_{11}, G_{21}, and G_{22} remained unchanged, but now

$$G_{12} = \widetilde{P}_s\widetilde{G}_{12} = \begin{bmatrix} P \\ P_s \end{bmatrix} = -B_{o1}A_{o1}^{-1} = -A_o^{-1}B_o.$$

Invoke (3.58) through (3.61) to establish that when P_s is stable that the system is still compatible and that $A_1^{-1}A_{o1}$ is still stable so that (3.36) of Theorem 3.1 again holds.

Solution Since the 1DOF system is admissible, $\widetilde{G}_{12}A_1$ is stable. Since P_s is stable, $\widetilde{P}_s$ is stable. Hence, one again gets that $G_{12}A_1 = \widetilde{P}_s\widetilde{G}_{12}A_1$ is stable. Also, $AG_{21} = A\widetilde{G}_{21}$ is unchanged and is, therefore, stable. In addition, it follows from $\widetilde{P}_s\widetilde{G}_{11} = \widetilde{G}_{11}$ and $\widetilde{P}_s\widetilde{T}_d = \widetilde{T}_d$ that

$$G_{11} + G_{12}A_1Y_1G_{21} = \widetilde{G}_{11} + \widetilde{P}_s\widetilde{G}_{12}A_1Y_1\widetilde{G}_{21} - \widetilde{T}_d = \widetilde{P}_s(\widetilde{G}_{11} + \widetilde{G}_{12}A_1Y_1\widetilde{G}_{21} - \widetilde{T}_d)$$

is stable because $\widetilde{P}_s$, $\widetilde{T}_d$, and $\widetilde{G}_{11} + \widetilde{G}_{12}A_1Y_1\widetilde{G}_{21}$ are stable. Thus, with the new choice for y_δ the system is compatible: $\widetilde{G}$ is admissible by assumption and $G_{22} = \widetilde{G}_{22}$. Also, from

$$-B_{o1} = G_{12}A_{o1} = \begin{bmatrix} PA_{o1} \\ P_sA_{o1} \end{bmatrix} = \begin{bmatrix} B_{p1}A_{p1}^{-1}A_{o1} \\ P_sA_{o1} \end{bmatrix}$$

it follows that $B_{p1}A_{p1}^{-1}A_{o1}$ and, therefore, $A_{p1}^{-1}A_{o1}$ are stable because B_{o1}, a polynomial matrix, is stable. Hence, $A_{o1} = A_{p1}\mathscr{M}$ where $\mathscr{M}$ is a stable matrix. It then follows that $A_1^{-1}A_{o1} = A_1^{-1}A_{p1}\mathscr{M}$ is stable by Lemma 2.10: $A_1^{-1}A_{p1}$ is stable since F is stable and it is necessary for admissibility that $\psi_{FP}(\psi_F\psi_P)^{-1}$ be stable.

Example* 3.15 Establish that $G = MM_*$ is analytic on Γ_s iff M is analytic on Γ_s where for the digital case Γ_s is $|\Lambda| = 1$ and for the analog case Γ_s is the finite $j\omega$-axis.

Solution For the digital case, $G(\lambda) = M(\lambda)M_*(\lambda)$ where $M_*(\lambda) = M^*(1/\bar{\lambda}) = M'(1/\lambda)$. Thus, $M_*(\lambda)$ can only have poles at $\lambda = \lambda_o^{-1}$ where λ_o is a pole of $M(\lambda)$. Hence, if $M(\lambda)$ is free of poles on $|\lambda| = 1$, so is $M_*(\lambda)$ and $G(\lambda)$. Conversely, if $G(\lambda)$ is free of poles on $|\lambda| = 1$, then when $M(\lambda)$ is $n_r \times n_c$ one gets that

$$\mathrm{Tr}[G(\varepsilon^{-j\theta})] = \mathrm{Tr}[M(\varepsilon^{-j\theta})M^*(\varepsilon^{-j\theta})] = \sum_{\mu=1}^{n_r}\sum_{\nu=1}^{n_c} |m_{\mu\nu}(\varepsilon^{-j\theta})|^2$$

must be finite or, equivalently, every element $m_{\mu\nu}(\varepsilon^{-j\theta})$ must be finite. That is, $M(\lambda)$ must be free of poles on $|\lambda| = 1$. Similarly for the analog case $G(s) = M(s)M_*(s)$ where $M_*(s) = M^*(-\bar{s}) = M'(-s)$. Thus, $M_*(s)$ can only have finite poles at $s = -s_o$ where s_o is a finite pole of $M(s)$. Hence, if $M(s)$ is free of poles on the finite $j\omega$-axis, so is $M_*(s)$ and $G(s)$. Conversely, when $G(s)$ is free of poles on the finite $j\omega$-axis and $M(s)$ is $n_r \times n_c$, one gets for finite ω that

$$\mathrm{Tr}\ [G(j\omega)] = \mathrm{Tr}\ [M(j\omega)M^*(j\omega)] = \sum_{\mu=1}^{n_r}\sum_{\nu=1}^{n_c} |m_{\mu\nu}(j\omega)|^2$$

must be finite or, equivalently, every element $m_{\mu\nu}(j\omega)$ must be finite. That is, $M(s)$ must be free of poles on the finite $j\omega$-axis.

Example 3.16 (a) For the 3DOF system treated in Sect. 3.8 establish that $\det(\nabla_{b*}\nabla_b) \neq 0$ on Γ_s when the parahermitian-positive matrix $kQ = W_{q*}W_q$ is positive definite on Γ_s where Γ_s is the contour described in Example 3.15. (b) It is clear from (3.349) that $\det(\nabla_c\nabla_{c*}) \neq 0$ on Γ_s iff $\det(\Omega_t\Omega_{t*}) \neq 0$ and $\det(\Omega_o\Omega_{o*}) \neq 0$ on Γ_s. When $(\Omega_r\Omega_{r*} + \Omega_{n_r}\Omega_{n_r*}) > 0$ on Γ_s and α_1 and α_2 are positive, establish that $\det(\Omega_o\Omega_{o*}) \neq 0$ on Γ_s. (c) An often encountered model is one in which

$P_d = P$ and d is a constant vector whose elements are zero-mean random variables. This corresponds to impulsive disturbances entering at the plant input and one can choose $A_d = I$ and B_d a diagonal matrix whose elements are the standard deviations of the respective elements of d. For this case with α_1 and α_2 positive establish that $\det(\Omega_t\Omega_{t*}) \neq 0$ when $\Omega_d\Omega_{d*} > 0$, $G_{n_m} > 0$, and $G_{n_l} > 0$ on Γ_s. **Comment:** It is important to recognize that under the circumstances of this example confirmation of $\det(\nabla_{b*}\nabla_b) \neq 0$ and $\det(\nabla_c\nabla_{c*}) \neq 0$ on Γ_s does not require a determination of a coprime matrix fraction description for FP.

Solution (a) Clearly,

$$\nabla_{b*}\nabla_b = A_{1*}(P_*P + kQ)A_1 = A_{1*}A_{p1*}^{-1}(B_{p1*}B_{p1} + kA_{p1*}QA_{p1})A_{p1}^{-1}A_1.$$

Hence,

$$\det(\nabla_{b*}\nabla_b) = \left(\frac{\det A_1}{\det A_{p1}}\right)_* \det\left(B_{p1*}B_{p1} + kA_{p1*}QA_{p1}\right)\left(\frac{\det A_1}{\det A_{p1}}\right).$$

Since the 3DOF system is admissible, $(\psi_{FP}/\psi_F\psi_P) \neq 0$ on Γ_s or, equivalently, $\left(\det A_1/\det A_{p1}\right) \neq 0$ on Γ_s because F is stable. Thus, $\det(\nabla_{b*}\nabla_b) \neq 0$ on Γ_s iff

$$\det(B_{p1*}B_{p1} + kA_{p1*}QA_{p1}) = \det\left\{\begin{bmatrix} B_{p1*} & A_{p1*}\end{bmatrix}\begin{bmatrix} I & 0 \\ 0 & kQ\end{bmatrix}\begin{bmatrix} B_{p1} \\ A_{p1}\end{bmatrix}\right\} \neq 0$$

on Γ_s. Since B_{p1}, A_{p1} is a right coprime polynomial matrix pair, $\begin{bmatrix} B'_{p1} & A'_{p1}\end{bmatrix}'$ has column rank throughout the complex plane and in particular on Γ_s. Hence, $\det(\nabla_{b*}\nabla_b) \neq 0$ on Γ_s if

$$\begin{bmatrix} I & 0 \\ 0 & kQ\end{bmatrix} > 0$$

on Γ_s. This is the case when $kQ > 0$ on Γ_s. (b) From (3.347) and (3.351),

$$\begin{aligned} G_o = \Omega_o\Omega_{o*} &= A_r(G_r + G_{n_r})A_{r*} = \alpha_1 B_r B_{r*} + \alpha_2 A_r(\Omega_r\Omega_{r*} + \Omega_{n_r}\Omega_{n_r*})A_{r*} \\ &= \begin{bmatrix} B_r & A_r\end{bmatrix}\begin{bmatrix} \alpha_1 I & 0 \\ 0 & \alpha_2(\Omega_r\Omega_{r*} + \Omega_{n_r}\Omega_{n_r*})\end{bmatrix}\begin{bmatrix} B_{r*} \\ A_{r*}\end{bmatrix}. \end{aligned}$$

Since $\begin{bmatrix} B_r & A_r\end{bmatrix}$ has row rank throughout the complex plane and in particular on Γ_s, it readily follows when $(\Omega_r\Omega_{r*} + \Omega_{n_r}\Omega_{n_r*}) > 0$ on Γ_s and α_1 and α_2 are positive that $G_o > 0$ and $\det(\Omega_o\Omega_{o*}) \neq 0$ on Γ_s. (c) From (3.350),

$$G_t = \Omega_t\Omega_{t*} = \begin{bmatrix} A & 0 \\ 0 & A_n\end{bmatrix}\begin{bmatrix} (FP_dG_dP_{d*}F_* + G_{n_m}) & FP_dG_dL_* \\ LG_dP_{d*}F_* & LG_dL_* + G_{n_l}\end{bmatrix}\begin{bmatrix} A_* & 0 \\ 0 & A_{n*}\end{bmatrix}.$$

It is easy to verify that

$$G_t = \Omega_t\Omega_{t*} = \begin{bmatrix} AFP_d \\ A_nL \end{bmatrix} G_d \begin{bmatrix} P_{d*}F_*A_* & L_*A_{n*} \end{bmatrix} + \begin{bmatrix} AG_{n_m}A_* & 0 \\ 0 & A_nG_{n_l}A_{n*} \end{bmatrix}$$

which is of the form

$$G_t = MG_{td}M_* + NG_{tn}N_* = \begin{bmatrix} M & N \end{bmatrix} \begin{bmatrix} G_{td} & 0 \\ 0 & G_{tn} \end{bmatrix} \begin{bmatrix} M_* \\ N_* \end{bmatrix}$$

where

$$M = \begin{bmatrix} AFP_d \\ A_nL \end{bmatrix} A_d^{-1} = \begin{bmatrix} AFP_dA_d^{-1} \\ A_nLA_d^{-1} \end{bmatrix} = \begin{bmatrix} AFP_dA_d^{-1} \\ B_n \end{bmatrix},$$

$$N = \begin{bmatrix} A & 0 \\ 0 & A_n \end{bmatrix}, G_{tn} = \begin{bmatrix} G_{n_m} & 0 \\ 0 & G_{n_l} \end{bmatrix},$$

and

$$G_{td} = \alpha_1 B_d B_{d*} + \alpha_2 A_d \Omega_d \Omega_{d*} A_{d*}.$$

By similar reasoning as for G_o it should be clear that $G_{td} > 0$ on Γ_s since α_1 and α_2 are positive and $\Omega_d\Omega_{d*} > 0$ on Γ_s. Obviously, $G_{tn} > 0$ on Γ_s because $G_{n_m} > 0$ and $G_{n_l} > 0$ on Γ_s. Now $G_t > 0$ on Γ_s iff $x = 0$ is the only solution for $x^*G_tx = 0$ on Γ_s. This is the case iff on Γ_s the matrix

$$\begin{bmatrix} M & N \end{bmatrix} = \begin{bmatrix} AFP_dA_d^{-1} & A & 0 \\ B_n & 0 & A_n \end{bmatrix}$$

has row rank or, equivalently, $x = 0$ is the only solution of

$$x^*\begin{bmatrix} M & N \end{bmatrix} = x^*\begin{bmatrix} AFP_dA_d^{-1} & A & 0 \\ B_n & 0 & A_n \end{bmatrix} = 0'.$$

When $A_d = I$ and $P_d = P$ one gets the following. First, since $LA_d^{-1} = L = A_n^{-1}B_n$ and L is stable, $\det A_n \neq 0$ on Γ_s. Hence, with $x' = [x_1' \quad x_2']$ it follows from the requirement that $x_2^*A_n = 0$ that $x_2 = 0$. Then, x_1 must satisfy

$$x_1^*\begin{bmatrix} AFP_dA_d^{-1} & A \end{bmatrix} = x_1^*\begin{bmatrix} AFP & A \end{bmatrix} = x_1^*\begin{bmatrix} B & A \end{bmatrix} = 0'.$$

But A, B is a left coprime polynomial matrix pair and so $x_1 = 0$ is the only solution. It immediately follows that $G_t > 0$ and $\det(\Omega_t\Omega_{t*}) \neq 0$ on Γ_s.

Example 3.17 (a) For the 1DOF system treated in Sect. 3.7 establish that $\det(\nabla_{b*}\nabla_b) \neq 0$ on Γ_s when the parahermitian-positive matrix $kQ = W_{q*}W_q$ is positive definite on Γ_s where Γ_s is the contour described in Example 3.15. (b) Establish that for (3.281) $\det(\nabla_c\nabla_{c*}) \neq 0$ on Γ_s when $\Omega_d\Omega_{d*} > 0$, $\Omega_r\Omega_{r*} > 0$, and $\Omega_n\Omega_{n*} > 0$ on Γ_s; α_1 and α_2 are positive; and $\det(AA_r^{-1}) \neq 0$ on Γ_s. **Comment**: It is important to recognize that under the circumstances of this example confirmation of $\det(\nabla_{b*}\nabla_b) \neq 0$ and $\det(\nabla_c\nabla_{c*}) \neq 0$ on Γ_s does not require a determination of a coprime matrix fraction description for FP since

$$\det(AA_r^{-1}) = (\det A)(\det A_r)^{-1} = \eta\psi_{FP}(\det A_r)^{-1}$$

where η is a constant and ψ_{FP} can be determined directly from the minors of FP.

Solution (a) The solution is the same as for Example 3.16(a). (b) For (3.281),

$$\begin{aligned}\nabla_c\nabla_{c*} &= AG_tA_* = A(P_oG_dP_{o*} + G_r + G_n)A_* \\ &= A\begin{bmatrix} -P_o & I & -I \end{bmatrix}\begin{bmatrix} G_d & 0 & 0 \\ 0 & G_r & 0 \\ 0 & 0 & G_n \end{bmatrix}\begin{bmatrix} -P_{o*} \\ I \\ -I \end{bmatrix}A_* = M\begin{bmatrix} G_{td} & 0 & 0 \\ 0 & G_{tr} & 0 \\ 0 & 0 & G_n \end{bmatrix}M_*\end{aligned}$$

where

$$\begin{aligned} M &= \begin{bmatrix} -AP_oA_d^{-1} & AA_r^{-1} & -A \end{bmatrix}, \\ G_{td} &= \alpha_1 B_dB_{d*} + \alpha_2 A_d\Omega_d\Omega_{d*}A_{d*}, \\ G_{tr} &= \alpha_1 B_rB_{r*} + \alpha_2 A_r\Omega_r\Omega_{r*}A_{r*}, \end{aligned}$$

and $G_n = \alpha_2\Omega_n\Omega_{n*}$. It readily follows from the same reasoning as in the solution to Example 3.16(b) that $G_{td} > 0$ and $G_{tr} > 0$ on Γ_s when $\Omega_d\Omega_{d*} > 0$ and $\Omega_r\Omega_{r*} > 0$ on Γ_s and α_1 and α_2 are positive. Also, $G_n > 0$ on Γ_s when $\Omega_n\Omega_{n*} > 0$ on Γ_s and $\alpha_2 > 0$. Hence, $AG_tA_* > 0$ on Γ_s iff $x^*M = 0$ on Γ_s implies that $x = 0$. Clearly, $x^*M = 0$ requires $x^*AA_r^{-1} = 0$ and, therefore, $x = 0$ when $\det(AA_r^{-1}) \neq 0$. That is, $AG_tA_* > 0$ on Γ_s is indeed the case and it follows that $\det(\nabla_c\nabla_{c*}) \neq 0$ on Γ_s.

Example* 3.18 (a) It is clear from (3.291) and (3.303) that for the $\mathcal{H}_2$ design of 1DOF systems the formula for K is of the form

$$K = \Lambda^{-1}(Z_a + \{\Lambda_*^{-1}\Gamma_i\Omega_*^{-1}\}_\infty)\Omega^{-1} - A_1^{-1}Y$$

for the digital case and $K = \Lambda^{-1}Z_a\Omega^{-1} - A_1^{-1}Y$ for the analog case where

$$Z_a = Z + \{\Lambda_*^{-1}\Gamma_i\Omega_*^{-1}\}_+ + \{\Lambda A_1^{-1}Y\Omega\}_-$$

Establish for the analog case that

$$C = (X_1 - KB)^{-1}(Y_1 + KA) = (I - A_1\Lambda^{-1}Z_a\Omega^{-1}B)^{-1}A_1\Lambda^{-1}Z_a\Omega^{-1}A$$

and show that one gets the same formula, but with Z_a replaced by $Z_d = Z_a + \{\Lambda_*^{-1}\Gamma_i\Omega_*^{-1}\}_\infty$, for the digital case. (b) Establish that when PA_1 is stable; the unstable poles of $FP = A^{-1}B = B_1A_1^{-1}$ all lie on the boundary Γ_s described in Example 3.15; $\det[A_{1*}(P_*P + kQ)A_1] \neq 0$ and $\det AG_tA \neq 0$ on Γ_s; Q and AG_tA are parahermitian-positive and good; and $\widetilde{\Lambda}$ and $\widetilde{\Omega}$ are square Wiener–Hopf factors satisfying $P_*P + kQ = \widetilde{\Lambda}_*\widetilde{\Lambda}$ and $G_t = \widetilde{\Omega}_*\widetilde{\Omega}$; then the Wiener–Hopf factor Λ satisfying (3.284) and (3.344) and the Wiener–Hopf factor Ω satisfying (3.284) can be chosen as $\Lambda = \widetilde{\Lambda}A_1$ and $\Omega = A\widetilde{\Omega}$. (c) When the conditions of part (b) prevail, establish that in the digital case the formula for C reduces to

$$C = (X_1 - KB)^{-1}(Y_1 + KA) = (I - \widetilde{\Lambda}^{-1}Z_d\widetilde{\Omega}^{-1}FP)^{-1}\widetilde{\Lambda}^{-1}Z_d\widetilde{\Omega}^{-1}$$

where with $\widetilde{\Gamma}_i = P_*(T_dG_r + P_dG_dP_{o*})$ one has

$$Z_d = Z + \{\widetilde{\Lambda}_*^{-1}\widetilde{\Gamma}_i\widetilde{\Omega}_*^{-1}\}_+ + \{\widetilde{\Lambda}_*^{-1}\widetilde{\Gamma}_i\widetilde{\Omega}_*^{-1}\}_\infty.$$

Also, establish that the same formula holds for the analog case with Z_d replaced by $Z_a = Z + \{\widetilde{\Lambda}_*^{-1}\widetilde{\Gamma}_i\widetilde{\Omega}_*^{-1}\}_+$. **Comment:** It is important to note that these reduced formulas for C do not require a determination of a coprime matrix fraction description for FP and the conditions $\det(\nabla_{b*}\nabla_b) \neq 0$ and $\det(\nabla_c\nabla_{c*}) \neq 0$ on Γ_s are automatically satisfied when the assumptions in Example 3.17 prevail.

Solution (a) For the analog case,

$$Y_1 + KA = Y_1 + \Lambda^{-1}Z_a\Omega^{-1}A - A_1^{-1}YA = \Lambda^{-1}Z_a\Omega^{-1}A$$

and

$$\begin{aligned} X_1 - KB &= X_1 - \Lambda^{-1}Z_a\Omega^{-1}B + A_1^{-1}YB = X_1 - \Lambda^{-1}Z_a\Omega^{-1}B + A_1^{-1}(I - A_1X_1) \\ &= A_1^{-1} - \Lambda^{-1}Z_a\Omega^{-1}B = A_1^{-1}(I - A_1\Lambda^{-1}Z_a\Omega^{-1}B). \end{aligned}$$

Hence,

$$C = (X_1 - KB)^{-1}(Y_1 + KA) = (I - A_1\Lambda^{-1}Z_a\Omega^{-1}B)^{-1}A_1\Lambda^{-1}Z_a\Omega^{-1}A.$$

Obviously, for the digital case the only thing that changes in the above is that Z_a is replaced by Z_d. (b) Since PA_1 is stable and Q is good,

$$A_{1*}(P_*P+kQ)A_1 = A_{1*}P_*PA_1 + kA_{1*}QA_1 = A_{1*}\widetilde{\Lambda}_*\widetilde{\Lambda}A_1$$

is good. Hence (see Example 3.15), $\Lambda = \widetilde{\Lambda}A_1$ is good or, equivalently, is analytic on Γ_s. Since $\det[A_{1*}(P_*P+kQ)A_1] \neq 0$ on Γ_s, it follows that $\det(P_*P+kQ) \neq 0$ on Γ_s. Thus, there does indeed exist a square Wiener–Hopf factor $\widetilde{\Lambda}$ such that $\widetilde{\Lambda}_*\widetilde{\Lambda} = P_*P+kQ$ and from det $(A_{1*}\widetilde{\Lambda}_*\widetilde{\Lambda}A) \neq 0$ on Γ_s one has det $(\widetilde{\Lambda}A) \neq 0$ on Γ_s. It follows that $\Lambda = \widetilde{\Lambda}A_1$ and $\Lambda^{-1} = A_1^{-1}\widetilde{\Lambda}^{-1}$ are good. Moreover, because $\widetilde{\Lambda}$ is a Wiener–Hopf factor, $\widetilde{\Lambda}$ and its inverse are analytic in $|\lambda| < 1$ for the digital case and in Re $s > 0$ for the analog case. The same is true for A_1^{-1} because all the unstable poles of FP lie on Γ_s. Thus, $\Lambda = \widetilde{\Lambda}A_1$ and $\Lambda^{-1} = A_1^{-1}\widetilde{\Lambda}^{-1}$ are analytic in $|\lambda| \leq 1$ for the digital case and in Re $s \geq 0$ for the analog case: i.e., $\Lambda = \widetilde{\Lambda}A_1$ is an acceptable Wiener–Hopf factor for $A_{1*}(P_*P+kQ)A_1 = \Lambda_*\Lambda$. Similarly, from $AG_tA_* = A\widetilde{\Omega}\widetilde{\Omega}_*A_*$ it follows that $A\widetilde{\Omega}$ is good and $\det(A\widetilde{\Omega}) \neq 0$ on Γ_s. That is, $\Omega = A\widetilde{\Omega}$ and $\Omega^{-1} = \widetilde{\Omega}^{-1}A^{-1}$ are good. Moreover, because $\widetilde{\Omega}$ is a Wiener–Hopf factor, $\widetilde{\Omega}$ and its inverse are analytic in $|\lambda| < 1$ for the digital case and in Re $s > 0$ for the analog case. The same is true for A^{-1} because all the unstable poles of FP lie on Γ_s. Thus, $\Omega = A\widetilde{\Omega}$ and $\Omega^{-1} = \widetilde{\Omega}^{-1}A^{-1}$ are analytic in $|\lambda| \leq 1$ for the digital case and in Re $s \geq 0$ for the analog case: i.e., $\Omega = A\widetilde{\Omega}$ is an acceptable Wiener–Hopf factor for $AG_tA_* = \Omega\Omega_*$. (c) It readily follows from the solution for part (a) that in the digital case

$$C = (I - A_1\Lambda^{-1}Z_d\Omega^{-1}B)^{-1}A_1\Lambda^{-1}Z_d\Omega^{-1}A.$$

So when the conditions of part (b) prevail,

$$C = (I - \widetilde{\Lambda}^{-1}Z_d\widetilde{\Omega}^{-1}A^{-1}B)^{-1}\widetilde{\Lambda}^{-1}Z_d\widetilde{\Omega}^{-1} = (I - \widetilde{\Lambda}^{-1}Z_d\widetilde{\Omega}^{-1}FP)^{-1}\widetilde{\Lambda}^{-1}Z_d\widetilde{\Omega}^{-1}$$

where

$$Z_d = Z_a + \{\Lambda_*^{-1}\Gamma_i\Omega_*^{-1}\}_\infty = Z + \{\Lambda_*^{-1}\Gamma_i\Omega_*^{-1}\}_+ + \{\Lambda_*^{-1}\Gamma_i\Omega_*^{-1}\}_\infty$$

because $\{\Lambda A_1^{-1}Y\Omega\}_- = 0$ when all the unstable poles of A_1^{-1} lie on Γ_s which is the case when all the unstable poles of FP lie on Γ_s. Moreover, from (3.289), $\Gamma_i = A_{1*}\widetilde{\Gamma}_iA_*$ and one gets

$$\begin{aligned} Z_d &= Z + \{\Lambda_*^{-1}A_{1*}\widetilde{\Gamma}_iA_*\Omega_*^{-1}\}_+ + \{\Lambda_*^{-1}A_{1*}\widetilde{\Gamma}_iA_*\Omega_*^{-1}\}_\infty \\ &= Z + \{\widetilde{\Lambda}_*^{-1}\widetilde{\Gamma}_i\widetilde{\Omega}_*^{-1}\}_+ + \{\widetilde{\Lambda}_*^{-1}\widetilde{\Gamma}_i\widetilde{\Omega}_*^{-1}\}_\infty. \end{aligned}$$

Example* 3.19 When K is given by (3.36), establish that (3.11) yields

$$C = (X_{s1} - K_g A_g B)^{-1}(Y_{s1} + K_g A_g A) = (Y_s + A_1 K_g A_g)(X_s - B_1 K_g A_g)^{-1}$$

where

$$K_s = A_1^{-1} A_{o1} \Gamma_o Y_{g1},$$
$$X_s = X - B_1 K_s, \quad X_{s1} = X_1 - K_s B,$$

and

$$Y_s = Y + A_1 K_s, \quad Y_{s1} = Y_1 + K_s A.$$

Moreover,

$$A X_s + B Y_s = I, \quad X_{s1} A_1 + Y_{s1} B_1 = I.$$

Solution From (3.11),

$$C = (X_1 - KB)^{-1}(Y_1 + KA) = (Y + A_1 K)(X - B_1 K)^{-1}.$$

From (3.36),

$$K = A_1^{-1} A_{o1} \Gamma_o Y_{g1} + K_g A_g = K_s + K_g A_g.$$

Hence,

$$\begin{aligned} C &= (X_1 - K_s B - K_g A_g B)^{-1}(Y_1 + K_s A + K_g A_g A) \\ &= (X_{s1} - K_g A_g B)^{-1}(Y_{s1} + K_g A_g A) \end{aligned}$$

and also

$$\begin{aligned} C &= (Y + A_1 K_s + A_1 K_g A_g)(X - B_1 K_s - B_1 K_g A_g)^{-1} \\ &= (Y_s + A_1 K_g A_g)(X_s - B_1 K_g A_g)^{-1}. \end{aligned}$$

Moreover,

$$A X_s + B Y_s = A(X - B_1 K_s) + B(Y + A_1 K_s) = AX + BY + (BA_1 - AB_1)K_s = I$$

and

$$X_{s1}A_1 + Y_{s1}B_1 = (X_1 - K_sB)A_1 + (Y_1 + K_sA)B_1 = X_1A_1 + Y_1B_1 - K_s(BA_1 - AB_1)$$
$$= I.$$

Example* **3.20** In Sects. 3.5 and 3.6, the first expression for K in (3.36) was used. The purpose of this example is to investigate for the analog case the use of the second expression for K instead. In particular, show that $\nabla_b = WB_{o1}A_{o1}^{-1}A_1 = -WG_{12}A_1$ is then the appropriate choice for ∇_b instead of $\nabla_{bo} = WB_{o1}$. Moreover, verify when $A_1^{-1}A_{o1}$ is stable and G is compatible that on the finite $j\omega$-axis, rank ∇_b = rank ∇_{bo}. In addition, establish that in place of (3.205) and (3.206) in Theorem 3.5 one gets

$$K = \Lambda^{-1}\Gamma_a\Omega^{-1}A_g - A_1^{-1}Y$$

where

$$\Gamma_a = Z + \{\Lambda_*^{-1}\nabla_{b*}WG_{11}G_w\nabla_{c*}\Omega_*^{-1}\}_+ + \{\Lambda A_1^{-1}(Y + A_{o1}\Gamma_oY_{g1})A_g^{-1}\Omega\}_-$$

with Λ now the Wiener–Hopf spectral factor satisfying

$$\nabla_{b*}\nabla_b = A_{1*}A_{o1*}^{-1}B_{o1*}W_*WB_{o1}A_{o1}^{-1}A_1 = \Lambda_*\Lambda.$$

Moreover, $\Lambda = \Lambda_oA_{o1}^{-1}A_1$ where Λ_o is the Wiener–Hopf spectral factor satisfying

$$\nabla_{bo*}\nabla_{bo} = B_{o1*}W_*WB_{o1} = \Lambda_{o*}\Lambda_o.$$

In addition, confirm that in place of (3.210) one gets

$$C = (I - A_1\Lambda^{-1}\Gamma_a\Omega^{-1}A_gB)^{-1}A_1\Lambda^{-1}\Gamma_a\Omega^{-1}A_gA.$$

Solution Using $A_1^{-1}A_{o1}K_m = K_g$ in (3.146) yields

$$\begin{aligned}\nabla &= WG_{11}G_w + WG_{12}(Y + A_{o1}\Gamma_oY_{g1})AG_{21}G_w + WG_{12}A_1K_gA_gAG_{21}G_w\\ &= WG_{11}G_w - WB_{o1}A_{o1}^{-1}(Y + A_{o1}\Gamma_oY_{g1})A_g^{-1}B_gB_w - WB_{o1}A_{o1}^{-1}A_1K_gA_gA_g^{-1}B_gB_w\\ &= WG_{11}G_w - WB_{o1}\tilde{\Gamma}_oB_gB_w - WB_{o1}A_{o1}^{-1}A_1K_gB_gB_w.\end{aligned}$$

This expression is identical with (3.147) except that $WB_{o1}K_mB_gB_w = \nabla_{bo}K_mB_gB_w$ is replaced with $\nabla_bK_gB_gB_w$ where $\nabla_b = -WG_{12}A_1 = WB_{o1}A_{o1}^{-1}A_1 = \nabla_{bo}A_{o1}^{-1}A_1$. Since $A_{o1}^{-1}A_1$ is stable when G is compatible and $A_1^{-1}A_{o1}$ is stable by assumption, it follows that $A_{o1}^{-1}A_1$ is nonsingular on the finite $j\omega$-axis. Thus, rank ∇_b = rank ∇_{bo}

on the finite $j\omega$-axis. Moreover, in terms of the spectral factorization $\nabla_{bo*}\nabla_{bo} = B_{o1*}W_*WB_{o1} = \Lambda_{o*}\Lambda_o$ one gets

$$\nabla_{b*}\nabla_b = A_{1*}A_{o1*}^{-1}B_{o1*}W_*WB_{o1}A_{o1}^{-1}A_1 = A_{1*}A_{o1*}^{-1}\Lambda_{o*}\Lambda_oA_{o1}^{-1}A_1$$

and it follows that now $\nabla_{b*}\nabla_b = \Lambda_*\Lambda$ where $\Lambda = \Lambda_oA_{o1}^{-1}A_1$ is a spectral factor because $A_{o1}^{-1}A_1$ and and $A_1^{-1}A_{o1}$ are stable. Since in Theorem 3.5, ∇_b is actually $\nabla_{bo} = \nabla_bA_1^{-1}A_{o1}$ and Λ is actually $\Lambda_o = \Lambda A_1^{-1}A_{o1}$, it follows from (3.205) and (3.206) that

$$K = A_1^{-1}(A_{o1}\Lambda_o^{-1}\Gamma_a\Omega^{-1}A_g - Y) = \Lambda^{-1}\Gamma_a\Omega^{-1}A_g - A_1^{-1}Y$$

where, with $Y_a = Y_o\Gamma_wY_g + YX_g$,

$$\begin{aligned}
\Gamma_a &= Z + \{\Lambda_{o*}^{-1}\nabla_{bo*}WG_{11}G_w\nabla_{c*}\Omega_*^{-1}\}_+ + \{\Lambda_oA_{o1}^{-1}Y_a\Omega\}_- \\
&= Z + \{\Lambda_*^{-1}\nabla_{b*}WG_{11}G_w\nabla_{c*}\Omega_*^{-1}\}_+ + \{\Lambda A_1^{-1}(Y_o\Gamma_wY_gA_g + YX_gA_g)A_g^{-1}\Omega\}_- \\
&= Z + \{\Lambda_*^{-1}\nabla_{b*}WG_{11}G_w\nabla_{c*}\Omega_*^{-1}\}_+ + \{\Lambda A_1^{-1}(Y_o\Gamma_wA_{g1}Y_{g1} + YX_gA_g)A_g^{-1}\Omega\}_- \\
&= Z + \{\Lambda_*^{-1}\nabla_{b*}WG_{11}G_w\nabla_{c*}\Omega_*^{-1}\}_+ + \{\Lambda A_1^{-1}(Y_o\Gamma_wA_{g1}Y_{g1} + Y - YB_{g1}Y_{g1})A_g^{-1}\Omega\}_- \\
&= Z + \{\Lambda_*^{-1}\nabla_{b*}WG_{11}G_w\nabla_{c*}\Omega_*^{-1}\}_+ + \{\Lambda A_1^{-1}(Y + A_{o1}\Gamma_oY_{g1})A_g^{-1}\Omega\}_-.
\end{aligned}$$

Finally, from (3.210) one gets

$$\begin{aligned}
C &= (I - A_{o1}\Lambda_o^{-1}\Gamma_a\Omega^{-1}A_gB)^{-1}A_{o1}\Lambda_o^{-1}\Gamma_a\Omega^{-1}A_gA \\
&= (I - A_1\Lambda^{-1}\Gamma_a\Omega^{-1}A_gB)^{-1}A_1\Lambda^{-1}\Gamma_a\Omega^{-1}A_gA.
\end{aligned}$$

Example 3.21 (a) When C in Fig. 3.2 is given by (3.11), establish that the transfer matrix from w to y_m is given by

$$G_m = (I - G_{22}C)^{-1}G_{21} = (X - B_1K)AG_{21}.$$

Use this result together with (3.14) and (3.15) to prove when G is compatible and A_w, B_w is a left coprime polynomial matrix pair that $G_w = A_w^{-1}B_w$ is strictly acceptable (see Definition 3.3) iff $\Gamma_g = AG_{21}A_w^{-1}$ and $\Gamma_h = (G_{11} + G_{12}A_1Y_1G_{21})A_w^{-1}$ are stable. (b) Establish when $A_1^{-1}A_{o1}$ is stable that an acceptable G_w is strictly acceptable iff $\Gamma_g = AG_{21}A_w^{-1}$ is stable. (c) When $A_1^{-1}A_{o1}$ is stable, show that Γ_g and Γ_h are stable iff $\Gamma_w = A_oG_{11}A_w^{-1}$ and $\Gamma_v = (G_{21} + G_{22}Y_oA_oG_{11})A_w^{-1}$ are stable.

Solution (a) It readily follows from $y_m = G_{21}w + G_{22}v = G_{21}w + G_{22}Cy_m$ that

$$\begin{aligned} y_m &= (I - G_{22}C)^{-1}G_{21}w = [I + A^{-1}B(Y + A_1K)(X - B_1K)^{-1}]^{-1}G_{21}w \\ &= (X - B_1K)[A(X - B_1K) + B(Y + A_1K)]^{-1}AG_{21}w = (X - B_1K)AG_{21}w = G_m w. \end{aligned}$$

Since G_δ, G_v, and G_m are stable, $G_w = A_w^{-1}B_w$ is strictly acceptable iff

$$T_w = \begin{bmatrix} G_\delta A_w^{-1} \\ G_v A_w^{-1} \\ G_m A_w^{-1} \end{bmatrix} = \begin{bmatrix} (G_{11} + G_{12}A_1Y_1G_{21}) \\ YAG_{21} \\ XAG_{21} \end{bmatrix} A_w^{-1} + \begin{bmatrix} G_{12}A_1 \\ A_1 \\ -B_1 \end{bmatrix} KAG_{21}A_w^{-1}$$

is stable. Hence, it is necessary that

$$\begin{aligned} [0 \quad B \quad A]T_w &= [B \quad A]\begin{bmatrix} G_v A_w^{-1} \\ G_m A_w^{-1} \end{bmatrix} = [(AX + BY) + (BA_1 - AB_1)K]AG_{21}A_w^{-1} \\ &= \Gamma_g \end{aligned}$$

be stable. It is also necessary that

$$G_\delta A_w^{-1} = (G_{11} + G_{12}A_1Y_1G_{21})A_w^{-1} + G_{12}A_1KAG_{21}A_w^{-1} = \Gamma_h + G_{12}A_1K\Gamma_g$$

be stable. Since $G_{12}A_1$ and K are stable, it follows that Γ_h must be stable. Obviously, Γ_g and Γ_h stable is sufficient for stable T_w. That is, G_w is strictly acceptable iff Γ_g and Γ_h are stable. (b) From Theorem 3.1, G_w is acceptable iff $\Gamma_w = A_oG_{11}A_w^{-1}$ and $\Gamma_o = A_{o1}^{-1}(Y_o\Gamma_wA_{g1} - YB_{g1})$ are stable. It is established in Example 3.3 when Γ_w is stable that Γ_o is stable iff $\Gamma_z = \Gamma_hA_{g1}$ is stable. So if $\Gamma_g = A_g^{-1}B_g = B_{g1}A_{g1}^{-1}$ is stable, $\Gamma_h = \Gamma_zA_{g1}^{-1}$ is stable and G_w is strictly acceptable. That Γ_g stable is necessary was established in part (a) of this example. (c) Using

$$G_{11} = \Gamma_hA_w - G_{12}A_1Y_1G_{21} = \Gamma_hA_w - G_{12}YAG_{21}$$

yields

$$\begin{aligned} \Gamma_v &= (G_{21} + G_{22}Y_oA_oG_{11})A_w^{-1} = G_{21}A_w^{-1} + G_{22}Y_oA_o(\Gamma_hA_w - G_{12}YAG_{21})A_w^{-1} \\ &= A^{-1}(\Gamma_g - BY_oA_o(\Gamma_h + A_o^{-1}B_oY\Gamma_g)) = A^{-1}(I - BY_oB_oY)\Gamma_g - A^{-1}BY_oA_o\Gamma_h \\ &= A^{-1}(I - B(I - A_{o1}X_{o1})Y)\Gamma_g - B_1A_1^{-1}A_{o1}Y_{o1}\Gamma_h \\ &= A^{-1}(I - BY + BA_{o1}X_{o1}Y)\Gamma_g - B_1A_1^{-1}A_{o1}Y_{o1}\Gamma_h \\ &= (X + B_1A_1^{-1}A_{o1}X_{o1}Y)\Gamma_g - B_1A_1^{-1}A_{o1}Y_{o1}\Gamma_h \end{aligned}$$

which is stable when Γ_g, Γ_h, and $A_1^{-1}A_{o1}$ are stable. Also, from

$$A_o\Gamma_h = A_oG_{11}A_w^{-1} + A_oG_{12}YAG_{21}A_w^{-1} = \Gamma_w - B_oY\Gamma_g$$

one gets that Γ_w is stable when Γ_g and Γ_h are stable. That is, Γ_w and Γ_v are stable when Γ_g and Γ_h are stable. Conversely, when Γ_w and Γ_v are stable it follows from

$$A\Gamma_v = AG_{21}A_w^{-1} + AG_{22}Y_oA_oG_{11}A_w^{-1} = \Gamma_g - BY_o\Gamma_w$$

that Γ_g is stable. Moreover, using

$$G_{21} = \Gamma_vA_w - G_{22}Y_oA_oG_{11}$$

yields

$$\begin{aligned}\Gamma_h &= (G_{11} + G_{12}YAG_{21})A_w^{-1} = G_{11}A_w^{-1} + G_{12}YA(\Gamma_v - G_{22}Y_oA_oG_{11}A_w^{-1})\\ &= (I - G_{12}YAG_{22}Y_oA_o)G_{11}A_w^{-1} + G_{12}A_1Y_1\Gamma_v = A_o^{-1}(I - B_oYBY_o)\Gamma_w + G_{12}A_1Y_1\Gamma_v\\ &= A_o^{-1}[I - B_o(I - A_1X_1)Y_o]\Gamma_w + G_{12}A_1Y_1\Gamma_v = (X_o - G_{12}A_1X_1Y_o)\Gamma_w + G_{12}A_1Y_1\Gamma_v\end{aligned}$$

which is stable when Γ_w and Γ_v are stable because $G_{12}A_1$ is stable when G is compatible. So Γ_g and Γ_h are stable when Γ_w and Γ_v are stable. That is, Γ_g and Γ_h are stable iff Γ_w and Γ_v are stable.

Example 3.22 The case in which the generalized persistent input standard configuration of Fig. 3.2 represents an analog system for which G_w is strictly acceptable is considered in this example. Prove the following parallel theorem to Theorem 3.5.

Theorem 3.5′ *When G is compatible, G_w is strictly acceptable for G, and W is stable; when on the finite $j\omega$-axis $\nabla_b = -WG_{12}A_1 = WB_{o1}A_{o1}^{-1}A_1$ has full column rank and $\nabla_c = AG_{21}G_w = A_g^{-1}B_gB_w$ has full row rank; and when the order relationships*

$$WG_{11}G_w \le 0(s^{-1}), (G_{12*}W_*WG_{12})^{-1} \le 0(s^{2k_1}), (G_{21}G_wG_{w*}G_{21*})^{-1} \le 0(s^{2k_2}), G_{22} \le 0(s^{k_3})$$

are satisfied with $k_1 + k_2 + k_3 \le 0$, then all controller transfer matrices C given by (3.11) *for which G_vG_w, $G_\delta G_w$, and G_mG_w are stable and the performance functional*

$$J = \frac{1}{2\pi j}\int_{-j\infty}^{j\infty} \text{Tr}\,[(WG_\delta G_w)(WG_\delta G_w)_*]\,ds$$

is finite are the ones for which

$$K = \Lambda^{-1}\Gamma_a\Omega^{-1} - A_1^{-1}Y$$

is chosen where

$$\Gamma_a = Z + \{\Lambda_*^{-1}\nabla_{b*}WG_{11}G_w\nabla_{c*}\Omega_*^{-1}\}_+ + \{\Lambda A_1^{-1}Y\Omega\}_-$$

with Z a strictly proper stable matrix. In the formulas for K and Γ_a*, the matrices* Λ *and* Ω *are the square stable Wiener–Hopf spectral factors with stable inverses satisfying*

$$\nabla_{b*}\nabla_b = \Lambda_*\Lambda,\ \nabla_c\nabla_{c*} = \Omega\Omega_*.$$

In this case, the controller transfer matrix is given by

$$C = (I - A_1\Lambda^{-1}\Gamma_a\Omega^{-1}B)^{-1}A_1\Lambda^{-1}\Gamma_a\Omega^{-1}A$$

and satisfies

$$C \le 0(s^{k_1+k_2-1}).$$

In addition,

$$J = \frac{1}{2\pi j}\int_{-j\infty}^{j\infty}\{\mathrm{Tr}\,[ZZ_* + \{\widetilde{\Gamma}\}_-\{\widetilde{\Gamma}\}_{-*}] + \rho_o\}ds$$

where

$$\rho_o = \mathrm{Tr}\,(\nabla_a\nabla_{a*}) - \mathrm{Tr}\,(\widetilde{\Gamma}\widetilde{\Gamma}_*)$$

and the minimum value for J is obtained when $Z \equiv 0$ *is chosen. In the formulas for J and* ρ_o*,*

$$\nabla_a = WG_{11}G_w - \nabla_b\Lambda^{-1}(\{\Lambda A_1^{-1}Y\Omega\}_+ + \{\Lambda A_1^{-1}Y\Omega\}_-)\Omega^{-1}\nabla_c$$

and

$$\widetilde{\Gamma} = \Lambda_*^{-1}\nabla_{b*}WG_{11}G_w\nabla_{c*}\Omega_*^{-1} - \{\Lambda A_1^{-1}Y\Omega\}_+ - \{\Lambda A_1^{-1}Y\Omega\}_-.$$

Solution From (3.15) and Example 3.21,

$$\begin{aligned}\nabla &= WG_\delta G_w = W(G_{11}+G_{12}A_1Y_1G_{21})G_w + WG_{12}A_1KAG_{21}G_w\\ &= W\Gamma_h B_w + WG_{12}A_1K\Gamma_g B_w = WG_{11}G_w - \nabla_b(K+Y_1A^{-1})\nabla_c\end{aligned}$$

where $G_{12}A_1$ is stable because G is compatible and, since G_w is strictly acceptable, Γ_g and Γ_h are stable. Hence, ∇ is stable for all stable K because W stable is assumed. In addition, ∇_b and ∇_c are stable. Moreover, on account of the rank assumptions on ∇_b and ∇_c, one can construct square Wiener–Hopf spectral factors Λ and Ω which together with their inverses are analytic in Re $s \geq 0$ and satisfy $\nabla_{b*}\nabla_b = \Lambda_*\Lambda$ and $\nabla_c\nabla_{c*} = \Omega\Omega_*$. (Clearly, A_g^{-1} and its inverse are stable. When it is also assumed that $A^{-1}A_{o1}$ is stable, then $A_{o1}^{-1}A_1$ and its inverse are stable. Hence, A_g^{-1} and $A_{o1}^{-1}A_1$ are nonsingular on the finite $j\omega$-axis and the rank assumptions on ∇_b and ∇_c are satisfied iff WB_{o1} has full column rank and B_gB_w has full row rank on the finite $j\omega$-axis.) Since $A_1Y_1 = YA$,

$$\nabla = WG_{11}G_w - \nabla_b\Lambda^{-1}(\Lambda K\Omega + \Lambda A_1^{-1}Y\Omega)\Omega^{-1}\nabla_c$$

and any stable K can be expressed as

$$K = \tilde{K} - \Lambda^{-1}\{\Lambda A_1^{-1}Y\Omega\}_\infty \Omega^{-1}$$

where $\tilde{K}$ is a stable matrix. Hence, $\nabla = \nabla_a - \nabla_b\tilde{K}\nabla_c$ where

$$\nabla_a = WG_{11}G_w - \nabla_b\Lambda^{-1}(\{\Lambda A_1^{-1}Y\Omega\}_+ + \{\Lambda A_1^{-1}Y\Omega\}_-)\Omega^{-1}\nabla_c.$$

Since $\nabla_b\Lambda^{-1}$ and $\Omega^{-1}\nabla_c$ are proper because they are paraunitary matrices and since $WG_{11}G_w \leq 0(s^{-1})$ is assumed, it immediately follows that ∇_a is strictly proper. Moreover, since ∇ is stable for any stable K or, equivalently for any stable $\tilde{K}$, one gets from $\nabla = \nabla_a$ when $\tilde{K} = 0$ that ∇_a is stable. In addition, on account of the rank assumptions on ∇_b and ∇_c, it is true that det $(\nabla_{b*}\nabla_b) \neq 0$ and det $(\nabla_c\nabla_{c*}) \neq 0$ on the finite $j\omega$-axis. Moreover, $\nabla_b = -WG_{12}A_1$ and $\nabla_c = A_g^{-1}B_gB_w = \Gamma_g B_w$ are stable. Thus, Theorem 3.4 applies with $\tilde{K}$ in place of $\tilde{K}_m$ and it follows that J is finite iff

$$K = \tilde{K} - \Lambda^{-1}\{\Lambda A_1^{-1}Y\Omega\}_\infty\Omega^{-1} = \Lambda^{-1}(Z + \{\tilde{\Gamma}\}_+ - \{\Lambda A_1^{-1}Y\Omega\}_\infty)\Omega^{-1}$$

where Z is a strictly-proper stable matrix and

$$\begin{aligned}\tilde{\Gamma} &= \Lambda_*^{-1}\nabla_{b*}\nabla_a\nabla_{c*}\Omega_*^{-1}\\ &= \Lambda_*^{-1}\nabla_{b*}WG_{11}G_w\nabla_{c*}\Omega_*^{-1} - \{\Lambda A_1^{-1}Y\Omega\}_+ - \{\Lambda A_1^{-1}Y\Omega\}_-\end{aligned}$$

is strictly proper and good because ∇_a is and because $\nabla_b \Lambda^{-1}$ and $\Omega^{-1}\nabla_c$ are proper and stable. Clearly,

$$\{\widetilde{\Gamma}\}_+ = \{\Lambda_*^{-1}\nabla_{b*}WG_{11}G_w\nabla_{c*}\Omega_*^{-1}\}_+ - \{\Lambda A_1^{-1}Y\Omega\}_+$$

and it readily follows that

$$\begin{aligned} K &= \Lambda^{-1}(Z + \{\Lambda_*^{-1}\nabla_{b*}WG_{11}G_w\nabla_{c*}\Omega_*^{-1}\}_+ - \{\Lambda A_1^{-1}Y\Omega\}_+ - \{\Lambda A_1^{-1}Y\Omega\}_\infty)\Omega^{-1} \\ &= \Lambda^{-1}(Z + \{\Lambda_*^{-1}\nabla_{b*}WG_{11}G_w\nabla_{c*}\Omega_*^{-1}\}_+ + \{\Lambda A_1^{-1}Y\Omega\}_- - \Lambda A_1^{-1}Y\Omega)\Omega^{-1} \\ &= \Lambda^{-1}\Gamma_a\Omega^{-1} - A_1^{-1}Y \end{aligned}$$

Also from Theorem 3.4,

$$J = \frac{1}{2\pi j}\int_{-j\infty}^{j\infty} \{\mathrm{Tr}\ [ZZ_* + \{\widetilde{\Gamma}\}_-\{\widetilde{\Gamma}\}_{-*}] + \rho_o\}\, ds$$

where

$$\rho_o = \mathrm{Tr}\ (\nabla_a\nabla_{a*}) - \mathrm{Tr}\ (\widetilde{\Gamma}\widetilde{\Gamma}_*)$$

and the minimum value for J is obtained when $Z \equiv 0$ is chosen. It remains to show that all controller transfer matrices C given by (3.11) exist when $K = \Lambda^{-1}\Gamma_a\Omega^{-1} - A_1^{-1}Y$. From (3.11),

$$\begin{aligned} C &= (X_1 - KB)^{-1}(Y_1 + KA) = [A_1(X_1 - KB)]^{-1}A_1(Y_1 + KA) \\ &= (A_1X_1 - A_1\Lambda^{-1}\Gamma_a\Omega^{-1}B - YB)^{-1}(A_1Y_1 + A_1\Lambda^{-1}\Gamma_a\Omega^{-1}A - YA) \\ &= (I - A_1\Lambda^{-1}\Gamma_a\Omega^{-1}B)^{-1}A_1\Lambda^{-1}\Gamma_a\Omega^{-1}A. \end{aligned}$$

Hence, the controllers exist provided $\det(I - A_1\Lambda^{-1}\Gamma_a\Omega^{-1}B) \not\equiv 0$. From the order assumptions,

$$A_1\Lambda^{-1}\Lambda_*^{-1}A_{1*} = A_1(\nabla_{b*}\nabla_b)^{-1}A_{1*} = (G_{12*}W_*WG_{12})^{-1} \leq 0(s^{2k_1})$$

and

$$A_*\Omega_*^{-1}\Omega^{-1}A = A_*(\nabla_c\nabla_{c*})^{-1}A = (G_{21}G_wG_{w*}G_{21*})^{-1} \leq 0(s^{2k_2}).$$

Invoking Lemma 3.2 yields $A_1\Lambda^{-1} \leq 0(s^{k_1})$, $\Omega^{-1}A \leq 0(s^{k_2})$, and

$$\Omega^{-1}B = -\Omega^{-1}AG_{22} \leq 0(s^{k_2+k_3}).$$

Clearly, Γ_a is strictly proper. It then follows that

$$A_1\Lambda^{-1}\Gamma_a\Omega^{-1}B \leq 0(s^{k_1+k_2+k_3-1})$$

is strictly proper when $k_1 + k_2 + k_3 \leq 0$. Thus,

$$\det(I - A_1\Lambda^{-1}\Gamma_a\Omega^{-1}B) \neq 0$$

is true and

$$A_1\Lambda^{-1}\Gamma_a\Omega^{-1}A \leq 0(k_1 + k_2 - 1) \Rightarrow C \leq 0(s^{k_1+k_2-1}).$$

Example* **3.23** (*An example of the optimal 1DOF $\mathcal{H}_2$ design for a strictly acceptable input*) Consider the 1DOF system in Fig. 3.3 in Sect. 3.7. The linearized plant model for submarine motion at 30 knots speed is represented in the form

$$\dot{x} = \Sigma x + Nv,\ v = [v_1 \quad v_2]',\ y = [y_1 \quad y_2]' = Hx,$$

$$\Sigma = \begin{bmatrix} -0.19003 & 4.4802 & 0 & 0.0014673 \\ 0.0085526 & -0.45988 & 0 & -0.0056095 \\ 1 & 0 & 0 & -15.433 \\ 0 & 1 & 0 & 0 \end{bmatrix},$$

$$N = \begin{bmatrix} -0.1855 & -0.57149 \\ 0.043308 & -0.055543 \\ 0 & 0 \\ 0 & 0 \end{bmatrix},$$

$$H = \begin{bmatrix} 0 & 0 & 1 & 0 \\ 0 & 0 & 0 & 1 \end{bmatrix},$$

where the outputs $y_1 = \delta D$ and $y_2 = \delta\theta$ indicate the changes of depth and pitch from the nominal values, respectively, and the control variables $v_1 = \delta b$ and $v_2 = \delta s$ indicate the changes in bow and stern plane angles, respectively (Grimble 2006, Sect. 10.4). That is, the transfer functions of the plant $P(s)$ in Fig. 3.3 are given by

$$\frac{\delta D}{\delta b} = \frac{-0.1855(s + 2.81906)(s + 0.19793)}{s(s + 0.55476)(s + 0.066663)(s + 0.028485)},$$

$$\frac{\delta D}{\delta s} = \frac{-0.57149(s-1.0113)(s+0.40668)}{s(s+0.55476)(s+0.066663)(s+0.028485)},$$

$$\frac{\delta\theta}{\delta b} = \frac{0.043308(s+0.153397)}{(s+0.55476)(s+0.066663)(s+0.028485)},$$

$$\frac{\delta\theta}{\delta s} = \frac{-0.055543(s+0.27803)}{(s+0.55476)(s+0.066663)(s+0.028485)}.$$

The disturbance $d(s)$, reference input $r(s)$ and measurement noise $n(s)$ in Fig. 3.3 are modeled as the following ones based on Sect. 10.4 of Grimble (2006);

$$d(s) = \Omega_d(s)\mu_d, r = A_r^{-1}B_r\sigma_r + \Omega_r\mu_r, n = \Omega_n(s)\mu_n$$

Here $\mu = [\mu_d' \quad \mu_r' \quad \mu_n']'$ is a column whose elements are independent zero-mean unit variance white-noise and σ_r is a vector whose elements are real random variables with zero mean and identity covariance matrix. The stable matrices Ω_d, Ω_n and Ω_r are given by

$$\Omega_d(s) = d_w^{-1}(s)\begin{bmatrix} \frac{0.4s}{(s+0.1)} & 0 \\ 0 & \frac{0.4s^2}{(s+0.1)(s+0.04)} \end{bmatrix},$$

$$\Omega_n = \begin{bmatrix} 0.02 & 0 & 4s^2/d_w & 0 \\ 0 & 0.02 & 0 & 0 \end{bmatrix},$$

$$d_w = s^4 + 8.555s^3 + 2.44s^2 + 1.884s + 0.15,$$

$$\Omega_r = \frac{0.02}{s+0.0001}\begin{bmatrix} 1 & 0 \\ 0 & 1 \end{bmatrix}.$$

The shape deterministic part of $r(s)$ is described by

$$A_r = \begin{bmatrix} s & 0 \\ 0 & s+0.02 \end{bmatrix}, \quad B_r = \begin{bmatrix} 1 & 0 \\ 0 & -1 \end{bmatrix},$$

where the -1 in B_r is chosen so that with a positive step increase command in depth the pitch command is negative. The effects of ocean currents, second-order wave and suction in Grimble (2006) are not considered here;

Design the optimal $\mathcal{H}_2$ controller of the 1DOF configuration in Fig. 3.3 in Sect. 3.7 with appropriate choices of the weighting matrices $W(s)$ and $G_w(s)$ (Assume that $L = 0$ and $F = P_d = T_d = I$ in the figure).

Solution The 1DOF configuration in Fig. 3.3 in Sect. 3.7 is considered with $L = 0$ and $F = I$. The matrices T_d and P_d in the figure are set as the identity ones; hence, $P_o = I$. The 1DOF system is admissible since $F = I$ and it is also compatible by

Lemma 3.4 since $T_d = I$. A coprime polynomial matrix fraction description of the plant $P(s) = H(sI - \Sigma)^{-1}N = A^{-1}(s)B(s)$ is given by

$$A(s) = \begin{bmatrix} -s(s+0.19003) & -10.95278s - 2.93127 \\ 0.008553s & -s^2 - 0.45988s + 0.12638 \end{bmatrix},$$

$$B(s) = \begin{bmatrix} 0.1855 & 0.57149 \\ -0.04331 & 0.055543 \end{bmatrix}.$$

This fractional description is obtained by using the command *ss2lmf* of the MATLAB Polynomial Toolbox. Invoking the standard model in Fig. 3.2 in Sect. 3.3 with the setup

$$w = \begin{bmatrix} d' & r' & n' \end{bmatrix}', y_\delta = \begin{bmatrix} \varepsilon' & v' \end{bmatrix}', y_m = u, v = v,$$

(the left-hand side variables are the ones in Fig. 3.2 and the right-hand side variables are the ones in Fig. 3.3.), one obtains the generalized matrices $G_{ij}(s)$ in Fig. 3.2 as

$$G_{11} = \begin{bmatrix} -I_2 & I_2 & 0 \\ 0 & 0 & 0 \end{bmatrix}, G_{12} = \begin{bmatrix} -P(s) \\ I_2 \end{bmatrix}, G_{21} = \begin{bmatrix} -I_2 & I_2 & -I_2 \end{bmatrix}, G_{22} = -P(s).$$

The weighting constants α_1 and α_2 are chosen as $\alpha_1 = \alpha_2 = 1$ and it follows that

$$G_w = \begin{bmatrix} 0 & 0 & 0 & \Omega_d & 0 & 0 \\ 0 & A_r^{-1}B_r & 0 & 0 & \Omega_r & 0 \\ 0 & 0 & 0 & 0 & 0 & \Omega_n \end{bmatrix},$$

with the setup $q = [\sigma_d' \; \sigma_r' \; \sigma_n' \; \mu_d' \; \mu_r' \; \mu_n']'$. The reason of taking the element at (2,2) position of A_r as $s + 0.02$, instead of s, is to meet the assumption that the matrix $\nabla_c = AG_{21}G_w$ has full row rank on the finite $j\omega$-axis.

It can be shown that a pair of left coprime polynomial matrices for $G_w = A_w^{-1}B_w$ is given by

$$A_w = diag\{D_d, d_r A_r, D_n\}, B_w = \begin{bmatrix} 0 & 0 & 0 & N_d & 0 & 0 \\ 0 & d_r B_r & 0 & 0 & 0.02A_r & 0 \\ 0 & 0 & 0 & 0 & 0 & N_n \end{bmatrix}$$

where

$$D_d = diag\{(s+0.1)d_w, (s+0.1)(s+0.04)d_w\}, D_n = diag\{d_w, 1\},$$
$$d_r = s + 0.0001,$$

$$N_d = diag\{0.4s, 0.4s^2\}, N_n = \begin{bmatrix} 0.02d_w & 0 & 4s^2 & 0 \\ 0 & 0.02 & 0 & 0 \end{bmatrix}.$$

Since $AG_{21}A_w^{-1} = \begin{bmatrix} -AD_d^{-1} & d_r^{-1}AA_r^{-1} & -AD_n^{-1} \end{bmatrix}$ is stable, the input weighting matrix G_w is strictly acceptable from the results in (b) of Example 3.21. In this case, it is convenient to exploit the optimal $\mathcal{H}_2$ controller formula in Theorem 3.5′ in Example 3.22 if the assumptions for the theorem are met.

Performance of the optimal $\mathcal{H}_2$ design is greatly affected by the choice of the weighting matrix $W(s)$. The matrix $W(s) = diag\{W_e, W_v\}$ in this example is chosen as

$$W_e = \begin{bmatrix} W_{e1} & 0 \\ 0 & W_{e2} \end{bmatrix} = \begin{bmatrix} \frac{300(s+1)}{s+30} & 0 \\ 0 & \frac{(s+1)}{s+0.01} \end{bmatrix}, W_v = \begin{bmatrix} \frac{30(s+1)}{s+30} & 0 \\ 0 & \frac{30(s+1)}{s+30} \end{bmatrix}.$$

The weighting functions W_{e1} is chosen as a high-pass filter in order to suppress the overshoot of the output $y_1 = \delta D$ stimulated by the actuator step input of δb. An efficient way to remove the steady-state errors for step inputs is to insert the integrators as a part of the controller $C(s)$. Such approach is not taken here. Instead, the weighting function W_{e2} is chosen as a low-pass filter in order for the output $y_2 = \delta\theta$ to track the step reference input δs with a small steady-state error. The output $y_1 = \delta D$ is to track the step reference input of δb or δs with zero steady-state error since the transfer functions $\delta D/\delta b$ and $\delta D/\delta s$ have integrator terms. The weighting function W_v is chosen as a high-pass filter to suppress the measurement noise. The filter gains are determined by observing the simulation results.

It is easy to verify that all assumptions in Theorem 3.5′ are satisfied. In fact, it can be shown that $A_{o1} = A_1$ and the matrix $\nabla_b = -WG_{12}A_1$ has full column rank on the finite $j\omega$-axis. The matrix $\nabla_c = AG_{21}G_w$ has full row rank on the finite $j\omega$-axis. Specifically, since

$$\nabla_c = \begin{bmatrix} -AD_d^{-1} & d_r^{-1}AA_r^{-1} & -AD_n^{-1} \end{bmatrix} \begin{bmatrix} 0 & 0 & 0 & N_d & 0 & 0 \\ 0 & d_rB_r & 0 & 0 & 0.02A_r & 0 \\ 0 & 0 & 0 & 0 & 0 & N_n \end{bmatrix}$$

$$= \begin{bmatrix} 0 & AA_r^{-1}B_r & 0 & -A\Omega_d & 0.02d_r^{-1}A & -A\Omega_n \end{bmatrix},$$

it has full row rank on the finite $j\omega$-axis when $AA_r^{-1}B_r$ does. It is not difficult to show that $\det(AA_r^{-1}B_r)$ has no zeros on the finite $j\omega$-axis, which implies that $AA_r^{-1}B_r$ has full row rank on the finite $j\omega$-axis. The order relationships in Theorem 3.5′ are satisfied with $k_1 = k_2 = 0$ and $k_3 = -1$. The optimal $\mathcal{H}_2$ controller is obtained by the formula

$$C = (I - A_1\Lambda^{-1}\Gamma_a\Omega^{-1}B)^{-1}A_1\Lambda^{-1}\Gamma_a\Omega^{-1}A,$$

where

$$\Gamma_a = \{\Lambda_*^{-1}\nabla_{b*}WG_{11}G_w\nabla_{c*}\Omega_*^{-1}\}_+ + \{\Lambda A_1^{-1}Y\Omega\}_-,$$

and the matrices Λ and Ω are the square stable Wiener–Hopf spectral factors with stable inverses satisfying

$$\nabla_{b*}\nabla_b = \Lambda_*\Lambda, \nabla_c\nabla_{c*} = \Omega\Omega_*.$$

It is noted here that the term $\{\Lambda A_1^{-1}Y\Omega\}_- = \{\Lambda Y_1 A^{-1}\Omega\}_-$ is zero since A^{-1} has no poles in $\text{Re}\, s < 0$ and thus the matrices X, Y, X_1, and Y_1 in (3.10) are not needed to obtain the optimal solution. For a strictly acceptable G_w, the optimal solution can be more conveniently obtained from the state-space formulas in Chap. 5. In fact, state-space formulas of the rational matrices $A_1\Lambda^{-1}$, $\Omega^{-1}A$, Γ_a, and

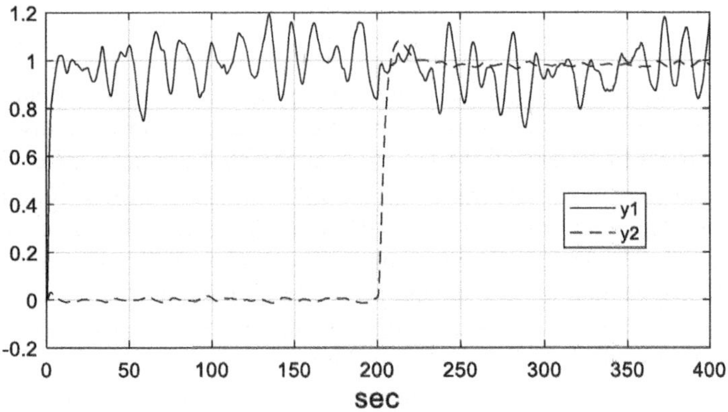

Fig. 3.4 Unit step responses of the outputs to reference changes in depth and pitch ($y_1 = \delta D, y_2 = \delta\theta$)

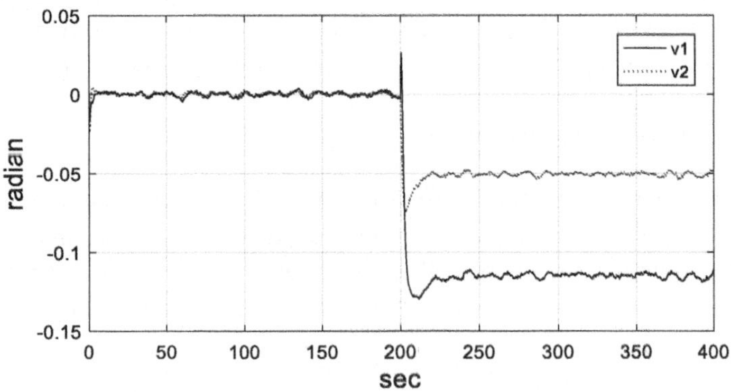

Fig. 3.5 Control activities due to reference changes in depth and pitch ($v_1 = \delta b, v_2 = \delta s$)

$$C = (I - A_1\Lambda^{-1}\Gamma_a\Omega^{-1}B)^{-1}A_1\Lambda^{-1}\Gamma_a\Omega^{-1}A$$

can be found in (5.66), (5.80), (5.83), and (5.89), respectively.

The optimal $\mathcal{H}_2$ controller is successfully obtained by using the formula in (5.89). The number of the states of the optimal controller is 26 and is not presented here. The unit step responses of y_1 and y_2 are shown in Fig. 3.4. The step reference inputs for y_1 and y_2 are exerted at $t = 0$ and $t = 200$, respectively. The actuating controls for these reference inputs are shown in Fig. 3.5. The actuating signals seem to change rather rapidly at $t = 200$ in Fig. 3.5. However, it is unlikely that a step input is a reference to the system in real operations; hence, the rapid actuating would not arise. The steady-state error of y_2 with respect to its step reference input is not exactly zero due to lack of an integrator in the corresponding loop and if one tries to make the error small, the overshoot of y_2 tends to increase.

Example* 3.24 (*An example of the optimal 2DOF $\mathcal{H}_2$ design for an acceptable input*) The Rosenbrock's process, whose transfer matrix is given below, is a typical multivariable system which shows a severe coupling phenomenon (Åström et al. 2002);

$$\begin{bmatrix} y_1 \\ y_2 \end{bmatrix} = P_0(s)\begin{bmatrix} v_1 \\ v_2 \end{bmatrix}, P_0(s) = \begin{bmatrix} \dfrac{1}{s+1} & \dfrac{2}{s+3} \\ \dfrac{1}{s+1} & \dfrac{1}{s+1} \end{bmatrix}.$$

The transfer matrix has a non-minimum phase zero at $s = 1$ and this yields fundamental limitations on control system performance. Design the optimal 2DOF $\mathcal{H}_2$ controller for the configuration in Fig. 1.1 in Sect. 1.2 with appropriate choices of the weighting matrices $W(s)$ and $G_w(s)$ (Remove the feedforward part and assume $F = I$ in the figure).

Solution This example concerns the design of an analog 2DOF control system which is the special case of the 3DOF system shown in Fig. 1.1 with $L = 0$, $n_l = 0$, and $\Omega_{n_l} = 0$. Hence, the design formulas for the 2DOF system can be obtained from those for the 3DOF system by removing from all matrices any block rows and block columns that contain the variables L, n_l, and Ω_{n_l}. In particular, for the 2DOF system one gets in place of (3.348), (3.350), (3.407), (3.408), (3.412), and (3.413) the reductions

$$\begin{aligned}
G_s &= \mathrm{diag}\{G_d, G_r, G_{n_r}, G_{n_m}\},\\
G_t &= \Omega_t\Omega_{t*} = A(FP_dG_dP_{d*}F_* + G_{n_m})A_*,\\
\widetilde{\Gamma}_w &= \Lambda_*^{-1}A_{1*}P_*P_dG_dP_{d*}F_*A_*\Omega_{t*}^{-1},\\
\widetilde{\Gamma}_a &= \Lambda Y_1 A^{-1}\Omega_t,\\
R_w &= A_1\Lambda^{-1}(Z_m + \{\widetilde{\Gamma}_w\}_+ + \{\widetilde{\Gamma}_a\}_-)\Omega_t^{-1}A,
\end{aligned}$$

and

$$C = [\,-C_w \quad C_u\,] = (X_1 - K_1B)^{-1}[\,-(Y_1 + K_1A) \quad H_1\,].$$

It is assumed that $P_d(s) = F(s) = T_d(s) = I$ for the 2DOF system; thus, it is admissible since $F = I$ and it is compatible since $T_d = I$. In order to eliminate the steady state errors for step reference inputs, the integrators $s^{-1}I_2$ are inserted in front of the plant and the augmented plant

$$P_{aug}(s) = P(s) = \begin{bmatrix} \dfrac{1}{s(s+1)} & \dfrac{2}{s(s+3)} \\ \dfrac{1}{s(s+1)} & \dfrac{1}{s(s+1)} \end{bmatrix}$$

is considered as the plant $P(s)$ in Fig. 1.1. The integrators are implemented as a part of the controller after the design of the optimal controller is carried out. The 2DOF system with the augmented plant is also admissible and compatible since $F = I$ and $T_d = I$. Coprime polynomial matrix fraction descriptions for the augmented plant are given by

$$\begin{aligned}
A(s) &= \begin{bmatrix} s(s+1)(s+3) & 0 \\ 0 & s(s+1) \end{bmatrix}, B(s) = \begin{bmatrix} s+3 & 2(s+1) \\ 1 & 1 \end{bmatrix},\\
A_1(s) &= \begin{bmatrix} s(s+1) & 0 \\ 0 & s(s+1)(s+3) \end{bmatrix}, B_1(s) = \begin{bmatrix} 1 & 2(s+1) \\ 1 & s+3 \end{bmatrix}.
\end{aligned}$$

With the choice of $\alpha_1 = \alpha_2 = 1$, the input weighting matrix G_w becomes

$$G_w = \begin{bmatrix} A_d^{-1}B_d & 0 & \Omega_d & 0 & 0 & 0 \\ 0 & A_r^{-1}B_r & 0 & \Omega_r & 0 & 0 \\ 0 & 0 & 0 & 0 & \Omega_{n_r} & 0 \\ 0 & 0 & 0 & 0 & 0 & \Omega_{n_m} \end{bmatrix},$$

and its element values are selected as

$$A_r = A_d = sI_2, B_r = B_d = I_2, \Omega_r = \Omega_d = 0, \Omega_{n_r} = \Omega_{n_m} = 0.1\,I_2.$$

The weighting matrices to the control variable are simply chosen as $kQ = I_2$. Then it is not difficult to verify that Assumptions 3.6–3.10 in Sect. 3.8 are satisfied. Thus, G_w is acceptable (note that $A_p = A$ since $F = I$). As for the optimal solution formulas, since the shape deterministic inputs in G_w only have a pole at $s = 0$, the more convenient formulas in (3.432) and (3.435) can be used instead of the ones in (3.404) and (3.411). Furthermore, $\{\widetilde{\Gamma}_a\}_- = \{\widetilde{\Gamma}_b\}_- = 0$ since A_1^{-1} has no poles in Re $s > 0$. It then follows, since $\Omega_t = \widetilde{\Omega}_t$ for the 2DOF case, that

$$R_u = A_1\Lambda^{-1}\{\widetilde{\Gamma}_r\}_+\widetilde{\Omega}_o^{-1}, R_w = A_1\Lambda^{-1}\{\widetilde{\Gamma}_w\}_+\widetilde{\Omega}_t^{-1}A,$$
$$\widetilde{\Gamma}_r = \Lambda_*^{-1}A_{1*}P_*A_r^{-1}B_rB_{r*}A_{r*}^{-1}\widetilde{\Omega}_{o*}^{-1}, \qquad \widetilde{\Gamma}_w = \Lambda_*^{-1}A_{1*}PA_d^{-1}B_dB_{d*}A_{d*}^{-1}A_*\widetilde{\Omega}_{t*}^{-1},$$

where Λ, $\widetilde{\Omega}_o$ and $\widetilde{\Omega}_t$ are Wiener–Hopf spectral factors of the equations

$$A_{1*}(P_*P + kQ)A_1 = \Lambda_*\Lambda, A_r^{-1}B_rB_{r*}A_{r*}^{-1} + \Omega_{n_r}\Omega_{n_r*} = \widetilde{\Omega}_o\widetilde{\Omega}_{o*},$$
$$A(A_d^{-1}B_dB_{d*}A_{d*}^{-1} + G_{n_m})A_* = \widetilde{\Omega}_t\widetilde{\Omega}_{t*}.$$

The computational results are given by

$$\Lambda = \begin{bmatrix} 0.5s^2 + 1.0576s + 1.036 & 0.40925s^2 + 2.0396s + 2.1088 \\ 0.40925s + 0.96262 & 0.5s^3 + 2.6104s^2 + 4.4816s + 2.9246 \end{bmatrix},$$
$$\widetilde{\Omega}_o = \begin{bmatrix} 1 + 0.1s & 0 \\ 0 & 1 + 0.1s \end{bmatrix}\frac{1}{s}, \qquad ,$$
$$\widetilde{\Omega}_t = \begin{bmatrix} 0.1s^3 + 1.4s^2 + 4.3s + 3 & 0 \\ 0 & 0.1s^2 + 1.1s + 1 \end{bmatrix}.$$

This example is a special case in that $\widetilde{\Omega}_o^{-1} = \widetilde{\Omega}_t^{-1}A$, and it therefore follows that $\{\widetilde{\Gamma}_r\}_+ = \{\widetilde{\Gamma}_w\}_+$, $R_u = R_w$, and $C_u = C_w$. The computational results are given by

$$\{\widetilde{\Gamma}_r\}_+ = \{\widetilde{\Gamma}_w\}_+ = \frac{1}{s}\begin{bmatrix} 0.99933 & 0.036708 \\ -0.036708 & 0.99933 \end{bmatrix},$$
$$R_w = R_u = \frac{1}{d_r(s)}\begin{bmatrix} r_{11}(s) & r_{12}(s) \\ r_{21}(s) & r_{22}(s) \end{bmatrix},$$

$$r_{11} = 19.987s(s+2.6979)(s^2+2.5529s+2.2254),$$
$$r_{12} = 0.73416s(s-21.941)(s+3.4322)(s+1.447),$$
$$r_{21} = -0.73416s(s+21.911)(s+3)(s+2.4866),$$
$$r_{22} = 19.987s(s+3)(s^2+2.0852s+2.0013),$$
$$d_r = (s+10)(s+2.6853)(s+0.36635)(s^2+3.2843s+4.066).$$

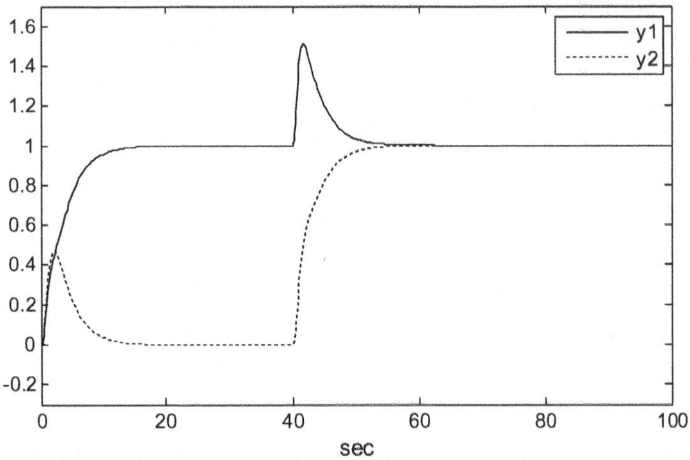

Fig. 3.6 Unit step responses of the outputs to reference changes

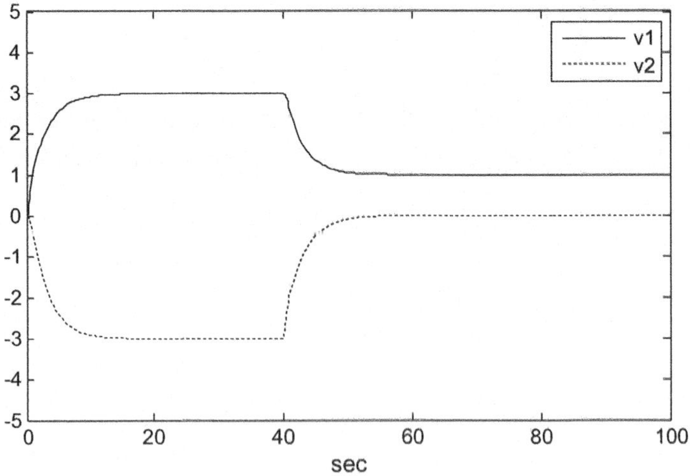

Fig. 3.7 Control activities due to reference changes

The controller matrices C_w and C_u are obtained by the formulas $C_w = C_u = (I - R_w P)^{-1} R_w$ and the computational results are given by

$$C_w(s) = C_u(s) = \frac{1}{d_c(s)} \begin{bmatrix} c_{11}(s) & c_{12}(s) \\ c_{21}(s) & c_{22}(s) \end{bmatrix},$$

$$\begin{aligned} c_{11} &= 19.987(s+9.7345)(s+1)(s^2+5.5164s+7.9224), \\ c_{12} &= 0.73416(s-20.476)(s+13.266)(s+1)(s+0.14786), \\ c_{21} &= -0.73416(s+3)(s+1)(s^2+34.398s+298.46), \\ c_{22} &= 19.987(s+9.7385)(s+3)(s+2.3467)(s+1), \\ d_c &= (s+10.135)(s+9.1359)(s+4.5579)(s^2+3.5067s+4.132). \end{aligned}$$

The McMillan degree of the controller $C(s) = [-C_w \quad C_u]$ is 5 and hence the final controller $C(s)\frac{1}{s}$ can be implemented as a state-space model with 7 states. The above optimal $\mathcal{H}_2$ controller can also be obtained by using the state-space formula in Park et al. (2002). It is confirmed that the formula in Park et al. (2002) gives the same results as the above ones.

The unit step responses of y_1 and y_2 are shown in Fig. 3.6. The step reference inputs for y_1 and y_2 are exerted at $t = 0$ and $t = 40$, respectively. The actuating controls for this reference inputs are shown in Fig. 3.7.

If the augmented plant $P_{aug}(s)$ is regarded as the plant $P(s)$ in Fig. 1.1 and the control variable $v(s)$ is regarded as the input to the inserted integrators, the cost $J = \|WG_\delta G_w\|_2^2$ for the $\mathcal{H}_2$ design in this example is 21.869.

Example* 3.25 Establish that the expressions for the functional J given by (3.129) and (3.130) in the digital case are equivalent. Also, show the same is true for (3.175) and (3.176) in the analog case.

Solution From Theorem 3.2 for the digital case, the expression for J in (3.129) is given by

$$J = \frac{1}{2\pi j} \oint_{|\lambda|=1} \{\mathrm{Tr}\,[ZZ_*] + \mathrm{Tr}(\nabla_a \nabla_{a*}) - \mathrm{Tr}\,(\widetilde{\Gamma}\widetilde{\Gamma}_*) + \mathrm{Tr}\,[\{\widetilde{\Gamma}\}_-\{\widetilde{\Gamma}\}_{-*}]\} \frac{d\lambda}{\lambda}$$

Thus, it only remains to show that

$$\begin{aligned} \widetilde{J} &= \frac{1}{2\pi j} \oint_{|\lambda|=1} \{\mathrm{Tr}\,(\widetilde{\Gamma}\widetilde{\Gamma}_*) - \mathrm{Tr}\,[\{\widetilde{\Gamma}\}_-\{\widetilde{\Gamma}\}_{-*}]\} \frac{d\lambda}{\lambda} \\ &= \frac{1}{2\pi j} \oint_{|\lambda|=1} \mathrm{Tr}\,[(\{\widetilde{\Gamma}\}_+ + \{\widetilde{\Gamma}\}_\infty)(\{\widetilde{\Gamma}\}_+ + \{\widetilde{\Gamma}\}_\infty)_*]\} \frac{d\lambda}{\lambda} \end{aligned}$$

Since $\widetilde{\Gamma}$ is good, $\{\widetilde{\Gamma}\}_+$ is stable. Moreover, $\widetilde{\Gamma} = \{\widetilde{\Gamma}\}_+ + \{\widetilde{\Gamma}\}_- + \{\widetilde{\Gamma}\}_\infty$. Hence,

$$2\pi j\widetilde{J} = \oint_{|\lambda|=1} \{\mathrm{Tr}\,[(\{\widetilde{\Gamma}\}_+ + \{\widetilde{\Gamma}\}_- + \{\widetilde{\Gamma}\}_\infty)(\{\widetilde{\Gamma}\}_+ + \{\widetilde{\Gamma}\}_- + \{\widetilde{\Gamma}\}_\infty)_*] - \mathrm{Tr}\,[\{\widetilde{\Gamma}\}_-\{\widetilde{\Gamma}\}_{-*}]\}\frac{d\lambda}{\lambda}$$

$$= \oint_{|\lambda|=1} \{\mathrm{Tr}\,[\{\widetilde{\Gamma}\}_+\{\widetilde{\Gamma}\}_{+*} + \{\widetilde{\Gamma}\}_+\{\widetilde{\Gamma}\}_{-*} + \{\widetilde{\Gamma}\}_+\{\widetilde{\Gamma}\}_{\infty *}]\}\frac{d\lambda}{\lambda}$$

$$+ \oint_{|\lambda|=1} \{\mathrm{Tr}\,[\{\widetilde{\Gamma}\}_-\{\widetilde{\Gamma}\}_{+*} + \{\widetilde{\Gamma}\}_-\{\widetilde{\Gamma}\}_{\infty *}]\}\frac{d\lambda}{\lambda}$$

$$+ \oint_{|\lambda|=1} \{\mathrm{Tr}\,[(\{\widetilde{\Gamma}\}_\infty\{\widetilde{\Gamma}\}_{+*} + \{\widetilde{\Gamma}\}_\infty\{\widetilde{\Gamma}\}_{-*} + \{\widetilde{\Gamma}\}_\infty\{\widetilde{\Gamma}\}_{\infty *}]\}\frac{d\lambda}{\lambda}.$$

$$= \oint_{|\lambda|=1} \{\mathrm{Tr}\,[\{\widetilde{\Gamma}\}_+\{\widetilde{\Gamma}\}_{+*} + \{\widetilde{\Gamma}\}_+\{\widetilde{\Gamma}\}_{-*} + \{\widetilde{\Gamma}\}_-\{\widetilde{\Gamma}\}_{+*}]\}\frac{d\lambda}{\lambda}$$

$$+ \oint_{|\lambda|=1} \{\mathrm{Tr}\,[(\{\widetilde{\Gamma}\}_+ + \{\widetilde{\Gamma}\}_-)\{\widetilde{\Gamma}\}_{\infty *}]\}\frac{d\lambda}{\lambda}$$

$$+ \oint_{|\lambda|=1} \{\mathrm{Tr}\,[\{\widetilde{\Gamma}\}_\infty(\{\widetilde{\Gamma}\}_{+*} + \{\widetilde{\Gamma}\}_{-*})]\}\frac{d\lambda}{\lambda} + \oint_{|\lambda|=1} \{\mathrm{Tr}\,[\{\widetilde{\Gamma}\}_\infty\{\widetilde{\Gamma}\}_{\infty *}]\}\frac{d\lambda}{\lambda}.$$

Now it follows from (3.121) with $Z = \{\widetilde{\Gamma}\}_+$ that

$$\oint_{|\lambda|=1} \{\mathrm{Tr}\,[\{\widetilde{\Gamma}\}_+\{\widetilde{\Gamma}\}_{-*} + \{\widetilde{\Gamma}\}_-\{\widetilde{\Gamma}\}_{+*}]\}\frac{d\lambda}{\lambda} = 0.$$

Moreover, one gets from (3.119) and (3.120) that

$$\oint_{|\lambda|=1} \{\mathrm{Tr}\,[\{\widetilde{\Gamma}\}_-)\{\widetilde{\Gamma}\}_{\infty *}]\}\frac{d\lambda}{\lambda} = \oint_{|\lambda|=1} \{\mathrm{Tr}\,[\{\widetilde{\Gamma}\}_\infty\{\widetilde{\Gamma}\}_{-*}]\}\frac{d\lambda}{\lambda} = 0$$

since $\{\widetilde{\Gamma}\}_\infty$ and $\lambda^{-1}\{\widetilde{\Gamma}\}_{-*}$ are analytic in $|\lambda| \le 1$. Thus,

$$2\pi j\widetilde{J} = \oint_{|\lambda|=1} \{\mathrm{Tr}\,[\{\widetilde{\Gamma}\}_+\{\widetilde{\Gamma}\}_{+*} + \{\widetilde{\Gamma}\}_+\{\widetilde{\Gamma}\}_{\infty *} + \{\widetilde{\Gamma}\}_\infty\{\widetilde{\Gamma}\}_{+*} + \{\widetilde{\Gamma}\}_\infty\{\widetilde{\Gamma}\}_{\infty *}]\}\frac{d\lambda}{\lambda}$$

$$= \oint_{|\lambda|=1} \{\mathrm{Tr}\,[(\{\widetilde{\Gamma}\}_+ + \{\widetilde{\Gamma}\}_\infty)(\{\widetilde{\Gamma}\}_+ + \{\widetilde{\Gamma}\}_\infty)_*]\}\frac{d\lambda}{\lambda}.$$

This completes the proof of Theorem 3.2.

Similarly, it follows from (3.175) and (3.176) for the analog case that

$$J = \frac{1}{2\pi j} \int_{-j\infty}^{j\infty} \{\mathrm{Tr}\,[ZZ_*] + \mathrm{Tr}\,(\nabla_a \nabla_{a*}) - \mathrm{Tr}\,(\widetilde{\Gamma}\widetilde{\Gamma}_*) + \mathrm{Tr}\,[\{\widetilde{\Gamma}\}_-\{\widetilde{\Gamma}\}_{-*}]\}ds.$$

Thus, it only remains to show that

$$\widetilde{J} = \frac{1}{2\pi j} \int_{-j\infty}^{j\infty} \{\mathrm{Tr}\,(\widetilde{\Gamma}\widetilde{\Gamma}_*) - \mathrm{Tr}\,[\{\widetilde{\Gamma}\}_-\{\widetilde{\Gamma}\}_{-*}]\}ds = \frac{1}{2\pi j} \int_{-j\infty}^{j\infty} \mathrm{Tr}\,[\{\widetilde{\Gamma}\}_+\{\widetilde{\Gamma}\}_{+*}]\}ds.$$

Since $\widetilde{\Gamma}$ is good and strictly proper, $\widetilde{\Gamma} = \{\widetilde{\Gamma}\}_+ + \{\widetilde{\Gamma}\}_-$. Moreover, $\{\widetilde{\Gamma}\}_+$ is stable. Hence,

$$\begin{aligned}\widetilde{J} &= \frac{1}{2\pi j} \int_{-j\infty}^{j\infty} \mathrm{Tr}\,[(\{\widetilde{\Gamma}\}_+ + \{\widetilde{\Gamma}\}_-)(\{\widetilde{\Gamma}\}_{+*} + \{\widetilde{\Gamma}\}_{-*}) - \mathrm{Tr}\,[\{\widetilde{\Gamma}\}_-\{\widetilde{\Gamma}\}_{-*}]\}ds \\ &= \frac{1}{2\pi j} \int_{-j\infty}^{j\infty} \mathrm{Tr}\,[\{\widetilde{\Gamma}\}_+\{\widetilde{\Gamma}\}_{+*} + \{\widetilde{\Gamma}\}_+\{\widetilde{\Gamma}\}_{-*} + \{\widetilde{\Gamma}\}_{+*}\{\widetilde{\Gamma}\}_-]\}ds = \frac{1}{2\pi j} \int_{-j\infty}^{j\infty} \mathrm{Tr}\,[\{\widetilde{\Gamma}\}_+\{\widetilde{\Gamma}\}_{+*}]\}ds\end{aligned}$$

because it follows from (3.169) with $Z = \{\widetilde{\Gamma}\}_+$ that

$$\frac{1}{2\pi j} \int_{-j\infty}^{j\infty} \mathrm{Tr}\,[\{\widetilde{\Gamma}\}_+\{\widetilde{\Gamma}\}_{-*} + \{\widetilde{\Gamma}\}_{+*}\{\widetilde{\Gamma}\}_-]\}ds = 0.$$

This completes the proof of Theorem 3.4.

Example* 3.26 For a square complex matrix $\mathscr{M}$ of order q with eigenvalues $\lambda_1, \lambda_2, \cdots, \lambda_q$ establish that when $|\lambda_k| \geq |\lambda_i|$, $i = 1 \to q$, then $|\lambda_k|$ is less than or equal to the largest singular value of $\mathscr{M}$ (i.e., the positive square root of the largest eigenvalue of the matrix $\mathcal{H} = \mathscr{M}\mathscr{M}^*$).

Solution The matrix $\mathcal{H} = \mathscr{M}\mathscr{M}^*$ is hermitian and nonnegative definite. Hence, its eigenvalues are real and nonnegative. Since λ_k is an eigenvalue of $\mathscr{M}$, it follows from

$$[\det(\lambda_k I - \mathscr{M})]^* = \det(\overline{\lambda}_k I - \mathscr{M}^*) = 0$$

that $\overline{\lambda}_k$ is an eigenvalue of $\mathcal{M}^*$. When x_k is an eigenvector associated with $\overline{\lambda}_k$, then

$$\mathcal{M}^* x_k = \overline{\lambda}_k x_k \Leftrightarrow x_k^* \mathcal{M} = \lambda_k x_k^*.$$

Thus,

$$x_k^* \mathcal{H} x_k = x_k^* \mathcal{M}\mathcal{M}^* x_k = \lambda_k \overline{\lambda}_k \|x_k\|^2 = |\lambda_k|^2 \|x_k\|^2.$$

Moreover, since $\mathcal{H}$ is hermitian, there exists a unitary matrix U satisfying $UU^* = U^*U = I$ for which

$$\mathcal{H} = U^* \Sigma U, \Sigma = \text{diag}\left\{\sigma_1^2, \sigma_2^2, \cdots, \sigma_q^2\right\}.$$

where the σ_i are the singular values for $\mathcal{M}$. Clearly, with $y = Ux_k$,

$$x_k^* \mathcal{H} x_k = y^* \Sigma y = \sum_{i=1}^{q} \sigma_i^2 |y_i|^2 \le \sigma_{max^2} \|y\|^2 = \sigma_{max^2} \|x_k\|^2.$$

Hence,

$$x_k^* \mathcal{H} x_k = |\lambda_k|^2 \|x_k\|^2 \le \sigma_{max^2} \|x_k\|^2 \Leftrightarrow |\lambda_k| \le \sigma_{max}$$

So when $\sigma_{max} < 1$, all eigenvalues of $\mathcal{M}$ are less than unity in magnitude.

Chapter 4
$\mathcal{H}_2$ Design of Multivariable Control Systems with Decoupling

4.1 Overview

In some linear multivariable control system designs, it is required for some set of plant outputs that each output in the set be controlled by only one of an equal number of reference inputs. In this case, the system is said to be decoupled and the transfer matrix from the reference inputs to the plant outputs, denoted by T, is a diagonal matrix. In this chapter, all stabilizing controllers for which T is diagonal are established and the $\mathcal{H}_2$ design of decoupled systems is treated for the standard model.

Any T, whether it is diagonal or not, that can be obtained using a stabilizing controller is called *realizable*. Further characterization of T is also needed to treat $\mathcal{H}_2$ design of decoupled systems. When T can be realized with a stabilizing controller that makes the regulated variables (y_δ in Fig. 3.1) and the control variables (v in Fig. 3.1) stable for a specific set of persistent inputs, then T and the inputs are called *acceptable*. The characterization of all realizable and acceptable T is treated in Sect. 4.2.

When decoupling design is required, the structure of T is confined to be diagonal and this property is used in Sect. 4.3 to parameterize all realizable and acceptable T for decoupled systems. In Sect. 4.4, minimization of a given cost functional is carried out over all diagonal acceptable T and the formula for all controllers, including the optimal one, that yield a finite cost functional is presented. In Sects. 4.5 and 4.6, $\mathcal{H}_2$ controller formulas for 1DOF and 3DOF decoupled systems are obtained from the general formulas presented in Sect. 4.4.

J. J. Bongiorno Jr. and K. Park, *Design of Linear Multivariable Feedback Control Systems*, https://doi.org/10.1007/978-3-030-44356-6_4

4.2 Characterization of Realizable and Acceptable *T*

The standard model for decoupling design is shown in Fig. 4.1. The variable r represents the reference inputs and e_o represents the exogenous inputs other than the reference inputs. With regard to Fig. 3.1, this corresponds to the choice $e' = [r' \quad e_o']$. The variable v represents the control variables. The variables y, y_z, and y_m denote the plant outputs, other controlled variables, and the measured variables, respectively. With regard to Fig. 3.1, this corresponds to the choice $y_c' = [y' \quad y_z']$. Dimensions of the vectors r and y are the same. The objective of decoupling design is to find stabilizing controllers that make T_{yr}, the transfer matrix from r to y, diagonal and invertible.

The transfer matrix $\widetilde{P}$ of the generalized plant follows from

$$\begin{bmatrix} y \\ y_z \\ y_m \end{bmatrix} = \widetilde{P} \begin{bmatrix} r \\ e_o \\ v \end{bmatrix}, \quad \widetilde{P} = \begin{bmatrix} P_{00} & P_{01} & P_{02} \\ P_{10} & P_{11} & P_{12} \\ P_{20} & P_{21} & P_{22} \end{bmatrix} \tag{4.1}$$

when the input–output relations are given by

$$y = P_{00}\, r + P_{01}\, e_o + P_{02}\, v \tag{4.2}$$

$$y_z = P_{10}\, r + P_{11}\, e_o + P_{12}\, v \tag{4.3}$$

and

$$y_m = P_{20}\, r + P_{21}\, e_o + P_{22}\, v. \tag{4.4}$$

The generalized plant $\widetilde{P}$ is assumed to be admissible which in accordance with Definition 2.2(*i*) in Sect. 2.3 is equivalent to

Assumption 4.1 The generalized plant $\widetilde{P}(p)$ is free of hidden poles in $\operatorname{Re} p \geq 0$ for the analog case and in $|p| \leq 1$ for the digital case and the polynomial $\psi_{\widetilde{P}}/\psi_{P_{22}}$ is stable.

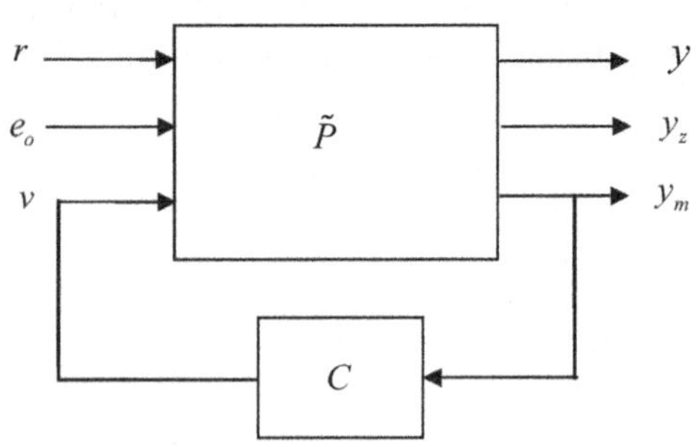

Fig. 4.1 The standard model for decoupling design (© 2009 IEEE. Reprinted with permission, from Park (2009))

Equation (4.2) accounts for the contributions from each input to the plant output y. In most cases, the plant output y is not impacted directly by the reference input r; hence, it is reasonable to assume that P_{00} is a null matrix. It is also typical that $y_m = \widetilde{F}y$, where $\widetilde{F}$ is the stable transfer matrix of a sensor or measuring device. It then follows from (4.2) to (4.4) that $P_{22} = \widetilde{F}P_{02}$. These observations are the basis for the following assumption.

Assumption 4.2 $P_{00} = 0$ and P_{22} is of the form $P_{22} = \widetilde{F}P_{02}$ with $\widetilde{F}$ stable.

With this assumption the target transfer matrix to be diagonalized is given by

$$T_{yr} = P_{02}C(I - P_{22}C)^{-1}P_{20} = P_{02}RP_{20} \tag{4.5}$$

with

$$R = C(I - P_{22}C)^{-1}. \tag{4.6}$$

Since the decoupling design requires that T_{yr} be invertible, it must have full rank (see Sect. B.2). It then follows from (4.5) that the next assumption is necessary.

Assumption 4.3 P_{02} has full row rank and P_{20} has full column rank.

Since the decoupling requirement is directly embodied in the imposition of the diagonal structure on the transfer matrix T_{yr}, it is convenient to define first all T_{yr} that can be obtained with stabilizing controllers.

Definition 4.1 A rational matrix T, whether diagonal or not, is called *realizable* for the plant $\widetilde{P}$ if there exists a stabilizing controller C that realizes the transfer matrix T_{yr} as the matrix T.

The characterization of all realizable T is established next. It is always possible to set $-P_{22} = A^{-1}B = B_1A_1^{-1}$, where $A^{-1}B$ and $B_1A_1^{-1}$ are coprime polynomial matrix fraction descriptions for which there always exist polynomial matrices X, Y, X_1, and Y_1 such that

$$\begin{bmatrix} A & B \\ Y_1 & -X_1 \end{bmatrix}\begin{bmatrix} X & B_1 \\ Y & -A_1 \end{bmatrix} = \begin{bmatrix} X & B_1 \\ Y & -A_1 \end{bmatrix}\begin{bmatrix} A & B \\ Y_1 & -X_1 \end{bmatrix} = \begin{bmatrix} I & 0 \\ 0 & I \end{bmatrix} \tag{4.7}$$

with $\det X \det X_1 \not\equiv 0$. (The coprime polynomial matrix fraction descriptions $A^{-1}B$ and $B_1A_1^{-1}$ are used to represent G, $-G_{22}$, or $-P_{22}$ in different sections of this book. It is always made clear which is intended so that no confusion arises.) The class of all stabilizing controllers for the plant $\widetilde{P}$ is characterized by the formula (see Lemma 2.2, Sect. 2.3)

$$C(p) = (X_1 - KB)^{-1}(Y_1 + KA) = (Y + A_1K)(X - B_1K)^{-1}, \tag{4.8}$$

where K is an arbitrary real rational stable matrix such that $\det(X - B_1K) \cdot \det(X_1 - KB) \not\equiv 0$. Inserting C in (4.8) into (4.6), one obtains

$$R = (Y + A_1K)A; \tag{4.9}$$

hence, the transfer matrix T_{yr} becomes

$$T_{yr} = P_{02}YAP_{20} + P_{02}A_1KAP_{20}. \tag{4.10}$$

By Assumption 4.1 and Lemma 2.3, the matrices $P_{02}YAP_{20}$, $P_{02}A_1$, and AP_{20} are stable. In view of (4.10), a rational matrix T is realizable for the plant $\widetilde{P}$ iff it is of the form

$$T = P_{02}YAP_{20} + P_{02}A_1KAP_{20}. \tag{4.11}$$

Since the matrices P_{02} and AP_{20} have row rank and column rank, respectively, there exist unimodular matrices V_a and V_b such that (see Theorem B.1)

$$P_{02}V_a = \begin{bmatrix} P_a & 0 \end{bmatrix} \text{ and } V_bAP_{20} = \begin{bmatrix} P_b \\ 0 \end{bmatrix} \tag{4.12}$$

with P_a and P_b square and invertible. Clearly, P_b is stable since AP_{20} is stable. One way to obtain the unimodular matrix V_a in (4.12) is to find a coprime polynomial matrix fraction description $P_{02} = A_{02}^{-1}B_{02}$ and to construct a row-Hermite form for B_{02}. Similarly, V_b can be obtained by introducing a coprime polynomial matrix fraction description of $AP_{20} = B_{120}A_{120}^{-1}$ and constructing a column-Hermite form for B_{120}. However, this is obviously more than is necessary.

For the system in Fig. 4.1, the tracking error is defined as

$$\varepsilon = T_d r - y, \tag{4.13}$$

where $T_d r$ can be regarded as the desired output. (In the previous chapters, the tracking error is defined by $\varepsilon = y - T_d r$. This discordance, however, does not affect the $\mathcal{H}_2$ design results in the sequel). The following assumption is critical in order to develop a tractable formula for realizable T.

Assumption 4.4 The matrix T_d is stable and there exists a stable matrix Q_o such that

$$P_{02}YAP_{20} = T_d + Q_oP_b. \tag{4.14}$$

Assumption 4.4 can be justified with regard to the practical requirement that the tracking error $\varepsilon = T_d r - y = (T_d - T_{yr})r$ be stable. First, it is natural to require that ε be stable for any stable r. Hence, $T_{\varepsilon r} = (T_d - T_{yr})$ must be stable which, since T_{yr} must be stable, requires that T_d be stable. Second, it is also natural from the standpoint of tracking persistent inputs when P_b^{-1} is unstable to require that ε be stable for any stable r_o when $r = P_b^{-1} r_o$. It then follows that

$$\begin{aligned} T_{\varepsilon r} P_b^{-1} &= (T_d - P_{02} Y A P_{20} - P_{02} A_1 K A P_{20}) P_b^{-1} \\ &= (T_d - P_{02} Y A P_{20}) P_b^{-1} - P_{02} A_1 K V_b^{-1} \begin{bmatrix} I \\ 0 \end{bmatrix} \end{aligned} \tag{4.15}$$

and, therefore,

$$-Q_o = (T_d - P_{02} Y A P_{20}) P_b^{-1} \tag{4.16}$$

must be stable. That is, Eq. (4.14) must hold.

Now it is possible to state the following theorem. The proof of the theorem is not given here since it can be obtained simply from the proof of Theorem 4.2 (see Remark 4.3 after the proof of Theorem 4.2).

Theorem 4.1 *Suppose that Assumptions* 4.1 through 4.4 *are satisfied. A rational matrix T is realizable for the plant $\widetilde{P}$ iff T, $(T_d - T)P_b^{-1}$ and $P_a^{-1} T P_b^{-1}$ are stable and there exists a stable matrix Q for which*

$$\det\left\{ I + \left[\widetilde{F} T P_b^{-1} \;\middle|\; P_{22} Y V_b^{-1} \begin{bmatrix} 0 \\ I \end{bmatrix} - B_1 Q \right] V_b A \right\} \not\equiv 0 . \tag{4.17}$$

For a realizable T, the corresponding R in (4.5) *is given by*

$$R = [\, R_{ro} \quad R_{vo} \,]\, V_b A \tag{4.18}$$

with

$$R_{ro} = V_a \begin{bmatrix} P_a^{-1} T P_b^{-1} \\ M_{21} \end{bmatrix}, \quad R_{vo} = Y V_b^{-1} \begin{bmatrix} 0 \\ I \end{bmatrix} + A_1 Q, \tag{4.19}$$

where M_{21} is an arbitrary stable matrix. For a realizable T, the corresponding stable K in (4.8) *is given by*

$$K = A_1^{-1} \left([\, R_{ro} \quad R_{vo} \,]\, V_b - Y \right). \tag{4.20}$$

Remark 4.1 When P_{20} is square, $V_b = I$ and $P_b = A P_{20}$. In this case, the formulas in (4.17) and (4.18) reduce to

$$\det(I + \widetilde{F} T P_{20}^{-1}) \not\equiv 0, \; R = V_a \begin{bmatrix} P_a^{-1} T P_{20}^{-1} \\ M_{21} A \end{bmatrix}. \tag{4.21}$$

When P_{02} is square, $V_a = I$ and $P_a = P_{02}$. In this case, the formula in (4.17) remains the same and R in (4.18) reduces to

$$R = \left[P_{02}^{-1} T P_b^{-1} \;\middle|\; Y V_b^{-1} \begin{bmatrix} 0 \\ I \end{bmatrix} + A_1 Q \right] V_b A. \tag{4.22}$$

When P_{20} and P_{02} are square, the formulas in (4.17) and (4.18) reduce to

$$\det(I + \widetilde{F} T P_{20}^{-1}) \not\equiv 0, \; R = P_{02}^{-1} T P_{20}^{-1}. \tag{4.23}$$

The characterization of all realizable T in Theorem 4.1 is enough to derive a parameterized formula of all realizable T in decoupled systems (see Theorem 4.3 in Sect. 4.3). However, further characterization of T is needed to treat $\mathcal{H}_2$ design of decoupled systems. In particular, when T is realizable and there exists a stabilizing controller for it that makes the regulated variables and the control variables stable for a specified set of persistent inputs, then T and the inputs are called *acceptable*. In order to be more precise with the definition of an acceptable T and acceptable inputs, it is first necessary to model the exogenous inputs which are treated in the $\mathcal{H}_2$ designs.

For the $\mathcal{H}_2$ design of the system in Fig. 4.1, the exogenous inputs r and e_o are modeled as done in Sect. 3.2. Moreover, as in Sects. 3.7 and 3.8, the $\mathcal{H}_2$ design problem of interest here is modeled in the form of Fig. 3.2 with the choices $w' = e' = [\, r' \quad e_o' \,]$ and $y_\delta = [\, \varepsilon \quad y_z \,]$. Specifically,

$$w = e = \begin{bmatrix} r \\ e_o \end{bmatrix} = [\, \widetilde{G}_{w1} \quad \widetilde{G}_{w2} \,] \begin{bmatrix} \sigma \\ \mu \end{bmatrix} = \widetilde{G}_w q, \; q = \begin{bmatrix} \sigma \\ \mu \end{bmatrix}, \tag{4.24}$$

where σ is a vector whose elements are real random variables with zero mean and covariance matrix $<\sigma\sigma'> = I$ and μ is a stochastic vector whose power spectral density matrix is given by $\Phi_{\mu\mu} = I$. It is assumed that σ and μ are independent processes. The matrix $\widetilde{G}_{w2}$ is assumed to be stable; hence, it follows for any constants $\alpha_1 > 0$ and $\alpha_2 \geq 0$ that according to Definition 3.2

$$G_w = \widetilde{G}_w \begin{bmatrix} \sqrt{\alpha_1} I & 0 \\ 0 & \sqrt{\alpha_2} I \end{bmatrix} = [\, \sqrt{\alpha_1} \widetilde{G}_{w1} \quad \sqrt{\alpha_2} \widetilde{G}_{w2} \,] \tag{4.25}$$

is acceptable iff $\widetilde{G}_{w1}$ is acceptable. As in Sect. 3.3, the matrix G_w is introduced here in anticipation of the optimization problem, where $\sqrt{\alpha_1} I$ and $\sqrt{\alpha_2} I$ can be viewed as the weightings on σ and μ, respectively. In many cases, some elements of the exogenous input e do not have shape-deterministic components and in this case one can assume with no loss in generality that

$$\widetilde{G}_{w1} = \begin{bmatrix} G_e \\ 0 \end{bmatrix}. \tag{4.26}$$

It should be noted that G_w is acceptable iff $G_{v1}G_e$ and $G_{\delta 1}G_e$ are both stable where $G_{v1} = G_v[I \quad 0]'$ and $G_{\delta 1} = G_\delta[I \quad 0]'$.

The regulated variable for $\mathcal{H}_2$ design considered here is

$$y_\delta = \begin{bmatrix} \varepsilon \\ y_z \end{bmatrix} = \begin{bmatrix} T_d r - y \\ y_z \end{bmatrix} \tag{4.27}$$

and the following generalized plant model for $\mathcal{H}_2$ design is used to exploit the results developed in Chap. 3 for Fig. 3.2:

$$\begin{bmatrix} y_\delta \\ y_m \end{bmatrix} = G\begin{bmatrix} e \\ v \end{bmatrix} = G\begin{bmatrix} w \\ v \end{bmatrix}, \; G = \begin{bmatrix} G_{11} & G_{12} \\ G_{21} & G_{22} \end{bmatrix}, \tag{4.28}$$

and

$$G_{11} = \begin{bmatrix} T_d & -P_{01} \\ P_{10} & P_{11} \end{bmatrix}, \; G_{12} = \begin{bmatrix} -P_{02} \\ P_{12} \end{bmatrix}, \; G_{21} = [P_{20} \; P_{21}], \; G_{22} = P_{22}. \tag{4.29}$$

Since $\widetilde{P}$ in (4.1) is admissible and T_d is stable under the assumptions made here, it is important to note that the matrices AG_{21}, $G_{12}A_1$, and $G_{11} + G_{12}YAG_{21}$ are stable; hence, the model G is compatible and this is implicit in the sequel. The vector y_z in (4.27) is the regulated variable other than the tracking error and often is the control variable itself which in view of the Eq. (4.3) corresponds to the case $P_{12} = I$, a stable matrix. It is reasonable, therefore, to make the following assumption.

Assumption 4.5 P_{12} is stable.

In order to characterize the acceptable G_w, the coprime polynomial matrix fraction descriptions

$$G_{12} = \begin{bmatrix} -P_{02} \\ P_{12} \end{bmatrix} = -A_o^{-1}B_o = -B_{o1}A_{o1}^{-1} \tag{4.30}$$

with

$$\begin{bmatrix} A_o & B_o \\ Y_{o1} & -X_{o1} \end{bmatrix}\begin{bmatrix} X_o & B_{o1} \\ Y_o & -A_{o1} \end{bmatrix} = \begin{bmatrix} X_o & B_{o1} \\ Y_o & -A_{o1} \end{bmatrix}\begin{bmatrix} A_o & B_o \\ Y_{o1} & -X_{o1} \end{bmatrix} = \begin{bmatrix} I & 0 \\ 0 & I \end{bmatrix} \tag{4.31}$$

are needed. It can be easily shown that $A_1^{-1}A_{o1}$ is stable. In fact, pre-multiplying (4.30) by $[-\widetilde{F}\ \ 0]$ yields, by Assumption 4.2, that $[-\widetilde{F}\ \ 0]G_{12} = \widetilde{F}P_{02} = P_{22} = G_{22}$ and, therefore, $[-\widetilde{F}\ \ 0]\ B_{o1}A_{o1}^{-1} = B_1A_1^{-1}$. Clearly, $[-\widetilde{F}\ 0]\ B_{o1} = B_1A_1^{-1}A_{o1}$ is stable since $\widetilde{F}$ is stable by Assumption 4.2. Then it can be concluded that $A_1^{-1}A_{o1}$ is stable since $(A_1,\ B_1)$ is a coprime pair.

In the sequel, Theorem 3.1 is applied to the model G in (4.28) to characterize the acceptable input $G_w = A_w^{-1}B_w$, where $(A_w,\ B_w)$ is a coprime polynomial pair. In Theorem 3.1, the acceptability of G_w is expressed in terms of A_w^{-1}. In this chapter, however, the acceptability condition is stated in terms of G_e which reduces computational efforts to check the acceptability of G_w. Recall that a matrix G_w is acceptable for G if there exists a stabilizing controller such that G_vG_w and $G_\delta G_w$ are both stable, where G_v and G_δ denote the stable transfer matrices from $w = e$ to v and y_δ, respectively. It then holds that G_w is acceptable for G when there exists a stabilizing controller such that $G_v\widetilde{G}_{w1} = G_v[G_e'\ \ 0']'$ and $G_\delta\widetilde{G}_{w1} = G_\delta[G_e'\ \ 0']'$ are both stable. Since $G_v = (Y + A_1K)AG_{21}$ and $G_\delta = G_{11} + G_{12}G_v$, the partitions

$$G_{11} = [G_{111}\quad G_{112}],\ G_{21} = [G_{211}\quad G_{212}] \tag{4.32}$$

that are compatible with the multiplications $G_{11}[G_e'\ \ 0']' = G_{111}G_e$ and $G_{21}[G_e'\ \ 0']' = G_{211}G_e$ are introduced. With $G_e = A_e^{-1}B_e$ a coprime polynomial matrix fraction description, G_w is acceptable for G iff

$$\Psi = G_v\begin{bmatrix} A_e^{-1} \\ 0 \end{bmatrix} = (Y + A_1K)AG_{211}A_e^{-1} \tag{4.33}$$

and

$$\Phi = G_\delta\begin{bmatrix} A_e^{-1} \\ 0 \end{bmatrix} = G_{111}A_e^{-1} + G_{12}\Psi \tag{4.34}$$

are stable. These equations have the same form as the ones in (3.20) and (3.21); hence, the same results are obtained as in Theorem 3.1 with the replacements $G_{11} \to G_{111}$, $G_{21} \to G_{211}$, and $A_w^{-1} \to A_e^{-1}$. This leads to the acceptability condition of G_w and the resulting K as follows. One can always set

$$AG_{211}A_e^{-1} = A_g^{-1}B_g = B_{g1}A_{g1}^{-1} \tag{4.35}$$

with

$$\begin{bmatrix} A_g & B_g \\ Y_{g1} & -X_{g1} \end{bmatrix}\begin{bmatrix} X_g & B_{g1} \\ Y_g & -A_{g1} \end{bmatrix} = \begin{bmatrix} X_g & B_{g1} \\ Y_g & -A_{g1} \end{bmatrix}\begin{bmatrix} A_g & B_g \\ Y_{g1} & -X_{g1} \end{bmatrix} = \begin{bmatrix} I & 0 \\ 0 & I \end{bmatrix}. \tag{4.36}$$

In this case, a matrix $G_w = A_w^{-1}B_w$ having the form shown in (4.25) and satisfying (4.26) with $G_e = A_e^{-1}B_e$ is acceptable for G iff $\Gamma_w = A_o G_{111} A_e^{-1}$ and $\Gamma_o = A_{o1}^{-1}(Y_o \Gamma_w A_{g1} - YB_{g1})$ are both stable and the associated controller can be realized in the form of (4.8) with the stable parameter K constrained by

$$K = A_1^{-1} A_{o1} \Gamma_o Y_{g1} + K_g A_g = K_s + K_g A_g, \tag{4.37}$$

where

$$K_s = A_1^{-1} A_{o1} \Gamma_o Y_{g1} \tag{4.38}$$

is stable and K_g is any stable rational matrix for which $\det(X - B_1 K) \cdot \det(X_1 - KB) \not\equiv 0$

It is convenient to introduce the following definitions for developing $\mathcal{H}_2$ design problems.

Definition 4.2 A stabilizing controller $C(p)$ in (4.8) is called *acceptable* for G and a given acceptable G_w if $G_\delta G_w$ and $G_v G_w$ are stable with this controller.

Definition 4.3 A realizable T is called *acceptable* for G and a given acceptable G_w if there exists an acceptable controller that realizes the transfer matrix T_{yr} as the matrix T.

As can be observed, the plant model $\widetilde{P}$ is used to formulate the decoupling problem and the model G is used to develop $\mathcal{H}_2$ design problems with an acceptable G_w. It is important to recognize that when the model G is given, the acceptable G_w and the acceptable C are completely identified. Specifically, all acceptable inputs G_w are implicitly identified as the ones for which $\Gamma_w = A_o G_{111} A_e^{-1}$ and $\Gamma_o = A_{o1}^{-1}(Y_o \Gamma_w A_{g1} - YB_{g1})$ are stable and for which there exist stable matrices K_g such that $\det(X - B_1 K) \cdot \det(X_1 - KB) \not\equiv 0$ when K is given by (4.37). All acceptable C for a given acceptable G_w are identified as in (4.8) with K being in the form of (4.37).

The characterization of all acceptable T is now sought. It is known that a realizable T always has the form $T = P_{02} Y A P_{20} + P_{02} A_1 K A P_{20}$ in (4.11). An acceptable T can be obtained by inserting the constrained formula of K in (4.37) into this equation. Hence, it can be concluded that a rational matrix T is acceptable for a given acceptable G_w iff it is of the form

$$T = P_{02}(Y + A_1 K_s) A P_{20} + P_{02} A_1 K_g A_g A P_{20}. \tag{4.39}$$

A useful equality for later use is

$$Y + A_1 K_s = Y + A_{o1} \Gamma_o Y_{g1} = Y + Y_o \Gamma_w A_{g1} Y_{g1} - YB_{g1} Y_{g1} \tag{4.40}$$

$$= Y + Y_o\Gamma_w Y_g A_g - Y(I - X_g A_g) \tag{4.41}$$

$$= Y_o\Gamma_w Y_g A_g + YX_g A_g = \Gamma_y A_g, \tag{4.42}$$

where

$$\Gamma_y = Y_o\Gamma_w Y_g + YX_g \tag{4.43}$$

is a stable matrix.

Since the matrix $A_g AP_{20}$ has full column rank, there exists a unimodular matrix V_g such that

$$V_g A_g AP_{20} = \begin{bmatrix} P_g \\ 0 \end{bmatrix} \tag{4.44}$$

with P_g square and invertible. Clearly, P_g is stable since AP_{20} is stable.

The following assumption is an extension of Assumption 4.4 which assures that a tractable formula for an acceptable T can be developed.

Assumption 4.4′ The matrix T_d is stable and there exists a stable matrix Q_g such that

$$P_{02}(Y + A_1 K_s)AP_{20} = T_d + Q_g P_g. \tag{4.45}$$

As in the case of Assumption 4.4, the above assumption can be justified with regard to the practical requirement that the tracking error $\varepsilon = T_d r - y = (T_d - T_{yr})r$ be stable. It is natural from the standpoint of tracking persistent inputs when P_g^{-1} is unstable to require that ε be stable for any stable r_o when $r = P_g^{-1} r_o$. From this reasoning it follows that

$$T_{\varepsilon r}P_g^{-1} = \{T_d - P_{02}(Y + A_1 K_s)AP_{20} - P_{02}A_1 K_g A_g AP_{20}\}P_g^{-1} \tag{4.46}$$

$$= \{T_d - P_{02}(Y + A_1 K_s)AP_{20}\}P_g^{-1} - P_{02}A_1 K_g V_g^{-1}\begin{bmatrix} I \\ 0 \end{bmatrix} \tag{4.47}$$

and, therefore,

$$-Q_g = \{T_d - P_{02}(Y + A_1 K_s)AP_{20}\}P_g^{-1} \tag{4.48}$$

must be stable. That is, Eq. (4.45) must hold. The matrix P_{20} is usually a minimum phase stable matrix; hence, the unstable poles of P_g^{-1} are a subset of the unstable poles of the plant and G_w. In particular, when $P_{20} = I$ and A_g^{-1} is stable, then

$P_g = A_g A$ and $r = P_g^{-1} r_o = A^{-1} A_g^{-1} r_o = A^{-1} \widetilde{r}_o$, where $\widetilde{r}_o$ is any stable vector. Since $\widetilde{F}$ is stable and the system is admissible, the unstable poles of A^{-1} are the same as the unstable poles of P_{02} which in most cases of interest corresponds to the actual plant transfer matrix. That is, the usual expectation that feedback systems can track persistent inputs with unstable poles the same as those of the plant is fulfilled in this case.

It should be noticed that when Assumption 4.4′ holds, then Assumption 4.4 is satisfied. In fact, it follows from (4.12) and (4.44) that

$$V_g A_g A P_{20} = V_g A_g V_b^{-1} \begin{bmatrix} P_b \\ 0 \end{bmatrix} = \begin{bmatrix} P_g \\ 0 \end{bmatrix}. \tag{4.49a}$$

Introducing the partition of the polynomial matrix

$$V_g A_g V_b^{-1} = \begin{bmatrix} L_{11} & L_{12} \\ L_{21} & L_{22} \end{bmatrix}. \tag{4.49b}$$

Then yields

$$V_g A_g V_b^{-1} \begin{bmatrix} P_b \\ 0 \end{bmatrix} = \begin{bmatrix} L_{11} P_b \\ L_{21} P_b \end{bmatrix} = \begin{bmatrix} P_g \\ 0 \end{bmatrix}; \tag{4.49c}$$

hence,

$$L_{11} P_b = P_g, \tag{4.49d}$$

where L_{11} is a polynomial matrix. When there exists a stable matrix Q_g such that $P_{02}(Y + A_1 K_s) A P_{20} = T_d + Q_g P_g$, then

$$P_{02} Y A P_{20} = T_d - P_{02} A_1 K_s A P_{20} + Q_g P_g = T_d - P_{02} A_1 K_s V_b^{-1} \begin{bmatrix} P_b \\ 0 \end{bmatrix} + Q_g P_g. \tag{4.50}$$

Thus, from (4.49d) and the partitioned stable matrix $P_{02} A_1 K_s V_b^{-1} = [\, L_{b1} \;\; L_{b2} \,]$ one gets

$$P_{02} Y A P_{20} = T_d - L_{b1} P_b + Q_g L_{11} P_b = T_d + (Q_g L_{11} - L_{b1}) P_b, \tag{4.51}$$

which shows that Assumption 4.4 is satisfied.

It is now possible to state and prove the following theorem.

Theorem 4.2 *Suppose that Assumptions* 4.1, 4.2, 4.3, 4.4′ *and* 4.5 *are satisfied. A rational matrix T is acceptable for a given acceptable G_w iff T, $(T_d - T)P_g^{-1}$ and $P_a^{-1}TP_g^{-1}$ are stable and there exists a stable matrix Q for which*

$$\det\left\{I+\left[\widetilde{F}TP_g^{-1} \;\middle|\; P_{22}\Gamma_y V_g^{-1}\begin{bmatrix}0\\ I\end{bmatrix}-B_1Q\right]V_gA_gA\right\}\not\equiv 0. \tag{4.52}$$

For any such T, the corresponding R in (4.5) *is given by*

$$R=\left[R_r \quad R_v\right]V_gA_gA \tag{4.53}$$

with

$$R_r = V_a\begin{bmatrix}P_a^{-1}TP_g^{-1}\\ M_{21}\end{bmatrix},\; R_v=\Gamma_yV_g^{-1}\begin{bmatrix}0\\ I\end{bmatrix}+A_1Q, \tag{4.54}$$

where M_{21} is an arbitrary stable matrix. For an acceptable T, the corresponding stable K_g in (4.37) *is given by*

$$K_g=A_1^{-1}(\left[R_r \quad R_v\right]V_g-\Gamma_y). \tag{4.55}$$

Remark 4.2 When P_{20} is square, $V_g = I$ and $P_g = A_gAP_{20}$. The formulas in (4.52) and (4.53) then reduce to

$$\det(I+\widetilde{F}TP_{20}^{-1})\not\equiv 0,\; R=V_a\begin{bmatrix}P_a^{-1}TP_{20}^{-1}\\ M_{21}A_gA\end{bmatrix}. \tag{4.56}$$

When P_{02} is square, $V_a = I$ and $P_a = P_{02}$. In this case, the formula in (4.52) remains the same and R in (4.53) reduces to

$$R=\left[P_{02}^{-1}TP_g^{-1} \;\middle|\; \Gamma_yV_g^{-1}\begin{bmatrix}0\\ I\end{bmatrix}+A_1Q\right]V_gA_gA. \tag{4.57}$$

When P_{20} and P_{02} are square, the formulas in (4.52) and (4.53) reduce to

$$\det(I+\widetilde{F}TP_{20}^{-1})\not\equiv 0,\; R=P_{02}^{-1}TP_{20}^{-1}. \tag{4.58}$$

Proof Necessity is considered first. Clearly, T in (4.39) must be stable since $P_{02}YAP_{20}$, $P_{02}A_1$, K_s and AP_{20} are stable. It follows from (4.39) and (4.12) that

$$T = [P_a \quad 0]V_a^{-1}(Y + A_1K_s)AP_{20} + [P_a \quad 0]V_a^{-1}A_1K_gA_gAP_{20}. \tag{4.59}$$

Hence,

$$P_a^{-1}T = [I \quad 0]V_a^{-1}(Y + A_1K_s)AP_{20} + [I \quad 0]V_a^{-1}A_1K_gA_gAP_{20} \tag{4.60}$$

must be stable. Next, it also follows from (4.39) and (4.44) that $T - P_{02}(Y + A_1K_s)AP_{20} = P_{02}A_1K_gV_g^{-1}[P_g' \quad 0']'$; hence,

$$\{T - P_{02}(Y + A_1K_s)AP_{20}\}P_g^{-1} = P_{02}A_1K_gV_g^{-1}\begin{bmatrix} I \\ 0 \end{bmatrix}. \tag{4.61}$$

Invoking Assumption 4.4′ then yields that

$$(T_d - T)P_g^{-1} = -P_{02}A_1K_gV_g^{-1}\begin{bmatrix} I \\ 0 \end{bmatrix} - Q_g \tag{4.62}$$

and this must be stable. Finally, rewriting (4.60) yields

$$P_a^{-1}T = [I \quad 0]V_a^{-1}(Y + A_1K_s + A_1K_gA_g)AP_{20}. \tag{4.63}$$

Hence, it follows from (4.42) to (4.63) that

$$P_a^{-1}T = [I \quad 0]V_a^{-1}(\Gamma_y + A_1K_g)A_gAP_{20} \tag{4.64}$$

$$= [I \quad 0]V_a^{-1}(\Gamma_y + A_1K_g)V_g^{-1}\begin{bmatrix} P_g \\ 0 \end{bmatrix}. \tag{4.65}$$

It can now be seen that $P_a^{-1}TP_g^{-1}$ must be stable.

It is important to recognize that if T is acceptable for a given acceptable G_w there must exist an $R = C(I - P_{22}C)^{-1}$ which has the structure of $R = (Y + A_1K)A$ with K of the form in (4.37). That is, there must exist an

$$R = (Y + A_1K)A = (Y + A_1K_s)A + A_1K_gA_gA \tag{4.66}$$

$$= \Gamma_yA_gA + A_1K_gA_gA = (\Gamma_y + A_1K_g)A_gA. \tag{4.67}$$

Clearly,

$$T = P_{02}RP_{20} = P_{02}R(A_gA)^{-1}A_gAP_{20} = [P_a \quad 0]V_a^{-1}R(A_gA)^{-1}V_g^{-1}\begin{bmatrix} P_g \\ 0 \end{bmatrix}; \tag{4.68}$$

hence, $P_a^{-1}TP_g^{-1} = [I \quad 0]\, V_a^{-1}R(A_gA)^{-1}V_g^{-1}[I \quad 0]'$ must be true. Therefore, $V_a^{-1}R(A_gA)^{-1}V_g^{-1}$ must be of the form

$$V_a^{-1}R(A_gA)^{-1}V_g^{-1} = \begin{bmatrix} P_a^{-1}TP_g^{-1} & M_{12} \\ M_{21} & M_{22} \end{bmatrix}. \tag{4.69}$$

In view of (4.67), $V_a^{-1}R(A_gA)^{-1}V_g^{-1}$ is stable; hence, $M_{12},\ M_{21}$, and M_{22} must be stable. Thus, for an acceptable T the matrix R must be of the form

$$R = V_a\begin{bmatrix} P_a^{-1}TP_g^{-1} & M_{12} \\ M_{21} & M_{22} \end{bmatrix}V_gA_gA = \left[V_a\begin{bmatrix} P_a^{-1}TP_g^{-1} \\ M_{21} \end{bmatrix} \quad V_a\begin{bmatrix} M_{12} \\ M_{22} \end{bmatrix}\right]V_gA_gA. \tag{4.70}$$

The stable matrices M_{12} and M_{22} in (4.70) are not completely arbitrary since R is constrained by the structure in (4.67). That is

$$R = \left[V_a\begin{bmatrix} P_a^{-1}TP_g^{-1} \\ M_{21} \end{bmatrix} \quad V_a\begin{bmatrix} M_{12} \\ M_{22} \end{bmatrix}\right]V_gA_gA = (\Gamma_y + A_1K_g)A_gA \tag{4.71}$$

is necessary. Multiplying (4.71) on the right by $(A_gA)^{-1}V_g^{-1}[0 \quad I]'$ yields

$$V_a\begin{bmatrix} M_{12} \\ M_{22} \end{bmatrix} = (\Gamma_y + A_1K_g)V_g^{-1}\begin{bmatrix} 0 \\ I \end{bmatrix} \tag{4.72}$$

$$= \Gamma_yV_g^{-1}\begin{bmatrix} 0 \\ I \end{bmatrix} + A_1Q, \tag{4.73}$$

where $Q = K_gV_g^{-1}[0 \quad I]'$ is a stable matrix. Hence, R must be of the form described by (4.53) and (4.54).

It remains to prove sufficiency. Solving for K_g in (4.67) and using (4.53), one obtains

$$K_g = A_1^{-1}(\ [R_r \quad R_v]\ V_g - \Gamma_y). \tag{4.74}$$

It has to be shown that this K_g is stable and that $\det(X - B_1K) \not\equiv 0$ when (4.52) is satisfied. Clearly, K_g is stable iff $[A_1' \quad B_1']'K_g$ is stable. Obviously A_1K_g is stable; hence, K_g is stable iff B_1K_g is stable. It follows from (4.74) that

$$-B_1K_g = P_{22}(\ [R_r \quad R_v]\ V_g - \Gamma_y) \tag{4.75}$$

$$= \widetilde{F}P_{02}\left(\left[V_a\begin{bmatrix} P_a^{-1}TP_g^{-1} \\ M_{21} \end{bmatrix} \middle| \ \Gamma_yV_g^{-1}\begin{bmatrix} 0 \\ I \end{bmatrix} + A_1Q \right]V_g - \Gamma_y\right) \tag{4.76}$$

$$= \tilde{F}\left(\left[P_{02}V_a\begin{bmatrix}P_a^{-1}TP_g^{-1}\\ M_{21}\end{bmatrix} \,\middle|\, P_{02}\Gamma_yV_g^{-1}\begin{bmatrix}0\\ I\end{bmatrix}+P_{02}A_1Q\right]V_g - P_{02}\Gamma_y\right) \tag{4.77}$$

$$= \tilde{F}\left(\left[TP_g^{-1} \,\middle|\, P_{02}\Gamma_yV_g^{-1}\begin{bmatrix}0\\ I\end{bmatrix}+P_{02}A_1Q\right]V_g - P_{02}\Gamma_y\right) \tag{4.78}$$

$$= \tilde{F}\left(\left[TP_g^{-1} \,\middle|\, P_{02}\Gamma_yV_g^{-1}\begin{bmatrix}0\\ I\end{bmatrix}+P_{02}A_1Q\right] - P_{02}\Gamma_yV_g^{-1}\right)V_g \tag{4.79}$$

$$= \tilde{F}\left(\left[TP_g^{-1} - P_{02}\Gamma_yV_g^{-1}\begin{bmatrix}I\\ 0\end{bmatrix} \,\middle|\, P_{02}\Gamma_yV_g^{-1}\begin{bmatrix}0\\ I\end{bmatrix}+P_{02}A_1Q - P_{02}\Gamma_yV_g^{-1}\begin{bmatrix}0\\ I\end{bmatrix}\right]\right)V_g \tag{4.80}$$

$$= \tilde{F}\left(\left[TP_g^{-1} - P_{02}\Gamma_yV_g^{-1}\begin{bmatrix}I\\ 0\end{bmatrix} \,\middle|\, P_{02}A_1Q\right]\right)V_g. \tag{4.81}$$

Since

$$A_gAP_{20}P_g^{-1} = V_g^{-1}\begin{bmatrix}I\\ 0\end{bmatrix}, \tag{4.82}$$

it follows from (4.81) that

$$-B_1K_g = \tilde{F}\left(\left[TP_g^{-1} - P_{02}\Gamma_yA_gAP_{20}P_g^{-1} \mid P_{02}A_1Q\right]\right)V_g. \tag{4.83}$$

Using the equality in (4.42) and the equality (4.45) in Assumption 4.4′, one gets

$$-B_1K_g = \tilde{F}\left(\left[TP_g^{-1} - P_{02}(Y + A_1K_s)AP_{20}P_g^{-1} \mid P_{02}A_1Q\right]\right)V_g \tag{4.84}$$

$$= \tilde{F}\left(\left[TP_g^{-1} - (T_d + Q_gP_g)P_g^{-1} \mid P_{02}A_1Q\right]\right)V_g \tag{4.85}$$

$$= \tilde{F}\left(\left[(T - T_d)P_g^{-1} - Q_g \mid P_{02}A_1Q\right]\right)V_g \tag{4.86}$$

which is stable. Hence, K_g is stable.

Finally, it is established that (4.52) is equivalent to $\det[(X - B_1K)A] \not\equiv 0$ which is the case iff $\det[(X - B_1K)] \not\equiv 0$. Clearly, from the expression for K_g in (4.74),

$$(X - B_1K)A = XA - B_1K_sA - B_1K_gA_gA \tag{4.87}$$

$$= XA - B_1K_sA - B_1A_1^{-1}([R_r \quad R_v]V_g - \Gamma_y)A_gA \tag{4.88}$$

$$= XA - B_1K_sA + \widetilde{F}P_{02}[R_r \quad R_v]V_gA_gA + B_1A_1^{-1}\Gamma_yA_gA \tag{4.89}$$

$$= XA - B_1K_sA + \widetilde{F}P_{02}[R_r \quad R_v]V_gA_gA + B_1A_1^{-1}(Y + A_1K_s)A \tag{4.90}$$

$$= XA + \widetilde{F}P_{02}[R_r \quad R_v]V_gA_gA + B_1A_1^{-1}YA \tag{4.91}$$

$$= I + \widetilde{F}P_{02}[R_r \quad R_v]V_gA_gA \tag{4.92}$$

$$= I + \widetilde{F}P_{02}\left[V_a\begin{bmatrix} P_a^{-1}TP_g^{-1} \\ M_{21}\end{bmatrix} \middle| \Gamma_yV_g^{-1}\begin{bmatrix}0 \\ I\end{bmatrix} + A_1Q\right]V_gA_gA \tag{4.93}$$

$$= I + \widetilde{F}\left[P_{02}V_a\begin{bmatrix} P_a^{-1}TP_g^{-1} \\ M_{21}\end{bmatrix} \middle| P_{02}\Gamma_yV_g^{-1}\begin{bmatrix}0 \\ I\end{bmatrix} + P_{02}A_1Q\right]V_gA_gA \tag{4.94}$$

$$= I + \widetilde{F}\left[TP_g^{-1} \middle| P_{02}\Gamma_yV_g^{-1}\begin{bmatrix}0 \\ I\end{bmatrix} + P_{02}A_1Q\right]V_gA_gA \tag{4.95}$$

$$= I + \left[\widetilde{F}TP_g^{-1} \middle| \widetilde{F}P_{02}(\Gamma_yV_g^{-1}\begin{bmatrix}0 \\ I\end{bmatrix} + A_1Q)\right]V_gA_gA \tag{4.96}$$

$$= I + \left[\widetilde{F}TP_g^{-1} \middle| P_{22}\Gamma_yV_g^{-1}\begin{bmatrix}0 \\ I\end{bmatrix} - B_1Q\right]V_gA_gA \tag{4.97}$$

and the proof of Theorem 4.2 is complete.

Remark 4.3 A comparison of the Eqs. (4.11) and (4.12) with (4.39) and (4.44) easily yields the proof of Theorem 4.2 from that of Theorem 4.1 using the substitutions of $V_g \to V_b$, $P_g \to P_b$, $A_g \to I$, and $Y + A_1K_s \to Y$ (hence $\Gamma_y \to Y$).

4.3 Characterization of Realizable Diagonal *T* and Acceptable Diagonal *T*

As a consequence of Theorem 4.1, a necessary condition for any rational matrix T to be realizable for the plant $\widetilde{P}$ is that T, $(T_d - T)P_b^{-1}$ and $P_a^{-1}TP_b^{-1}$ be stable. As a consequence of Theorem 4.2, a necessary condition for any rational matrix T to be acceptable is that T, $(T_d - T)P_g^{-1}$ and $P_a^{-1}TP_g^{-1}$ be stable. These necessary conditions apply whether T is diagonal or not. When decoupling design is considered, the structure of T is confined to be diagonal and this property is used in the sequel to parameterize all realizable T and all acceptable T for a decoupled system.

In a decoupled system, it is natural to choose a diagonal T_d in defining the tracking error $\varepsilon = T_d r - y$. It is also convenient to impose the restriction that T_d^{-1} is stable. Thus, these additional restrictions on T_d are included in

Assumption 4.6 T_d is diagonal and T_d^{-1} is stable.

Let

$$T = diag\{t_i\},\ t_i = \frac{h_i}{g_i}, \tag{4.98}$$

where h_i and g_i are polynomials. Since T must be stable, the polynomial g_i must be stable (that is, it must be Hurwitz for the analog case and Schur for digital case). Since P_b and P_g are stable, the stability of $P_a^{-1}TP_b^{-1}$ and $P_a^{-1}TP_g^{-1}$ implies the stability of $P_a^{-1}T$. For the stability of $P_a^{-1}T$, it is required that h_i cancel all unstable poles of i-th column of P_a^{-1}. Let θ_i denote the unique monic polynomial of the minimal degree such that $\{i\text{-th } column\ of\ P_a^{-1}\} \times \theta_i$ is stable and define $\Delta_\theta = diag\{\theta_i\}$. Then h_i must be of the form $h_i = m_i\theta_i$, m_i polynomial. That is, t_i must be of the form $t_i = \theta_i d_{1i}$, where $d_{1i} = m_i/g_i$ is an arbitrary stable rational function. This leads to

$$T = \Delta_\theta D_1, \tag{4.99}$$

where $D_1 = diag\{d_{1i}\}$ is an arbitrary stable diagonal matrix. For the same reason, the stability of $(T_d - T)P_b^{-1}$ for a realizable T requires that

$$T_d - T = \Delta_\varphi D_2, \tag{4.100}$$

where φ_i denotes the unique monic polynomial of the minimal degree such that $\varphi_i \times \{i\text{-th } row\ of\ P_b^{-1}\}$ is stable, $\Delta_\varphi = diag\{\varphi_i\}$, and D_2 is an arbitrary stable diagonal matrix. It follows from (4.99) and (4.100) that

$$T_d = \Delta_\theta D_1 + \Delta_\varphi D_2. \tag{4.101}$$

Notice that any common zero of θ_i and φ_i, being a zero in the unstable region, is a zero of the i-th diagonal element of T_d but this violates the assumption that T_d^{-1} is stable. Thus, it is necessary that the pair θ_i, φ_i be coprime. When the pair θ_i, φ_i is coprime, Eq. (4.101) has the same form as (2.242). Hence, the reasoning in Sect. 2.6 can be used to obtain the general solution formulas for D_1 and D_2 in (4.101) which are

$$D_1 = \Delta_\alpha T_d + \Delta\Delta_\varphi \tag{4.102}$$

and

$$D_2 = \Delta_\beta T_d - \Delta\Delta_\theta. \tag{4.103}$$

Here Δ is an arbitrary stable diagonal matrix, and Δ_α and Δ_β are diagonal matrices defined by

$$\Delta_\alpha = diag\{\hat{\alpha}_i\},\ \Delta_\beta = diag\{\beta_i\}, \tag{4.104}$$

where $\hat{\alpha}_i$ and β_i are unique polynomials of minimal degree satisfying (see Sect. B.9)

$$\hat{\alpha}_i\theta_i + \beta_i\varphi_i = 1. \tag{4.105}$$

It now follows from (4.99) to (4.102) that necessarily any realizable diagonal T must have the form

$$T = \Delta_\theta(\Delta_\alpha T_d + \Delta\Delta_\varphi). \tag{4.106}$$

Moreover,

$$P_a^{-1}TP_b^{-1} = P_a^{-1}\Delta_\theta(\Delta_\alpha T_d + \Delta\Delta_\varphi)P_b^{-1} = P_a^{-1}\Delta_\theta\Delta_\alpha T_d P_b^{-1} + P_a^{-1}\Delta_\theta\Delta\Delta_\varphi P_b^{-1} \tag{4.107}$$

must be a stable matrix. Since the second term in (4.107) is stable by the definitions of θ_i and φ_i, it is necessary that $\Sigma = P_a^{-1}\Delta_\theta\Delta_\alpha T_d P_b^{-1}$ be stable. Conversely, when Assumptions 4.1 through 4.4 and Assumption 4.6 are satisfied it is easy to show that the T given by (4.106) satisfies T, $(T_d - T)P_b^{-1}$, and $P_a^{-1}TP_b^{-1}$ stable if Σ is stable. It then follows from Theorem 4.1 that

Theorem 4.3 *When Assumptions* 4.1 through 4.4 *and* 4.6 *are satisfied, a diagonal rational T is realizable (equivalently, a decoupling controller C exists) iff (1) θ_i and φ_i are coprime (equivalently, no column of P_a^{-1} and the corresponding number row in P_b^{-1} possess coincident poles in the unstable region); (2) the data construct $\Sigma = P_a^{-1}\Delta_\theta\Delta_\alpha T_d P_h^{-1}$ is stable; and (3) there exists a diagonal stable rational matrix Δ and a stable matrix Q for which (4.17) holds with*

$$T = \Delta_\theta\Delta_\alpha T_d + \Delta_\theta\Delta\Delta_\varphi. \tag{4.108}$$

The results in Theorem 4.3 are very important in the sense that the conditions for the existence of a decoupling controller are described and the characterization of all realizable T in decoupled systems is explicitly expressed. Also,

Remark 4.4 Assumption 4.4 is critical to derive the parameterized formula for T in (4.108) of Theorem 4.3. When Assumption 4.4 is not met, a different method to parameterize the acceptable and diagonal T is possible (Park 2012). This method, however, requires vector operations which lead to a dimensional inflation that complicates design calculations.

Attention is now turned to the characterization of acceptable T for $\mathcal{H}_2$ design of decoupled systems with acceptable persistent inputs. A necessary condition for a rational matrix T to be acceptable for a given acceptable G_w is that T, $(T_d - T)P_g^{-1}$, and $P_a^{-1}TP_g^{-1}$ be stable. These conditions are the same as the ones required for realizable T except for the replacement of P_b by P_g; hence, by slight modification of the derivations leading to Theorem 4.3 for realizable T one gets Theorem 4.4 below. In this theorem, θ_i is the same one defined previously and φ_{gi} denotes the unique monic polynomial of the minimal degree such that $\varphi_{gi} \times \{i\text{-th } \textit{row of } P_g^{-1}\}$ is stable, where P_g is defined as in (4.44).

Theorem 4.4 *When Assumptions* 4.1, 4.2, 4.3, 4.4′, 4.5, *and* 4.6 *are satisfied, a diagonal rational T is acceptable (equivalently, an acceptable decoupling controller C exists) for a given acceptable G_w iff (1) θ_i and φ_{gi} are coprime (equivalently, no column of P_a^{-1} and the corresponding number row in P_g^{-1} possess coincident poles in the unstable region); (2) the data construct $\Sigma_g = P_a^{-1}\Delta_\theta\Delta_{\alpha g} T_d P_g^{-1}$ is stable, where $\Delta_{\alpha g} = diag\{\alpha_{gi}\}$ and α_{gi}, β_{gi} are unique polynomials of minimal degrees satisfying*

$$\alpha_{gi}\theta_i + \beta_{gi}\varphi_{gi} = 1; \tag{4.109}$$

and (3) there exists a diagonal stable rational matrix Δ and a stable matrix Q for which (4.52) holds with

$$T = \Delta_\theta\Delta_{\alpha g}T_d + \Delta_\theta\Delta\Delta_{\varphi g}, \tag{4.110}$$

where $\Delta_{\varphi g} = diag\{\varphi_{gi}\}$.

When a diagonal T is determined by Theorem 4.3 or Theorem 4.4, the corresponding controller C can be obtained from the formula

$$C = (I + RP_{22})^{-1}R, \tag{4.111}$$

where the matrix R is given by (4.18) for Theorem 4.3 and by (4.53) for Theorem 4.4, respectively.

Remark 4.5 When a diagonal T exits which is acceptable in accordance with Theorem 4.4, it is likely that a diagonal T exits which is realizable in accordance with Theorem 4.3. The following derivation shows that this is indeed the case. First it is shown that φ_i in Theorem 4.3 divides φ_{gi} in Theorem 4.4. Since $P_b^{-1} = P_g^{-1}L_{11}$ from (4.49d) and L_{11} is a polynomial matrix, the highest multiplicity of any unstable pole in the i-th row of P_g^{-1} is greater than or equal to its highest multiplicity in the i-th row of P_b^{-1}. Hence, φ_i divides φ_{gi} and $\varphi_{gi} = \varphi_i l_i$, where l_i is a polynomial. That is, if θ_i, φ_{gi} is a coprime pair, then so is θ_i, φ_i. It follows from (4.109) that $\alpha_{gi}\theta_i + \beta_{gi}\varphi_i l_i = 1$ and, therefore, $\alpha_{gi}\theta_i + \beta_{gi}\varphi_i l_i - 1 = \alpha_{gi}\theta_i + \beta_{gi}\varphi_i l_i - (\alpha_i\theta_i + \beta_i\varphi_i) = 0$ which leads to the equalities

$$\alpha_{gi}\theta_i = \alpha_i\theta_i + \beta_i\varphi_i - \beta_{gi}\varphi_i l_i \tag{4.112}$$

and

$$\alpha_{gi} - \alpha_i = \theta_i^{-1}(\beta_i - \beta_{gi}l_i)\varphi_i. \tag{4.113}$$

Since θ_i, φ_i is a coprime pair, no cancelation happens between the product of θ_i^{-1} and φ_i; hence, $\theta_i^{-1}(\beta_i - \beta_{gi}l_i)$ must be a polynomial. Next suppose that $\Sigma_g = P_a^{-1}\Delta_\theta\Delta_{\alpha g}T_d P_g^{-1} = P_a^{-1}\Delta_\theta\Delta_{\alpha g}T_d P_b^{-1}L_{11}^{-1}$ is stable (L_{11} is invertible since $P_g = L_{11}P_b$). Then, $\Sigma_g L_{11} = P_a^{-1}\Delta_\theta\Delta_{\alpha g}T_d P_b^{-1}$ is stable. The equality in (4.112) yields that $\Delta_{\alpha g}\Delta_\theta = \Delta_\alpha\Delta_\theta + \Delta_\beta\Delta_\varphi - \Delta_{\beta g}\Delta_\varphi\Delta_l$, with $\Delta_l = diag\{l_i\}$, and it follows from

$$\Sigma_g L_{11} = P_a^{-1}\Delta_\theta\Delta_{\alpha g}T_d P_b^{-1} = P_a^{-1}\Delta_\alpha\Delta_\theta T_d P_b^{-1} + P_a^{-1}(\Delta_\beta\Delta_\varphi - \Delta_{\beta g}\Delta_\varphi\Delta_l)T_d P_b^{-1} \tag{4.114}$$

that

$$\Sigma_g L_{11} = P_a^{-1}\Delta_\alpha\Delta_\theta T_d P_b^{-1} + P_a^{-1}\Delta_\theta\Delta_\theta^{-1}(\Delta_\beta - \Delta_{\beta g}\Delta_l)\Delta_\phi T_d P_b^{-1}. \tag{4.115}$$

Since $P_a^{-1}\Delta_\theta$, $\Delta_\varphi T_d P_b^{-1} = T_d\Delta_\varphi P_b^{-1}$, and $\Delta_\theta^{-1}(\Delta_\beta - \Delta_{\beta g}\Delta_l)$ are all stable, the stability of $\Sigma_g L_{11}$ implies the stability of $P_a^{-1}\Delta_\alpha\Delta_\theta T_d P_b^{-1}$. Therefore, if a diagonal T is acceptable in accordance with Theorem 4.4, then it is realizable in accordance with Theorem 4.3 provided there exists a stable matrix Q for which (4.17) holds.

It is also useful to scrutinize the case when a decoupling controller exists by Theorem 4.3, but the conditions of Theorem 4.4 are not all met. When one encounters this situation, $\mathcal{H}_2$ design cannot proceed unless it is practical to consider an alternative G_w for which an acceptable diagonal T exists. (It is not difficult to show as in Example 4.9 that when a realizable diagonal T exists, it is also an acceptable diagonal T for any strictly acceptable G_w.) In order to avoid a trial and error approach to find an acceptable G_w, one might try to identify all the matrices G_{wd} for which there exists a realizable and diagonal T such that $G_\delta G_{wd}$ and $G_v G_{wd}$ are stable and to parameterize such T. This would be similar to the approach taken in Sect. 3.3 to differentiate acceptable controllers from stabilizing controllers. Although this may be possible, it is expected that for decoupled systems little or no advantage over the trial and error method would be gained. Hence, Theorem 4.4 provides a practical way to find an acceptable T and acceptable G_w for the decoupled G.

4.4 $\mathcal{H}_2$ Design of Decoupled Systems for Acceptable Inputs

In this section, minimization of a given cost functional over the diagonal acceptable T in Theorem 4.4 is carried out when decoupling controllers exist. In the plant model G in (4.28), y_δ is the regulated variable and the transfer matrix from e to y_δ is G_δ. As explained in Sects. 3.5 and 3.6, a cost functional which incorporates the transient responses and the steady-state stochastic performance for the generalized plant in (4.28) is given by

$$J = \|WG_\delta G_w\|_2^2 = \frac{1}{2\pi j} \int_{-j\infty}^{j\infty} Tr\,[(WG_\delta G_w)\,(WG_\delta G_w)_*]ds, \tag{4.116}$$

in the analog case and

$$J = \|WG_\delta G_w\|_2^2 = \frac{1}{2\pi j} \oint_{|\lambda|=1} Tr\,[(WG_\delta G_w)(WG_\delta G_w)_*]\frac{d\lambda}{\lambda}, \tag{4.117}$$

in the digital case. The weighting matrix W is stable and G_w is an acceptable matrix for which a diagonal acceptable T in Theorem 4.4 exists. The transfer matrix G_δ is given by

$$G_\delta = G_{11} + G_{12}C(I - G_{22}C)^{-1}G_{21} = G_{11} + G_{12}RG_{21}. \tag{4.118}$$

The $\mathcal{H}_2$ decoupling problem for the system in Fig. 4.1 is to find the decoupling controller which minimizes the quadratic cost functional J. Employing the formula in (4.53) for R, one obtains

$$E = WG_\delta G_w = WG_{11}G_w + WG_{12}[R_r \quad R_v]V_g A_g A G_{21} G_w. \tag{4.119}$$

Introducing the compatible partition

$$V_g A_g A G_{21} G_w = V_g[\sqrt{\alpha_1}B_g B_e \quad \sqrt{\alpha_2}A_g A G_{21}\widetilde{G}_{w2}] = \begin{bmatrix} N_{g1} \\ N_{g2} \end{bmatrix} \tag{4.120}$$

leads to

$$E = WG_{11}G_w + WG_{12}(R_rN_{g1} + R_vN_{g2}) \tag{4.121}$$

$$= WG_{11}G_w + WG_{12}\left\{ V_a \begin{bmatrix} P_a^{-1}TP_g^{-1} \\ M_{21} \end{bmatrix} N_{g1} + (\Gamma_y V_g^{-1} \begin{bmatrix} 0 \\ I \end{bmatrix} + A_1 Q) N_{g2} \right\} \tag{4.122}$$

$$= WG_{11}G_w + W\begin{bmatrix} -P_{02} \\ P_{12} \end{bmatrix} V_a \begin{bmatrix} P_a^{-1}TP_g^{-1} \\ M_{21} \end{bmatrix} N_{g1} + WG_{12}\Gamma_y V_g^{-1} \begin{bmatrix} 0 \\ I \end{bmatrix} N_{g2} + WG_{12}A_1QN_{g2} \tag{4.123}$$

$$= WG_{11}G_w + WG_{12}\Gamma_y V_g^{-1} \begin{bmatrix} 0 \\ I \end{bmatrix} N_{g2} + W\begin{bmatrix} [-P_a \quad 0] \\ P_{12}V_a \end{bmatrix} \begin{bmatrix} P_a^{-1}TP_g^{-1} \\ M_{21} \end{bmatrix} N_{g1} + WG_{12}A_1QN_{g2}. \tag{4.124}$$

One can now introduce the compatible partition

$$P_{12}V_a = [N_{a1} \quad N_{a2}] \tag{4.125}$$

to obtain

$$E = W\left(G_{11}G_w + G_{12}\Gamma_y V_g^{-1} \begin{bmatrix} 0 \\ I \end{bmatrix} N_{g2} + \left[\begin{array}{c|c} -P_a & 0 \\ N_{a1} & N_{a2} \end{array}\right] \begin{bmatrix} P_a^{-1}TP_g^{-1} \\ M_{21} \end{bmatrix} N_{g1} + G_{12}A_1QN_{g2}\right) \tag{4.126}$$

$$= W\left(G_{11}G_w + G_{12}\Gamma_y V_g^{-1} \begin{bmatrix} 0 \\ I \end{bmatrix} N_{g2} + \begin{bmatrix} -P_a \\ N_{a1} \end{bmatrix} P_a^{-1}TP_g^{-1}N_{g1} + \begin{bmatrix} 0 \\ N_{a2} \end{bmatrix} M_{21}N_{g1} + G_{12}A_1QN_{g2} \right). \tag{4.127}$$

Inserting the diagonal acceptable T of (4.110) into (4.127) then yields

$$E = \Phi_1 + W\begin{bmatrix} -\Delta_\theta \\ N_{a1}P_a^{-1}\Delta_\theta \end{bmatrix} \Delta\Delta_{\varphi g} P_g^{-1} N_{g1} + W\begin{bmatrix} 0 \\ N_{a2} \end{bmatrix} M_{21}N_{g1} + WG_{12}A_1QN_{g2}, \tag{4.128}$$

where

$$\Phi_1 = WG_{11}G_w + WG_{12}\Gamma_y V_g^{-1} \begin{bmatrix} 0 \\ I \end{bmatrix} N_{g2} + W\begin{bmatrix} -P_a \\ N_{a1} \end{bmatrix} P_a^{-1}\Delta_\theta \Delta_{\alpha g} T_d P_g^{-1} N_{g1}. \tag{4.129}$$

It easily follows from (4.120) and (4.125) that N_{a1}, N_{a2}, N_{g1}, and N_{g2} are stable. Hence, Φ_1 is a stable matrix since $E = WG_\delta G_w$ is stable and the last three terms in (4.128) are all stable. It is also important to note for the structure of E in (4.128) that the free design parameters available to minimize the cost functional $J = \|E\|_2^2$ are Δ, M_{21}, and Q. Among them, Δ is restricted to be diagonal and this requirement prevents one from directly applying the method developed in Chap. 3 to obtain the optimal $\mathcal{H}_2$ solution. However, the structure of E can be transformed into the solvable standard form treated in Chap. 3 with the aid of the following vector-matrix operations and notations.

As already noted in the next to last paragraph of Sect. 1.4, the Kronecker product $G \otimes R$ is the matrix whose ij-block is $g_{ij}R$ and $vec\, G = [g_1' \quad g_2' \quad \cdots \quad g_n']'$ is the vector obtained by stacking the n columns g_j, $j = 1 \to n$ of G. Also needed is the Khatri-Rao product $G \bullet R$ which is the matrix whose i-th column is given by $g_i \otimes r_i$, where g_i and r_i are the respective i-th columns of G and R. In addition, for a diagonal matrix G, the vector formed by stacking all the diagonal elements of G is denoted by $vecd\, G = [g_{11} \quad g_{22} \quad \cdots \quad g_{nn}]'$. Moreover, $vec\,(AVD) = (D' \otimes A) vec\, V$ and when V is a diagonal matrix then $vec\,(AVD) = (D' \bullet A) vecd\, V$ is easily verified. Finally (see Example 4.1), $\|E\|_2 = \|vec\, E\|_2$: i.e., the vec operation does not alter the value of the $\mathcal{H}_2$ norm. It now follows from (4.128) that

$$\begin{aligned} vec\, E = \phi_1 &+ \left\{(\Delta_{\varphi g} P_g^{-1} N_{g1})' \bullet \left(W \begin{bmatrix} -\Delta_\theta \\ N_{a1} P_a^{-1} \Delta_\theta \end{bmatrix}\right)\right\} \tilde{d} \\ &+ \left\{N_{g1}' \otimes \left(W \begin{bmatrix} 0 \\ N_{a2} \end{bmatrix}\right)\right\} m_{21} + \left\{N_{g2}' \otimes (W G_{12} A_1)\right\} \tilde{q}, \end{aligned} \tag{4.130}$$

where

$$\tilde{d} = vecd\Delta, \; m_{21} = vec\, M_{21}, \; \tilde{q} = vec\, Q, \tag{4.131}$$

and

$$\begin{aligned} \phi_1 = vec\, \Phi_1 &= vec\,(W G_{11} G_w) + \left\{N_{g2}' \otimes (W G_{12} A_1)\right\} vec\left(A_1^{-1} \Gamma_y V_g^{-1} \begin{bmatrix} 0 \\ 1 \end{bmatrix}\right) \\ &+ \left\{(\Delta_{\varphi g} P_g^{-1} N_{g1})' \bullet \left(W \begin{bmatrix} -\Delta_\theta \\ N_{a1} P_a^{-1} \Delta_\theta \end{bmatrix}\right)\right\} vecd(\Delta_{\alpha g} T_d \Delta_{\varphi g}^{-1}). \end{aligned} \tag{4.132}$$

Rearranging (4.130), one obtains

$$vec\, E = \phi_1 + \Pi\, q_L, \tag{4.133}$$

where $q_L = [\tilde{d}' \quad m_{21}' \quad \tilde{q}']'$ is an arbitrary stable vector and

$$\Pi = \left[(\Delta_{\varphi g} P_g^{-1} N_{g1})' \bullet (W \begin{bmatrix} -\Delta_\theta \\ N_{a1} P_a^{-1} \Delta_\theta \end{bmatrix}) \;\middle|\; N'_{g1} \otimes (W \begin{bmatrix} 0 \\ N_{a2} \end{bmatrix}) \;\middle|\; N'_{g2} \otimes (W G_{12} A_1) \right] \tag{4.134}$$

is a stable matrix. Actually, q_L is not completely arbitrary because (4.52) must be satisfied. Relaxing this requirement initially, however, makes the optimization problem treated in the sequel tractable. Moreover, optimum solutions obtained in practice usually lead to controllers with a K matrix that satisfies $\det(X - B_1 K) \cdot \det(X_1 - KB) \not\equiv 0$ That is, the solutions are ones which can be achieved with a stabilizing controller. The form in (4.133) fits the standard one treated in $\mathcal{H}_2$ problems and the procedures developed in Sects. 3.5 and 3.6 to obtain the optimal solution can now be used again here.

The Wiener–Hopf spectral factor of $\Pi_* \Pi$ plays an important role in the sequel and as in Sects. 3.5 and 3.6 it is assumed that Π has full column rank on $|\lambda| = 1$ for the digital case and on the finite part of the $s = j\omega$ axis for the analog case. In order to treat the digital and analog cases together within a unified frame, the terminology of "*the stability boundary*" is used to denote the unit circle of $|\lambda| = 1$ for the digital case and the finite part of the $s = j\omega$ axis for the analog case. Some reasonable assumptions guaranteeing that Π has full column rank are established in the sequel with the aid of the following lemmas and identities.

Lemma 4.1 *For the set of matrices G_i, $i = 1 \to n$, whose elements are complex numbers it is true that if G_1 has column rank, then $rank(G_1 G_2) = rank(G_2)$ and if G_i has full column rank for $i = 1 \to n - 1$, then $rank(G_1 G_2 \cdots G_n) = rank(G_n)$. (See Example* 4.2 *for the proof.)*

Lemma 4.2 *M and N are matrices whose elements are complex numbers. If M and N each have full column rank, then $M \otimes N$ has full column rank. (See Example* 4.3 *for the proof.)*

Also useful are the identities $(A \otimes B)(C \otimes D) = (AC) \otimes (BD)$, $(A \otimes B)(C \bullet D) = (AC \bullet BD)$, and $[A \,\vdots\, B] \otimes C = [A \otimes C \,\vdots\, B \otimes C]$.

Starting with (4.134) one obtains that

$$\Pi = \left[(N'_{g1} (\Delta_{\varphi g} P_g^{-1})') \bullet \left(W \begin{bmatrix} -\Delta_\theta \\ N_{a1} P_a^{-1} \Delta_\theta \end{bmatrix}\right) \;\middle|\; N'_{g1} \otimes \left(W \begin{bmatrix} 0 \\ N_{a2} \end{bmatrix}\right) \;\middle|\; N'_{g2} \otimes (W G_{12} A_1) \right] \tag{4.135}$$

$$= \left[(N'_{g1} \otimes W) \left\{ (\Delta_{\varphi g} P_g^{-1})') \bullet \begin{bmatrix} -\Delta_\theta \\ N_{a1} P_a^{-1} \Delta_\theta \end{bmatrix} \right\} \;\middle|\; (N'_{g1} \otimes W) \left(I \otimes \begin{bmatrix} 0 \\ N_{a2} \end{bmatrix}\right) \;\middle|\; (N'_{g2} \otimes W)(I \otimes G_{12} A_1) \right] \tag{4.136}$$

$$= \left[(N'_{g1} \otimes W) \mid (N'_{g2} \otimes W) \right] \left[\begin{array}{c|c|c} (\Delta_{\varphi g} P_g^{-1})' \bullet \begin{bmatrix} -\Delta_\theta \\ N_{a1} P_a^{-1} \Delta_\theta \end{bmatrix} & I \otimes \begin{bmatrix} 0 \\ N_{a2} \end{bmatrix} & 0 \\ \hline 0 & 0 & I \otimes (G_{12} A_1) \end{array} \right] \tag{4.137}$$

$$= \{ (V_g A_g A G_{21} G_w)' \otimes W \} \left[\begin{array}{c|c|c} (\Delta_{\varphi g} P_g^{-1})' \bullet \begin{bmatrix} -\Delta_\theta \\ N_{a1} P_a^{-1} \Delta_\theta \end{bmatrix} & I \otimes \begin{bmatrix} 0 \\ N_{a2} \end{bmatrix} & 0 \\ \hline 0 & 0 & I \otimes (G_{12} A_1) \end{array} \right] \tag{4.138}$$

$$= \{ (V_g A_g A G_{21} G_w)' \otimes W \} \left[\begin{array}{c|c|c} (\Delta_{\varphi g} P_g^{-1})' \bullet \begin{bmatrix} I & 0 \\ 0 & N_{a1} \end{bmatrix} \begin{bmatrix} -\Delta_\theta \\ P_a^{-1} \Delta_\theta \end{bmatrix} & I \otimes \begin{bmatrix} 0 \\ N_{a2} \end{bmatrix} & 0 \\ \hline 0 & 0 & I \otimes (G_{12} A_1) \end{array} \right] \tag{4.139}$$

$$= \{ (V_g A_g A G_{21} G_w)' \otimes W \} \left[\begin{array}{c|c|c} \left(I \otimes \begin{bmatrix} I & 0 \\ 0 & N_{a1} \end{bmatrix} \right) \left((\Delta_{\varphi g} P_g^{-1})' \bullet \begin{bmatrix} -\Delta_\theta \\ P_a^{-1} \Delta_\theta \end{bmatrix} \right) & I \otimes \begin{bmatrix} 0 \\ N_{a2} \end{bmatrix} & 0 \\ \hline 0 & 0 & I \otimes (G_{12} A_1) \end{array} \right] \tag{4.140}$$

$$= \Pi_1 \, \Pi_2 \, \Pi_3, \tag{4.141}$$

where

$$\Pi_1 = (V_g A_g A G_{21} G_w)' \otimes W \;,\; \Pi_2 = \left[\begin{array}{c|c|c} I \otimes \begin{bmatrix} I & 0 \\ 0 & N_{a1} \end{bmatrix} & I \otimes \begin{bmatrix} 0 \\ N_{a2} \end{bmatrix} & 0 \\ \hline 0 & 0 & I \end{array} \right], \tag{4.142}$$

and

$$\Pi_3 = \left[\begin{array}{c|c|c} (\Delta_{\varphi g} P_g^{-1})' \bullet \begin{bmatrix} -\Delta_\theta \\ P_a^{-1} \Delta_\theta \end{bmatrix} & 0 & 0 \\ \hline 0 & I & 0 \\ \hline 0 & 0 & I \otimes (G_{12} A_1) \end{array} \right]. \tag{4.143}$$

Hence, Π has full column rank on the stability boundary if Π_1, Π_2, and Π_3 each have full column rank on the stability boundary (Lemma 4.1). The following assumptions assure that this is the case.

Assumption 4.7 The stable weighting matrix W has full column rank on the stability boundary.

Assumption 4.8 $A_g A G_{21} G_w = A_g A[\, P_{20}\ P_{21}\,] G_w$ has full row rank on the stability boundary.

Assumption 4.9 P_{12} has full column rank on the stability boundary.

Assumption 4.10 $G_{12}A_1 = \begin{bmatrix} -P_{02} \\ P_{12} \end{bmatrix} A_1$ has full column rank on the stability boundary.

Assumption 4.11 $\begin{bmatrix} \Delta_\theta \\ P_a^{-1}\Delta_\theta \end{bmatrix}$ has full column rank on the stability boundary.

Assumption 4.12 Each row of $\Delta_{\varphi g} P_g^{-1}$ has row rank (i.e., rank of 1) on the stability boundary (or, each row of $\Delta_{\varphi g} P_g^{-1}$ does not vanish on the stability boundary).

Assumptions 4.7 and 4.8 assure, by Lemma 4.2, that Π_1 has full column rank on the stability boundary. Assumption 4.9 assures that Π_2 has full column rank on the stability boundary. In fact, there exist constant permutation matrices E_1, E_2, and E_3 (Brewer 1978) such that

$$\Pi_2 = \left[\begin{array}{c|c|c} I \otimes \begin{bmatrix} I & 0 \\ 0 & N_{a1} \end{bmatrix} & I \otimes \begin{bmatrix} 0 \\ N_{a2} \end{bmatrix} & 0 \\ \hline 0 & 0 & I \end{array}\right] = \left[\begin{array}{c|c|c} E_1\left(\begin{bmatrix} I & 0 \\ 0 & N_{a1} \end{bmatrix} \otimes I\right) E_2 & E_1\left(\begin{bmatrix} 0 \\ N_{a2} \end{bmatrix} \otimes I\right) E_3 & 0 \\ \hline 0 & 0 & I \end{array}\right] \tag{4.144}$$

$$= \begin{bmatrix} E_1 & 0 \\ 0 & I \end{bmatrix} \left[\begin{array}{c|c|c} \begin{bmatrix} I & 0 \\ 0 & N_{a1} \end{bmatrix} \otimes I & \begin{bmatrix} 0 \\ N_{a2} \end{bmatrix} \otimes I & 0 \\ \hline 0 & 0 & I \end{array}\right] \begin{bmatrix} E_2 & 0 & 0 \\ 0 & E_3 & 0 \\ 0 & 0 & I \end{bmatrix} \tag{4.145}$$

$$= \begin{bmatrix} E_1 & 0 \\ 0 & I \end{bmatrix} \left[\begin{array}{c|c} \begin{bmatrix} I & 0 & 0 \\ 0 & N_{a1} & N_{a2} \end{bmatrix} \otimes I & 0 \\ \hline 0 & I \end{array}\right] \left[\begin{array}{cc|c} E_2 & 0 & 0 \\ 0 & E_3 & 0 \\ \hline 0 & 0 & I \end{array}\right] \tag{4.146}$$

$$= \begin{bmatrix} E_1 & 0 \\ 0 & I \end{bmatrix} \left[\begin{array}{c|c} \begin{bmatrix} I & 0 \\ 0 & P_{12}V_a \end{bmatrix} \otimes I & 0 \\ \hline 0 & I \end{array}\right] \left[\begin{array}{cc|c} E_2 & 0 & 0 \\ 0 & E_3 & 0 \\ \hline 0 & 0 & I \end{array}\right]. \tag{4.147}$$

This shows that if P_{12} has full column rank on the stability boundary, then Π_2 has full column rank on the stability boundary. Assumptions 4.10, 4.11, and 4.12

assure that Π_3 has full column rank on the stability boundary. In fact, when $G_{12}A_1$ satisfies Assumption 4.10, Π_3 has full column rank on the stability boundary if

$$\Pi_4 = (\Delta_{\varphi g} P_g^{-1})' \bullet \begin{bmatrix} -\Delta_\theta \\ P_a^{-1}\Delta_\theta \end{bmatrix} = \left(I \otimes \begin{bmatrix} -\Delta_\theta \\ P_a^{-1}\Delta_\theta \end{bmatrix} \right) \{ (\Delta_{\varphi g} P_g^{-1})' \bullet I \} \quad (4.148)$$

has full column rank on the stability boundary. When Assumption 4.11 holds, then $rank(\Pi_4) = rank\{(\Delta_{\varphi g}P_g^{-1})' \bullet I\} = rank\{E_4(I \bullet (\Delta_{\varphi g}P_g^{-1})')\} = rank\{(I \bullet (\Delta_{\varphi g}P_g^{-1})'\}$, where E_4 is a permutation matrix. When Assumption 4.12 is satisfied, then $I \bullet (\Delta_{\varphi g}P_g^{-1})'$ has full column rank on the stability boundary. Thus, Π_4 and, therefore, Π have full column rank on the stability boundary.

With Θ a Wiener–Hopf spectral factor in the equation

$$\Pi_* \Pi = \Theta_* \Theta \quad (4.149)$$

it is convenient to define

$$U = \Pi \Theta^{-1}. \quad (4.150)$$

By the property of Wiener–Hopf factors, Θ and Θ^{-1} are analytic in $|\lambda| < 1$ for the digital case and in $\mathrm{Re}\, s > 0$ for the analog case. The matrix Θ is good since Π is stable. Moreover, since Π has full column rank on the stability boundary, Θ^{-1} is analytic on the stability boundary. Hence, Θ and Θ^{-1} are stable.

The matrix $U = \Pi\Theta^{-1}$ is paraunitary (i.e., $U_* U = \Theta_*^{-1} \Pi_* \Pi \Theta^{-1} = I$), stable and proper (Lemma 3.2). Since $U_* U = I$, it is clear that $UU_*U = U$ and $U - UU_*U = (I - UU_*)U = 0$. Thus, $(I - UU_*)\Pi = 0$ and it readily follows from (4.134) that

$$(I - UU_*) \left\{ (\Delta_{\varphi g} P_g^{-1} N_{g1})' \bullet \left(W \begin{bmatrix} -\Delta_\theta \\ N_{a1} P_a^{-1} \Delta_\theta \end{bmatrix} \right] \right\} = 0 \quad (4.151)$$

and

$$(I - UU_*)\{N'_{g2} \otimes (WG_{12}A_1)\} = 0. \quad (4.152)$$

Since $(I - UU_*)_* = (I - UU_*)$, it follows from (4.151) and (4.152) that

$$\left\{ (\Delta_{\varphi g} P_g^{-1} N_{g1})' \bullet \left(W \begin{bmatrix} -\Delta_\theta \\ N_{a1} P_a^{-1} \Delta_\theta \end{bmatrix} \right] \right\}_* (I - UU_*) = 0 \quad (4.153)$$

and

$$\{N'_{g2} \otimes (WG_{12}A_1)\}_*(I - UU_*) = 0. \tag{4.154}$$

4.4.1 $\mathcal{H}_2$ Design for the Digital Case

It is convenient at this point to treat $\mathcal{H}_2$ designs for the digital and analog cases separately. The digital case is treated first in this section and the analog case is treated in Sect. 4.4.2. The objective is to find all q_L that yield a finite cost functional $J = \|vec\, E\|_2^2$. It follows from (4.133) and (4.150) that

$$(vec\, E)_* vec\, E = (\phi_1 + \Pi q_L)_*(\phi_1 + \Pi q_L) = (\phi_1 + U\Theta\, q_L)_*(\phi_1 + U\Theta\, q_L) \tag{4.155}$$

$$= (U_*\phi_1 + \Theta\, q_L)_*(U_*\phi_1 + \Theta\, q_L) + \rho_o, \tag{4.156}$$

where $\rho_o = \phi_{1*}\phi_1 - \phi_{1*}UU_*\phi_1$. It can be shown from (4.132), and (4.151) to (4.154) that

$$\rho_0 = \phi_{1*}(I - UU_*)\phi_1 = (vec\,(WG_{11}G_w))_*(I - UU_*)vec\,(WG_{11}G_w). \tag{4.157}$$

Since $\{U_*\phi_1\}_+$, $\{U_*\phi_1\}_\infty$, Θ, and Θ^{-1} are stable and q_L is an arbitrary stable vector, it follows that

$$\delta_f = \{U_*\phi_1\}_+ + \{U_*\phi_1\}_\infty + \Theta\, q_L \tag{4.158}$$

is an arbitrary stable vector. Using (4.158) in (4.156) then yields

$$(vec\, E)_* vec\, E = (\delta_f + \{U_*\phi_1\}_-)_*(\delta_f + \{U_*\phi_1\}_-) + \rho_o. \tag{4.159}$$

By the same reasoning as in Sect. 3.5 ((3.118)–(3.122)), it follows that

$$J = \|vec\, E\|_2^2 = \|\{U_*\phi_1\}_- + \delta_f\|_2^2 + \frac{1}{2\pi j}\oint_{|\lambda|=1} \rho_o \frac{d\lambda}{\lambda} \tag{4.160}$$

$$= \|\delta_f\|_2^2 + \|\{U_*\phi_1\}_-\|_2^2 + \frac{1}{2\pi j}\oint_{|\lambda|=1} \rho_o \frac{d\lambda}{\lambda} \tag{4.161}$$

It now becomes clear that all $q_L = \Theta^{-1}(-\{U_*\phi_1\}_+ - \{U_*\phi_1\}_\infty + \delta_f)$ yield finite cost for J and minimization of J is achieved when $\delta_f = 0$. So with no further proof necessary the following theorem can be stated.

Theorem 4.5 *When Assumptions* 4.1, 4.2, 4.3, 4.4′, *and* 4.5 through 4.12 *hold and when the existence condition for a decoupling controller in Theorem* 4.4 *is met, the T and R matrices associated with the decoupling acceptable controllers that yield finite cost J are given as in* (4.110) *and* (4.53)*, respectively, with*

$$q_L = -\Theta^{-1}(\{U_*\phi_1\}_\infty + \{U_*\phi_1\}_+ - \delta_f), \tag{4.162}$$

where δ_f is an arbitrary stable rational vector. The optimal q_L minimizing the cost functional is given by

$$q_{opt} = -\Theta^{-1}(\{U_*\phi_1\}_\infty + \{U_*\phi_1\}_+). \tag{4.163}$$

When J_{opt} denotes the cost functional associated with the optimal solution q_{opt}, the cost functional J yielded by q_L in (4.162) is given by

$$J = J_{opt} + \|\delta_f\|_2^2,\ J_{opt} = \|\{U_*\phi_1\}_-\|_2^2 + \frac{1}{2\pi j}\oint_{|\lambda|=1} \rho_o \frac{d\lambda}{\lambda}. \tag{4.164}$$

At this point there is no guarantee that using (4.162) for q_L in (4.53) yields a K determined by (4.37) and (4.55) such that $\det(X_1 - KB) \not\equiv 0$. However, this is typically the case on account of Remark 3.3.

Remark 4.6 For the stable paraunitary matrix U in (4.150), it is possible to find a left Wiener–Hopf factor $U_\perp$ such that $I - UU_* = U_\perp U_{\perp *}$ (see Example 4.4). In this case, the integral term in (4.164) is given by

$$\frac{1}{2\pi j}\oint_{|\lambda|=1} \rho_o \frac{d\lambda}{\lambda} = \|vec\,(U_{\perp *}WG_{11}G_w)\|_2^2. \tag{4.165}$$

4.4.2 $\mathcal{H}_2$ Design for the Analog Case

The feature that distinguishes $\mathcal{H}_2$ design for the analog case from that for the digital case is the fact that for the cost functional $J = \|vec\,E\|_2^2$ to be finite in the analog case, it is required that $vec\,E \le O(s^{-1})$. Then $U_*\,vec\,E \le O(s^{-1})$ and it follows from $U_*\,vec\,E = U_*\phi_1 + U_*\Pi q_L = U_*\,\phi_1 + \Theta q_L = \{U_*\,\phi_1\}_\infty + \{U_*\,\phi_1\}_+ + \{U_*\,\phi_1\}_- + \Theta q_L$ that the matrix Θq_L must have the form

$$\Theta q_L = -\{U_*\phi_1\}_\infty + \delta_z, \tag{4.166}$$

where δ_z is an arbitrary stable and strictly proper vector. Inserting (4.166) into $vec\, E = \phi_1 + \Pi\, q_L = \phi_1 + U\Theta\, q_L$, one obtains

$$vec\, E = \phi_1 - U\{U_*\phi_1\}_\infty + U\delta_z = \phi_1 - U(U_*\phi_1 - \{U_*\phi_1\}_+ - \{U_*\phi_1\}_-) + U\delta_z \quad (4.167)$$

$$= (I - UU_*)\phi_1 + U\{U_*\phi_1\}_+ + U\{U_*\phi_1\}_- + U\delta_z. \quad (4.168)$$

It follows from (4.132), (4.151), and (4.152) that

$$vec\, E = (I - UU_*)vec\,(WG_{11}G_w) + U\{U_*\phi_1\}_+ + U\{U_*\phi_1\}_- + U\delta_z. \quad (4.169)$$

Since U and ϕ_1 are stable, the vector $\{U_*\phi_1\}_+$ is also stable. Hence, it is possible to parameterize δ_z as $\delta_z = -\{U_*\phi_1\}_+ + \delta_f$, where δ_f is an arbitrary stable and strictly proper vector. With this parameterization, it follows from (4.169) that

$$vec\, E = (I - UU_*)vec\,(WG_{11}G_w) + U(\{U_*\phi_1\}_- + \delta_f) \quad (4.170)$$

and from (4.166) that

$$q_L = \Theta^{-1}(-\{U_*\phi_1\}_\infty + \delta_z) = -\Theta^{-1}(\{U_*\phi_1\}_\infty + \{U_*\phi_1\}_+ - \delta_f). \quad (4.171)$$

In view of (4.170), $vec\, E$ is strictly proper when the following assumption is met.

Assumption 4.13 The matrix $WG_{11}G_w = W\begin{bmatrix} T_d & -P_{01} \\ P_{10} & P_{11} \end{bmatrix}G_w$ is strictly proper.

It is also clear from (4.168) that $vec\ E$ is good; hence, J is finite. Moreover, it is not difficult to show using (4.170) that

$$\begin{aligned}(vec\, E)_*(vec\, E) &= vec\,(WG_{11}G_w)_*(I - UU_*)vec\,(WG_{11}G_w) \\ &\quad + (\{U_*\phi_1\}_- + \delta_f)_*(\{U_*\phi_1\}_- + \delta_f).\end{aligned} \quad (4.172)$$

Thus,

$$\begin{aligned}J = \|vec\, E\|_2^2 &= \frac{1}{2\pi j}\int_{-j\infty}^{j\infty} vec\,(WG_{11}G_w)_*(I - UU_*)vec\,(WG_{11}G_w)ds \\ &\quad + \frac{1}{2\pi j}\int_{-j\infty}^{j\infty} (\{U_*\phi_1\}_- + \delta_f)_*(\{U_*\phi_1\}_- + \delta_f)ds.\end{aligned} \quad (4.173)$$

The same reasoning as in Sect. 3.6 ((3.168)–(3.171)) now leads to

$$J = \|E\|_2^2 = \|vec\, E\|_2^2 = J_{opt} + \left\|\delta_f\right\|^2, \tag{4.174}$$

where

$$J_{opt} = \left\|\{U_*\phi_1\}_-\right\|_2^2 + \frac{1}{2\pi j}\int_{-j\infty}^{j\infty} vec\,(WG_{11}G_w)_*(I - UU_*)vec\,(WG_{11}G_w)ds. \tag{4.175}$$

Therefore, minimization of the cost functional J can be achieved by taking $\delta_f = 0$ and the cost functional associated with the optimal solution $q_{opt} = -\Theta^{-1}(\{U_*\phi_1\}_\infty + \{U_*\phi_1\}_+)$ is given by J_{opt} (see Remark 4.6 for another expression for the integral term in (4.175)).

When Assumption 4.13 holds, $\{U_* vec\,(WG_{11}G_w)\}_\infty = 0$. It then follows from (4.129) and (4.132) that

$$\{U_*\phi_1\}_\infty = \{U_*\phi_g\}_\infty, \tag{4.176}$$

where

$$\phi_g = vec\,(WG_{12}\Gamma_y V_g^{-1}\begin{bmatrix}0\\ I\end{bmatrix}N_{g2} + W\begin{bmatrix}-P_a\\ N_{a1}\end{bmatrix}P_a^{-1}\Delta_\theta\Delta_{\alpha g}T_d P_g^{-1}N_{g1}) \tag{4.177}$$

$$= \{N_{g2}' \otimes (WG_{12}A_1)\}vec\,(A_1^{-1}\Gamma_y V_g^{-1}\begin{bmatrix}0\\ I\end{bmatrix}) + \{(\Delta_{\varphi g}P_g^{-1}N_{g1})' \bullet W\begin{bmatrix}-\Delta_\theta\\ N_{a1}P_a^{-1}\Delta_\theta\end{bmatrix}\}vecd(\Delta_{\alpha g}T_d\Delta_{\varphi g}^{-1}). \tag{4.178}$$

It is next established that the controller C associated with the q_L in (4.171) is strictly proper when an order assumption is imposed on the model G. It has been shown that the stable vector q_L in (4.171) yields finite cost $J = \|WG_\delta G_w\|_2$. Thus,

$$WG_\delta G_w = WG_{11}G_w + WG_{12}C(I - G_{22}C)^{-1}G_{21}G_w = WG_{11}G_w + WG_{12}RG_{21}G_w \tag{4.179}$$

is strictly proper when q_L in (4.171) is used. Since $WG_{11}G_w$ is strictly proper, $WG_\delta G_w$ is strictly proper iff $WG_{12}RG_{21}G_w$ is also. The implication of this requirement on the controller transfer matrix C is explored next under

Assumption 4.14 $P_{22} = G_{22}$ is proper. The matrices WG_{12} and $G_{21}G_w$ behave as $WG_{12} \to E_{12}\,s^k$ and $G_{21}G_w \to E_{21}\,s^l$ as $s \to \infty$ with the condition that the constant matrices E_{12} and E_{21} have full column rank and full row rank, respectively, and $k + l \geq 0$. (In most applications, $k = l = 0$.)

Since $WG_{12}RG_{21}G_w$ is strictly proper, it follows from Assumption 4.14 that $\lim_{s\to\infty} WG_{12}RG_{21}G_w = \lim_{s\to\infty}(WG_{12}s^{-k})(s^kRs^l)(G_{21}G_ws^{-l}) = E_{12}\lim_{s\to\infty}(s^kRs^l)E_{21} = 0$. This implies that $R \le O(s^{-1})$ since E_{12} and E_{21} have full column and full row rank, respectively, and $k+l \ge 0$. Since $R \le O(s^{-1})$ and $P_{22} = G_{22} \le O(s^0)$, the controller $C = (I+RP_{22})^{-1}R$ is well defined and is strictly proper. In addition, since $S - RP_{22} = I$ with $S = (I - CP_{22})^{-1} = A_1(X_1 - KB)$, it follows that $S \to I$, $s \to \infty$. Hence, for any K meeting this requirement one gets $\det S = (\det A_1)$ $\det(X_1 - KB) \to 1$, $s \to \infty$ and $\det(X_1 - KB) \not\equiv 0$. It is now possible to state without the need of further proof

Theorem 4.6 *When Assumptions* 4.1, 4.2, 4.3, 4.4′, *and* 4.5 through4.14 *hold and when the existence condition for a decoupling controller in Theorem* 4.4 *is met, the T and R matrices associated with the decoupling acceptable controllers that yield finite cost J are given as in* (4.110) *and in* (4.52)*, respectively, with*

$$q_L = -\Theta^{-1}(\{U_*\phi_g\}_\infty + \{U_*\phi_1\}_+ - \delta_f), \tag{4.180}$$

where δ_f is an arbitrary stable strictly proper vector. The optimal q_L that minimizes the cost functional is given by

$$q_{opt} = -\Theta^{-1}(\{U_*\phi_g\}_\infty + \{U_*\phi_1\}_+). \tag{4.181}$$

The cost functional J yielded by q_L in (4.180) *is given by*

$$J = J_{opt} + \|\delta_f\|^2, \tag{4.182}$$

where J_{opt} in (4.175) is the cost functional associated with the optimal solution q_{opt}. *The associated controller $C(s)$ with q_L in* (4.180) *is strictly proper.*

4.4.3 $\mathcal{H}_2$ Design of Decoupled Systems for Strictly Acceptable Inputs

A matrix $G_w = A_w^{-1}B_w$, $(A_w,\ B_w)$ being a coprime polynomial pair, is called strictly acceptable for the plant G in (4.28) if there exists a stabilizing controller C that makes $G_\delta A_w^{-1}$, $G_v A_w^{-1}$, and $G_m A_w^{-1}$ stable, where G_δ, G_v, and G_m are the transfer matrices from $e = [r'\ e_o']'$ to y_δ, v and y_m, respectively (see Definition 3.3). Note that a stable G_w is always strictly acceptable. A stabilizing controller C for which $G_\delta A_w^{-1}$, $G_v A_w^{-1}$, and $G_m A_w^{-1}$ are stable is called *strictly acceptable* for the plant G. It can be shown that G_w is strictly acceptable for a compatible G iff $AG_{21}A_w^{-1}$ and $(G_{11} + G_{12}YAG_{21})A_w^{-1}$ are stable (see Example 3.21(a)) and the class of all strictly acceptable controllers is given by the formula in (4.8) with K an arbitrary real rational stable matrix such that $\det(X - B_1K) \cdot \det(X_1 - KB) \not\equiv 0$. That is, the class of all strictly acceptable controllers is equivalent to the class of all stabilizing controllers (this follows indirectly from the result in Remark 3.5 or directly from the formula for T_w in Solution 3.21(a)). In other words, when G_w is strictly acceptable and C is any stabilizing controller, the three matrices $G_\delta A_w^{-1}$, $G_v A_w^{-1}$, and $G_m A_w^{-1}$ are stable. It is important to recognize that the R matrices associated with decoupling controllers that yield finite cost J for a given strictly acceptable G_w are given as in (4.18) of Theorem 4.1 with the diagonal T in (4.108) of Theorem 4.3.

In this section, minimization of a given cost functional over the diagonal realizable T in Theorem 4.3 is treated when decoupling controllers exist. The cost functional to be minimized is $J = \|WG_\delta G_w\|_2^2$, where W is a stable weighting matrix and G_w is a strictly acceptable matrix. The transfer matrix G_δ is given by

$$G_\delta = G_{11} + G_{12}C(I - G_{22}C)^{-1}G_{21} = G_{11} + G_{12}RG_{21}. \tag{4.183}$$

Employing the formula in (4.18) for R, one gets

$$WG_\delta G_w = WG_{11}G_w + WG_{12}\begin{bmatrix} R_{ro} & R_{vo} \end{bmatrix} V_b A G_{21} G_w. \tag{4.184}$$

The steps needed to find $\mathcal{H}_2$ solutions here are the same as the ones in the previous section with minor modifications and the approach is simply outlined as follows.

First, the compatible partition of the stable matrix

$$V_b A G_{21} G_w = V_b A \begin{bmatrix} P_{20} & P_{21} \end{bmatrix} G_w = \begin{bmatrix} N_{b1} \\ N_{b2} \end{bmatrix} \tag{4.185}$$

is introduced. Clearly, (4.184) and (4.185) follow from (4.119) and (4.120) with the substitutions $V_g A_g \to V_b$, $N_{g1} \to N_{b1}$, $N_{g2} \to N_{b2}$, $R_r \to R_{ro}$, and $R_v \to R_{vo}$. The latter two are obtained with the substitutions $P_g \to P_b$ and $\Gamma_y V_g^{-1} \to YV_b^{-1}$. Moreover, the expression for T needed here is given by (4.108) which is obtained from (4.110) with the substitutions $\Delta_{\alpha g} \to \Delta_\alpha$ and $\Delta_{\varphi g} \to \Delta_\varphi$. Hence, in place of (4.129) one gets

$$\Phi_{o1} = WG_{11}G_w + WG_{12}YV_b^{-1}\begin{bmatrix} 0 \\ I \end{bmatrix} N_{b2} + W\begin{bmatrix} -P_a \\ N_{a1} \end{bmatrix} P_a^{-1}\Delta_\theta \Delta_\alpha T_d P_b^{-1} N_{b1} \tag{4.186}$$

and it is convenient to introduce

$$\phi_{o1} = vec\,\Phi_{o1}, \tag{4.187}$$

and

$$\phi_{ob} = vec\,(WG_{12}YV_b^{-1}\begin{bmatrix} 0 \\ I \end{bmatrix} N_{b2} + W\begin{bmatrix} -P_a \\ N_{a1} \end{bmatrix} P_a^{-1}\Delta_\theta\Delta_\alpha T_d P_b^{-1}N_{b1}). \tag{4.188}$$

Moreover, in place of (4.134) one has

$$\Pi_o = \left[(\Delta_\varphi P_b^{-1}N_{b1})' \bullet (W\begin{bmatrix} -\Delta_\theta \\ N_{a1}P_a^{-1}\Delta_\theta \end{bmatrix}) \middle| N_{b1}' \otimes (W\begin{bmatrix} 0 \\ N_{a2} \end{bmatrix}) \middle| N_{b2}' \otimes (WG_{12}A_1) \right] \tag{4.189}$$

where all variables are defined as in Sects. 4.2 and 4.3. So for the case of interest here it follows that in place of (4.133) one has $vec\,E = \phi_{o1} + \Pi_o q_L$. Also, analogous to (4.149) and (4.150), the Wiener–Hopf spectral factor Θ_o of the equation

$$\Pi_{o*}\Pi_o = \Theta_{o*}\Theta_o \tag{4.190}$$

is introduced and

$$U_o = \Pi_o\Theta_o^{-1} \tag{4.191}$$

is defined. Finally, it is necessary that Assumptions 4.8 and 4.12, respectively, be replaced by the following two:

Assumption 4.8′ $AG_{21}G_w = A[\,P_{20} \quad P_{21}\,]\,G_w$ has full row rank on the stability boundary.

and

Assumption 4.12′ Each row of $\Delta_\varphi P_b^{-1}$ has row rank (i.e., rank of 1) on the stability boundary (or, each row of $\Delta_\varphi P_b^{-1}$ does not vanish on the stability boundary).

It now follows in straightforward fashion with the assumptions in Theorems 4.1 and 4.3 and those introduced in Sect. 4.4 that the theorems corresponding to Theorems 4.5 and 4.6 for the case in which G_w is strictly acceptable rather than acceptable are the two that follow:

Theorem 4.7 (Digital case) *When Assumptions* 4.1 through 4.4, 4.6, 4.7, 4.8′, 4.9, 4.10, 4.11, *and* 4.12′ *hold and when the existence condition for a decoupling controller in Theorem* 4.3 *is met, and the T and R matrices associated with decoupling controllers that yield finite cost J for a given strictly acceptable* G_w *are given as in* (4.108) *and in* (4.18), *respectively, with*

$$q_L = -\Theta_o^{-1}(\{U_{o*}\phi_{o1}\}_\infty + \{U_{o*}\phi_{o1}\}_+ - \delta_f), \tag{4.192}$$

where δ_f *is an arbitrary stable rational vector. The optimal* q_L *minimizing the cost functional is given by*

$$q_{opt} = -\Theta_o^{-1}(\{U_{o*}\phi_{o1}\}_\infty + \{U_{o*}\phi_{o1}\}_+). \tag{4.193}$$

When J_{opt} denotes the cost functional associated with the optimal solution q_{opt}, the cost functional J yielded by q_L in (4.192) is given by

$$J = J_{opt} + \|\delta_f\|_2^2,\; J_{opt} = \|\{U_{o*}\phi_{o1}\}_-\|_2^2 + \frac{1}{2\pi j}\oint_{|\lambda|=1} \rho_o \frac{d\lambda}{\lambda}, \tag{4.194}$$

where

$$\rho_o = \phi_{o1*}(I - U_o U_{o*})\phi_{o1} = (vec\,(WG_{11}G_w))_*(I - U_o U_{o*})vec\,(WG_{11}G_w). \tag{4.195}$$

Theorem 4.8 (Analog case) *When Assumptions* 4.1 through 4.4, 4.6, 4.7, 4.8′, 4.9, 4.10, 4.11, 4.12′, 4.13, *and* 4.14 *hold and when the existence condition for a decoupling controller in Theorem* 4.3 *is met, the T and R matrices associated with decoupling controllers that yield finite cost J for a given strictly acceptable G_w are given as in (4.108) and in (4.18), respectively, with*

$$q_L = -\Theta_o^{-1}(\{U_{o*}\phi_{ob}\}_\infty + \{U_{o*}\phi_{o1}\}_+ - \delta_f), \tag{4.196}$$

where δ_f is an arbitrary stable strictly proper rational vector. The optimal q_L that minimizes the cost functional is given by

$$q_{opt} = -\Theta_o^{-1}(\{U_{o*}\phi_{ob}\}_\infty + \{U_{o*}\phi_{o1}\}_+). \tag{4.197}$$

When J_{opt} denotes the cost functional associated with the optimal solution q_{opt}, the cost functional J yielded by q_L in (4.196) is given by

$$J = J_{opt} + \|\delta_f\|^2, \tag{4.198}$$

where

$$J_{opt} = \|\{U_{o*}\phi_{o1}\}_-\|_2^2 + \frac{1}{2\pi j}\int_{-j\infty}^{j\infty} vec\,(WG_{11}G_w)_*(I - U_o U_{o*})vec\,(WG_{11}G_w)ds. \tag{4.199}$$

The controller C associated with q_L in (4.196) is strictly proper.

In the next two sections, the solution formulas for the $\mathcal{H}_2$ design of decoupled systems with 1DOF and 3DOF controller configurations are obtained from the results derived in this section by simple substitutions. This shows the merit of developing a theory for a generalized model when several configurations, which are special cases of this model are of interest.

4.5 $\mathcal{H}_2$ Design of 1DOF Decoupled Systems

In this section, $\mathcal{H}_2$ design of decoupled systems for the 1DOF system in Fig. 3.3 is treated for strictly acceptable inputs (it is explained later in Remark 4.8 that the acceptable inputs G_w considered in Sect. 3.7 for $\mathcal{H}_2$ design of 1DOF systems are actually strictly acceptable when F in Fig. 3.3 is stable). For the 1DOF system in Fig. 3.3, the design objective is to find stabilizing controllers for which the transfer matrix from r to y is diagonal and the $\mathcal{H}_2$ cost functional penalized on the error $\varepsilon = T_d r - y$ and the control variable v is finite. The $\mathcal{H}_2$ solution for 1DOF decoupled systems is obtained by using the formulas derived in Sect. 4.4 for the standard model in Fig. 4.1. Specifically, the results in Theorems 4.7 and 4.8 are used. For this purpose, the plant model $\widetilde{P}$ in (4.1) and G in (4.28) for the 1DOF system in Fig. 3.3 are obtained first.

When $r = r$, $e_o = \begin{bmatrix} d' & n' \end{bmatrix}'$, $v = v$, $y = y$, $y_z = v$, and $y_m = u$, where the left-hand side variables are the ones in Fig. 4.1 and the right-hand side variables are the ones in Fig. 3.3, it then follows that

$$\widetilde{P} = \begin{bmatrix} P_{00} & P_{01} & P_{02} \\ P_{10} & P_{11} & P_{12} \\ P_{20} & P_{21} & P_{22} \end{bmatrix} = \left[\begin{array}{c|c|c|c} 0 & P_d & 0 & P \\ \hline 0 & 0 & 0 & I \\ \hline I & -P_o & -I & -FP \end{array}\right], \; P_o = FP_d + L. \tag{4.200}$$

The regulated variable y_δ and the exogenous input e for $\mathcal{H}_2$ design are set as $y_\delta = [\varepsilon' \;\; v']'$ and $e = [r' \;\; e_o']'$. One then gets from (4.29) that the plant G in (4.28) is given by

$$G_{11} = \begin{bmatrix} T_d & -P_d & 0 \\ 0 & 0 & 0 \end{bmatrix}, \; G_{12} = \begin{bmatrix} -P \\ I \end{bmatrix}, \tag{4.201}$$

$$G_{21} = [I \;\; -P_o \;\; -I], \; G_{22} = -FP. \tag{4.202}$$

The exogenous input $e = [r' \;\; e_o']'$ is modeled by

$$e = \begin{bmatrix} r \\ e_o \end{bmatrix} = \begin{bmatrix} r \\ d \\ n \end{bmatrix} = \begin{bmatrix} r_1 \\ d_1 \\ 0 \end{bmatrix} + \begin{bmatrix} r_2 \\ d_2 \\ n \end{bmatrix} = e_1 + e_2, \tag{4.203}$$

where r_1 and d_1 are the shape-deterministic components of r and d and where r_2, d_2, and n are purely stochastic. Consider the fractional descriptions for the shape-deterministic signals $r_1 = A_r^{-1}B_r\sigma_r$, $d_1 = A_d^{-1}B_d\sigma_d$, where A_r, B_r and A_d, B_d are left coprime polynomial matrix pairs. Then

$$e = e_1 + e_2 = [\,\widetilde{G}_{w1} \quad \widetilde{G}_{w2}\,]\begin{bmatrix} \sigma \\ \mu \end{bmatrix} = \widetilde{G}_{w1}\sigma + \widetilde{G}_{w2}\mu = \begin{bmatrix} G_e \\ 0 \end{bmatrix}\sigma + \widetilde{G}_{w2}\mu, \tag{4.204}$$

$$\sigma = [\,\sigma_r' \quad \sigma_d'\,]',\ \mu = [\,\mu_r' \ \mu_d' \ \mu_n'\,]', \tag{4.205}$$

$$G_e = A_e^{-1}B_e,\ A_e = \begin{bmatrix} A_r & 0 \\ 0 & A_d \end{bmatrix},\ B_e = \begin{bmatrix} B_r & 0 \\ 0 & B_d \end{bmatrix}, \tag{4.206}$$

and

$$\widetilde{G}_{w2} = \Omega_s,\ \Omega_s = \text{diag}\{\Omega_r \quad \Omega_d \quad \Omega_n\,\}, \tag{4.207}$$

where $\sigma = [\,\sigma_r' \quad \sigma_d'\,]'$ is a vector whose elements are random variables satisfying $\langle\sigma\sigma'\rangle = I$; the elements of μ are in the temporal domain zero-mean independent white-noise processes with unit variance; and Ω_r, Ω_d, and Ω_n are stable left Wiener–Hopf factors associated with the power spectral density matrices for the stochastic signals r_2, d_2, and n, respectively.

Clearly, A_e, B_e is a left coprime polynomial matrix pair and the matrices G_{111} and G_{211} in (4.32) are given by

$$G_{111} = \begin{bmatrix} T_d & -P_d \\ 0 & 0 \end{bmatrix},\ G_{211} = [\,I \quad -P_o\,]. \tag{4.208}$$

Finally, the coprime polynomial matrix fraction description for FP used in the 1DOF system of Sect. 3.7 is used again. Specifically,

$$-G_{22} = FP = A^{-1}B = B_1A_1^{-1} \tag{4.209}$$

and

$$\begin{bmatrix} A & B \\ Y_1 & -X_1 \end{bmatrix}\begin{bmatrix} X & B_1 \\ Y & -A_1 \end{bmatrix} = \begin{bmatrix} X & B_1 \\ Y & -A_1 \end{bmatrix}\begin{bmatrix} A & B \\ Y_1 & -X_1 \end{bmatrix} = \begin{bmatrix} I & 0 \\ 0 & I \end{bmatrix}, \tag{4.210}$$

where all the submatrices in (4.210) are polynomial matrices.

The cost functional to be minimized is $J = \|WG_\delta G_w\|_2^2$, where G_δ is the transfer matrix from e to y_δ, W is a stable weighting matrix, and $G_w = [\,\sqrt{\alpha_1}\widetilde{G}_{w1} \quad \sqrt{\alpha_2}\widetilde{G}_{w2}\,]$ with $\alpha_1 > 0$ and $\alpha_2 \geq 0$. The solution formula for the optimization problem is contained in Theorems 4.7 and 4.8 under Assumptions 4.1 through 4.4,

Assumptions 4.6, 4.7, 4.8′, 4.9, 4.10, 4.11, 4.12′, 4.13, and 4.14 for the standard model. The corresponding assumptions for the 1DOF system in Fig. 3.3 are now enumerated. In place of Assumption 4.1 one naturally gets

Assumption 4.15 The 1DOF system in Fig. 3.3 is admissible or, equivalently,

(1) The subsystems L, $[P_d \quad P]$, and F have no unstable hidden poles (i.e., h_o in (3.241) is a stable polynomial),
(2) $\psi_{[P_d\ P]}/\psi_P$ is a stable polynomial (i.e., h_1 in (3.242) is a stable polynomial),
(3) L is a stable matrix, and
(4) $\psi_F\psi_P/\psi_{FP}$ is a stable polynomial.

When the 1DOF system in Fig. 3.3 is admissible, it easily follows from Lemma 3.3 that AP_o, PA_1, and $P_d - PA_1Y_1P_o$ are stable. It also follows from Lemma 3.4 that $G = [G_{ij}]$ in (4.201) and (4.202) is compatible when the 1DOF system is admissible and T_d is stable (the stability of T_d is assumed later).

It is clear from (4.200) that $P_{02} = P$ and $P_{22} = -FP$; hence, $\widetilde{F} = -F$ and Assumption 4.2 is equivalent

Assumption 4.16 F is stable.

It also follows from (4.200) that $P_{20} = I$; thus, Assumption 4.3 is equivalent to

Assumption 4.17 The plant P has full row rank.

When P has full row rank, there exists as in Sect. 4.2 a unimodular matrix V_a such that

$$PV_a = [P_a \quad 0]. \tag{4.211}$$

The conditions which assure that Assumption 4.4 is met are developed next. Since $AP_{20} = A$, it follows from (4.12) that $V_b = I$ and $P_b = A$. Hence, Assumption 4.4 is satisfied when T_d is stable and there exists a stable matrix Q_o such that $PYA = T_d + Q_oA$. The latter condition is satisfied iff $(T_d - PYA)A^{-1}$ is stable. Since $(T_d - PYA) = (T_d - PA_1Y_1)$ is stable, this is the case iff $(T_d - PYA)A^{-1}B$ is stable. Now $(T_d - PYA)A^{-1}B = (T_d - PYA)FP = = (T_dF - I)P + P(I - YB) = (T_dF - I)P + PA_1X_1$ and the matrix $(T_d - PYA)A^{-1}B$ is stable iff $(T_dF - I)P$ is stable. Thus, Assumption 4.4 holds when the following assumption is made.

Assumption 4.18 T_d is stable and $(T_dF - I)P$ is stable.

Assumption 4.5 is trivially satisfied and Assumption 4.6 is retained as

Assumption 4.19 T_d is diagonal and T_d^{-1} is stable.

Now in order to proceed with the $\mathcal{H}_2$ decoupling design for strictly acceptable inputs, a decoupling controller for the 1DOF system is assumed to exist. In this regard, it is convenient to introduce the set $\{\theta_i\}$ of unique monic polynomials of minimal degree such that the $\{i\text{-th } \textit{column of } P_a^{-1}\} \times \theta_i$ is stable for all columns of P_a^{-1} and the set $\{\varphi_i\}$ of unique monic polynomials of minimal degree such that $\varphi_i \times \{i\text{-th } row\ of\ A^{-1}\}$ is stable for all rows of $P_b^{-1} = A^{-1}$. It is also convenient to introduce $\Delta_\theta = diag\{\theta_i\}$, $\Delta_\varphi = diag\{\varphi_i\}$, $\Delta_\alpha = diag\{\hat{\alpha}_i\}$, and $\Delta_\beta = diag\{\beta_i\}$, where $\hat{\alpha}_i$ and β_i are unique polynomials of minimal degrees satisfying $\hat{\alpha}_i\theta_i + \beta_i\varphi_i = 1$ when θ_i, φ_i are coprime. Then (see Theorem 4.3), the existence of a decoupling controller requires

Assumption 4.20 (1) θ_i and φ_i are coprime (equivalently, no column of P_a^{-1} and the corresponding number row in A^{-1} possess coincident poles in the unstable region) and (2) the data construct $\Sigma = P_a^{-1}\Delta_\theta\Delta_\alpha T_d A^{-1}$ is stable.

The satisfaction of Assumption 4.7 is guaranteed by

Assumption 4.21 The weighting matrix W is of the form $W = diag\{W_e,\ W_v\}$. The matrices W_e and W_v are stable and have full column rank on the stability boundary.

Assumption 4.8′ is changed to the following one since $\alpha_1 > 0$ and

$$G_{21}G_w = [\,\sqrt{\alpha_1}A_r^{-1}B_r \quad -\sqrt{\alpha_1}P_oA_d^{-1}B_d \quad \sqrt{\alpha_2}\Omega_r \quad -\sqrt{\alpha_2}P_o\Omega_d \quad -\sqrt{\alpha_2}\Omega_n\,]. \tag{4.212}$$

Assumption 4.22 $A\,[A_r^{-1}B_r \quad P_oA_d^{-1}B_d \quad \sqrt{\alpha_2}\Omega_r \quad \sqrt{\alpha_2}P_o\Omega_d \quad \sqrt{\alpha_2}\Omega_n]$ has full row rank on the stability boundary.

Assumption 4.9 is trivially met since $P_{12} = I$. Assumptions 4.10, 4.11, and 4.12′ are, respectively, simply replaced with the following three assumptions:

Assumption 4.23 $\begin{bmatrix} -P \\ I \end{bmatrix} A_1$ has full column rank on the stability boundary.

Assumption 4.24 $\begin{bmatrix} \Delta_\theta \\ P_a^{-1}\Delta_\theta \end{bmatrix}$ has full column rank on the stability boundary.

Assumption 4.25 No row of $\Delta_\varphi A^{-1}$ vanishes on the stability boundary.

Since

$$WG_{11}G_w = \begin{bmatrix} \sqrt{\alpha_1}W_eT_dA_r^{-1}B_r & -\sqrt{\alpha_1}W_eP_dA_d^{-1}B_d & \sqrt{\alpha_2}W_eT_d\Omega_r & -\sqrt{\alpha_2}W_eP_d\Omega_d & 0 \\ 0 & 0 & 0 & 0 & 0 \end{bmatrix}, \tag{4.213}$$

Assumption 4.13 is equivalent to

Assumption 4.26 The matrices $W_e\,[\,T_dA_r^{-1}B_r \;\; P_dA_d^{-1}B_d\,]$ and $\sqrt{\alpha_2}\,W_e\,[\,T_d\Omega_r$ $P_d\,\Omega_d\,]$ are strictly proper.

Assumption 4.14 is simply changed to

Assumption 4.27 *FP* is proper. In addition, as $s \to \infty$, the following behavior holds: $\begin{bmatrix} W_e P \\ W_v \end{bmatrix} \to E_{12}\, s^k$ and $[\, A_r^{-1}B_r \;\; P_o A_d^{-1}B_d \;\; \sqrt{\alpha_2}\Omega_r \;\; \sqrt{\alpha_2}P_o\Omega_d \;\; \sqrt{\alpha_2}\Omega_n \,] \to E_{21}s^l$, where E_{12} and E_{21} have full column rank and full row rank, respectively, and $k + l \geq 0$.

Having established assumptions for the 1DOF case which assure that the assumptions made in Theorems 4.7 and 4.8 are met, one can invoke these theorems to find the solution for the 1DOF case. In particular, the formula $q_L = -\Theta_o^{-1}(\{U_{o*}\phi_{o1}\}_\infty + \{U_{o*}\phi_{o1}\}_+ - \delta_f)$ can be used. Calculation of $\Pi_{o*}\Pi_o$ is done first to obtain Θ_o in (4.190). Since $P_{20} = I$ is a square matrix, the matrix R is given by (4.21) in Remark 4.1. Thus, $[R_{ro} \;\; R_{vo}]$ in (4.184) is replaced with R_{ro} and there is no need to introduce the partition (4.185) for the 1DOF case. Instead, one can use $V_b A G_{21} G_w = A[I \;\; -P_o \;\; -I]G_w = N_{b1}$. That is, for the 1DOF case one can set $N_{b2} = 0$ in (4.185) and the equations following it. In particular, the third block column of Π_o in (4.189) disappears and accordingly, $q_L = [\widetilde{d}' \;\; m'_{21}]'$. Moreover,

$$\begin{aligned} N_{b1} &= A G_{21} G_w \\ &= A\,[\,\sqrt{\alpha_1}A_r^{-1}B_r \;\; -\sqrt{\alpha_1}P_o A_d^{-1}B_d \;\; \sqrt{\alpha_2}\Omega_r \;\; -\sqrt{\alpha_2}P_o\Omega_d \;\; -\sqrt{\alpha_2}\Omega_n\,]. \end{aligned} \tag{4.214}$$

When V_a is partitioned as $V_a = [\, V_{a1} \;\; V_{a2} \,]$ corresponding to the column sizes of the partition $[-P_a \;\; 0\,]$, the partition matrices N_{a1} and N_{a2} in (4.125) are given by the pair

$$N_{a1} = V_{a1},\; N_{a2} = V_{a2}. \tag{4.215}$$

A number of helpful matrix formulas for the calculation of $\Pi_{o*}\Pi_o$ are now presented. (For easy reference the definition of the Schur product stated in Sect. 1.4 is repeated here: the *Schur* product of two equi-size matrices G and R is denoted by $G \circ R$ and is the matrix whose *i*-row, *j*-column entry is $g_{ij}r_{ij}$). It is not difficult to confirm that

$$(AB)_* = B_*A_*,\; [A \;\; B]_* = \begin{bmatrix} A_* \\ B_* \end{bmatrix},\; (A \otimes B)_* = A_* \otimes B_*, \tag{4.216}$$

$$(A^{-1})_* = (A_*)^{-1},\; (A \bullet B)_* = (A'_* \bullet B'_*)', \tag{4.217}$$

$$(F \otimes G)(L \bullet M) = (FL) \bullet (GM),\; (F \bullet G)'(L \bullet M) = (F'L) \circ (G'M), \tag{4.218}$$

$$(F \bullet G)_*(F \bullet G) = (F_*F) \circ (G_*G),\; (F \circ G)_* = F_* \circ G_*. \tag{4.219}$$

A straightforward calculation now yields

$$\Pi_{o*}\Pi_o = \begin{bmatrix} \Psi & \{(\Delta_\varphi G_{rd} A_*)' \bullet (V_{a2*} W_{v*} W_v V_{a1} P_a^{-1} \Delta_\theta)\}_* \\ (\Delta_\varphi G_{rd} A_*)' \bullet (V_{a2*} W_{v*} W_v V_{a1} P_a^{-1} \Delta_\theta) & (A G_{rd} A_*)' \otimes (V_{a2*} W_{v*} W_v V_{a2}) \end{bmatrix}, \tag{4.220}$$

where

$$\Psi = (\Delta_\varphi G_{rd} \Delta_{\varphi*})' \circ \{\Delta_{\theta*}(W_{e*} W_e + P_{a*}^{-1} V_{a1*} W_{v*} W_v V_{a1} P_a^{-1}) \Delta_\theta\}, \tag{4.221a}$$

and

$$G_{rd} = \alpha_1 (A_r^{-1} B_r B_{r*} A_{r*}^{-1} + P_o A_d^{-1} B_d B_{d*} A_{d*}^{-1} P_{o*}) + \alpha_2 (\Omega_r \Omega_{r*} + P_o \Omega_d \Omega_{d*} P_{o*} + \Omega_n \Omega_{n*}). \tag{4.221b}$$

The Wiener–Hopf factor Θ_o in (4.190) is usually calculated numerically after the values of the variables are inserted into the expression for $\Pi_{o*}\Pi_o$ given by (4.220). Formulas for $U_{o*}\phi_{o1} = \Theta_{o*}^{-1}\Pi_{o*}\phi_{o1}$ and $U_{o*}\phi_{ob} = \Theta_{o*}^{-1}\Pi_{o*}\phi_{ob}$ in the 1DOF case are easy (although tedious) to derive beginning with (4.186)–(4.188). The vectors $\phi_{o1} = vec\, \Phi_{o1}$ and ϕ_{ob} become

$$\phi_{o1} = vec\,(W G_{11} G_w) + \phi_{ob} \tag{4.222}$$

and

$$\phi_{ob} = \left\{ (\Delta_\varphi G_{21} G_w)' \bullet \left(\begin{bmatrix} -W_e \\ W_v V_{a1} P_a^{-1} \end{bmatrix} \Delta_\theta \right\} vec\, d(\Delta_\alpha T_d \Delta_\varphi^{-1}), \tag{4.223}$$

where $W G_{11} G_w$ and $G_{21} G_w$ are given in (4.213) and (4.212), respectively. Straightforward calculations then yield

$$U_{o*}\phi_{o1} = \Theta_{o*}^{-1} \begin{bmatrix} \Psi\, vecd(\Delta_\alpha T_d \Delta_\varphi^{-1}) - (\Delta_\varphi \bullet \Delta_\theta)_* vec\,(W_{e*} W_e G_{tp}) \\ vec\,(V_{a2*} W_{v*} W_v V_{a1} P_a^{-1} \Delta_\theta \Delta_\alpha T_d G_{rd} A_*) \end{bmatrix} \tag{4.224}$$

with

$$G_{tp} = \alpha_1 (T_d A_r^{-1} B_r B_{r*} A_{r*}^{-1} + P_d A_d^{-1} B_d B_{d*} A_{d*}^{-1} P_{o*}) + \alpha_2 (T_d \Omega_r \Omega_{r*} + P_d \Omega_d \Omega_{d*} P_{o*}). \tag{4.225}$$

When Assumption 4.26 is satisfied, $U_{o*} vec\,(W G_{11} G_w)$ is strictly proper and in this case, $\{U_{o*}\phi_{o1}\}_\infty = \{U_{o*}\phi_{ob}\}_\infty$ and $U_{o*}\phi_{ob}$ is given by

$$U_{o*}\phi_{ob} = \Theta_{o*}^{-1}\begin{bmatrix} \Psi\, vecd(\Delta_\alpha T_d \Delta_\varphi^{-1}) \\ vec\,(V_{a2*}W_{v*}W_v V_{a1} P_a^{-1}\Delta_\theta \Delta_\alpha T_d G_{rd} A_*) \end{bmatrix}. \tag{4.226}$$

The preparation needed for the application of Theorems 4.7 and 4.8 to the 1DOF system in Fig. 3.3 is now complete.

Before stating the resulting theorems, it is convenient to identify the strictly acceptable inputs G_w (or, equivalently, G_e) for the 1DOF system in Fig. 3.3. The strictly acceptable inputs in the standard model are ones for which $AG_{21}G_w$ and $(G_{11} + G_{12}YAG_{21})G_w$ (equivalently, $AG_{211}G_e$ and $(G_{111} + G_{12}YAG_{211})G_e$) are stable. When F is stable, it is shown later in Remark 4.7 that these conditions are met and a persistent input G_e is strictly acceptable for the 1DOF system iff $A_pP_dA_d^{-1}$, LA_d^{-1}, $A_pT_dA_r^{-1}$, and $(I - FT_d)A_r^{-1}$ are stable, where A_p is the denominator matrix of the coprime polynomial matrix fraction description

$$P = A_p^{-1}B_p. \tag{4.227}$$

These conditions are the same as the ones in Sect. 3.7 (see the conditions in Lemma 3.4).

Theorem 4.9 (Digital case) *When Assumptions* 4.15 through 4.25 *hold and* $A_pP_dA_d^{-1}$, LA_d^{-1}, $A_pT_dA_r^{-1}$, *and* $(I - FT_d)A_r^{-1}$ *are stable, decoupling controllers for the 1DOF system in Fig.* 3.3 *exist. Moreover, the set of all loop transfer matrices* $R = C(I - P_{22}C)^{-1}$ *associated with the ones for which the cost* $J = \|WG_\delta G_w\|_2^2$ *is finite is given by*

$$R = V_a\begin{bmatrix} P_a^{-1}T \\ M_{21}A \end{bmatrix} = V_a\begin{bmatrix} P_a^{-1}\Delta_\theta(\Delta_\alpha T_d + \Delta\Delta_\varphi) \\ M_{21}A \end{bmatrix}, \tag{4.228}$$

$$\tilde{d} = vecd\Delta, \quad m_{21} = vec\, M_{21}, \tag{4.229}$$

and

$$\begin{bmatrix} \tilde{d} \\ m_{21} \end{bmatrix} = -\Theta_o^{-1}(\{U_{o*}\phi_{o1}\}_+ + \{U_{o*}\phi_{o1}\}_\infty - \delta_f), \tag{4.230}$$

where δ_f *is an arbitrary stable vector and* $U_{o*}\phi_{o1}$ *is obtained from (4.224). All finite values of the cost functional J associated with any stable choice for* δ_f *in (4.230) are given by*

$$J = J_{opt} + \|\delta_f\|_2^2,\ J_{opt} = \|\{U_{o*}\phi_{o1}\}_-\|_2^2 + \frac{1}{2\pi j}\oint_{|\lambda|=1} \rho_o \frac{d\lambda}{\lambda}, \tag{4.231}$$

where $WG_{11}G_w$ is given by (4.213) and

$$\rho_o = (vec\,(WG_{11}G_w))_*(I - U_oU_{o*})vec\,(WG_{11}G_w). \tag{4.232}$$

The choice for δ_f which yields the minimum value of J is called the optimal choice and is given by $\delta_f = 0$.

Theorem 4.10 (Analog case) *When Assumptions* 4.15 through 4.27 *hold and $A_pP_dA_d^{-1}$, LA_d^{-1}, $A_pT_dA_r^{-1}$, and $(I - FT_d)A_r^{-1}$ are stable, decoupling controllers for the 1DOF system in Fig. 3.3 exist. Moreover, the set of all loop transfer matrices $R = C(I - P_{22}C)^{-1}$ associated with the ones for which the cost $J = \|WG_\delta G_w\|_2^2$ is finite is given by*

$$R = V_a \begin{bmatrix} P_a^{-1}T \\ M_{21}A \end{bmatrix} = V_a \begin{bmatrix} P_a^{-1}\Delta_\theta(\Delta_\alpha T_d + \Delta\Delta_\varphi) \\ M_{21}A \end{bmatrix}, \tag{4.233}$$

$$\tilde{d} = vec d\Delta, \quad m_{21} = vec\,M_{21}, \tag{4.234}$$

and

$$\begin{bmatrix} \tilde{d} \\ m_{21} \end{bmatrix} = -\Theta_o^{-1}(\{U_{o*}\phi_{ob}\}_\infty + \{U_{o*}\phi_{o1}\}_+ - \delta_f), \tag{4.235}$$

where δ_f is an arbitrary stable strictly proper vector and $U_{o*}\phi_{o1}$ and $U_{o*}\phi_{ob}$ are given by (4.224) and (4.226), respectively. All finite values of the cost functional J associated with any stable strictly proper choice for δ_f in (4.230) are given by

$$J = J_{opt} + \|\delta_f\|_2^2, \tag{4.236}$$

where

$$J_{opt} = \|\{U_{o*}\phi_{o1}\}_-\|_2^2 + \frac{1}{2\pi j}\int_{-j\infty}^{j\infty} vec\,(WG_{11}G_w)_*(I - U_oU_{o*})vec\,(WG_{11}G_w)ds \tag{4.237}$$

and $WG_{11}G_w$ is given by (4.213). The choice for δ_f which yields the minimum value of J is called the optimal choice and is given by $\delta_f = 0$. The controller C associated with R in (4.233) is strictly proper.

Remark 4.7 It remains to establish for the 1DOF system that when F is stable the strictly acceptable G_e are those for which $A_pP_dA_d^{-1}$, LA_d^{-1}, $A_pT_dA_r^{-1}$, and $(I - FT_d)A_r^{-1}$ are stable. In order to identify strictly acceptable inputs, coprime pairs A_o, B_o and A_{o1}, B_{o1} for $-G_{12} = [\,P' \;\; -I\,]' = A_o^{-1}B_o = B_{o1}A_{o1}^{-1}$ in (4.30) are obtained first. The coprime polynomial matrix fraction description for P given by (3.60) and (3.61) is used again for the 1DOF system:

$$P = A_p^{-1}B_p = B_{p1}A_{p1}^{-1} \tag{4.238}$$

with

$$\begin{bmatrix} A_p & B_p \\ Y_{p1} & -X_{p1} \end{bmatrix}\begin{bmatrix} X_p & B_{p1} \\ Y_p & -A_{p1} \end{bmatrix} = \begin{bmatrix} X_p & B_{p1} \\ Y_p & -A_{p1} \end{bmatrix}\begin{bmatrix} A_p & B_p \\ Y_{p1} & -X_{p1} \end{bmatrix} = \begin{bmatrix} I & 0 \\ 0 & I \end{bmatrix}. \tag{4.239}$$

Then the desired coprime polynomial pairs A_o, B_o and A_{o1}, B_{o1} and associated polynomial matrices X_o, Y_o, X_{o1}, and Y_{o1} satisfying $A_oX_o + B_oY_o = I$, $X_{o1}A_{o1} + Y_{o1}B_{o1} = I$, and $Y_{o1}X_o = X_{o1}Y_o$ are given by

$$A_o = \begin{bmatrix} A_p & 0 \\ 0 & I \end{bmatrix} B_o = \begin{bmatrix} B_p \\ -I \end{bmatrix},\; B_o = \begin{bmatrix} B_p \\ -I \end{bmatrix},\; X_o = \begin{bmatrix} X_p & 0 \\ Y_p & I \end{bmatrix},\; Y_o = [\,Y_p \;\; 0\,] \tag{4.240}$$

and

$$A_{o1} = A_{p1},\; B_{o1} = \begin{bmatrix} B_{p1} \\ -A_{p1} \end{bmatrix},\; X_{o1} = X_{p1},\; Y_{o1} = [\,Y_{p1} \;\; 0\,]. \tag{4.241}$$

As explained in the statement after Eq. (4.31), $A_1^{-1}A_{o1}$ is stable since F is stable. In Example 3.21(c), it is proven for the system in (4.28) that when $A_1^{-1}A_{o1}$ is stable $G_w = A_w^{-1}B_w$, (A_w, B_w) being a coprime polynomial pair, is strictly acceptable iff $\Gamma_w = A_oG_{11}A_w^{-1}$(equivalently $\widetilde{\Gamma}_w = A_oG_{111}A_e^{-1}$) and $\Gamma_v = (G_{21} + G_{22}Y_oA_oG_{11})A_w^{-1}$ (equivalently $\widetilde{\Gamma}_v = (G_{211} + G_{22}Y_oA_oG_{111})A_e^{-1}$) are stable. In the 1DOF system, $\widetilde{\Gamma}_w = A_oG_{111}A_e^{-1} = \begin{bmatrix} A_pT_dA_r^{-1} & -A_pP_dA_d^{-1} \end{bmatrix}$; hence $\widetilde{\Gamma}_w$ is stable iff $A_pP_dA_d^{-1}$ and $A_pT_dA_r^{-1}$ are stable. Next,

$$\widetilde{\Gamma}_v = (G_{211} + G_{22}Y_oA_oG_{111})A_e^{-1} \tag{4.242}$$

$$= \left[(I - FPY_pA_pT_d)A_r^{-1} \mid -(P_o - FPY_pA_pP_d)A_d^{-1} \right] \tag{4.243}$$

$$=\left[(I-FT_d)A_r^{-1}+F(I-PY_pA_p)T_dA_r^{-1} \mid -F(I-PY_pA_p)P_dA_d^{-1}-LA_d^{-1}\right] \tag{4.244}$$

$$=\left[(I-FT_d)A_r^{-1}+FX_pA_pT_dA_r^{-1} \mid -FX_pA_pP_dA_d^{-1}-LA_d^{-1}\right] \tag{4.245}$$

is stable iff $(I - FT_d)A_r^{-1}$ and LA_d^{-1} are stable since F, $A_pP_dA_d^{-1}$, and $A_pT_dA_r^{-1}$ are stable. In summary, when F is stable a persistent input G_e in the 1DOF system is strictly acceptable iff $A_pP_dA_d^{-1}$, LA_d^{-1}, $A_pT_dA_r^{-1}$, and $(I - FT_d)A_r^{-1}$ are stable.

Remark 4.8 The persistent input G_w considered in Sect. 3.7 for the $\mathcal{H}_2$ design of the 1DOF system shown in Fig. 3.3 is acceptable for all stabilizing controllers. It is proven in Lemma 3.4 that G_w is acceptable for all stabilizing controllers, provided F, T_d, and $\widetilde{G}_{w2}$ are stable, iff $A_pP_dA_d^{-1}$, LA_d^{-1}, $A_pT_dA_r^{-1}$, and $(I - FT_d)A_r^{-1}$ are stable. As can be seen in Remark 4.7, this condition is the same as the one for strictly acceptable inputs. In fact, it is proven in Example 3.21(b) for the system in Fig. 3.2 that when $A_1^{-1}A_{o1}$ is stable an acceptable G_w is strictly acceptable iff $\Gamma_g = AG_{21}G_w$ is stable. Since the G_w considered in Sect. 3.7 satisfies this condition, it is strictly acceptable when F is stable.

Remark 4.9 It is important to note that for the 1DOF system explicit necessary and sufficient conditions for G_w to be strictly acceptable are easily derived as in Remark 4.7. This is not the case, however, for acceptable G_w. For the standard model shown in Fig. 3.2, all the acceptable inputs G_w are ones for which $\Gamma_w = A_oG_{11}A_w^{-1}$ and $\Gamma_o = A_{o1}^{-1}(Y_o\Gamma_wA_{g1} - YB_{g1})$ (or $\Gamma_z = (G_{11} + G_{12}A_1Y_1G_{21})A_w^{-1}A_{g1}$, see Example 3.3) are stable, where $AG_{21}A_w^{-1} = B_{g1}A_{g1}^{-1}$ is a coprime polynomial matrix fraction description. In the case of the 1DOF system shown in Fig. 3.3, this condition is difficult to express explicitly. It is, however, possible to find a sufficient condition for an acceptable G_w which has a rather simple form (see Example 4.5 (a)). In Example 2.12(c), it is shown that there can exist an acceptable G_w that is not strictly acceptable for a specific 1DOF system. The following is another illustration of this phenomenon in connection with decoupling design. Consider the system in Fig. 3.3 with

$$P=\begin{bmatrix} \frac{\frac{1}{s}}{\frac{s+1}{s-1}} & \frac{s+2}{\frac{\frac{s}{3}}{s-1}} \end{bmatrix},\ F=T_d=P_d=I_2,\ L=\begin{bmatrix} 0 & 1 \\ 0 & 0 \end{bmatrix}, \tag{4.246}$$

and

$$G_e = \begin{bmatrix} A_r^{-1}B_r & 0 \\ 0 & A_d^{-1}B_d \end{bmatrix} = \left[\begin{array}{cc|cc} \frac{1}{s} & 0 & 0 & 0 \\ 0 & \frac{1}{s-1} & 0 & 0 \\ \hline 0 & 0 & \frac{1}{s} & 0 \\ 0 & 0 & 0 & \frac{1}{s-1} \end{array}\right]. \tag{4.247}$$

Since LA_d^{-1} (equivalently, $LA_d^{-1}B_d$) is not stable, the matrix G_e is not strictly acceptable but it can be shown that G_e is acceptable (see Example 4.5(b)). In this case, the formula in Theorem 4.10 cannot be used; however, the general formula in Theorem 4.6 can be used to obtain the $\mathcal{H}_2$ optimal decoupling solution.

4.6 $\mathcal{H}_2$ Design of 3DOF Decoupled Systems

In this section, $\mathcal{H}_2$ design of decoupled systems for the 3DOF system in Fig. 1.1 is treated for acceptable inputs. For the 3DOF system in Fig. 1.1, the design objective is to find stabilizing controllers for which the transfer matrix from r to y is diagonal and the $\mathcal{H}_2$ cost functional penalized on the error $\varepsilon = T_d r - y$ and the control variable v is finite. The $\mathcal{H}_2$ solution for 3DOF decoupled systems is obtained by using the formulas derived in Sect. 4.4 for the standard model in Fig. 4.1. Specifically, the results in Theorems 4.5 and 4.6 are used. For this purpose, the plant model $\widetilde{P}$ in (4.1) and G in (4.28) for the 3DOF system in Fig. 1.1 are obtained first.

When $r = r$, $e_o = [d' \;\; n_r' \;\; n_m' \;\; n_l']'$, $v = v$, $y = y$, $y_z = v$, and $y_m = [w' \;\; z' \;\; u']'$, where the left-hand side variables are the ones in Fig. 4.1 and the right-hand side variables are the ones in Fig. 1.1 (for consistency with the notation in Sect. 3.8, n_r is used here for the noise n in Fig. 1.1), then

$$\tilde{P} = \begin{bmatrix} P_{00} & P_{01} & P_{02} \\ P_{10} & P_{11} & P_{12} \\ P_{20} & P_{21} & P_{22} \end{bmatrix} = \left[\begin{array}{c|cccc|c} 0 & P_d & 0 & 0 & 0 & P \\ \hline 0 & 0 & 0 & 0 & 0 & I \\ 0 & FP_d & 0 & I & 0 & FP \\ \hline 0 & L & 0 & 0 & I & 0 \\ I & 0 & I & 0 & 0 & 0 \end{array}\right]. \tag{4.248}$$

The regulated variable y_δ and the exogenous input e for $\mathcal{H}_2$ design are set as $y_\delta = [\varepsilon' \;\; v']'$ and $e = [r' \;\; e_o']'$. One then gets from (4.29) that the plant G in (4.28) is given by

$$G_{11} = \begin{bmatrix} T_d & -P_d & 0 & 0 & 0 \\ 0 & 0 & 0 & 0 & 0 \end{bmatrix}, \; G_{12} = \begin{bmatrix} -P \\ I \end{bmatrix}, \tag{4.249}$$

$$G_{21} = \begin{bmatrix} 0 & FP_d & 0 & I & 0 \\ 0 & L & 0 & 0 & I \\ I & 0 & I & 0 & 0 \end{bmatrix}, \; G_{22} = \begin{bmatrix} FP \\ 0 \\ 0 \end{bmatrix}. \tag{4.250}$$

As in Sect. 3.8, the model for the exogenous input $e = [r' \;\; e_o']'$ is

$$e = \begin{bmatrix} r \\ e_o \end{bmatrix} = \begin{bmatrix} r \\ d \\ n_r \\ n_m \\ n_l \end{bmatrix} = \begin{bmatrix} r_1 \\ d_1 \\ 0 \\ 0 \\ 0 \end{bmatrix} + \begin{bmatrix} r_2 \\ d_2 \\ n_r \\ n_m \\ n_l \end{bmatrix} = e_1 + e_2, \tag{4.251}$$

where r_1 and d_1 are the shape-deterministic components of r and d and where r_2, d_2, n_r, n_m, and n_l are purely stochastic. The fractional descriptions for the shape-deterministic signals are $r_1 = A_r^{-1}B_r\sigma_r$, $d_1 = A_d^{-1}B_d\sigma_d$, where A_r, B_r and A_d, B_d are left coprime polynomial matrix pairs. Then

$$e = e_1 + e_2 = [\,\widetilde{G}_{w1} \;\; \widetilde{G}_{w2}\,]\begin{bmatrix} \sigma \\ \mu \end{bmatrix} = \widetilde{G}_{w1}\sigma + \widetilde{G}_{w2}\mu = \begin{bmatrix} G_e \\ 0 \end{bmatrix}\sigma + \widetilde{G}_{w2}\mu, \tag{4.252}$$

$$\sigma = [\,\sigma_r' \;\; \sigma_d'\,]', \; \mu = [\,\mu_r' \;\; \mu_d' \;\; \mu_{n_r}' \;\; \mu_{n_m}' \;\; \mu_{n_l}'\,]', \tag{4.253}$$

$$G_e = A_e^{-1}B_e, \; A_e = \begin{bmatrix} A_r & 0 \\ 0 & A_d \end{bmatrix}, \; B_e = \begin{bmatrix} B_r & 0 \\ 0 & B_d \end{bmatrix}, \tag{4.254}$$

$$\widetilde{G}_{w2} = \Omega_s, \; \Omega_s = \text{diag}\,\{\,\Omega_r \;\; \Omega_d \;\; \Omega_{n_r} \;\; \Omega_{n_m} \;\; \Omega_{n_l}\,\}, \tag{4.255}$$

where $\sigma = [\,\sigma_r' \;\; \sigma_d'\,]'$ is a vector whose elements are random variables satisfying $\langle \sigma\sigma' \rangle = I$; where the elements $\hat{\mu}_r$, $\hat{\mu}_d$, $\hat{\mu}_{n_r}$, $\hat{\mu}_{n_m}$, and $\hat{\mu}_{n_l}$ are zero-mean independent white-noise processes with unit variance; and where Ω_d, Ω_r, Ω_{n_r}, Ω_{n_m}, and Ω_{n_l} are stable left Wiener–Hopf factors associated with the power spectral density matrices for the stochastic signals r_2, d_2, n_r, n_m, and n_l, respectively. Clearly, A_e, B_e is a left coprime polynomial matrix pair and the matrices G_{111} and G_{211} in (4.32) are given by

$$G_{111} = \begin{bmatrix} T_d & -P_d \\ 0 & 0 \end{bmatrix}, \; G_{211} = \begin{bmatrix} 0 & FP_d \\ 0 & L \\ I & 0 \end{bmatrix}. \tag{4.256}$$

The coprime polynomial matrix fraction description for FP used in the 3DOF system of Sect. 3.8 is employed again here. Specifically,

$$FP = A^{-1}B = B_1A_1^{-1} \tag{4.257}$$

with polynomial submatrices satisfying

$$\begin{bmatrix} A & B \\ Y_1 & -X_1 \end{bmatrix}\begin{bmatrix} X & B_1 \\ Y & -A_1 \end{bmatrix} = \begin{bmatrix} X & B_1 \\ Y & -A_1 \end{bmatrix}\begin{bmatrix} A & B \\ Y_1 & -X_1 \end{bmatrix} = \begin{bmatrix} I & 0 \\ 0 & I \end{bmatrix}. \tag{4.258}$$

With $\widetilde{A}^{-1}\widetilde{B} = \widetilde{B}_1\widetilde{A}_1^{-1}$ coprime polynomial matrix fraction descriptions for $-G_{22} = -P_{22} = [\,(-FP)' \quad 0' \quad 0'\,]'$ (here $\widetilde{A}$, $\widetilde{B}$, $\widetilde{A}_1$, and $\widetilde{B}_1$ are used for consistency with the notation in Sects. 3.4 and 3.8) it follows that $\widetilde{A}$, $\widetilde{B}$, $\widetilde{A}_1$, $\widetilde{B}_1$, $\widetilde{X}$, $\widetilde{Y}$, $\widetilde{X}_1$, and $\widetilde{Y}_1$ satisfying (4.7) are given by

$$\widetilde{A} = \begin{bmatrix} A & 0 & 0 \\ 0 & I & 0 \\ 0 & 0 & I \end{bmatrix}, \; \widetilde{B} = \begin{bmatrix} -B \\ 0 \\ 0 \end{bmatrix}, \; \widetilde{X} = \begin{bmatrix} X & 0 & 0 \\ 0 & I & 0 \\ 0 & 0 & I \end{bmatrix}, \; \widetilde{Y} = [\,-Y \quad 0 \quad 0\,] \tag{4.259}$$

and

$$\widetilde{A}_1 = A_1, \; \widetilde{B}_1 = \begin{bmatrix} -B_1 \\ 0 \\ 0 \end{bmatrix}, \; \widetilde{X}_1 = X_1, \; \widetilde{Y}_1 = [\,-Y_1 \quad 0 \quad 0\,]. \tag{4.260}$$

The cost functional to be minimized is $J = \|WG_\delta G_w\|_2^2$, where G_δ is the transfer matrix from e to y_δ, W is a stable weighting matrix, and $G_w = [\,\sqrt{\alpha_1}\widetilde{G}_{w1} \quad \sqrt{\alpha_2}\widetilde{G}_{w2}\,]$ with $\alpha_1 > 0$ and $\alpha_2 \geq 0$. The solution formula for this optimization problem is contained in Theorems 4.5 and 4.6 under Assumptions 4.1 through 4.14 for the standard model. The corresponding assumptions for the 3DOF system are now enumerated.

Exploiting the results of Lemma 2.8 yields in place of Assumption 4.1 (here, P_a in Fig. 2.1 is assumed to be I)

Assumption 4.28 The 3DOF system in Fig. 1.1 is admissible or, equivalently,

(1) The subsystems L, $[\,P_d \quad P\,]$, and F have no unstable hidden poles,
(2) $\psi_{[P_d \; P]}/\psi_P$ is a stable polynomial,
(3) L is a stable matrix, and
(4) $\psi_F\psi_P/\psi_{FP}$ is a stable polynomial.

When the 3DOF system in Fig. 1.1 is admissible, it easily follows from Lemma 2.7 that AFP_d, L, PA_1, and $(I - PA_1Y_1F)P_d$ are all stable. It is also not difficult to verify that $G = [G_{ij}]$ in (4.249) and (4.250) is compatible when the 3DOF system is admissible and T_d is stable (the stability of T_d is assumed later).

It is obvious from (4.248) that $P_{22} = \widetilde{F}P_{02}$, where $\widetilde{F} = [\,F' \;\; 0 \;\; 0\,]'$; hence, Assumption 4.2 is equivalent to

Assumption 4.29 F is stable.

It is also clear from (4.248) that $P_{02} = P$ and $P_{20} = [0\;0\;I]'$; hence, Assumption 4.3 is equivalent to

Assumption 4.30 The plant P has full row rank.

When P has full row rank, there exists as in Sect. 4.2 a unimodular matrix V_a such that

$$PV_a = [P_a \quad 0]. \tag{4.261}$$

It is established next that a stable Q_g satisfying (4.45) in Assumption 4.4′ always exists for the 3DOF system when T_d is stable. In this regard, the polynomial matrices A_o, B_o, A_g, and B_g are needed. First, the coprime polynomial matrix pairs A_g, B_g and A_{g1}, B_{g1} for which

$$\widetilde{A}G_{211}A_e^{-1} = \begin{bmatrix} 0 & AFP_dA_d^{-1} \\ 0 & LA_d^{-1} \\ A_r^{-1} & 0 \end{bmatrix} = A_g^{-1}B_g = B_{g1}A_{g1}^{-1} \tag{4.262}$$

are determined. This can be done simply (as in Sect. 3.4) in terms of the coprime polynomial matrix fraction descriptions associated with

$$\begin{bmatrix} AFP_dA_d^{-1} \\ LA_d^{-1} \end{bmatrix} = A_q^{-1}B_q = B_{q1}A_{q1}^{-1}, \tag{4.263}$$

where the polynomial matrices satisfy

$$\begin{bmatrix} A_q & B_q \\ Y_{q1} & -X_{q1} \end{bmatrix}\begin{bmatrix} X_q & B_{q1} \\ Y_q & -A_{q1} \end{bmatrix} = \begin{bmatrix} X_q & B_{q1} \\ Y_q & -A_{q1} \end{bmatrix}\begin{bmatrix} A_q & B_q \\ Y_{q1} & -X_{q1} \end{bmatrix} = \begin{bmatrix} I & 0 \\ 0 & I \end{bmatrix}. \tag{4.264}$$

Clearly,

$$\begin{aligned} \widetilde{A}G_{211}A_e^{-1} &= \begin{bmatrix} 0 & A_q^{-1}B_q \\ A_r^{-1} & 0 \end{bmatrix} = \begin{bmatrix} A_q & 0 \\ 0 & A_r \end{bmatrix}^{-1}\begin{bmatrix} 0 & B_q \\ I & 0 \end{bmatrix} \\ &= \begin{bmatrix} 0 & B_{q1} \\ I & 0 \end{bmatrix}\begin{bmatrix} A_r & 0 \\ 0 & A_{q1} \end{bmatrix}^{-1}. \end{aligned} \tag{4.265}$$

It is easy to confirm that the equalities in (4.36) are satisfied with

$$A_g = \begin{bmatrix} A_q & 0 \\ 0 & A_r \end{bmatrix},\ B_g = \begin{bmatrix} 0 & B_q \\ I & 0 \end{bmatrix},\ A_{g1} = \begin{bmatrix} A_r & 0 \\ 0 & A_{q1} \end{bmatrix},\ B_{g1} = \begin{bmatrix} 0 & B_{q1} \\ I & 0 \end{bmatrix} \tag{4.266}$$

and

$$X_g = \begin{bmatrix} X_q & 0 \\ 0 & 0 \end{bmatrix},\ Y_g = \begin{bmatrix} 0 & I \\ Y_q & 0 \end{bmatrix},\ X_{g1} = \begin{bmatrix} 0 & 0 \\ 0 & X_{q1} \end{bmatrix},\ Y_{g1} = \begin{bmatrix} 0 & I \\ Y_{q1} & 0 \end{bmatrix}. \tag{4.267}$$

Next, coprime pairs A_o, B_o and A_{o1}, B_{o1} are derived for which $-G_{12} = [\,P' \;\; -I\,]' = A_o^{-1}B_o = B_{o1}A_{o1}^{-1}$. Using the coprime polynomial matrix fraction description for P in the 3DOF system one has

$$P = A_p^{-1}B_p = B_{p1}A_{p1}^{-1} \tag{4.268}$$

with

$$\begin{bmatrix} A_p & B_p \\ Y_{p1} & -X_{p1} \end{bmatrix}\begin{bmatrix} X_p & B_{p1} \\ Y_p & -A_{p1} \end{bmatrix} = \begin{bmatrix} X_p & B_{p1} \\ Y_p & -A_{p1} \end{bmatrix}\begin{bmatrix} A_p & B_p \\ Y_{p1} & -X_{p1} \end{bmatrix} = \begin{bmatrix} I & 0 \\ 0 & I \end{bmatrix}. \tag{4.269}$$

It then easily follows that

$$A_o = \begin{bmatrix} A_p & 0 \\ 0 & I \end{bmatrix},\ B_o = \begin{bmatrix} B_p \\ -I \end{bmatrix},\ X_o = \begin{bmatrix} X_p & 0 \\ Y_p & I \end{bmatrix},\ Y_o = [\,Y_p \;\; 0\,] \tag{4.270}$$

and

$$A_{o1} = A_{p1},\ B_{o1} = \begin{bmatrix} B_{p1} \\ -A_{p1} \end{bmatrix},\ X_{o1} = X_{p1},\ Y_{o1} = [\,Y_{p1} \;\; 0\,] \tag{4.271}$$

satisfy (4.30) and (4.31).

It is established in Sect. 3.4 that when F and T_d are stable and the 3DOF system is admissible then $A_e^{-1}B_e$ is acceptable iff $A_pP_dA_d^{-1}$ and $A_pT_dA_r^{-1}$ are both stable. Also,

$$A_g\tilde{A}P_{20} = \left[\begin{array}{c|c} A_q & 0 \\ \hline 0 & A_r \end{array}\right]\left[\begin{array}{cc|c} A & 0 & 0 \\ 0 & I & 0 \\ \hline 0 & 0 & I \end{array}\right]\left[\begin{array}{c} 0 \\ 0 \\ \hline I \end{array}\right] = \begin{bmatrix} 0 \\ 0 \\ A_r \end{bmatrix}$$; hence, (4.44) is satisfied with

$$V_g = \begin{bmatrix} 0 & 0 & I \\ I & 0 & 0 \\ 0 & I & 0 \end{bmatrix}$$ and $P_g = A_r$. Clearly,

$$\Gamma_w = A_oG_{111}A_e^{-1} = \begin{bmatrix} A_pT_dA_r^{-1} & -A_pP_dA_d^{-1} \\ 0 & 0 \end{bmatrix} \tag{4.272}$$

and (4.43) yields

$$\Gamma_y = Y_o\Gamma_w Y_g + \tilde{Y}X_g = \left[-Y_pA_pP_dA_d^{-1}Y_q + [-Y\ 0]X_q \mid Y_pA_pT_dA_r^{-1}\right]. \quad (4.273)$$

Thus,

$$P_{02}(\tilde{Y} + \tilde{A}_1K_o)\tilde{A}P_{20} = P_{02}\Gamma_yA_g\tilde{A}P_{20} = P\Gamma_y\begin{bmatrix}0\\0\\\hline A_r\end{bmatrix} = PY_pA_pT_d\,. \quad (4.274a)$$

Since $A_pX_p + B_pY_p = I$ implies $X_p + PY_p = A_p^{-1}$ and $X_pA_pT_d + PY_pA_pT_d = T_d$, it follows that $PY_pA_pT_d = T_d - X_pA_pT_d = T_d - X_pA_pT_dA_r^{-1}A_r$. Hence, (4.274a) yields

$$Q_g = [P_{02}(\widetilde{Y} + \widetilde{A}_1K_o)\widetilde{A}P_{20} - T_d]P_g^{-1} = (PY_pA_pT_d - T_d)A_r^{-1} = -X_pA_pT_dA_r^{-1} \quad (4.274b)$$

which is stable for acceptable $A_e^{-1}B_e$, the only case of interest. That is, Assumption 4.4′ simply becomes

Assumption 4.31 The matrix T_d is stable.

Assumption 4.5 is met since $P_{12} = I$ and Assumption 4.6 remains the same as

Assumption 4.32 T_d is diagonal and T_d^{-1} is stable.

It is of interest to examine whether or not the 3DOF system has a decoupling controller before attempting an $\mathcal{H}_2$ design. The existence of a decoupling controller can be checked by Theorem 4.3. Since $\left[\begin{array}{cc|c}A & 0 & 0\\0 & I & 0\\\hline 0 & 0 & I\end{array}\right]\begin{bmatrix}0\\0\\I\end{bmatrix} = \begin{bmatrix}0\\0\\I\end{bmatrix}$, it follows that $V_b\widetilde{A}P_{20} = \begin{bmatrix}P_b\\0\\0\end{bmatrix}$ is satisfied with $V_b = \begin{bmatrix}0 & 0 & I\\I & 0 & 0\\0 & I & 0\end{bmatrix}$ and $P_b = I$. Moreover, $P_{02}\widetilde{Y}\widetilde{A}P_{20} = 0$. Therefore, (4.14) is satisfied with $Q_o = -T_d$, a stable matrix, and Assumption 4.4 is met. In addition, for the 3DOF system, $\varphi_i = 1$ for each i since $P_b = I$; hence, the pair θ_i, φ_i is coprime and it follows that $\hat{\alpha}_i = 0$, $\beta_i = 1$ in (4.105) and $\Delta_\alpha = 0$. Clearly, $\Sigma = P_a^{-1}\Delta_\theta\Delta_\alpha T_dP_b^{-1} = 0$ is stable and it is not difficult to show that with $Q = 0$ Eq. (4.17) is satisfied when B_1 and Y are replaced with $\widetilde{B}_1$ and $\widetilde{Y}$. Therefore, under Theorem 4.3 the 3DOF system always admits a decoupling controller.

Now in order to proceed with the $\mathcal{H}_2$ decoupling design for acceptable inputs, an acceptable decoupling controller for the 3DOF system is assumed to exist. In this regard it is convenient to introduce the set $\{\theta_i\}$ of unique monic polynomials of minimal degree such that the $\{$ *i*-th *column of* $P_a^{-1}\} \times \theta_i$ is stable for all columns of P_a^{-1} and the set $\{\varphi_{gi}\}$ of unique monic polynomials of minimal degree such that $\varphi_{gi} \times \{$*i*-th *row of* $A_r^{-1}\}$ is stable for all rows of $P_g^{-1} = A_r^{-1}$. That $P_g = A_r$ can be used is clear from the expressions for V_g and P_g given in the paragraph containing (4.272). It is also convenient to introduce $\Delta_\theta = diag\{\theta_i\}$, $\Delta_{\varphi g} = diag\{\varphi_{gi}\}$, $\Delta_{\alpha g} = diag\{\alpha_{gi}\}$, and $\Delta_{\beta g} = diag\{\beta_{gi}\}$, where α_{gi} and β_{gi} are unique polynomials of minimal degrees satisfying $\alpha_{gi}\theta_i + \beta_{gi}\varphi_{gi} = 1$ when θ_i, φ_{gi} are coprime. Then (see Theorem 4.4), the existence of an acceptable decoupling controller requires

Assumption 4.33 (1) θ_i and φ_{gi} are coprime (equivalently, no column of P_a^{-1} and the corresponding number row in A_r^{-1} possess coincident poles in the unstable region) and (2) the data construct $\Sigma_g = P_a^{-1}\Delta_\theta\Delta_{\alpha g}T_d A_r^{-1}$ is stable.

As for Assumption 4.7, the weighting matrix W is specified as follows:

Assumption 4.34 The weighting matrix W is of the form $W = diag\{I,\ W_v\}$ and W_v is stable and has full column rank on the stability boundary.

Since

$$A_g\widetilde{A}G_{21}G_w = A_g\widetilde{A}G_{21}\left[\sqrt{\alpha_1}\,\widetilde{G}_{w1} \quad \sqrt{\alpha_2}\,\widetilde{G}_{w2}\right]$$

$$= \begin{bmatrix} 0 & \sqrt{\alpha_1}\,B_qB_d & 0 & \sqrt{\alpha_2}\,B_qA_d\Omega_d & 0 & \sqrt{\alpha_2}\,A_q diag\{A\Omega_{n_m},\ \Omega_{n_l}\} \\ \sqrt{\alpha_1}B_r & 0 & \sqrt{\alpha_2}A_r\Omega_r & 0 & \sqrt{\alpha_2}A_r\Omega_{n_l} & 0 \end{bmatrix}, \tag{4.275}$$

Assumption 4.8 is replaced by

Assumption 4.35 The matrices $[B_qB_d \quad \sqrt{\alpha_2}B_qA_d\Omega_d \quad \sqrt{\alpha_2}A_q diag\{A\Omega_{n_m},\ \Omega_{n_l}\}]$ and $[B_r \ \sqrt{\alpha_2}A_r\Omega_r \ \sqrt{\alpha_2}A_r\Omega_{n_l}]$ have full row rank on the stability boundary (see Example 4.8 for an equivalent condition).

Assumption 4.9 is trivially met since $P_{12} = I$. Assumptions 4.10 through 4.12 are, respectively, simply replaced with the following three assumptions:

Assumption 4.36 $\begin{bmatrix} -P \\ I \end{bmatrix} A_l$ has full column rank on the stability boundary.

Assumption 4.37 $\begin{bmatrix} \Delta_\theta \\ P_a^{-1}\Delta_\theta \end{bmatrix}$ has full column rank on the stability boundary.

Assumption 4.38 No row of $\Delta_{\varphi g}A_r^{-1}$ vanishes on the stability boundary.

Since

$$WG_{11}G_w = \begin{bmatrix} \sqrt{\alpha_1}T_dA_r^{-1}B_r & -\sqrt{\alpha_1}P_dA_d^{-1}B_d & \sqrt{\alpha_2}T_d\Omega_r & -\sqrt{\alpha_2}P_d\Omega_d & 0 & 0 & 0 \\ 0 & 0 & 0 & 0 & 0 & 0 & 0 \end{bmatrix}, \quad (4.276)$$

Assumption 4.13 is equivalent to

Assumption 4.39 The matrices $T_dA_r^{-1}B_r$, $P_d A_d^{-1}B_d$, $\sqrt{\alpha_2}T_d\,\Omega_r$, and $\sqrt{\alpha_2}P_d\Omega_d$ are strictly proper.

The assumption corresponding to Assumption 4.14 is simply

Assumption 4.40 *FP* is proper and when $s \to \infty$, then $\begin{bmatrix} P \\ W_v \end{bmatrix} \to E_{12}\, s^k$

and

$$\begin{bmatrix} A_r^{-1}B_r & \sqrt{\alpha_2}\Omega_r & 0 & 0 & \sqrt{\alpha_2}\Omega_{n_r} & 0 & 0 \\ 0 & 0 & FP_dA_d^{-1}B_d & \sqrt{\alpha_2}FP_d\Omega_d & 0 & \sqrt{\alpha_2}\Omega_{n_m} & 0 \\ 0 & 0 & LA_d^{-1}B_d & \sqrt{\alpha_2}L\Omega_d & 0 & 0 & \sqrt{\alpha_2}\Omega_{n_l} \end{bmatrix}$$
$$\to E_{21}s^l,$$

where E_{12} and E_{21} have full column rank and full row rank, respectively, and $k+l \geq 0$.

Having established assumptions for the 3DOF case which assure that the assumptions made in Theorems 4.5 and 4.6 are met, one can invoke these theorems to find the solution for the 3DOF case. In particular, the formula $q_L = -\Theta^{-1}(\{U_*\phi_1\}_\infty + \{U_*\phi_1\}_+ - \delta_f)$ can be used.

For the 3DOF system, it follows from (4.53) that the acceptable $R = C(I - P_{22}C)^{-1}$ is of the form

$$R = [\, R_r \;\; R_v \,]V_g A_g \tilde{A} = [R_r \mid R_v]\left[\begin{array}{cc|c} 0 & 0 & I \\ \hline I & 0 & 0 \\ 0 & I & 0 \end{array}\right]\left[\begin{array}{c|c} A_q & 0 \\ \hline 0 & A_r \end{array}\right]\left[\begin{array}{cc|c} A & 0 & 0 \\ 0 & I & 0 \\ \hline 0 & 0 & I \end{array}\right] \quad (4.277a)$$

$$= \left[R_vA_q\begin{bmatrix} A & 0 \\ 0 & I \end{bmatrix} \;\middle|\; R_rA_r \right] = [-R_w \;\; -R_z \mid R_u]. \quad (4.277b)$$

The minus sign in the definition of R_w and R_z in (4.277b) is for consistency with the definition in (2.228). The separate roles of $[\, R_w \; R_z \,]$ and R_u in feedback design are well explained in Youla and Bongiorno (1985) and Park and Bongiorno (1990). In decoupling design, only R_u is relevant since it follows from (4.5) and (4.277b) that

$$T_{yr} = P_{02}RP_{20} = P[-R_w \quad -R_z \quad R_u][0 \quad 0 \quad I]' = PR_u = PR_rA_r. \tag{4.278}$$

In view of the expressions for R_v and R_r in (4.54), $[\,R_w\ R_z\,]$ in the 3DOF decoupled system is determined by the selection of the free parameter Q (equivalently, $\widetilde{q}$) and R_u is determined by the selection of the free parameters of T (equivalently, $\widetilde{d}$) and M_{21} (equivalently, m_{21}). In the theorems of this section, solution formulas are expressed in terms of R_u and $[\,R_w\ R_z\,]$.

The evaluation of $q_L = -\Theta^{-1}(\{U_*\phi_1\}_\infty + \{U_*\phi_1\}_+ - \delta_f)$ begins with the calculation of $\Pi_*\Pi$ so that Θ in (4.149) can be obtained. The submatrices N_{g1} and N_{g2} are determined by (4.120) and (4.275) which yield

$$N_{g1} = [\,\sqrt{\alpha_1}B_r\ 0\ \sqrt{\alpha_2}A_r\Omega_r\ 0\ \sqrt{\alpha_2}A_r\Omega_{n_l}\ 0\ \ 0\,] \tag{4.279}$$

and

$$N_{g2} = [\,0 \quad \sqrt{\alpha_1}\,B_qB_d \ \ 0 \quad \sqrt{\alpha_2}\,B_qA_d\Omega_d \quad 0 \quad \sqrt{\alpha_2}\,A_q diag\{A\Omega_{n_m}, \Omega_{n_l}\}\,]. \tag{4.280}$$

When V_a is partitioned as $V_a = [\,V_{a1} \quad V_{a2}\,]$ corresponding to the column sizes of the partition $[-P_a \quad 0\,]$, the partition matrices N_{a1} and N_{a2} in (4.125) are given by the pair

$$N_{a1} = V_{a1},\ N_{a2} = V_{a2}. \tag{4.281}$$

It now easily follows from (4.134) that $\Pi = [\,\Pi_{g1} \quad \Pi_{g2}\,]$ with

$$\Pi_{g1} = \left[(\Delta_{\varphi g}A_r^{-1}N_{g1})' \bullet (W\begin{bmatrix} -\Delta_\theta \\ V_{a1}P_a^{-1}\Delta_\theta \end{bmatrix}) \;\middle|\; N'_{g1} \otimes (W\begin{bmatrix} 0 \\ V_{a2} \end{bmatrix}) \right] \tag{4.282}$$

$$= \left[(\Delta_{\varphi g}A_r^{-1}N_{g1})' \bullet \begin{bmatrix} -\Delta_\theta \\ W_vV_{a1}P_a^{-1}\Delta_\theta \end{bmatrix} \;\middle|\; N'_{g1} \otimes \begin{bmatrix} 0 \\ W_vV_{a2} \end{bmatrix} \right] \tag{4.283}$$

and

$$\Pi_{g2} = N'_{g2} \otimes (W\begin{bmatrix} -P \\ I \end{bmatrix}A_1) = N'_{g2} \otimes (\begin{bmatrix} -P \\ W_v \end{bmatrix}A_1). \tag{4.284}$$

It can be readily shown that $N_{g1}N_{g2*} = 0$. It then follows with the aid of the identities given after Lemma 4.2 and those contained in (4.216)–(4.218) that $\Pi_{g1*}\Pi_{g2} = 0$. Hence,

$$\Pi_*\Pi = \begin{bmatrix} \Pi_{g1*} \\ \Pi_{g2*} \end{bmatrix}[\,\Pi_{g1} \quad \Pi_{g2}\,] = \begin{bmatrix} \Pi_{g1*}\Pi_{g1} & 0 \\ 0 & \Pi_{g2*}\Pi_{g2} \end{bmatrix}. \tag{4.285}$$

Straightforward calculations using the properties in (4.216)–(4.219) yield

$$\Pi_{g1*}\Pi_{g1} = \begin{bmatrix} \Psi_g & \{(\Delta_{\varphi g}G_{rl}A_{r*})' \bullet (V_{a2*}W_{v*}W_vV_{a1}P_a^{-1}\Delta_\theta)\}_* \\ (\Delta_{\varphi g}G_{rl}A_{r*})' \bullet (V_{a2*}W_{v*}W_vV_{a1}P_a^{-1}\Delta_\theta) & (A_rG_{rl}A_{r*})' \otimes (V_{a2*}W_{v*}W_vV_{a2}) \end{bmatrix}, \tag{4.286}$$

where

$$\Psi_g = (\Delta_{\varphi g}G_{rl}\Delta_{\varphi g*})' \circ \{\Delta_{\theta *}(I + P_{a*}^{-1}V_{a1*}W_{v*}W_vV_{a1}P_a^{-1})\Delta_\theta\} \tag{4.287a}$$

and

$$G_{rl} = \alpha_1 A_r^{-1}B_rB_{r*}A_{r*}^{-1} + \alpha_2(\Omega_r\Omega_{r*} + \Omega_{n_r}\Omega_{n_r*}). \tag{4.287b}$$

In view of the diagonal form in (4.285), it is natural to introduce Wiener–Hopf factors Θ_{g1}, Θ_{g2} satisfying

$$\Pi_{g1*}\Pi_{g1} = \Theta_{g1*}\Theta_{g1},\ \Pi_{g2*}\Pi_{g2} = \Theta_{g2*}\Theta_{g2}. \tag{4.288}$$

Now it follows from (4.285) and (4.288) that

$$\Theta = \begin{bmatrix} \Theta_{g1} & 0 \\ 0 & \Theta_{g2} \end{bmatrix},\ U = \Pi\Theta^{-1} = [\ U_{g1}\ \ U_{g2}\ ] \tag{4.289}$$

with

$$U_{g1} = \Pi_{g1}\Theta_{g1}^{-1},\ U_{g2} = \Pi_{g2}\Theta_{g2}^{-1}. \tag{4.290}$$

Clearly, $U_{g1*}U_{g1} = I$ and $U_{g2*}U_{g2} = I$; hence, $U_{g1} \leq O(s^0)$ and $U_{g2} \leq O(s^0)$ by Lemma 3.2.

For the calculation of $U_*\phi_1 = \begin{bmatrix} U_{g1*}\phi_1 \\ U_{g2*}\phi_1 \end{bmatrix}$ it is convenient to denote Φ_1 in (4.129) as

$$\Phi_1 = \Phi_{11} + \Phi_{g1} + \Phi_{g2}, \tag{4.291}$$

where

$$\Phi_{11} = WG_{11}G_w,\ \Phi_{g1} = W\begin{bmatrix} -P_a \\ V_{a1} \end{bmatrix} P_a^{-1}\Delta_\theta\Delta_{\alpha g}T_dA_r^{-1}N_{g1}, \tag{4.292}$$

and

$$\Phi_{g2} = \begin{bmatrix} -P \\ W_v \end{bmatrix} \Gamma_y V_g^{-1} \begin{bmatrix} 0 \\ I \end{bmatrix} N_{g2}. \tag{4.293}$$

Then, $\phi_1 = vec\,\Phi_1 = \phi_{11} + \phi_{g1} + \phi_{g2}$ with

$$\phi_{11} = vec\,(WG_{11}G_w),\ \phi_{g2} = vec\,\Phi_{g2} = \{N'_{g2} \otimes (\begin{bmatrix} -P \\ W_v \end{bmatrix} A_1)\} vec\,(A_1^{-1}\Gamma_y V_g^{-1} \begin{bmatrix} 0 \\ I \end{bmatrix}), \tag{4.294}$$

and

$$\phi_{g1} = vec\,\Phi_{g1} = \{(\Delta_{\varphi g} A_r^{-1} N_{g1})' \bullet (W \begin{bmatrix} -\Delta_\theta \\ V_{a1} P_a^{-1} \Delta_\theta \end{bmatrix})\} vecd(\Delta_{\alpha g} T_d \Delta_{\varphi g}^{-1}) \tag{4.295}$$

$$= (N'_{g1} \otimes W)\{(\Delta_{\varphi g} A_r^{-1})' \bullet \begin{bmatrix} -\Delta_\theta \\ V_{a1} P_a^{-1} \Delta_\theta \end{bmatrix}\} vecd(\Delta_{\alpha g} T_d \Delta_{\varphi g}^{-1}). \tag{4.296}$$

In the expression for ϕ_{g2} in (4.294) the term in braces is Π_{g2} and in the expression for ϕ_{g1} in (4.295) the term in braces is $\Pi_{g1}[I \quad 0]'$. It then easily follows from $\Pi_{g1*}\Pi_{g2} = 0$, that $U_{g1*}\phi_1 = U_{g1*}(\phi_{11} + \phi_{g1})$ and $U_{g2*}\phi_1 = U_{g2*}(\phi_{11} + \phi_{g2})$. That is, the solution for $q_L = [\tilde{d}' \ m'_{21} \ \tilde{q}']'$ given by

$$q_L = -\begin{bmatrix} \Theta_{g1}^{-1} & 0 \\ 0 & \Theta_{g2}^{-1} \end{bmatrix} \left(\left\{ \begin{bmatrix} U_{g1*}\phi_1 \\ U_{g2*}\phi_1 \end{bmatrix} \right\}_\infty + \left\{ \begin{bmatrix} U_{g1*}\phi_1 \\ U_{g2*}\phi_1 \end{bmatrix} \right\}_+ - \begin{bmatrix} \delta_{f1} \\ \delta_{f2} \end{bmatrix} \right) \tag{4.297}$$

with $\delta_f = [\ \delta'_{f1} \ \ \delta'_{f2}\ ]'$ yields

$$\begin{bmatrix} \tilde{d} \\ m_{21} \end{bmatrix} = -\Theta_{g1}^{-1}(\{\,U_{g1*}(\phi_{11} + \phi_{g1})\,\}_\infty + \{\,U_{g1*}(\phi_{11} + \phi_{g1})\,\}_+ - \delta_{f1}) \tag{4.298}$$

and

$$\tilde{q} = -\Theta_{g2}^{-1}(\{\,U_{g2*}(\phi_{11} + \phi_{g2})\,\}_\infty + \{\,U_{g2*}(\phi_{11} + \phi_{g2})\,\}_+ - \delta_{f2}). \tag{4.299}$$

when it is recognized that the square matrix Θ_{g2} has the same number of columns as Π_{g2} which in turn equals the number of rows of $\tilde{q}$. Straightforward calculations (see Example 4.6) now lead to

$$U_{g1*}(\phi_{11}+\phi_{g1}) = \Theta_{g1*}^{-1}\left[\frac{\Psi_g\, vecd(\Delta_{\alpha g}T_d\Delta_{\varphi g}^{-1}) - (\Delta_{\varphi g}\bullet\Delta_\theta)_* vec(T_dG_r)}{vec(V_{a2*}W_{v*}W_vV_{a1}P_a^{-1}\Delta_\theta\Delta_{\alpha g}T_dG_{rl}A_{r*})}\right], \tag{4.300}$$

where

$$G_r = \alpha_1A_r^{-1}B_rB_{r*}A_{r*}^{-1} + \alpha_2\Omega_r\Omega_{r*}. \tag{4.301}$$

When Assumption 4.39 is satisfied, $U_{g1*}\phi_{11}$ is strictly proper and hence $\{U_{g1*}\phi_{11}\}_\infty = 0$. In this case, $\{U_{g1*}(\phi_{11}+\phi_{g1})\}_\infty = \{U_{g1*}\phi_{g1}\}_\infty$ and $U_{g1*}\phi_{g1}$ is given by

$$U_{g1*}\phi_{g1} = \Theta_{g1*}^{-1}\left[\frac{\Psi_g\, vecd(\Delta_{\alpha g}T_d\Delta_{\varphi g}^{-1})}{vec\,(V_{a2*}W_{v*}W_vV_{a1}P_a^{-1}\Delta_\theta\Delta_{\alpha g}T_dG_{rl}A_{r*})}\right]. \tag{4.302}$$

Simplification of the formula for $\tilde{q}$ in (4.299) is also possible. To begin with,

$$\Pi_{g2*}\Pi_{g2} = \{N'_{g2}\otimes(P_wA_1)\}_*\{N'_{g2}\otimes(P_wA_1)\} = (N_{g2}N_{g2*})'\otimes\{(P_wA_1)_*(P_wA_1)\}, \tag{4.303}$$

where

$$P_w = \begin{bmatrix} -P \\ W_v \end{bmatrix}. \tag{4.304}$$

It is also convenient to introduce the Wiener–Hopf factors Ω_{qd} and Λ which satisfy

$$N_{g2}N_{g2*} = A_q\left(\begin{bmatrix} AFP_d \\ L \end{bmatrix} G_d \begin{bmatrix} AFP_d \\ L \end{bmatrix}_* + diag\{AG_{n_m}A_*,\ G_{n_l}\}\right)A_{q*} = \Omega_{qd}\Omega_{qd*} \tag{4.305}$$

and

$$(P_wA_1)_*(P_wA_1) = A_{1*}(P_*P + W_{v*}W_v)A_1 = \Lambda_*\Lambda, \tag{4.306}$$

where

$$G_d = \alpha_1A_d^{-1}B_dB_{d*}A_{d*}^{-1} + \alpha_2\Omega_d\Omega_{d*},\ G_{n_m} = \alpha_2\Omega_{n_m}\Omega_{n_m*},\ G_{n_l} = \alpha_2\Omega_{n_l}\Omega_{n_l*}. \tag{4.307}$$

Assumption 4.35 assures that Ω_{qd}^{-1} and Assumptions 4.34 and 4.36 assure that Λ^{-1} are analytic on the stability boundary and, therefore, stable. Clearly,

$$\Pi_{g2*}\Pi_{g2} = (\Omega'_{qd*}\Omega'_{qd}) \otimes (\Lambda_*\Lambda) = (\Omega'_{qd*} \otimes \Lambda_*)(\Omega'_{qd} \otimes \Lambda). \tag{4.308}$$

Now for two square nonsingular matrices M and N it is not difficult to confirm that $(M \otimes N)^{-1}$ is given by $M^{-1} \otimes N^{-1}$. Hence,

$$\Theta_{g2} = \Omega'_{qd} \otimes \Lambda \tag{4.309}$$

is a stable Wiener–Hopf factor for (4.288) whose inverse is also stable. Moreover,

$$U_{g2} = \Pi_{g2}\Theta_{g2}^{-1} = \{N'_{g2} \otimes (P_w A_1)\}\,(\Omega_{qd}'^{-1} \otimes \Lambda^{-1}) = (\Omega_{qd}^{-1} N_{g2})' \otimes (P_w A_1 \Lambda^{-1}) \tag{4.310}$$

and

$$U_{g2*}\phi_1 = U_{g2*}(\phi_{11} + \phi_{g2}) = \{(\Omega_{qd}^{-1}N_{g2})'_* \otimes (P_w A_1 \Lambda^{-1})_*\}(\phi_{11} + \phi_{g2}) \tag{4.311}$$

$$= \{(\Omega_{qd}^{-1}N_{g2})'_* \otimes (P_w A_1 \Lambda^{-1})_*\}\, vec\,(\Phi_{11} + \Phi_{g2}) \tag{4.312}$$

$$= vec\,\{(P_w A_1 \Lambda^{-1})_*(\Phi_{11} + \Phi_{g2})\,(\Omega_{qd}^{-1}N_{g2})_*\} = vec\,(\widetilde{\Phi}_{11} + \widetilde{\Phi}_{g2}), \tag{4.313}$$

where

$$\widetilde{\Phi}_{11} = (P_w A_1 \Lambda^{-1})_* \Phi_{11}\,(\Omega_{qd}^{-1}N_{g2})_* \tag{4.314}$$

$$= \Lambda_*^{-1} A_{1*} P_* P_d G_d A_{d*} B_{q*} \Omega_{qd*}^{-1} \tag{4.315}$$

and

$$\widetilde{\Phi}_{g2} = (P_w A_1 \Lambda^{-1})_* \Phi_{g2}\,(\Omega_{qd}^{-1}N_{g2})_* \tag{4.316}$$

$$= \Lambda A_1^{-1} \Gamma_y V_g^{-1} \begin{bmatrix} 0 \\ I \end{bmatrix} \Omega_{yd}. \tag{4.317}$$

It then follows from (4.299) that

$$\widetilde{q} = vec\,Q = -\Theta_{g2}^{-1}(\{\,U_{g2*}(\phi_{11} + \phi_{g2})\,\}_\infty + \{\,U_{g2*}(\phi_{11} + \phi_{g2})\,\}_+ - \delta_{f2}) \tag{4.318}$$

$$= -(\Omega_{qd}'^{-1} \otimes \Lambda^{-1}) vec\,(\{\widetilde{\Phi}_{11} + \widetilde{\Phi}_{g2}\}_{\infty,+} - Z_{f2}) \tag{4.319}$$

$$= -vec\,[\Lambda^{-1}(\{\widetilde{\Phi}_{11}+\widetilde{\Phi}_{g2}\}_{\infty,+} - Z_{f2})\,\Omega_{qd}^{-1}\,], \tag{4.320}$$

where Z_{f2} is uniquely determined by its size from $vec\,Z_{f2} = \delta_{f2}$ and $\{G\}_{\infty,+}$ denotes $\{G\}_{\infty} + \{G\}_{+}$. The formula (4.320) for $\widetilde{q}$ represents a significant simplification since no Kronecker products are involved. Obviously,

$$Q = -\Lambda^{-1}(\{\widetilde{\Phi}_{11}+\widetilde{\Phi}_{g2}\}_{\infty,+} - Z_{f2})\,\Omega_{qd}^{-1} \tag{4.321}$$

$$= -\Lambda^{-1}\Big(\{\widetilde{\Phi}_{11}\}_{\infty,+} + \widetilde{\Phi}_{g2} - \{\widetilde{\Phi}_{g2}\}_{-} - Z_{f2}\Big)\Omega_{qd}^{-1} \tag{4.322}$$

$$= -\Lambda^{-1}\Big(\{\widetilde{\Phi}_{11}\}_{\infty,+} - \{\widetilde{\Phi}_{g2}\}_{-} - Z_{f2}\Big)\Omega_{qd}^{-1} - \Lambda^{-1}\widetilde{\Phi}_{g2}\Omega_{qd}^{-1} \tag{4.323}$$

$$= -\Lambda^{-1}\Big(\{\widetilde{\Phi}_{11}\}_{\infty,+} - \{\widetilde{\Phi}_{g2}\}_{-} - Z_{f2}\Big)\Omega_{qd}^{-1} - A_1^{-1}\Gamma_y V_g^{-1}\begin{bmatrix}0\\ I\end{bmatrix}. \tag{4.324}$$

Hence, (4.277b) and (4.53) yield

$$[\,-R_w \quad -R_z\,] = R_y A_q\begin{bmatrix}A & 0\\ 0 & I\end{bmatrix} = (\Gamma_y V_g^{-1}\begin{bmatrix}0\\ I\end{bmatrix} + A_1 Q)A_q\begin{bmatrix}A & 0\\ 0 & I\end{bmatrix} \tag{4.325}$$

$$= -A_1\Lambda^{-1}\Big(\{\widetilde{\Phi}_{11}\}_{\infty,+} - \{\widetilde{\Phi}_{g2}\}_{-} - Z_{f2}\Big)\Omega_{qd}^{-1}A_q\begin{bmatrix}A & 0\\ 0 & I\end{bmatrix}. \tag{4.326}$$

Finally, in the cost functional given by (4.164), one can use $\|\delta_f\|_2^2 = \|\delta_{f1}\|_2^2 + \|Z_{f2}\|_2^2$ since $\delta_f = [\,\delta_{f1}' \;\; \delta_{f2}'\,]'$ and $vec\,Z_{f2} = \delta_{f2}$. One can also use $\|\{U_*\phi_1\}_-\|_2^2 = \|\{U_{g1*}\phi_1\}_-\|_2^2 + \|\{U_{g2*}\phi_1\}_-\|_2^2 = \Big\|\{U_{g1*}(\phi_{11}+\phi_{g1})\}_-\Big\|_2^2 + \Big\|\{U_{g2*}(\phi_{11}+\phi_{g2})\}_-\Big\|_2^2$ since $U_*\phi_1 = [\,(U_{g1*}\phi_1)' \mid (U_{g2*}\phi_1)'\,]'$.

The solution of the digital $\mathcal{H}_2$ design problem for the decoupled 3DOF system in Fig. 1.1 now easily follows from Theorem 4.5. Specifically,

Theorem 4.11 (Digital case) *When Assumptions* 4.28 through 4.38 *hold and both* $A_pP_dA_d^{-1}$ *and* $A_pT_dA_r^{-1}$ *are stable, decoupling controllers for the 3DOF system in Fig.* 1.1 *exist. Moreover, the set of all loop transfer matrices* $R = C(I - P_{22}C)^{-1} = [-R_w - R_z\; R_u]$ *associated with the ones for which* $J = \|WG_\delta G_w\|_2^2$ *is finite is given by*

$$R_u = R_r A_r = V_a\begin{bmatrix}P_a^{-1}T\\ M_{21}A_r\end{bmatrix} = V_a\begin{bmatrix}P_a^{-1}\Delta_\theta(\Delta_{ag}T_d + \Delta\Delta_{\varphi g})\\ M_{21}A_r\end{bmatrix}, \tag{4.327}$$

$$\widetilde{d} = vecd\Delta, m_{21} = vec\, M_{21}, \begin{bmatrix} \widetilde{d} \\ m_{21} \end{bmatrix} = -\Theta_{g1}^{-1}\left(\left\{U_{g1*}(\phi_{11}+\phi_{g1})\right\}_{\infty,+} - \delta_{f1}\right), \tag{4.328}$$

and

$$[R_w\ R_z] = A_1\Lambda^{-1}\left(\left\{\widetilde{\Phi}_{11}\right\}_{\infty,+} - \left\{\widetilde{\Phi}_{g2}\right\}_{-} - Z_{f2}\right)\Omega_{qd}^{-1}A_q\begin{bmatrix} A & 0 \\ 0 & I \end{bmatrix}, \tag{4.329}$$

where δ_{f1} *is an arbitrary stable vector and* Z_{f2} *is an arbitrary stable rational matrix. The matrices* $U_{g1^*}(\phi_{11}+\phi_{g1})$, $\widetilde{\Phi}_{11}$ and $\widetilde{\Phi}_{g2}$ *are given by (4.300), (4.315), and (4.317), respectively. The optimal* R_u and $[R_w\quad R_z]$ *that minimize the cost functional are the ones given by (4.327) and (4.329), respectively, with* δ_{f1} *and* Z_{f2} *zero. When* J_{opt} *denotes the cost functional associated with the optimal solution, the cost functional* J *yielded by* R_u and $[R_w\, R_z]$ *in (4.327) and (4.329) is given by*

$$J = J_{opt} + \left\|\delta_{f1}\right\|_2^2 + \left\|Z_{f2}\right\|_2^2, \tag{4.330}$$

where

$$J_{opt} = \frac{1}{2\pi j}\oint_{|\lambda|=1}(vec\,(WG_{11}G_w))_*\left(I - U_{g1}U_{g1*} - U_{g2}U_{g2*}\right)(vec\,(WG_{11}G_w))\frac{d\lambda}{\lambda}$$
$$+\left\|\left\{U_{g1*}(\phi_{11}+\phi_{g1})\right\}_{-}\right\|_2^2 + \left\|\left\{U_{g2*}(\phi_{11}+\phi_{g2})\right\}_{-}\right\|_2^2, \tag{4.331}$$

with $U_{g2*}(\phi_{11}+\phi_{g2})$ *and* $WG_{11}G_w$ *given by (4.313) and (4.276), respectively.*

The solution of the analog $\mathcal{H}_2$ design problem for the decoupled 3DOF system in Fig. 1.1 is obtained from Theorem 4.6. Specifically,

Theorem 4.12 (Analog case) *When Assumptions* 4.28 through 4.40 *hold and both* $A_pP_dA_d^{-1}$ *and* $A_pT_dA_r^{-1}$ *are stable, decoupling controllers for the 3DOF system in Fig. 1.1 exist. Moreover, the set of all loop transfer matrices* $R = C(I - P_{22}C)^{-1} = [-R_w - R_z\ R_u]$ *associated with the ones for which* $J = \|WG_\delta G_w\|_2^2$ *is finite is given by*

$$R_u = R_rA_r = V_a\begin{bmatrix} P_a^{-1}T \\ M_{21}A_r \end{bmatrix} = V_a\begin{bmatrix} P_a^{-1}\Delta_\theta\left(\Delta_{\alpha g}T_d + \Delta\Delta_{\varphi g}\right) \\ M_{21}A_r \end{bmatrix}, \tag{4.332}$$

$$\widetilde{d} = vecd\,\Delta, m_{21} = vec\,M_{21}, \tag{4.333}$$

$$\begin{bmatrix} \widetilde{d} \\ m_{21} \end{bmatrix} = -\Theta_{g1}^{-1}\left(\{U_{g1*}\phi_{g1})\}_{\infty} + \left\{U_{g1*}(\phi_{11}+\phi_{g1})\right\}_{+} - \delta_{f1}\right), \tag{4.334}$$

and

$$[R_w\ R_z] = A_1\Lambda^{-1}\left(\left\{\widetilde{\Phi}_{11}\right\}_{+} - \left\{\widetilde{\Phi}_{g2}\right\}_{-} - Z_{f2}\right)\Omega_{qd}^{-1}A_q\begin{bmatrix} A & 0 \\ 0 & I \end{bmatrix}, \tag{4.335}$$

where δ_{f1} *is an arbitrary stable strictly proper vector and* Z_{f2} *is an arbitrary stable strictly proper rational matrix. The matrices* $U_{g1*}(\phi_{11}+\phi_{g1}), U_{g1*}\phi_{g1}$, $\widetilde{\Phi}_{11}$, and $\widetilde{\Phi}_{g2}$ *are given by* (4.300), (4.302), (4.315), *and* (4.317), *respectively. The optimal* R_u *and* $[R_w\,R_z]$ *that minimize the cost functional are the ones given by* (4.332) *and* (4.335) *with* δ_{f1} *and* Z_{f2} *zero, respectively. When* J_{opt} *denotes the cost functional associated with the optimal solution, the cost functional J yielded by* R_u *and* $[R_w\,R_z]$ *in* (4.332) *and* (4.335) *is given by*

$$J = J_{opt} + \left\|\delta_{f1}\right\|_2^2 + \left\|Z_{f2}\right\|_2^2, \tag{4.336}$$

where

$$\begin{aligned} J_{opt} = \frac{1}{2\pi j}\int_{-j\infty}^{j\infty} & vec\,(WG_{11}G_w)_*(I - U_{g1}U_{g1*} - U_{g2}U_{g2*})vec\,(WG_{11}G_w)ds \\ & + \left\|\{U_{g1*}(\phi_{11}+\phi_{g1})\}_{-}\right\|_2^2 + \left\|\{U_{g2*}(\phi_{11}+\phi_{g2})\}_{-}\right\|_2^2, \end{aligned} \tag{4.337}$$

with $U_{g2*}(\phi_{11}+\phi_{g2})$ *and* $WG_{11}G_w$ *given by* (4.313) *and* (4.276), *respectively. The controller C associated with* R_u *and* $[R_w\,R_z]$ *in* (4.332) *and* (4.335) *is strictly proper.*

Remark 4.10 As stated previously, $[R_w\,R_z]$ is irrelevant to the decoupling design and its $\mathcal{H}_2$ solution is not affected by the decoupling constraint. In fact it can be shown that if $W_{v*}W_v$ in (4.306) is replaced with kQ as in (3.344), the $\mathcal{H}_2$ solution formula for $[R_w\,R_z]$ in (4.329) is the same as the one for $[R_w\,R_z]$ in (3.376) in Sect. 3.8, where the ordinary $\mathcal{H}_2$ problem is treated (see Example 4.7). That is,

$$[R_w\ R_z] = A_1\Lambda^{-1}\left(\left\{\widetilde{\Phi}_{11}\right\}_{\infty,+} - \left\{\widetilde{\Phi}_{g2}\right\}_{-} - Z_{f2}\right)\Omega_{qd}^{-1}A_q\begin{bmatrix} A & 0 \\ 0 & I \end{bmatrix} \tag{4.338}$$

$$= A_1\Lambda^{-1}(Z_{mn} + \{[\widetilde{\Gamma}_w \quad \widetilde{\Gamma}_z]\}_+ + \{[\widetilde{\Gamma}_w \quad \widetilde{\Gamma}_z]\}_\infty + \{[\widetilde{\Gamma}_a \quad \widetilde{\Gamma}_b]\}_-)\Omega_t^{-1}\begin{bmatrix} A & 0 \\ 0 & A_n \end{bmatrix} \tag{4.339}$$

where the variables $\widetilde{\Gamma}_w, \widetilde{\Gamma}_z, \widetilde{\Gamma}_a, \widetilde{\Gamma}_b$, and Ω_t are defined as in Sect. 3.8. It can be shown similarly that the formula in (4.335) is the same as the one in (3.411).

4.7 Historical Perspective and Commentary

One of the main characteristics of multivariable systems is interaction between input and output variables. Efforts to eliminate cross-coupling between these variables require controllers that make the transfer matrix from the inputs to the outputs diagonal, or more generally, block diagonal. The decoupling design problem dates back to Voznesenskij (1936), Kavanagh (1957), and Strejc (1960) (see also the references in Wang (2003) and Kučera (2011)). The early approaches are based on transfer matrix methods without imposing the stability constraint.

The first meaningful results are presented in Morgan (1964) where decoupling design with static state feedback is treated. Falb and Wolovich (1967) present a necessary and sufficient condition for the existence of a decoupling controller for state-space models with a characterization of the class of feedback matrices which decouple a system. Decoupling with dynamic state feedback is investigated using the geometric approach in Morse and Wonham (1970) where it is claimed that if a linear constant multivariable system can be decoupled at all, then it can always be decoupled using linear dynamic compensation. Hautus and Heymann (1983) exploit the transfer matrix methods for decoupling design and show that dynamic state feedback is equivalent to combined dynamic output feedback and feedforward reference compensation, often referred to as a 2DOF controller.

For the 2DOF configuration, Doseor and Gündes (1986) and Lee and Bongiorno (1993a, b) show that the decoupling and stability problems can be treated independently and decoupling design is always possible if the plant is internally stabilizable. On the other hand, decoupling and stability properties are not independent in the 1DOF configuration; hence, if decoupling is a design requirement, it is more restrictive for the 1DOF configuration than for the 2DOF configuration. So when decoupling is required, one might be tempted to exclude 1DOF systems from consideration. However, Howze and Bhattacharyya (1997) point out that the asymptotic tracking property of 2DOF controllers is fragile with respect to the controller parameters in the absence of error feedback. This is not the case for the unity-feedback 1DOF controller configuration. With respect to this issue then a 1DOF system could be preferable and should be considered as well.

Unlike the 2DOF case, a 1DOF decoupling controller does not always exist. A sufficient condition for the existence of one is suggested in Vardulakis (1987) and necessary and sufficient conditions are established in Wang (1992), Lin (1997), Youla and Bongiorno (2000), and Gómez and Goodwin (2000). An algebraic approach based on coprime factorizations is adopted by Gómez and Goodwin (2000) to treat both triangular and diagonal decoupling design problems. A notable feature in Youla and Bongiorno (2000) is the parameterization of all realizable decoupled transfer matrices in an explicit affine form which makes it very convenient to treat performance and/or robust stability in the design of decoupled systems. In the papers by Wang (1992) and Lin (1997), the conventional model with unity feedback is treated. In Youla and Bongiorno (2000), the conventional model with non-unity feedback is addressed.

In Park (2012), a parameterization of diagonal decoupling controllers is presented for the generalized plant model which accommodates 1, 2, and 3DOF configurations with nonsquare plant and non-unity feedback cases. However, the method used to obtain the realizable diagonal T requires vector operations which inflate the dimensionality of the problem. Thus, the benefit gained from the general formulation in Park (2012) lies more in a unified treatment rather than in any simplification of the solution. In this regard, the formula of the realizable diagonal T in Theorem 4.3 is an improved one which provides a unified treatment along with a simpler solution for the generalized plant model.

Though the existence and the structure of realizable decoupled systems have been fairly well analyzed, not many papers treat the issue of performance or robust stability. The first to treat the robust stability problem with decoupling controllers is Safonov and Chen (1982). In this paper, decoupling controllers are described for which stability margin in the $\mathcal{H}_\infty$ norm context is maximized and output regulation constraints are taken into consideration. The $\mathcal{H}_2$ decoupling controller formulas for analog 2DOF and 3DOF systems appear first in Lee and Bongiorno (1993a, b). In these two papers, the tracking error and plant input are penalized in the quadratic cost functional and persistent inputs are treated whose only unstable poles lie on the $j\omega$ axis. Corrêa, Soares, and Sales (2001) treat the optimal tracking problem with decoupling constraints for 2DOF analog systems. This paper allows persistent deterministic exogenous inputs with poles in $\mathrm{Re}\,s \geq 0$ where the plant has none and insists on asymptotic tracking with zero steady-state error. In this case, the plant input must have poles in $\mathrm{Re}\,s \geq 0$ and the $\mathcal{H}_2$ norm of the plant input is not finite. This difficulty is circumvented by incorporating only the stable part of the plant input into the system performance measure and is the same approach taken in Corrêa and Da Silveira (1995). As a consequence of this approach, however, plant input saturation can become a problem. In Park, Choi, and Kuc (2002), the results contained in Lee and Bongiorno (1993a) are extended to the 2DOF standard configuration and computational algorithms based on state-variable representations are also given. Kučera (2011) treats the optimal $\mathcal{H}_2$ block decoupling problem by adopting a general setting for the 2DOF controller configuration. The class of all controllers that decouple and stabilize the system is determined in parametric form and the parameter is used to obtain the optimal controller which minimizes the $\mathcal{H}_2$ norm of the transfer matrix from the reference input to the error. In formulating the cost functional, however, the plant input is not considered and neither are persistent inputs. The $\mathcal{H}_2$ problem for 1DOF systems is considerably more difficult to solve than that of 2DOF systems. Brinsmead and Goodwin (2001) investigate the fundamental limits on a decoupled 1DOF system employing an $\mathcal{H}_2$ measure of the tracking error. Youla and Bongiorno (2000) and Bongiorno and Youla (2001) treat $\mathcal{H}_2$ decoupling problems in which both tracking error and plant saturation for the 1DOF configuration are considered. Park (2008) treats the $\mathcal{H}_2$ decoupling design problem for the generalized plant model, but many restrictive assumptions are made; hence, the merit of the generalized plant model is not fully exploited. The $\mathcal{H}_2$ solution formulas for decoupling design derived in Sect. 4.4 are applicable for the most general setting and cover 1DOF, 2DOF, and 3DOF systems with acceptable persistent inputs as special cases. In this regard, Sect. 4.4 represents a generalization of the results in Youla and Bongiorno (2000), Bongiorno and Youla (2001), and Park (2008). Moreover, many of the results contained in this chapter are new ones appearing here for the first time.

That the Schur matrix product plays a key role in the Wiener–Hopf design of optimal decoupled systems was first reported in Lee and Bongiorno (1993a). The parameterization of all closed-loop transfer matrices each of which can be realized with a 1DOF stabilizing decoupling controller for a plant with a square transfer matrix first appeared in Youla and Bongiorno (2000). The parameterization for a plant with rectangular transfer matrix appears in Bongiorno and Youla (2001).

4.8 Examples

Example 4.1 Show that the vector operation *vec* does not alter the $\mathcal{H}_2$ norm of a matrix.

Solution Let $E = \left[e_{ij}\right]$ and suppose that the size of E is $p \times q$. It can be readily shown then that

$$Tr(E_* E) = \sum_{j=1}^{q} \sum_{i=1}^{p} \left(e_{ij^*} e_{ij}\right) = (vec\, E)_* (vec\, E);$$

hence, $\|E\|_2 = \|vecE\|_2$.

Example 4.2 (*Lemma* 4.1) Prove the following when $G_i, i = 1 \to n$, are matrices whose elements are complex numbers. If G_1 has column rank, then $\text{rank}(G_1G_2) = \text{rank}(G_2)$. If G_i has full column rank for $i = 1 \to n-1$ then $rank(G_1G_2\cdots G_n) = rank(G_n)$.

Solution Employing the Sylvester inequality for G_1 and G_2, one obtains

$$rank(G_1) + rank(G_2) - \begin{pmatrix} \textit{the number of the} \\ \textit{columns of } G_1 \end{pmatrix} \leq rank(G_1G_2) \leq rank(G_2).$$

Since G_1 has full column rank it follows that $rank(G_2) \leq rank(G_1G_2) \leq rank(G_2)$; hence, $rank(G_1G_2) = rank(G_2)$. When G_2 has full column rank, G_1G_2 has full column rank too. When G_i has full column rank, for $i = 1 \to n-1$, one easily obtains by induction that $rank(G_1G_2\cdots G_n) = rank(G_n)$.

Example 4.3 (*Lemma* 4.2) Prove the following when M and N are matrices whose elements are complex numbers. If M and N each have full column rank, then $M \otimes N$ has full column rank.

Solution There always exist permutation matrices E_5 and E_6 such that $M \otimes N = E_5(N \otimes M)E_6$ (Brewer 1978). Since permutation matrices are invertible, it follows that $rank(M \otimes N) = rank(N \otimes M)$. When the sizes of M and N are $n \times m$ and $p \times q$, respectively, the Sylvester inequality and the identity $M \otimes N = (M \otimes I_p)(I_m \otimes N)$ yield

$$rank(M \otimes I_p) + rank(I_m \otimes N) - mp \leq rank(M \otimes N) \leq rank(I_m \otimes N) = mq.$$

Since $rank(M \otimes I_p) = rank(I_p \otimes M) = mp$, it follows that $rank(M \otimes N) = mq$ which is the number of the columns of $M \otimes N$.

Example 4.4 A real rational matrix $U(s)$ is called inner if it is stable and $U_*(s)U(s) = I$ (paraunitary). Show that for an inner matrix $U(s)$ there always exists a left Wiener–Hopf factor $U_\perp(s)$, called complementary inner, such that $I - U(s)U_*(s) = U_\perp(s)U_{\perp*}(s)$. In this case $I = U(s)U_*(s) + U_\perp(s)U_{\perp*}(s) = \widetilde{U}(s)\widetilde{U}_*(s)$, where $\widetilde{U}(s) = [U(s)U_\perp(s)]$ is a square inner matrix so that $\widetilde{U}^{-1}(s) = \widetilde{U}_*(s)$; hence, $\widetilde{U}_*(s)\widetilde{U}(s) = I$. An equivalent result for a digital system inner $U(\lambda)$ also holds.

Solution When the size of $U(s)$ is $l \times m \quad (l \geq m))$ and when $Q(s) = I_l - U(s)U_*(s)$ is introduced, one can first establish that the normal rank of $Q(s)$ is $l - m$ and then that $Q(j\omega) \geq 0$ for all ω. Since $U_*(s)U(s) = I_m$, the normal rank of both $U(s)$ and $U(s)U_*(s)$ is m. Clearly, from the equality $I_l - U(s)U_*(s) + U(s)U_*(s) = I_l$ one gets

$$\begin{aligned} l &= rank(I_l - U(s)U_*(s) + U(s)U_*(s)) \leq rank(Q(s)) + rank(U(s)U_*(s)) \\ &= rank(Q(s)) + m. \end{aligned}$$

Thus, $l - m \leq rank(Q(s))$. On the other hand, $Q(s)U(s) = (I_l - U(s)U_*(s)) U(s) = 0$ leads to $rank(Q(s)) + rank(U(s)) - l = rank(Q(s)) + m - l \leq 0$ by the Sylvester inequality. Hence, $rank(Q(s)) = l - m$. Next, $Q(j\omega)$ is obviously Hermitian. Moreover, since $U^*(j\omega)U(j\omega) = I$, an eigenvalue $\lambda(j\omega)$ of $U(j\omega) U^*(j\omega)$ is 0 or 1. Thus, an eigenvalue $1 - \lambda(j\omega)$ of $Q(j\omega) = I_l - U(j\omega)U^*(j\omega)$ is 0 or 1. Since $x^*Q(j\omega)x \geq \lambda_{\min}(Q(j\omega))x^*x$ for any x, it is concluded that $Q(j\omega) \geq 0$. That is, the matrix $Q(s)$ is parahermitian-positive and it follows from Theorem C.1 in Appendix C that there exists a left Wiener–Hopf factor $U_\perp(s)$ whose size is $l \times (l - m)$ which satisfies $Q(s) = U_\perp(s)U_{\perp *}(s)$. Following the same procedures for a digital system inner $U(\lambda)$, one obtains identical results for the digital case.

Example 4.5 Suppose that F, T_d, T_d^{-1} and $(T_dF - I)P$ are stable and the 1DOF system described by (4.200)–(4.208) is admissible. (a) Prove that the persistent input G_e in (4.206) is acceptable if $A_pT_dA_r^{-1}$, $A_pP_dA_d^{-1}$ and $LA_d^{-1}A_{n1}$ are stable, where A_p and A_{n1} are the denominator matrices of the coprime polynomial matrix fraction descriptions $P = A_p^{-1}B_p$ and $AP_oA_d^{-1} = B_{n1}A_{n1}^{-1}$, respectively. (b) Show that the G_e in (4.247) for the example in Remark 4.9 is acceptable. **Hint**: For part (a) first apply the concepts discussed in the paragraph containing Eqs. (4.32)–(4.36) to Example 3.3. In particular, establish that G_e is acceptable iff $\widetilde{\Gamma}_w = A_oG_{111}A_e^{-1}$ and $\widetilde{\Gamma}_z = (G_{111} + G_{12}A_1Y_1G_{211})A_e^{-1}A_{g1}$ are stable.

Solution (a) As noted between (4.34) and (4.35), all one has to do is make the replacements $G_{11} \to G_{111}$, $G_{21} \to G_{211}$, and $A_w^{-1} \to A_e^{-1}$ to get from Example 3.3 that $G_e = A_e^{-1}B_e$ is acceptable iff $\widetilde{\Gamma}_w$ and $\widetilde{\Gamma}_z$ are stable. For the 1DOF system in (4.200)–(4.208), $\widetilde{\Gamma}_w = A_oG_{111}A_e^{-1} = \begin{bmatrix} A_pT_dA_r^{-1} & -A_pP_dA_d^{-1} \end{bmatrix}$; hence, $\widetilde{\Gamma}_w$ is stable iff $A_pP_dA_d^{-1}$ and $A_pT_dA_r^{-1}$ are stable. In addition,

$$\tilde{\Gamma}_z = (G_{111} + G_{12}A_1Y_1G_{211})A_e^{-1}A_{g1} = \left(\begin{bmatrix} T_d & -P_d \\ 0 & 0 \end{bmatrix} + \begin{bmatrix} -P \\ I \end{bmatrix} A_1Y_1 \begin{bmatrix} I & -P_o \end{bmatrix} \right) \begin{bmatrix} A_r^{-1} & 0 \\ 0 & A_d^{-1} \end{bmatrix} A_{g1}$$

$$= \begin{bmatrix} T_d - PA_1Y_1 & -(P_d - PA_1Y_1P_o) \\ A_1Y_1 & -A_1Y_1P_o \end{bmatrix} \begin{bmatrix} A_r^{-1} & 0 \\ 0 & A_d^{-1} \end{bmatrix} A_{g1}$$

$$= \begin{bmatrix} T_d - PYA & -(P_d - PA_1Y_1P_o) \\ YA & -YAP_o \end{bmatrix} \begin{bmatrix} A_r^{-1} & 0 \\ 0 & A_d^{-1} \end{bmatrix} A_{g1}$$

$$= \left[\begin{array}{c|c} T_d(I - FPYA) + (T_dF - I)PYA & -(I - PA_1Y_1F)P_d + PA_1Y_1L \\ \hline YA & -YAFP_d - YAL \end{array} \right] \begin{bmatrix} A_r^{-1} & 0 \\ 0 & A_d^{-1} \end{bmatrix} A_{g1}$$

$$= \left[\begin{array}{c|c} T_dXA + (T_dF - I)PYA & PA_1Y_1L \\ \hline YA & -YAL \end{array} \right] \begin{bmatrix} A_r^{-1} & 0 \\ 0 & A_d^{-1} \end{bmatrix} A_{g1} + \left[\begin{array}{c|c} 0 & -(I - PA_1Y_1F)P_dA_d^{-1} \\ \hline 0 & -YAFP_dA_d^{-1} \end{array} \right] A_{g1}$$

$$= \left[\begin{array}{c|c} T_dX + (T_dF - I)PY & PA_1Y_1 \\ \hline Y & -YA \end{array} \right] \begin{bmatrix} AA_r^{-1} & 0 \\ 0 & LA_d^{-1} \end{bmatrix} A_{g1} + \left[\begin{array}{c|c} 0 & -(I - PA_1Y_1F)P_dA_d^{-1} \\ \hline 0 & -YAFP_dA_d^{-1} \end{array} \right] A_{g1}.$$

Clearly, when $A_pP_dA_d^{-1}$ is stable, $(I - PA_1Y_1F)P_dA_d^{-1} = (I - PA_1Y_1F)A_p^{-1}A_p P_dA_d^{-1}$ and $YAFP_dA_d^{-1} = YAFA_p^{-1}A_pP_dA_d^{-1}$ are stable because $(I - PA_1Y_1F)A_p^{-1}$ (equivalently, $(I - PA_1Y_1F)P = PA_1X_1$) and AFA_p^{-1} (equivalently, $AFP = B$) are stable. Therefore, when $A_pP_dA_d^{-1}$ is stable, $\tilde{\Gamma}_z$ is stable iff

$$\tilde{\Gamma}_{zo} = \left[\begin{array}{c|c} T_dX + (T_dF - I)PY & PA_1Y_1 \\ \hline Y & -YA \end{array} \right] \begin{bmatrix} AA_r^{-1} & 0 \\ 0 & LA_d^{-1} \end{bmatrix} A_{g1}$$

is stable. Since T_d, PA_1, and $(T_dF - I)P$ are stable by assumption, $\tilde{\Gamma}_{zo}$ is stable if $\begin{bmatrix} AA_r^{-1} & 0 \\ 0 & LA_d^{-1} \end{bmatrix} A_{g1}$ is stable. It can be shown that the stability of $(T_dF - I)P$ assures the stability of $(I - FT_d)A_r^{-1}$ when F, T_d, and T_d^{-1} are stable. In fact, $(T_dF - I)P$ is stable iff $(T_dF - I)A_p^{-1}$ is stable. It then follows since T_d^{-1} and $A_pT_dA_r^{-1}$ are stable that the matrix $T_d^{-1}(T_dF - I)A_p^{-1}A_pT_dA_r^{-1} = T_d^{-1}(T_dF - I)T_dA_r^{-1} = (FT_d - I)A_r^{-1}$ is stable. When $(I - FT_d)A_r^{-1}$ and $A_pT_dA_r^{-1}$ are stable, it can be shown that AA_r^{-1} is stable (see (3.261)). Moreover, the denominator matrix A_{g1} of the coprime polynomial matrix fraction description $AG_{211}A_e^{-1} = \begin{bmatrix} AA_r^{-1} & -AP_oA_d^{-1} \end{bmatrix} = B_{g1}A_{g1}^{-1}$ is easily parameterized. With $AA_r^{-1} = B_{m1}A_{m1}^{-1}$ and

$AP_oA_d^{-1} = B_{n1}A_n^{-1}$ coprime polynomial matrix fraction descriptions; $\widetilde{A}_{g1} = \begin{bmatrix} A_{m1} & 0 \\ 0 & A_{n1} \end{bmatrix}$; and $\widetilde{B}_{g1} = \begin{bmatrix} B_{m1} & -B_{n1} \end{bmatrix}$, it is clear that $\begin{bmatrix} AA_r^{-1} & -AP_oA_d^{-1} \end{bmatrix} = \widetilde{B}_{g1}\widetilde{A}_{g1}^{-1}$ is a polynomial matrix fraction description for which $\widetilde{A}_{g1}, \widetilde{B}_{g1}$ may not be a coprime pair. In this case, since $B_{g1}A_{g1}^{-1}$ is a left-coprime polynomial matrix fraction description for $\begin{bmatrix} AA_r^{-1} & -AP_oA_d^{-1} \end{bmatrix}$, there exists a square polynomial matrix M_{g1} which is not unimodular such that $\widetilde{A}_{g1} = A_{g1}M_{g1}$ and $\widetilde{B}_{g1} = B_{g1}M_{g1}$ (see Definition B.1 and Lemma B.3). Since the matrix AA_r^{-1} is stable, $\det A_{m1}$ is Hurwitz and it can be concluded that $\det M_{g1}$ is also Hurwitz for the following reason. Suppose that p_o is a finite unstable pole of

$$H_g = \begin{bmatrix} AA_r^{-1} & -AP_oA_d^{-1} \end{bmatrix} = \begin{bmatrix} AA_r^{-1} & 0 \end{bmatrix} + \begin{bmatrix} 0 & -AP_oA_d^{-1} \end{bmatrix}.$$

Then, the equality $\delta(H;p_o) = \delta\left(AP_oA_d^{-1};p_o\right)$ holds by the property B.3 of the McMillan degree in Sect. B.7 of Appendix B. This property holds for every finite unstable pole of H_g; hence, $\det A_{g1}$ and $\det A_{n1}$ possess the same unstable zeros with the same multiplicities. Since $\det \widetilde{A}_{g1} = \det A_{m1} \det A_{n1} = \det A_{g1} \det M_{g1}$, it is concluded that $\det M_{g1}$ must be Hurwitz. Clearly, $A_{g1} = \widetilde{A}_{g1}M_{g1}^{-1} = \begin{bmatrix} A_{m1} & 0 \\ 0 & A_{n1} \end{bmatrix} M_{g1}^{-1}$ with M_{g1}^{-1} stable and it immediately follows that $\begin{bmatrix} AA_r^{-1} & 0 \\ 0 & LA_d^{-1} \end{bmatrix}$ $A_{g1} = \begin{bmatrix} AA_r^{-1}A_{m1} & 0 \\ 0 & LA_d^{-1}A_{n1} \end{bmatrix} M_{g1}^{-1}$ is stable because AA_r^{-1} and $LA_d^{-1}A_{n1}$ are stable. Thus, $\widetilde{\Gamma}_w$ and $\widetilde{\Gamma}_z$ are stable and G_e in (4.206) is acceptable.

(b) For the system in (4.246) and (4.247), one gets $A = A_p = \begin{bmatrix} s & 0 \\ 0 & s-1 \end{bmatrix}$, $B = B_p = \begin{bmatrix} 1 & s+2 \\ s+1 & 3 \end{bmatrix}$, $P_o = \begin{bmatrix} 1 & 1 \\ 0 & 1 \end{bmatrix}$, $A_r = A_d = \begin{bmatrix} s & 0 \\ 0 & s-1 \end{bmatrix}$, $B_r = B_d = I_2$, $AP_oA_d^{-1} = \begin{bmatrix} 1 & \frac{s}{s-1} \\ 0 & 1 \end{bmatrix}$, $A_{n1} = \begin{bmatrix} 1 & 0 \\ 0 & s-1 \end{bmatrix}$, and $B_{n1} = \begin{bmatrix} 1 & s \\ 0 & s-1 \end{bmatrix}$. Now it can be readily shown that $A_pT_dA_r^{-1}$, $A_pP_dA_d^{-1}$, and $LA_d^{-1}A_{n1}$ are stable; hence, it follows from part (a) that G_e in (4.247) is acceptable.

Example 4.6 Fill in the details of the calculations for $U_{g1*}\left(\phi_{11} + \phi_{g1}\right)$ in (4.300).

Solution It is first important to recall the identity $vec\,(AVD) = (D' \otimes A)vec\,V$; the useful identities cited after Lemma 4.2; and (4.216)–(4.219). It then follows from (4.282) that

$$\Pi_{g1} = \left[(N'_{g1} \otimes W)\{(\Delta_{\varphi g} A_r^{-1})' \bullet \begin{bmatrix} -\Delta_\theta \\ V_{a1} P_a^{-1} \Delta_\theta \end{bmatrix}\} \;\middle|\; (N'_{g1} \otimes W)(I \otimes \begin{bmatrix} 0 \\ V_{a2} \end{bmatrix}) \right]$$
$$= (N'_{g1} \otimes W)\left[(\Delta_{\varphi g} A_r^{-1})' \bullet \begin{bmatrix} -\Delta_\theta \\ V_{a1} P_a^{-1} \Delta_\theta \end{bmatrix} \;\middle|\; I \otimes \begin{bmatrix} 0 \\ V_{a2} \end{bmatrix} \right].$$

Hence,

$$U_{g1*}(\phi_{11} + \phi_{g1}) = \Theta_{g1*}^{-1} \Pi_{g1*}(\phi_{11} + \phi_{g1})$$

$$= \Theta_{g1*}^{-1} \left[(\Delta_{\varphi g} A_r^{-1})' \bullet \begin{bmatrix} -\Delta_\theta \\ V_{a1} P_a^{-1} \Delta_\theta \end{bmatrix} \;\middle|\; I \otimes \begin{bmatrix} 0 \\ V_{a2} \end{bmatrix} \right]_* (N'_{g1*} \otimes W_*) vec(\Phi_{11} + \Phi_{g1})$$

$$= \Theta_{g1*}^{-1} \left[(\Delta_{\varphi g} A_r^{-1})' \bullet \begin{bmatrix} -\Delta_\theta \\ V_{a1} P_a^{-1} \Delta_\theta \end{bmatrix} \;\middle|\; I \otimes \begin{bmatrix} 0 \\ V_{a2} \end{bmatrix} \right]_* \{vec(W_* \Phi_{11} N_{g1*}) + (N'_{g1*} \otimes W_*)\phi_{g1}\}$$

$$= \Theta_{g1*}^{-1} \begin{bmatrix} \{(\Delta_{\varphi g} A_r^{-1})' \bullet \begin{bmatrix} -\Delta_\theta \\ V_{a1} P_a^{-1} \Delta_\theta \end{bmatrix}\}_* \{vec(W_* \Phi_{11} N_{g1*}) + (N'_{g1*} \otimes W_*)\phi_{g1}\} \\ \hline (I \otimes [0 \quad V_{a2*}]) vec\{W_*(\Phi_{11} + \Phi_{g1}) N_{g1*}\} \end{bmatrix} = \Theta_{g1*}^{-1} \begin{bmatrix} \widetilde{\Phi}_1 \\ \hline \widetilde{\Phi}_2 \end{bmatrix}.$$

It easily follows from (4.276) and (4.279) that

$$\begin{aligned} \widetilde{\Phi}_1 &= \left\{ (\Delta_{\varphi g} A_r^{-1})' \bullet \begin{bmatrix} -\Delta_\theta \\ V_{a1} P_a^{-1} \Delta_\theta \end{bmatrix} \right\}_* \left\{ vec(W_* \Phi_{11} N_{g1*}) + \left(N'_{g1*} \otimes W_*\right) \phi_{g1} \right\} \\ &= \left\{ (\Delta_{\varphi g} A_r^{-1})' \bullet \begin{bmatrix} -\Delta_\theta \\ V_{a1} P_a^{-1} \Delta_\theta \end{bmatrix} \right\}_* vec \begin{bmatrix} T_d G_r A_{r*} \\ 0 \end{bmatrix} + \left\{ (\Delta_{\varphi g} A_r^{-1})' \bullet \begin{bmatrix} -\Delta_\theta \\ V_{a1} P_a^{-1} \Delta_\theta \end{bmatrix} \right\}_* \left\{ \left(N'_{g1*} \otimes W_*\right) \phi_{g1} \right\}, \end{aligned} \tag{E4.1}$$

where

$$G_r = \alpha_1 A_r^{-1} B_r B_{r*} A_{r*}^{-1} + \alpha_2 \Omega_r \Omega_{r*}.$$

The first term of (E4.1) simplifies to

$$\begin{aligned} \left\{ (\Delta_{\varphi g} A_r^{-1})' \bullet \begin{bmatrix} -\Delta_\theta \\ V_{a1} P_a^{-1} \Delta_\theta \end{bmatrix} \right\}_* vec \begin{bmatrix} T_d G_r A_{r*} \\ 0 \end{bmatrix} &= \left\{ \left(A_r'^{-1} \otimes \begin{bmatrix} -I \\ V_{a1} P_a^{-1} \end{bmatrix} \right) (\Delta_{\varphi g} \bullet \Delta_\theta) \right\}_* vec \begin{bmatrix} T_d G_r A_{r*} \\ 0 \end{bmatrix} \\ &= (\Delta_{\varphi g} \bullet \Delta_\theta)_* \left(A_r'^{-1} \otimes \begin{bmatrix} -I \\ V_{a1} P_a^{-1} \end{bmatrix} \right)_* vec \begin{bmatrix} T_d G_r A_{r*} \\ 0 \end{bmatrix} \\ &= (\Delta_{\varphi g} \bullet \Delta_\theta)_* vec \left(\begin{bmatrix} -I & P_{a^*}^{-1} V_{a1*} \end{bmatrix} \begin{bmatrix} T_d G_r A_{r*} \\ 0 \end{bmatrix} A_{r*}^{-1} \right) \\ &= (\Delta_{\varphi g} \bullet \Delta_\theta)_* vec(-T_d G_r) = -(\Delta_{\varphi g} \bullet \Delta_\theta)_* vec(T_d G_r). \end{aligned} \tag{E4.2}$$

Using the expression for ϕ_{g1} in (4.295), the second term of (E4.1) reduces to

$$\begin{aligned}
&\left\{(\Delta_{\varphi g}A_r^{-1})'\bullet\begin{bmatrix}-\Delta_\theta\\ V_{a1}P_a^{-1}\Delta_\theta\end{bmatrix}\right\}_*\left\{\left(N'_{g1*}\otimes W_*\right)\phi_{g1}\right\} = \left\{(\Delta_{\varphi g}A_r^{-1}N_{g1})'\bullet\left(W\begin{bmatrix}-\Delta_\theta\\ V_{a1}P_a^{-1}\Delta_\theta\end{bmatrix}\right)\right\}_*\phi_{g1}\\
&= \left\{(\Delta_{\varphi g}A_r^{-1}N_{g1})'\bullet\left(W\begin{bmatrix}-\Delta_\theta\\ V_{a1}P_a^{-1}\Delta_\theta\end{bmatrix}\right)\right\}_*\left\{(\Delta_{\varphi g}A_r^{-1}N_{g1})'\bullet\left(W\begin{bmatrix}-\Delta_\theta\\ V_{a1}P_a^{-1}\Delta_\theta\end{bmatrix}\right)\right\}\\
&\quad vecd\left(\Delta_{\alpha g}T_d\Delta_{\varphi g}^{-1}\right)\\
&= \left\{\left[(\Delta_{\varphi g}A_r^{-1}N_{g1})'_*(\Delta_{\varphi g}A_r^{-1}N_{g1})'\right]\circ\begin{bmatrix}-\Delta_\theta\\ V_{a1}P_a^{-1}\Delta_\theta\end{bmatrix}_* W_*W\begin{bmatrix}-\Delta_\theta\\ V_{a1}P_a^{-1}\Delta_\theta\end{bmatrix}\right\}\\
&\quad vecd\left(\Delta_{\alpha g}T_d\Delta_{\varphi g}^{-1}\right)\\
&= \left\{(\Delta_{\varphi g^*}G'_{rl}\Delta_{\varphi g})\circ\left[\Delta_{\theta^*}\left(I+P_{a^*}^{-1}V_{a1*}W_{v*}W_vV_{a1}P_a^{-1}\right)\Delta_\theta\right]\right\}\\
&\quad vecd\left(\Delta_{\alpha g}T_d\Delta_{\varphi g}^{-1}\right).
\end{aligned} \tag{E4.3}$$

Next,

$$\begin{aligned}
\widetilde{\Phi}_2 &= (I\otimes[0\quad V_{a2^*}])vec\{W_*(\Phi_{11}+\Phi_{g1})N_{g1*}\} = vec\{[0\quad V_{a2^*}]W_*(\Phi_{11}+\Phi_{g1})N_{g1^*}\}\\
&= vec\{[0\quad V_{a2^*}]W_*\Phi_{g1}N_{g1*}\} = vec\left(V_{a2^*}W_{v^*}W_vV_{a1}P_a^{-1}\Delta_\theta\Delta_{\alpha g}T_dG_{rl}A_{r*}\right).
\end{aligned} \tag{E4.4}$$

Equation (4.300) now easily follows from (E4.2), (E4.3), and (E4.4).

Example 4.7 Show that the formula in (4.329) is the same as the one in (3.376).

Solution Let $AFP_dA_d^{-1} = A_m^{-1}B_m$ and $LA_d^{-1} = A_n^{-1}B_n$ be coprime polynomial matrix fraction descriptions. Define $\widetilde{A}_q = \begin{bmatrix}A_m & 0\\ 0 & A_n\end{bmatrix}$ and $\widetilde{B}_q = \begin{bmatrix}B_m\\ B_n\end{bmatrix}$. Clearly, $H_d = \begin{bmatrix}AFP_dA_d^{-1}\\ LA_d^{-1}\end{bmatrix} = \widetilde{A}_q^{-1}\widetilde{B}_q$ is a polynomial matrix fraction description, where $\widetilde{A}_q, \widetilde{B}_q$ may not be a coprime pair. Since $A_q^{-1}B_q$ is a left-coprime fraction description for H_d, there exists a polynomial matrix M_q such that $\widetilde{A}_q = M_qA_q$. Since the matrix $AFP_dA_d^{-1}$ is stable (see (3.90)), $\det A_m$ is Hurwitz and it can be concluded that $\det M_q$ is also Hurwitz for the following reasons. Suppose that p_o is a finite unstable pole of

$$H_d = \begin{bmatrix}AFP_dA_d^{-1}\\ LA_d^{-1}\end{bmatrix} = \begin{bmatrix}AFP_dA_d^{-1}\\ 0\end{bmatrix} + \begin{bmatrix}0\\ LA_d^{-1}\end{bmatrix}.$$

Then, the equality $\delta(H_d;p_o) = \delta\left(LA_d^{-1};p_o\right)$ holds by the property B.3 of the McMillan degree in Sect. B.7 of Appendix B. This property holds for every finite unstable pole of H_d; hence, $\det A_q$ and $\det A_n$ possess the same unstable zeros with the same multiplicities. Since $\det\widetilde{A}_q = \det A_m\det A_n = \det M_q\det A_q$, it is concluded that $\det M_q$ must be Hurwitz. Clearly,

$$A_q = M_q^{-1}\widetilde{A}_q = M_q^{-1}\begin{bmatrix} A_m & 0 \\ 0 & I \end{bmatrix}\begin{bmatrix} I & 0 \\ 0 & A_n \end{bmatrix} \quad \text{(E4.5)}$$

with M_q^{-1} stable and it follows from (4.305) that

$$N_{g2}N_{g2^*} = A_q\left(\begin{bmatrix} AFP_d \\ L \end{bmatrix} G_d \begin{bmatrix} AFP_d \\ L \end{bmatrix}_* + diag\{AG_{n_m}A_*, G_{n_l}\}\right)A_{q*} \quad \text{(E4.6)}$$

$$= M_q^{-1}\begin{bmatrix} A_m & 0 \\ 0 & I \end{bmatrix}\begin{bmatrix} I & 0 \\ 0 & A_n \end{bmatrix}\left(\begin{bmatrix} AFP_d \\ L \end{bmatrix} G_d \begin{bmatrix} AFP_d \\ L \end{bmatrix}_* + diag\{AG_{n_m}A_*, G_{n_l}\}\right)\begin{bmatrix} I & 0 \\ 0 & A_{n*} \end{bmatrix}\begin{bmatrix} A_{m*} & 0 \\ 0 & I \end{bmatrix}M_{q*}^{-1}, \quad \text{(E4.7)}$$

where G_d is given by (4.307). Since G_{n_m} and G_{n_l} are good and since $AFP_dA_d^{-1} = A_m^{-1}B_m$ and $LA_d^{-1} = A_n^{-1}B_n$ are stable, it follows that there exists a stable Wiener–Hopf factor Ω_t satisfying

$$\begin{aligned} G_t &= \begin{bmatrix} I & 0 \\ 0 & A_n \end{bmatrix}\left(\begin{bmatrix} AFP_d \\ L \end{bmatrix} G_d \begin{bmatrix} AFP_d \\ L \end{bmatrix}_* + diag\{AG_{n_m}A_*, G_{n_l}\}\right)\begin{bmatrix} I & 0 \\ 0 & A_{n*} \end{bmatrix} \\ &= \Omega_t\Omega_{t*}. \end{aligned} \quad \text{(E4.8)}$$

One then gets from (E4.6), (E4.7), and (E4.8) that

$$N_{g2}N_{g2*} = M_q^{-1}\begin{bmatrix} A_m & 0 \\ 0 & I \end{bmatrix}\Omega_t\Omega_{t*}\begin{bmatrix} A_{m*} & 0 \\ 0 & I \end{bmatrix}M_{q*}^{-1} = \Omega_{qd}\Omega_{qd*},$$

where

$$\Omega_{qd} = M_q^{-1}\begin{bmatrix} A_m & 0 \\ 0 & I \end{bmatrix}\Omega_t \quad \text{(E4.9)}$$

is a stable matrix. Moreover, on account of Assumption 4.35, N_{g2} has full row rank on the stability boundary so that

$$\det(N_{g2}N_{g2^*}) = \left[\frac{(\det A_m)(\det \Omega_t)}{(\det M_q)}\right]\left[\frac{(\det A_m)(\det \Omega_t)}{(\det M_q)}\right]_* \neq 0$$

on the stability boundary. Hence, Ω_t^{-1} is analytic on the stability boundary and, therefore, stable. It then follows that the Ω_{qd} given by (E4.9) is a stable Wiener–Hopf factor for $N_{g2}N_{g2^*}$ whose inverse is also stable. In addition, it follows from (E4.5) and (E4.9) that

$$\Omega_{qd} = A_q \begin{bmatrix} I & 0 \\ 0 & A_n^{-1} \end{bmatrix} \Omega_t \tag{E4.10}$$

and

$$\Omega_{qd}^{-1} A_q \begin{bmatrix} A & 0 \\ 0 & I \end{bmatrix} = \Omega_t^{-1} \begin{bmatrix} A & 0 \\ 0 & A_n \end{bmatrix}. \tag{E4.11}$$

Equations (3.369), (4.315), and (E4.10) then yield

$$\begin{aligned} \widetilde{\Phi}_{11} &= \Lambda_*^{-1} A_{1*} P_* P_d G_d A_{d^*} B_{q^*} A_{q^*}^{-1} \begin{bmatrix} I & 0 \\ 0 & A_{n*} \end{bmatrix} \Omega_{t^*}^{-1} \\ &= \Lambda_*^{-1} A_{1*} P_* P_d G_d A_{d^*} \begin{bmatrix} A_{d^*}^{-1} P_{d^*} F_* A_* & A_{d^*}^{-1} L_* A_{n^*} \end{bmatrix} \Omega_{t^*}^{-1} = \begin{bmatrix} \widetilde{\Gamma}_w & \widetilde{\Gamma}_z \end{bmatrix}. \end{aligned} \tag{E4.12}$$

The remaining task is to show that $-\left\{\widetilde{\Phi}_{g2}\right\}_- = \{[\widetilde{\Gamma}_a \;\; \widetilde{\Gamma}_b]\}_-$. First notice that if $W_{v*} W_v$ in (4.306) is taken as kQ in (3.344), then Λ in (4.306) is equal to Λ in (3.344). It follows from (4.273), (4.317), and (E4.10) that

$$\begin{aligned} \widetilde{\Phi}_{g2} &= \Lambda A_1^{-1} \Gamma_y V_g^{-1} \begin{bmatrix} 0 \\ I \end{bmatrix} \Omega_{qd} = \Lambda A_1^{-1} \left(-Y_p A_p P_d A_d^{-1} Y_q + [-Y \;\; 0] X_q \right) \Omega_{qd} \\ &= \Lambda A_1^{-1} \left(-Y_p A_p P_d A_d^{-1} Y_q + [-Y \;\; 0] X_q \right) A_q \begin{bmatrix} I & 0 \\ 0 & A_n^{-1} \end{bmatrix} \Omega_t. \end{aligned}$$

Since $A_q X_q + B_q Y_q = I$, it follows that $X_q A_q = I - A_q^{-1} B_q Y_q A_q = I - \begin{bmatrix} AFP_d A_d^{-1} \\ LA_d^{-1} \end{bmatrix} Y_q A_q$; hence, $[-Y \;\; 0] X_q A_q = [-Y \;\; 0] + YAFP_d A_d^{-1} Y_q A_q$. That is,

$$\widetilde{\Phi}_{g2} = \Lambda A_1^{-1} [-Y \;\; 0] \Omega_t + \Lambda A_1^{-1} (YAF - Y_p A_p) P_d A_d^{-1} Y_q A_q \begin{bmatrix} I & 0 \\ 0 & A_n^{-1} \end{bmatrix} \Omega_t. \tag{E4.13}$$

It can be shown that the second term on the right-hand side of (E4.13) is stable. In fact, it follows from the equalities $A_p X_p + B_p Y_p = I$ and $AX + BY = I$ that $FPY_p A_p = F - FX_p A_p$ and $FPYAF = F - XAF$, respectively. Thus, $FP(YAF - Y_p A_p) = FX_p A_p - XAF$ and $B_1 A_1^{-1} (YAF - Y_p A_p) A_p^{-1} = FX_p - XAFA_p^{-1}$. Since AFA_p^{-1} and FX_p are stable, $B_1 A_1^{-1} (YAF - Y_p A_p) A_p^{-1}$ is stable; hence, $A_1^{-1} (YAF - Y_p A_p) A_p^{-1}$ is stable. Therefore, the second term on the right-hand side of (E4.13) is stable and one gets from (3.370) that

$$-\left\{\widetilde{\Phi}_{g2}\right\}_{-} = \{\Lambda A_1^{-1}[Y \quad 0]\Omega_t\}_{-} = \{[\widetilde{\Gamma}_a \quad \widetilde{\Gamma}_b]\}_{-}. \tag{E4.14}$$

Setting $Z_{f2} = -Z_{mn}$ and using (E4.11), (E4.12), and (E4.14), one can show that the expression for $[R_w \quad R_z]$ in (4.329) is equal to the one in (3.376).

Example 4.8 Show when $A_q^{-1}B_q$ is given by (4.263) that the full rank condition on the stability boundary of the matrix

$$\left[B_q B_d \ \ \sqrt{\alpha_2}\ B_q A_d \Omega_d \ \ \sqrt{\alpha_2} A_q diag\{A\Omega_{n_m}, \Omega_{n_l}\}\right]$$

is equivalent to that of the matrix

$$\begin{bmatrix} AFP_d A_d^{-1} B_d & \sqrt{\alpha_2} AFP_d \Omega_d & \sqrt{\alpha_2} A\Omega_{n_m} & 0 \\ B_n B_d & \sqrt{\alpha_2} B_n A_d \Omega_d & 0 & \sqrt{\alpha_2} A_n \Omega_{n_l} \end{bmatrix},$$

where $LA_d^{-1} = A_n^{-1}B_n$ is a coprime polynomial matrix fraction description.

Solution Using the relationship in (E4.5) of the previous example, one obtains

$$\begin{aligned}
&\left[B_q B_d \ \ \sqrt{\alpha_2} B_q A_d \Omega_d \ \ \sqrt{\alpha_2} A_q diag\{A\Omega_{n_m}, \Omega_{n_l}\}\right] \\
&= A_q\left[A_q^{-1} B_q B_d \ \ \sqrt{\alpha_2} A_q^{-1} B_q A_d \Omega_d \ \ \sqrt{\alpha_2} diag\{A\Omega_{n_m}, \Omega_{n_l}\}\right] \\
&= M_q^{-1}\begin{bmatrix} A_m & 0 \\ 0 & I \end{bmatrix}\begin{bmatrix} I & 0 \\ 0 & A_n \end{bmatrix}\begin{bmatrix} AFP_d A_d^{-1} B_d & \sqrt{\alpha_2} AFP_d \Omega_d & \sqrt{\alpha_2} A\Omega_{n_m} & 0 \\ LA_d^{-1} B_d & \sqrt{\alpha_2} L\Omega_d & 0 & \sqrt{\alpha_2}\Omega_{n_l} \end{bmatrix} \\
&= M_q^{-1}\begin{bmatrix} A_m & 0 \\ 0 & I \end{bmatrix}\begin{bmatrix} AFP_d A_d^{-1} B_d & \sqrt{\alpha_2} AFP_d \Omega_d & \sqrt{\alpha_2} A\Omega_{n_m} & 0 \\ B_n B_d & \sqrt{\alpha_2} B_n A_d \Omega_d & 0 & \sqrt{\alpha_2} A_n \Omega_{n_l} \end{bmatrix}.
\end{aligned}$$

Since M_q^{-1}, M_q, A_m^{-1}, and A_m have full rank on the stability boundary, the third matrix in the last expression above has the same rank on the stability boundary as the original one.

Example 4.9 Establish that any rational diagonal T which is realizable is also acceptable for any strictly acceptable G_w when G is compatible. **Hint**: Make use of the results obtained in Example 3.21.

Solution When a rational diagonal T is realizable, then there exists a stable K and associated stabilizing controller C given by (4.8) for which $T_{yr} = T$. With this same controller it follows from Example 3.21 when G is compatible and G_w is strictly acceptable that $G_v G_w = (Y + A_1 K) A G_{21} A_w^{-1} B_w$ and $G_\delta G_w = [(G_{11} + G_{12} A_1 Y_1 G_{21}) + G_{12} A_1 K A G_{21}] A_w^{-1} B_w$ are stable because $G_{12} A_1, A G_{21} A_w^{-1}$, and $(G_{11} + G_{12} A_1 Y_1 G_{21}) A_w^{-1}$ are stable. Hence, the stabilizing controller C and the diagonal T it realizes are acceptable.

Example 4.10 (An example of the optimal decoupling 2DOF $\mathcal{H}_2$ design for an acceptable input) For the Rosenbrock's process in Example 3.24, design the optimal decoupling 2DOF $\mathcal{H}_2$ controller of the configuration in Fig. 1.1 in Sect. 1.2 with appropriate choices of the weighting matrices $W(s)$ and $G_w(s)$ (Remove the feedforward part and assume $F = I$ in the figure).

Solution The 3DOF configuration in Fig. 1.1 is considered with the feedforward part removed and $P_d(s) = F(s) = T_d(s) = I$. As done in Solution 3.24, the integrators $\frac{1}{s}I_2$ are inserted in front of the plant $P_0(s)$ and the augmented plant $P_{aug}(s) = \frac{1}{s}P_0(s)$ is regarded as the plant $P(s)$ in Fig. 1.1. This 2DOF system with the augmented plant is admissible and compatible since $F = I$ and $T_d = I$. Coprime polynomial matrix fraction descriptions for the augmented plant are given in Solution 3.24. With the choice of $\alpha_1 = \alpha_2 = 1$, the input weighting matrix G_w becomes

$$G_w = \begin{bmatrix} A_d^{-1}B_d & 0 & \Omega_d & 0 & 0 & 0 \\ 0 & A_r^{-1}B_r & 0 & \Omega_r & 0 & 0 \\ 0 & 0 & 0 & 0 & \Omega_{n_r} & 0 \\ 0 & 0 & 0 & 0 & 0 & \Omega_{n_m} \end{bmatrix},$$

and its elements values are selected as

$$A_r = A_d = sI_2, B_r = B_d = I_2, \Omega_r = \Omega_d = 0, \Omega_{n_r} = \Omega_{n_m} = 0.1I_2.$$

It is explained that G_w is acceptable in Solution 3.24. The weighting matrix to the control variable are chosen as $W_v = I_2$. The inverse of the plant is given by

$$P^{-1} = \frac{s(s+1)}{s-1}\begin{bmatrix} -(s+3) & 2(s+1) \\ s+3 & -(s+3) \end{bmatrix},$$

and it follows that

$$\Delta_\theta = (s-1)I_2, \Delta_{\varphi g} = sI_2, \Delta_{\alpha g} = -I_2, \Delta_{\beta g} = I_2.$$

Then it can be shown that Assumptions 4.28 through 4.40 in Sect. 4.6 hold and both $A_pA_d^{-1}$ and $A_pA_r^{-1}$ are stable (note that $A_p = A$ since $F = I$). Thus the optimal decoupling solution for the 2DOF system is given by the formulas in Theorem 4.12. Reminding that the matrix L is empty, one obtains

$$R_u = P^{-1}\Delta_\theta(\Delta_{\alpha g} + \Delta\Delta_{\varphi g}),$$
$$\tilde{d} = vecd\Delta = -\Theta_{g1}^{-1}\left(\{U_{g1*}\phi_{g1}\}_\infty + \{U_{g1*}(\phi_{11} + \phi_{g1})\}_+\right).$$

Here, Θ_{g1} is a Wiener–Hopf spectral factor of the equation $\Psi_g = \Theta_{g1*}\Theta_{g1}$ with

$$\Psi_g = \left(\Delta_{\varphi g} G_{rl} \Delta_{\varphi g*}\right)' \circ \left\{\Delta_{\theta *}\left(I + P_*^{-1}P^{-1}\right)\Delta_\theta\right\},$$

where

$$G_{rl} = A_r^{-1}B_rB_{r*}A_{r*}^{-1} + \Omega_{n_r}\Omega_{n_r*},\ U_{g1*}\phi_{g1} = \Theta_{g1*}^{-1}\Psi_g vecd\left(\Delta_{\alpha g}\Delta_{\varphi g}^{-1}\right),$$

$$U_{g1*}(\phi_{11} + \phi_{g1}) = \Theta_{g1*}^{-1}\left(\Psi_g vecd\left(\Delta_{\alpha g}\Delta_{\varphi g}^{-1}\right) - \left(\Delta_{\varphi g} \bullet \Delta_\theta\right)_* vec(G_r)\right)$$

$$G_r = A_r^{-1}B_rB_{r^*}A_{r^*}^{-1}.$$

The computational results obtained by MATLAB Polynomial Toolbox are given by

$$\Theta_{g1} = \begin{bmatrix} \Theta_{g11} & 0 \\ 0 & \Theta_{g12} \end{bmatrix},$$

$$\Theta_{g11} = 0.14142s^4 + 2.012s^3 + 6.5344s^2 + 5.6638s + 1,$$

$$\Theta_{g12} = 0.22361s^4 + 2.8777s^3 + 6.934s^2 + 5.28s + 1,$$

$$\left\{U_{g1^*}\phi_{g1}\right\}_\infty = \begin{bmatrix} -(0.14142s^3 + 2.012s^2 + 6.5344s + 5.6638) \\ -(0.22361s^3 + 2.8777s^2 + 6.934s + 5.28) \end{bmatrix},$$

$$\left\{U_{g1^*}(\phi_{11} + \phi_{g1})\right\}_+ = 0,$$

$$\widetilde{d} = vecd\Delta = \begin{bmatrix} \dfrac{s^3 + 14.277s^2 + 46.205s + 40.049}{s^4 + 14.277s^3 + 46.205s^2 + 40.049s + 7.0711} \\ \dfrac{s^3 + 12.869s^2 + 31.01s + 23.613}{s^4 + 12.869s^3 + 31.01s^2 + 23.613s + 4.4721} \end{bmatrix},$$

$$R_u = \frac{1}{r_{ud}(s)}\begin{bmatrix} r_{u11}(s) & r_{u12}(s) \\ r_{u21}(s) & r_{u22}(s) \end{bmatrix},$$

$$r_{u11} = 7.0711s(s+3)(s+1.5877)(s+0.28168),$$

$$r_{u12} = -8.9443s(s+2.9907)(s+1)(s+0.23644),$$

$$r_{u21} = -7.0711s(s+3)(s+1.5877)(s+0.28168),$$

$$r_{u22} = 4.4721s(s+3)(s+2.9907)(s+0.23644),$$

$$r_{ud} = (s+10)(s+2.9907)(s+1.5877)(s+0.28168)(s+0.23644).$$

The optimal solutions R_w and C_w are the same as the ones in Solution 3.24 by Theorem 4.12. The optimal controller C_u is obtained by the formula $C_u = (I - R_w P)^{-1} R_u$ and it is given by

$$C_u(s) = \frac{1}{c_{ud}} \begin{bmatrix} c_{u11} & c_{u12} \\ c_{u21} & c_{u22} \end{bmatrix},$$

$$\begin{aligned} c_{u11} = {} & 7.0711(s+8.6187)(s+5.808)(s+2.6158)(s+2.2632) \\ & \times (s+1.5877)(s+0.28168)\left(s^2+1.0303s+0.52012\right), \end{aligned}$$

$$\begin{aligned} c_{u12} = {} & -8.9443(s+9.376)(s+3.7468)(s+2.9907)(s+2.8387) \\ & \times (s+0.23644)(s+0.008201)\left(s^2+2.3663s+1.8027\right), \end{aligned}$$

$$\begin{aligned} c_{u21} = {} & -7.0711(s+9.3755)(s+3)(s+1.5877)(s+1.1184) \\ & \times (s+0.28168)(s+0.18583)\left(s^2+6.6563s+11.246\right), \end{aligned}$$

$$\begin{aligned} c_{u22} = {} & 4.4721(s+8.603)(s+5.8243)(s+3)(s+2.9907) \\ & \times (s+1.5444)(s+0.23644)\left(s^2+1.3643s+0.59025\right), \end{aligned}$$

$$\begin{aligned} c_{ud} = {} & (s+10.135)(s+9.1359)(s+4.5579)(s+2.9907)(s+1.5877) \\ & \times (s+0.28168)(s+0.23644)\left(s^2+3.5067s+4.132\right). \end{aligned}$$

The McMillan degree of the controller $C(s) = [-C_w \quad C_u]$ is 16 and hence the final controller $C(s)\frac{1}{s}$ can be implemented as a state-space model with 18 states.

The unit step responses of y_1 and y_2 are shown in Fig. 4.2. The step reference inputs for y_1 and y_2 are exerted at $t = 0$ and $t = 40$, respectively. The actuating controls for these reference inputs are shown in Fig. 4.3.

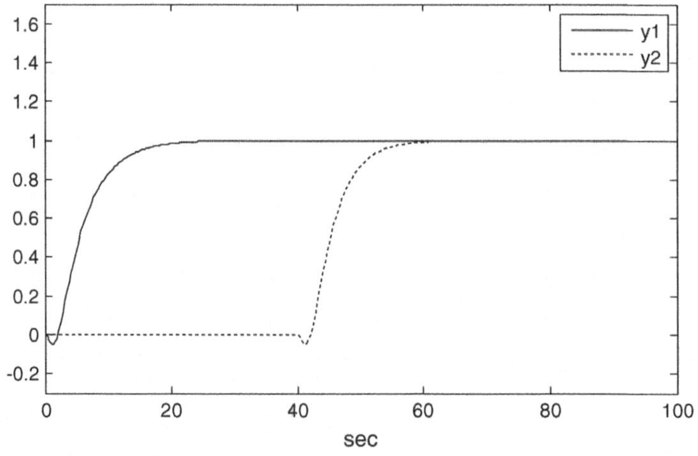

Fig. 4.2 Unit step responses of the outputs to reference changes

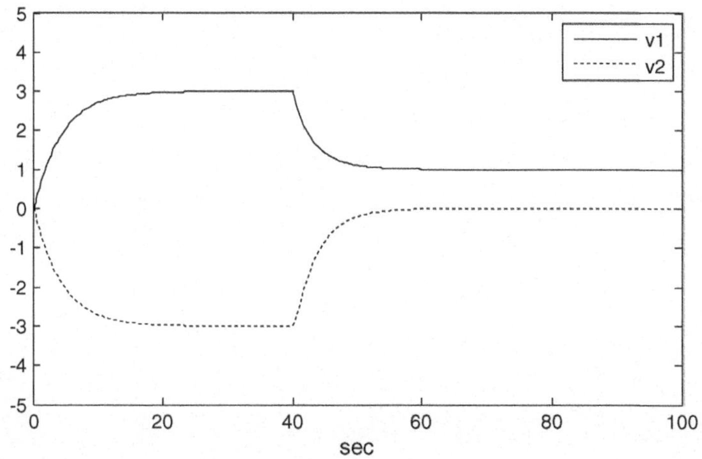

Fig. 4.3 Control activities due to reference changes

I If the augmented plant $P_{aug}(s)$ is regarded as the plant $P(s)$ in Fig. 1.1 and the control variable $v(s)$ is regarded as the input to the inserted integrators, the cost $J = \|WG_\delta G_w\|_2^2$ for the $\mathcal{H}_2$ design in this example is 23.878. Compared to the non-decoupling $\mathcal{H}_2$ design in Example 3.24 where the cost J is 21.869, the decoupling design yields 9.19% increase in its cost.

Chapter 5
Numerical Calculation of Wiener–Hopf Controllers

5.1 Overview

A state-variable representation of the Wiener–Hopf controllers determined in Chap. 3 for strictly acceptable persistent inputs is developed in this chapter for both the analog and digital cases. First, a generalized Bezout identity (GBI) in stable rational matrices is derived from a polynomial GBI. Then the Wiener–Hopf controllers in Chap. 3, which are described in terms of the parameters in a polynomial GBI, are described instead by the parameters in a stable rational GBI. A state-space representation for the stable rational GBI description is then obtained from two algebraic Riccati equations making it possible to find a state-space representation for the Wiener–Hopf controllers. It is shown that the frequency-domain assumptions which guarantee the existence of the Wiener–Hopf controllers correspond to equivalent ones in the state-space domain.

5.2 Basic Machinery for Analog Systems

A collection of useful relationships between state-space and frequency-domain representations needed in the sequel is provided in this section for easy reference and review. These basic properties can be found in Kailath (1980), Chen (1984), and Zhou et al. (1996). A linear time-invariant system described by

$$\dot{\hat{x}}(t) = \Sigma\hat{x}(t) + N\hat{u}(t), \hat{y}(t) = H\hat{x}(t) + D\hat{u}(t) \tag{5.1}$$

is considered where the elements of the vectors $\hat{x}(t), \hat{u}(t)$, and $\hat{y}(t)$ are, respectively, the state, the input, and the output variables. The transfer matrix of the system is given by $G(s) = H(sI - \Sigma)^{-1}N + D$. A constant matrix Σ is called stable if all eigenvalues of Σ lie in Re $s < 0$ and a rational matrix $G(s)$ is called stable if it is

J. J. Bongiorno Jr. and K. Park, *Design of Linear Multivariable Feedback Control Systems*, https://doi.org/10.1007/978-3-030-44356-6_5

analytic in $\mathrm{Re}\, s \geq 0$. A triple (Σ, N, H) is minimal iff it is controllable and observable (Kailath 1980). In this case, the characteristic polynomial $\Delta(s) = \det(sI - \Sigma)$ of the system and the characteristic denominator $\psi_G(s)$ of the transfer matrix $G(s) = H(sI - \Sigma)^{-1}N + D$ are identical. When the triple (Σ, N, H) is not minimal, some poles of $(sI - \Sigma)^{-1}$ are canceled out in the product of $H(sI - \Sigma)^{-1}N$. In this case, the polynomials $\Delta(s)$ and $\psi_G(s)$ are related by the equation

$$\det(sI - \Sigma) = h(s)\psi_G(s), \tag{5.2}$$

where the polynomial $h(s)$ accounts for the hidden poles of the system. This is easily established from a well-known result on system equivalence. When (Σ, N, H) is uncontrollable and unobservable, there exists a nonsingular matrix L such that (see Theorem 3.10 in Zhou et al. 1996)

$$HL = [\widetilde{H}_1 \quad 0 \mid \widetilde{H}_3 \quad 0],\ L^{-1}\Sigma L = \left[\begin{array}{cc|cc} \widetilde{\Sigma}_{11} & 0 & \widetilde{\Sigma}_{13} & 0 \\ \widetilde{\Sigma}_{21} & \widetilde{\Sigma}_{22} & \widetilde{\Sigma}_{23} & \widetilde{\Sigma}_{24} \\ \hline 0 & 0 & \widetilde{\Sigma}_{33} & 0 \\ 0 & 0 & \widetilde{\Sigma}_{43} & \widetilde{\Sigma}_{44} \end{array}\right],\ L^{-1}N = \left[\begin{array}{c} \widetilde{N}_1 \\ \widetilde{N}_2 \\ \hline 0 \\ 0 \end{array}\right] \tag{5.3}$$

where the triple $(\widetilde{\Sigma}_{11}, \widetilde{N}_1, \widetilde{H}_1)$ is controllable and observable. Hence, the transfer matrix of the system is also equivalent to

$$G(s) = \widetilde{H}_1(sI - \widetilde{\Sigma}_{11})^{-1}\widetilde{N}_1 + D = HL(sI - L^{-1}\Sigma L)^{-1}L^{-1}N + D. \tag{5.4}$$

Clearly, the characteristic polynomial of the system is given by

$$\begin{aligned} \Delta &= \det[L^{-1}(sI - \Sigma)L] = \det(sI - \widetilde{\Sigma}) \\ &= \det(sI - \widetilde{\Sigma}_{11})\det(sI - \widetilde{\Sigma}_{22})\det(sI - \widetilde{\Sigma}_{33})\det(sI - \widetilde{\Sigma}_{44}). \end{aligned} \tag{5.5}$$

Since the triple $(\widetilde{\Sigma}_{11}, \widetilde{N}_1, \widetilde{H}_1)$ is minimal, it can be concluded from (5.4) that $\psi_G = \det(sI - \widetilde{\Sigma}_{11})$ and it follows from (5.2) and (5.5) that

$$h(s) = \det(sI - \widetilde{\Sigma}_{22})\det(sI - \widetilde{\Sigma}_{33})\det(sI - \widetilde{\Sigma}_{44}). \tag{5.6}$$

When (Σ, N, H) is controllable and unobservable, or when (Σ, N, H) is uncontrollable and observable, a similar approach can show that the equality in (5.2) still holds.

When the triple (Σ, N, H) is stabilizable and detectable, the matrices $\widetilde{\Sigma}_{22}$, $\widetilde{\Sigma}_{33}$, and $\widetilde{\Sigma}_{44}$ in (5.3) are stable and the polynomial $h(s)$ is Hurwitz. In this case, the system in (5.1) is free of unstable hidden poles and its transfer matrix $G(s)$ is stable iff $\widetilde{\Sigma}_{11}$ is stable. In general, one can always write (Lemma B.5)

$$\begin{aligned} G(s) &= \widetilde{H}_1(sI - \widetilde{\Sigma}_{11})^{-1}\widetilde{N}_1 + D = H(sI - \Sigma)^{-1}N + D = A^{-1}(s)B(s) \\ &= B_1(s)A_1^{-1}(s), \end{aligned} \tag{5.7}$$

where the last two expressions are, respectively, left and right coprime polynomial matrix fraction descriptions for $G(s)$. It then follows that $\psi_G(s) = \eta \det A(s) = \eta_1 \det A_1(s)$, where η and η_1 are constants chosen so that $\psi_G(s)$ is monic (Lemma B.7). Obviously,

$$\det(sI - \Sigma) = h(s)\psi_G(s) = \eta h(s) \det A(s) = \eta_1 h(s) \det A_1(s). \tag{5.8}$$

When $G(s) = B_1(s)A_1^{-1}(s)$ denotes a coprime polynomial matrix fraction description, a complex number s_o is a transmission zero of $G(s)$ if the rank of the matrix $B_1(s)$ is less than its normal rank at s_o. If s_o is a transmission zero of $G(s)$ which is not a pole, then the rank of $G(s)$ is less than its normal rank at s_o.

The pair (Σ, N) is controllable iff the matrix $[\, sI - \Sigma \quad N \,]$ has full row rank for all s. The pair (Σ, H) is observable iff the matrix $[\, sI - \Sigma' \quad H' \,]'$ has full column rank for all s. The pair (Σ, N) is stabilizable iff the matrix $[\, sI - \Sigma \quad N \,]$ has full row rank in $\mathrm{Re}\, s \geq 0$. The pair (Σ, H) is detectable iff the matrix $[\, sI - \Sigma' \quad H' \,]'$ has full column rank in $\mathrm{Re}\, s \geq 0$. Using these facts it is easy to show that if (Σ, N_1) or (Σ, N_2) is controllable (stabilizable) then $(\Sigma, [N_1 \; N_2])$ is controllable (stabilizable), and if (Σ, H_1) or (Σ, H_2) is observable (detectable) then $(\Sigma, [\, H_1' \quad H_2' \,]')$ is observable (detectable).

For the linear time-invariant system in (5.1), it is convenient to define the system matrix

$$S(s) = \begin{bmatrix} \Sigma - sI & N \\ H & D \end{bmatrix}. \tag{5.9}$$

The rank of the system matrix $S(s)$ can only be less than its normal rank at those points in the complex plane which are eigenvalues of Σ corresponding to uncontrollable or unobservable modes of the triple (Σ, N, H) or transmission zeros of $G(s) = H(sI - \Sigma)^{-1}N + D$. When (Σ, N, H) is minimal, the system matrix $S(s)$ is Smith-equivalent to $\mathrm{diag}\{I, B_1(s)\}$. That is, the two matrices $S(s)$ and $\mathrm{diag}\{I, B_1(s)\}$ have the same Smith form (see Sects. 6.5.3 and 8.3.2 in Kailath 1980 and Example 5.9). When (Σ, N, H) is stabilizable and detectable (i.e., the region $\Re$ in Example 5.9 is the half-plane $\mathrm{Re}\, s \geq 0$), then two matrices $S(s)$ and $\mathrm{diag}\{I, B_1(s)\}$ have the same rank in $\mathrm{Re}\, s \geq 0$.

In Sects. 5.2 and 5.3, it is convenient to represent the transfer matrix $G(s)$ associated with the state-space parameters (Σ, N, H, D) as a block matrix distinguished by dashed partition lines. Specifically,

$$G(s) = H(sI - \Sigma)^{-1}N + D = \left[\begin{array}{c|c} \Sigma & N \\ \hline H & D \end{array}\right]. \tag{5.10}$$

When

$$G(s) = \left[\begin{array}{c|c} \Sigma & N \\ \hline H & D \end{array}\right],\ G_1(s) = \left[\begin{array}{c|c} \Sigma_1 & N_1 \\ \hline H_1 & D_1 \end{array}\right],\ G_2(s) = \left[\begin{array}{c|c} \Sigma_2 & N_2 \\ \hline H_2 & D_2 \end{array}\right], \tag{5.11}$$

and D^{-1} denotes both the inverse of D when D is square and nonsingular and the right inverse of D when D has full row rank, it is not difficult to verify that

$$G(s) = \left[\begin{array}{c|c} \Sigma & N \\ \hline H & D \end{array}\right] = \left[\begin{array}{c|c} L^{-1}\Sigma L & L^{-1}N \\ \hline HL & D \end{array}\right],\ G_*(s) = \left[\begin{array}{c|c} -\Sigma' & H' \\ \hline -N' & D' \end{array}\right], \tag{5.12}$$

$$G(s)G^{-1}(s) = \left[\begin{array}{c|c} \Sigma & N \\ \hline H & D \end{array}\right]\left[\begin{array}{c|c} \Sigma - ND^{-1}H & ND^{-1} \\ \hline -D^{-1}H & D^{-1} \end{array}\right] = I, \tag{5.13}$$

$$G_1(s)G_2(s) = \left[\begin{array}{cc|c} \Sigma_1 & N_1H_2 & N_1D_2 \\ 0 & \Sigma_2 & N_2 \\ \hline H_1 & D_1H_2 & D_1D_2 \end{array}\right] = \left[\begin{array}{cc|c} \Sigma_2 & 0 & N_2 \\ N_1H_2 & \Sigma_1 & N_1D_2 \\ \hline D_1H_2 & H_1 & D_1D_2 \end{array}\right], \tag{5.14}$$

and

$$G_1(s) + G_2(s) = \left[\begin{array}{cc|c} \Sigma_1 & 0 & N_1 \\ 0 & \Sigma_2 & N_2 \\ \hline H_1 & H_2 & D_1 + D_2 \end{array}\right]. \tag{5.15}$$

In particular, (5.13) can be confirmed using (5.14) with $G_1(s) = G(s)$ and $G_2(s) = G^{-1}(s)$.

For a compact description of transfer matrices it is convenient to write

$$U(s) = \begin{bmatrix} U_{11}(s) & U_{12}(s) \\ U_{21}(s) & U_{22}(s) \end{bmatrix},\ V(s) = \begin{bmatrix} V_{11}(s) & V_{12}(s) \\ V_{21}(s) & V_{22}(s) \end{bmatrix}, \tag{5.16}$$

and to introduce the following transformations. The linear fractional transformation (LFT) of $U(s)$ with respect to $\phi(s)$, $U(s)\Box\phi(s)$, is defined as

$$U(s)\Box\phi(s) = \begin{bmatrix} U_{11} & U_{12} \\ U_{21} & U_{22} \end{bmatrix}\Box\phi(s) = U_{11} + U_{12}\phi(I - U_{22}\phi)^{-1}U_{21}, \tag{5.17}$$

and the homographic transformation of $V(s)$ with respect to $\phi(s)$, $V(s)\Diamond\phi(s)$, is defined as

$$V(s)\Diamond\phi(s) = \begin{bmatrix} V_{11} & V_{12} \\ V_{21} & V_{22} \end{bmatrix} \Diamond\phi(s) = (V_{11}\phi + V_{12})(V_{21}\phi + V_{22})^{-1}. \tag{5.18}$$

A homographic transformation with respect to $\phi(s)$ can be obtained from a LFT with respect to $\phi(s)$ through the relationship $V(s)\Diamond\phi(s) = U(s)\Box\phi(s)$ when

$$U(s) = \begin{bmatrix} U_{11}(s) & U_{12}(s) \\ U_{21}(s) & U_{22}(s) \end{bmatrix} = \begin{bmatrix} V_{12}V_{22}^{-1} & V_{11} - V_{12}V_{22}^{-1}V_{21} \\ V_{22}^{-1} & -V_{22}^{-1}V_{21} \end{bmatrix}, \tag{5.19}$$

where the existence of V_{22}^{-1} is assumed. When the state-space parameters of $V(s)$ are given by

$$V = \begin{bmatrix} V_{11} & V_{12} \\ V_{21} & V_{22} \end{bmatrix} = \left[\begin{array}{c|cc} \Sigma & N_1 & N_2 \\ \hline H_1 & D_{11} & D_{12} \\ H_2 & D_{21} & D_{22} \end{array}\right], \tag{5.20}$$

it follows that (see Example 5.10(a))

$$U(s) = \begin{bmatrix} U_{11}(s) & U_{12}(s) \\ U_{21}(s) & U_{22}(s) \end{bmatrix} = \left[\begin{array}{c|cc} \Sigma - N_2D_{22}^{-1}H_2 & N_2D_{22}^{-1} & N_1 - N_2D_{22}^{-1}D_{21} \\ \hline H_1 - D_{12}D_{22}^{-1}H_2 & D_{12}D_{22}^{-1} & D_{11} - D_{12}D_{22}^{-1}D_{21} \\ -D_{22}^{-1}H_2 & D_{22}^{-1} & -D_{22}^{-1}D_{21} \end{array}\right] \tag{5.21}$$

describes a set of state-space parameters for $U(s)$ in (5.19). When the state-space parameters of $U(s)$ and $\phi(s)$ are given by

$$U(s) = \begin{bmatrix} U_{11}(s) & U_{12}(s) \\ U_{21}(s) & U_{22}(s) \end{bmatrix} = \left[\begin{array}{c|cc} \Sigma & N_1 & N_2 \\ \hline H_1 & D_{11} & D_{12} \\ H_2 & D_{21} & D_{22} \end{array}\right], \quad \phi(s) = \left[\begin{array}{c|c} \Sigma_\phi & N_\phi \\ \hline H_\phi & D_\phi \end{array}\right], \tag{5.22}$$

a set of state-space parameters for $U(s)\Box\phi(s)$ is given by (see Example 5.10(b))

$$U\Box\phi = \left[\begin{array}{cc|c} \Sigma + N_2E_1^{-1}D_\phi H_2 & N_2E_1^{-1}H_\phi & N_1 + N_2E_1^{-1}D_\phi D_{21} \\ N_\phi E_2^{-1}H_2 & \Sigma_\phi + N_\phi E_2^{-1}D_{22}H_\phi & N_\phi E_2^{-1}D_{21} \\ \hline H_1 + D_{12}D_\phi E_2^{-1}H_2 & D_{12}E_1^{-1}H_\phi & D_{11} + D_{12}D_\phi E_2^{-1}D_{21} \end{array}\right] \tag{5.23}$$

where

$$E_1 = I - D_\phi D_{22}, \ E_2 = I - D_{22}D_\phi. \tag{5.24}$$

For a positive definite matrix E, the notation $E^{\frac{1}{2}}$ denotes a symmetric positive definite matrix satisfying $E^{\frac{1}{2}}E^{\frac{1}{2}} = E$.

5.3 State-Space Representation of Wiener–Hopf Controllers: The Analog Case

In this section state-space algorithms for one of the Wiener–Hopf controllers in Chap. 3 are developed for the analog case. In particular, the generalized persistent input standard configuration of Fig. 3.2 with G compatible and proper is considered for the case in which G_w, a weighted $\widetilde{G}_w$, is strictly acceptable and proper.

As explained in Sect. 3.3, when the coprime polynomial matrix fraction descriptions

$$G_{22} = -A^{-1}B = -B_1A_1^{-1} \tag{5.25}$$

associated with the generalized Bezout identity

$$\begin{bmatrix} A & B \\ Y_1 & -X_1 \end{bmatrix}\begin{bmatrix} X & B_1 \\ Y & -A_1 \end{bmatrix} = \begin{bmatrix} I & 0 \\ 0 & I \end{bmatrix} = \begin{bmatrix} X & B_1 \\ Y & -A_1 \end{bmatrix}\begin{bmatrix} A & B \\ Y_1 & -X_1 \end{bmatrix} \tag{5.26}$$

are given and G is compatible, the class of all stabilizing controller transfer matrices is characterized by

$$C = (X_1 - KB)^{-1}(Y_1 + KA) = (Y + A_1K)(X - B_1K)^{-1}, \tag{5.27}$$

where K is a stable rational matrix such that $\det(X - B_1K) \cdot \det(X_1 - KB) \not\equiv 0$. (The coprime polynomial matrix fraction description $A^{-1}B = B_1A_1^{-1}$ is used to represent G, $-G_{22}$, or $-P_{22}$ in different sections of this book. It is always made clear which is intended so that no confusion arises.) When the system G is compatible, the matrices $G_{12}A_1, AG_{21}$, and $G_{11} + G_{12}A_1Y_1G_{21}$ are stable. A matrix G_w is strictly acceptable for a compatible G if there exists a stable matrix K such that G_vG_w, $G_\delta G_w$, and G_mG_w are stable, where G_δ, G_v, and G_m are the transfer matrices from w to y_δ, v, and y_m, respectively (see Definition 3.3). When G is compatible, a rational matrix G_w is strictly acceptable iff $AG_{21}G_w$ and $(G_{11} + G_{12}A_1Y_1G_{21})G_w$ are stable (see Example 3.21).

The $\mathcal{H}_2$ design problems solved in Chap. 3 with a compatible G in Fig. 3.2 yield all stabilizing controllers that make the performance functional

$$J = \frac{1}{2\pi j}\int_{-j\infty}^{j\infty} \mathrm{Tr}[(WG_\delta G_w)(WG_\delta G_w)_*]ds \tag{5.28}$$

finite when the weighting matrix W is stable and G_w is acceptable or strictly acceptable. The results for the case of strictly acceptable G_w are given by Theorem 3.5′ in Example 3.22 under the following assumptions:

Assumption 5.1 G is compatible and G_w is strictly acceptable.

Assumption 5.2 $WG_{11}G_w$ is strictly proper and W is stable.

Assumption 5.3 The following order relationships are satisfied:

$$G_{22} \leq O(s^0), \tag{5.29}$$

$$(G_{12*}W_*WG_{12})^{-1} \leq O(s^0), \tag{5.30}$$

$$(G_{21}G_wG_{w*}G_{21*})^{-1} \leq O(s^0). \tag{5.31}$$

(Attention is restricted here to the particular case of $k_1 = k_2 = k_3 = 0$ in Theorem 3.5′.)

Assumption 5.4 $\nabla_b = -WG_{12}A_1$ has full column rank on the finite part of the $s = j\omega$ axis.

Assumption 5.5 $\nabla_c = AG_{21}G_w$ has full row rank on the finite part of the $s = j\omega$ axis.

The following corollary is obtained from Theorem 3.5′ with $k_1 = k_2 = k_3 = 0$.

Corollary 5.1 *Suppose that Assumptions 5.1 through 5.5 are satisfied and that the matrices Λ and Ω are square stable Wiener–Hopf spectral factors with stable inverses satisfying*

$$\nabla_{b*}\nabla_b = A_{1*}G_{12*}W_*WG_{12}A_1 = \Lambda_*\Lambda \tag{5.32}$$

and

$$\nabla_c\nabla_{c*} = AG_{21}G_wG_{w*}G_{21*}A_* = \Omega\Omega_*. \tag{5.33}$$

Then all controller transfer matrices C given by (5.27) for which G_yG_w, $G_\delta G_w$, and G_mG_w are stable and the performance functional J in (5.28) is finite are the ones for which

$$K = \Lambda^{-1}\Gamma_a\Omega^{-1} - A_1^{-1}Y \tag{5.34}$$

is chosen where

$$\Gamma_a = Z + \{\Gamma_\phi\}_+ + \{\Lambda A_1^{-1} Y\Omega\}_-, \Gamma_\phi = \Lambda_*^{-1} \nabla_{b*} W G_{11} G_w \nabla_{c*} \Omega_*^{-1} \tag{5.35}$$

with Z a strictly proper stable matrix. In this case, the controller transfer matrix in (5.27) *satisfies* $C \leq 0(s^{-1})$. *Moreover, the performance functional J is minimized when* $Z = 0$.

The stabilizing controllers in (5.27) are described in terms of the polynomial matrices in the generalized Bezout identity (GBI) in (5.26); hence, the same is true for the Wiener–Hopf controllers given by Corollary 5.1. In order to obtain state-space representations for these controllers, it is therefore convenient to first establish a stable proper rational GBI associated with G_{22} using the polynomial GBI in (5.26). The basic ways to do this are explained in Sect. 2.7. Here, the method using the Wiener–Hopf approach (Park et al. 1999) is taken. First, notice that

$$\begin{bmatrix} \Omega^{-1} & 0 \\ \Gamma_\theta \Omega^{-1} & \Lambda \end{bmatrix} \begin{bmatrix} \Omega & 0 \\ -\Lambda^{-1}\Gamma_\theta & \Lambda^{-1} \end{bmatrix} = I = \begin{bmatrix} \Omega & 0 \\ -\Lambda^{-1}\Gamma_\theta & \Lambda^{-1} \end{bmatrix} \begin{bmatrix} \Omega^{-1} & 0 \\ \Gamma_\theta \Omega^{-1} & \Lambda \end{bmatrix} \tag{5.36}$$

and consider the case

$$\Gamma_\theta = \{\Gamma_\phi - \Lambda A_1^{-1} Y\Omega\}_+ - \{\Lambda A_1^{-1} Y\Omega\}_\infty = \{\Gamma_\phi\}_+ + \{\Lambda A_1^{-1} Y\Omega\}_- - \Lambda A_1^{-1} Y\Omega. \tag{5.37}$$

Clearly, the matrix Γ_θ is stable iff $\{\Gamma_\phi - \Lambda A_1^{-1} Y\Omega\}_+$ is stable. That this is the case follows from (5.32) and (5.33) which give

$$\Lambda A_1^{-1} = \Lambda_*^{-1} A_{1*} G_{12*} W_* W G_{12}, \Omega = A G_{21} G_w G_{w*} G_{21*} A_* \Omega_*^{-1}. \tag{5.38}$$

Hence,

$$\Gamma_\phi - \Lambda A_1^{-1} Y\Omega = -\Lambda_*^{-1} A_{1*} G_{12*} W_* W (G_{11} + G_{12} Y A G_{21}) G_w G_{w*} G_{21*} A_* \Omega_*^{-1}. \tag{5.39}$$

Since $\Lambda^{-1}, G_{12}A_1, W, \Omega^{-1}, (G_{11} + G_{12}YAG_{21})G_w$, and $AG_{21}G_w$ are stable, the matrix $\Gamma_\phi - \Lambda A_1^{-1} Y\Omega$ is analytic on the finite part of the $j\omega$-axis. Thus, $\{\Gamma_\phi - \Lambda A_1^{-1} Y\Omega\}_+$ is indeed stable and it follows that $\Gamma_\theta = \{\Gamma_\phi - \Lambda A_1^{-1} Y\Omega\}_+ - \{\Lambda A_1^{-1} Y\Omega\}_\infty$ is also stable.

Using the equalities in (5.36), one gets from (5.26) that

$$\begin{bmatrix} \Omega^{-1} & 0 \\ \Gamma_\theta \Omega^{-1} & \Lambda \end{bmatrix} \begin{bmatrix} A & B \\ Y_1 & -X_1 \end{bmatrix} \begin{bmatrix} X & B_1 \\ Y & -A_1 \end{bmatrix} \begin{bmatrix} \Omega & 0 \\ -\Lambda^{-1}\Gamma_\theta & \Lambda^{-1} \end{bmatrix} = \begin{bmatrix} I & 0 \\ 0 & I \end{bmatrix}, \quad (5.40)$$

and

$$\begin{bmatrix} I & 0 \\ 0 & I \end{bmatrix} = \begin{bmatrix} X & B_1 \\ Y & -A_1 \end{bmatrix} \begin{bmatrix} \Omega & 0 \\ -\Lambda^{-1}\Gamma_\theta & \Lambda^{-1} \end{bmatrix} \begin{bmatrix} \Omega^{-1} & 0 \\ \Gamma_\theta \Omega^{-1} & \Lambda \end{bmatrix} \begin{bmatrix} A & B \\ Y_1 & -X_1 \end{bmatrix}, \quad (5.41)$$

which leads to

$$\begin{bmatrix} \widetilde{A} & \widetilde{B} \\ \widetilde{Y}_1 & -\widetilde{X}_1 \end{bmatrix} \begin{bmatrix} \widetilde{X} & \widetilde{B}_1 \\ \widetilde{Y} & -\widetilde{A}_1 \end{bmatrix} = \begin{bmatrix} I & 0 \\ 0 & I \end{bmatrix} = \begin{bmatrix} \widetilde{X} & \widetilde{B}_1 \\ \widetilde{Y} & -\widetilde{A}_1 \end{bmatrix} \begin{bmatrix} \widetilde{A} & \widetilde{B} \\ \widetilde{Y}_1 & -\widetilde{X}_1 \end{bmatrix}, \quad (5.42)$$

where

$$\widetilde{A} = \Omega^{-1}A, \widetilde{B} = \Omega^{-1}B, \widetilde{A}_1 = A_1\Lambda^{-1}, \widetilde{B}_1 = B_1\Lambda^{-1}, \quad (5.43)$$

$$\widetilde{X} = X\Omega - B_1\Lambda^{-1}\Gamma_\theta = A^{-1}\Omega - B_1\Lambda^{-1}(\{\Gamma_\phi\}_+ + \{\Lambda A_1^{-1}Y\Omega\}_-), \quad (5.44)$$

$$\widetilde{Y} = Y\Omega + A_1\Lambda^{-1}\Gamma_\theta = A_1\Lambda^{-1}(\{\Gamma_\phi\}_+ + \{\Lambda A_1^{-1}Y\Omega\}_-), \quad (5.45)$$

$$\widetilde{X}_1 = \Lambda X_1 - \Gamma_\theta\Omega^{-1}B = \Lambda A_1^{-1} - (\{\Gamma_\phi\}_+ + \{\Lambda A_1^{-1}Y\Omega\}_-)\Omega^{-1}B, \quad (5.46)$$

and

$$\widetilde{Y}_1 = \Lambda Y_1 + \Gamma_\theta\Omega^{-1}A = (\{\Gamma_\phi\}_+ + \{\Lambda A_1^{-1}Y\Omega\}_-)\Omega^{-1}A. \quad (5.47)$$

It is also clear that

$$G_{22} = -\widetilde{A}^{-1}\widetilde{B} = -\widetilde{B}_1\widetilde{A}_1^{-1}. \quad (5.48)$$

Lemma 5.1 *When Assumptions* 5.1, 5.3, 5.4, *and* 5.5 *are satisfied, then the expression in (5.42) is a stable proper rational GBI associated with* G_{22}.

Proof By Assumptions 5.4 and 5.5, the matrices Λ^{-1} and Ω^{-1} are analytic in $\mathrm{Re}\,s \geq 0$; therefore, $\widetilde{A}, \widetilde{B}, \widetilde{A}_1$, and $\widetilde{B}_1$ are stable. The matrices $\widetilde{A}_1 = A_1\Lambda^{-1}$ and $\widetilde{A} = \Omega^{-1}A$ are proper. In fact, it follows from (5.32) and (5.33) that

$$G_{12*}W_*WG_{12} = A_{1*}^{-1}\Lambda_*\Lambda A_1^{-1}, \quad G_{21}G_wG_{w*}G_{21*} = A^{-1}\Omega\Omega_*A_*^{-1}. \quad (5.49)$$

By the order assumptions in (5.30) and (5.31) it can be concluded that $\widetilde{A}_1 = A_1\Lambda^{-1} \le O(s^0)$ and $\widetilde{A} = \Omega^{-1}A \le O(s^0)$ (Lemma 3.2). The properness of $\widetilde{B}_1 = B_1\Lambda^{-1}$ and $\widetilde{B} = \Omega^{-1}B$ is obvious since $\widetilde{B}_1 = -G_{22}\widetilde{A}_1$ and $\widetilde{B} = -\widetilde{A}G_{22}$, where G_{22} is proper by the assumption in (5.29). The properness of $\widetilde{X}_1, \widetilde{X}, \widetilde{Y}_1$, and $\widetilde{Y}$ is obvious since $\widetilde{A}_1 = A_1\Lambda^{-1}$, $\widetilde{A} = \Omega^{-1}A$, $\widetilde{B}_1 = B_1\Lambda^{-1}$, $\widetilde{B} = \Omega^{-1}B$, and $\{\Gamma_\phi\}_+ + \{\Lambda A_1^{-1}Y\Omega\}_-$ are all proper. That $\widetilde{X}_1, \widetilde{X}, \widetilde{Y}_1$, and $\widetilde{Y}$ are stable is obvious since $\Lambda, \Lambda^{-1}, \Omega, \Omega^{-1}$ and Γ_θ are all stable.

Now it is possible to describe the Wiener–Hopf controllers of Corollary 5.1 in terms of the stable proper rational matrices in (5.42). First, it should be noticed from (5.34) and (5.35) that

$$\begin{aligned} K &= \Lambda^{-1}(\Gamma_a - \Lambda A_1^{-1}Y\Omega)\Omega^{-1} \\ &= \Lambda^{-1}(Z + \{\Gamma_\phi\}_+ + \{\Lambda A_1^{-1}Y\Omega\}_- - \Lambda A_1^{-1}Y\Omega)\Omega^{-1} \end{aligned} \tag{5.50}$$

$$= \Lambda^{-1}(Z + \{\Gamma_\phi - \Lambda A_1^{-1}Y\Omega\}_+ - \{\Lambda A_1^{-1}Y\Omega\}_\infty)\Omega^{-1} = \Lambda^{-1}(Z+\Gamma_\theta)\Omega^{-1}. \tag{5.51}$$

It now follows from (5.27) and (5.34) that

$$C = \left(Y + A_1\Lambda^{-1}(Z+\Gamma_\theta)\Omega^{-1}\right)\left(X - B_1\Lambda^{-1}(Z+\Gamma_\theta)\Omega^{-1}\right)^{-1} \tag{5.52}$$

$$= \left(Y\Omega + A_1\Lambda^{-1}(Z+\Gamma_\theta)\right)\left(X\Omega - B_1\Lambda^{-1}(Z+\Gamma_\theta)\right)^{-1} \tag{5.53}$$

$$= (\widetilde{Y} + \widetilde{A}_1 Z)(\widetilde{X} - \widetilde{B}_1 Z)^{-1} = \begin{bmatrix} \widetilde{A}_1 & \widetilde{Y} \\ -\widetilde{B}_1 & \widetilde{X} \end{bmatrix} \diamond Z. \tag{5.54}$$

An alternative form for the controller transfer matrix in (5.54) is given by

$$C = (\widetilde{Y} + \widetilde{A}_1 Z)(\widetilde{X} - \widetilde{B}_1 Z)^{-1} = (\widetilde{X}_1 - Z\widetilde{B})^{-1}(\widetilde{Y}_1 + Z\widetilde{A}). \tag{5.55}$$

This can be shown by using the method in Lemma 2.2 or can be shown directly by multiplying $[Z \quad I]$ on the left and $[I \; -Z']'$ on the right to the first equality in (5.42) to get

$$[Z \quad I]\begin{bmatrix} \widetilde{A} & \widetilde{B} \\ \widetilde{Y}_1 & -\widetilde{X}_1 \end{bmatrix}\begin{bmatrix} \widetilde{X} & \widetilde{B}_1 \\ \widetilde{Y} & -\widetilde{A}_1 \end{bmatrix}\begin{bmatrix} I \\ -Z \end{bmatrix} = [Z \quad I]\begin{bmatrix} I \\ -Z \end{bmatrix} = 0. \tag{5.56}$$

This leads to $(Z\widetilde{A} + \widetilde{Y}_1)(\widetilde{X} - \widetilde{B}_1 Z) + (Z\widetilde{B} - \widetilde{X}_1)(\widetilde{Y} + \widetilde{A}_1 Z) = 0$ from which the equality in (5.55) follows.

The next step is the derivation of a state-space realization for the controller. Specifically, given

$$G_N = \begin{bmatrix} WG_{11}G_w & WG_{12} \\ G_{21}G_w & G_{22} \end{bmatrix} = \left[\begin{array}{c|cc} \Sigma & N_1 & N_2 \\ \hline H_1 & D_{11} & D_{12} \\ H_2 & D_{21} & D_{22} \end{array}\right] \tag{5.57}$$

the objective is the determination of a state-space realization free of any unstable hidden poles for the controller transfer matrix in (5.54). It is of course necessary that G_N be proper and this is assured when $W, G,$ and G_w are proper. Moreover, since G is compatible, the response to initial conditions associated with the physical components within G is stabilized by C. So there is no need to be concerned about this issue in regard to (5.57) although it makes good sense that the state-space model for G_N be consistent with $G_{22} = H_2(sI - \Sigma)^{-1}N_2 + D_{22}$ being stabilized by C. Specifically, with $\Delta_\Sigma = \det(sI - \Sigma)$, it is assumed that $h_2 = \Delta_\Sigma/\psi_{G_{22}}$ is a stable polynomial or, equivalently, (Σ, N_2) is stabilizable and (Σ, H_2) is detectable. This equivalence is established in the solution of Example 5.9(b). It is shown in the sequel for this case that the derived state-space realization for C is indeed free of unstable hidden poles. In addition, it is shown later that the order of the state-space realization for the optimal Wiener–Hopf controller is equal to that of the matrix Σ. Therefore, choosing the lowest order possible for Σ is desirable in this regard. However, the constraint that the controller realization be minimal is not imposed here for more flexibility.

Constraints on the realization in (5.57) imposed by Assumptions 5.1 through 5.5 are summarized in the following lemma.

Lemma 5.2 *The frequency-domain conditions in Assumptions* 5.1 *through* 5.5 *impose constraints on the state-space parameters in (5.57). In particular, the following assumptions must be satisfied:*

Assumption 5.1′ (Σ, N_2) is stabilizable and (Σ, H_2) is detectable.

Assumption 5.2′ $D_{11} = 0$.

Assumption 5.3′ D_{12} and D_{21} have full column and full row rank, respectively.

Assumption 5.4′ The matrix $S_{12}(j\omega) = \begin{bmatrix} \Sigma - j\omega I & N_2 \\ H_1 & D_{12} \end{bmatrix}$ has full column rank for all ω.

Assumption 5.5′ The matrix $S_{21}(j\omega) = \begin{bmatrix} \Sigma - j\omega I & N_1 \\ H_2 & D_{21} \end{bmatrix}$ has full row rank for all ω.

Proof The need for Assumption 5.1′ has already been discussed. Clearly, Assumption 5.2 requires that $WG_{11}G_w$ be strictly proper and this is possible in (5.57) iff Assumption 5.2′ holds. Assumption 5.3′ is a consequence of the following considerations. Employing the formulas in (5.12) and (5.14) one gets from $W(s)G_{12}(s) = H_1(sI - \Sigma)^{-1}N_2 + D_{12}$ that

$$G_{12*}W_*WG_{12} = \left[\begin{array}{cc|c} -\Sigma' & H_1'H_1 & H_1'D_{12} \\ 0 & \Sigma & N_2 \\ \hline -N_2' & D_{12}'H_1 & D_{12}'D_{12} \end{array}\right]. \tag{5.58}$$

Assumption 5.3 implies that $(G_{12*}W_*WG_{12})^{-1}$ exists and is proper. It then follows from (5.13) and (5.58) that this is the case iff the inverse of the matrix $D_{12}'D_{12}$ exists or, equivalently, the matrix D_{12} has full column rank. A similar argument can be used to establish that $(G_{21}G_wG_{w*}G_{21*})^{-1} \leq O(s^0)$ iff D_{21} has full row rank. Assumption 5.4′ is considered next. In particular, it is now established under Assumption 5.1′ that Assumption 5.4′ holds iff Assumption 5.4 holds or, equivalently, $\det(A_{1*}G_{12*}W_*WG_{12}A_1) \neq 0$ on the finite part of the $s = j\omega$ axis. One can always write

$$W(s)G_{12}(s) = H_1(sI - \Sigma)^{-1}N_2 + D_{12} = B_{p1}A_{p1}^{-1} \tag{5.59}$$

and

$$G_{22}(s) = H_2(sI - \Sigma)^{-1}N_2 + D_{22} = -B_1A_1^{-1}, \tag{5.60}$$

with A_{p1}, B_{p1} and A_1, B_1 right coprime polynomial matrix pairs. It then follows that $\det(sI - \Sigma) = h_{p1}\det A_{p1} = h_1 \det A_1$ and the polynomial h_1 must be Hurwitz by Assumption 5.1′. Hence,

$$\det(A_{1*}G_{12*}W_*WG_{12}A_1) = \det A_{1*}\det(B_{p1*}B_{p1})\det A_1/(\det A_{p1*}\det A_{p1}) \tag{5.61}$$

$$= h_{p1*}\det(B_{p1*}B_{p1})h_{p1}/(h_{1*}h_1). \tag{5.62}$$

Since h_1 is Hurwitz, $\det(A_{1*}G_{12*}W_*WG_{12}A_1)$ has no $j\omega$-axis zeros iff both h_{p1} and $\det(B_{p1*}B_{p1})$ have no $j\omega$-axis zeros. Since (Σ, N_2) is stabilizable, $[(\Sigma - j\omega I) \;\; N_2]$ has full row rank on the $s = j\omega$ axis. In this case, h_{p1} has no $j\omega$-axis zeros iff $[(\Sigma' - j\omega I) \;\; H_1']$ has full row rank on the $j\omega$-axis (follows from Example 5.9(b) with $\Re$ the $j\omega$-axis). Also, $\det(B_{p1*}B_{p1})$ has no $j\omega$-axis zeros iff $B_{p1}(s)$ has full column rank on the finite part of the $j\omega$-axis or, equivalently, $W(s)G_{12}(s)$ has no transmission zeros on the $s = j\omega$ axis. It then follows (from Example 5.9(a) with $\mathfrak{R}$ the $j\omega$-axis) that diag $\{I, B_{p1}(j\omega)\}$ and $S_{12}(j\omega)$ have the same rank; hence, $S_{12}(j\omega)$ has full column rank for all ω. That is, under Assumption 5.1′, the Assumptions 5.4 and 5.4′ are equivalent. The proof of the equivalence of Assumptions 5.5 and 5.5′ is similar and omitted.

In the next lemma, a state-space representation of the stable proper rational matrix $\widetilde{A}_1$ is developed under Assumptions 5.1′ through 5.5′ (or, equivalently, 5.1 through 5.5).

Lemma 5.3 *Suppose that Assumptions 5.1′ through 5.5′ are satisfied and that*

$$(sI - \Sigma)^{-1} N_2 = B_2(s) A_2^{-1}(s) \tag{5.63}$$

denotes a right coprime polynomial fraction description for $(sI - \Sigma)^{-1} N_2$. *Then the Wiener–Hopf factor* $\Lambda(s)$ *in (5.32) is given by*

$$\Lambda(s) = E_1^{\frac{1}{2}} (A_2(s) + L_1 B_2(s)) V_1^{-1}(s) = E_1^{\frac{1}{2}} (A_1(s) + L_1 B_2(s) V_1^{-1}(s)), \tag{5.64}$$

where $V_1(s) = A_1^{-1}(s) A_2(s)$ *is a polynomial matrix whose determinant is Hurwitz and* E_1 *and* L_1 *are given below in (5.67). Moreover, state-space parameters for the matrix* $\widetilde{A}_1(s) = A_1(s)\Lambda^{-1}(s)$ *are obtained from the formula*

$$\widetilde{A}_1(s) = A_1(s)\Lambda^{-1}(s) = \left(I + L_1 (sI - \Sigma)^{-1} N_2\right)^{-1} E_1^{-\frac{1}{2}} \tag{5.65}$$

$$= \left(I - L_1 (sI - \Sigma + N_2 L_1)^{-1} N_2\right) E_1^{-\frac{1}{2}}, \tag{5.66}$$

in which

$$E_1 = D_{12}' D_{12}, L_1 = E_1^{-1}(D_{12}' H_1 + N_2' M_1). \tag{5.67}$$

The matrix M_1 *is the unique positive semi-definite solution of the algebraic Riccati equation (ARE)*

$$\begin{aligned}(\Sigma - N_2 E_1^{-1} D_{12}' H_1)' M_1 + M_1 (\Sigma - N_2 E_1^{-1} D_{12}' H_1) - M_1 N_2 E_1^{-1} N_2' M_1 \\ + H_1' (I - D_{12} E_1^{-1} D_{12}') H_1 = 0\end{aligned} \tag{5.68}$$

with the property that $\Sigma - N_2 L_1$ *is stable.*

Proof Since $G_{22} = H_2(sI-\Sigma)^{-1}N_2 + D_{22} = -B_1A_1^{-1}$, $(sI-\Sigma)^{-1}N_2 = B_2(s)A_2^{-1}(s)$, the pair (Σ, N_2) is stabilizable, and the pair (Σ, H_2) is detectable, it follows that $\det(sI-\Sigma) = h_1(s)\det A_1(s)$ and $\det(sI-\Sigma) = h_2(s)\det A_2(s)$, where $h_1(s)$ and $h_2(s)$ are Hurwitz polynomials. In addition, it is clear that $-B_1A_1^{-1}A_2 = G_{22}A_2 = H_2B_2 + D_{22}A_2$ is a polynomial matrix. Hence,

$$V_1(s) = A_1^{-1}A_2 = (X_1 + Y_1B_1A_1^{-1})A_2 = X_1A_2 + Y_1B_1A_1^{-1}A_2 \tag{5.69}$$

is a polynomial matrix and $\det V_1 = \det A_2/\det A_1 = h_1/h_2$ is a Hurwitz polynomial. Moreover, since (Σ, N_2) is stabilizable and Assumption 5.4′ is satisfied, the ARE in (5.68) has a unique solution $M_1 \geq 0$ with the property that $\Sigma - N_2L_1$ is stable (Zhou et al. 1996; Corollary 13.10). A little algebra from (5.68) yields (see Example 5.1)

$$G_{12*}(s)W_*(s)W(s)G_{12}(s) = \left(L_1(sI-\Sigma)^{-1}N_2 + I\right)_* D_{12}'D_{12}\left(L_1(sI-\Sigma)^{-1}N_2 + I\right) \tag{5.70}$$

$$= (L_1B_2A_2^{-1} + I)_* D_{12}'D_{12}(L_1B_2A_2^{-1} + I) \tag{5.71}$$

and this leads to

$$A_{1*}G_{12*}W_*WG_{12}A_1 = (A_1(s) + L_1B_2(s)V_1^{-1})_* D_{12}'D_{12}(A_1(s) + L_1B_2(s)V_1^{-1}) \tag{5.72}$$

$$= V_{1*}^{-1}(A_2(s) + L_1B_2(s))_* D_{12}'D_{12}(A_2(s) + L_1B_2(s))V_1^{-1}. \tag{5.73}$$

It is easy to prove that $(A_2 + L_1B_2)^{-1}$ is analytic in $\text{Re}\, s \geq 0$. In fact, $\det(I + L_1(sI-\Sigma)^{-1}N_2) = \det(I + (sI-\Sigma)^{-1}N_2L_1) = \det(sI - \Sigma + N_2L_1)/\det(sI-\Sigma)$ and it is also true that $\det(I + L_1(sI-\Sigma)^{-1}N_2) = \det(I + L_1B_2A_2^{-1}) = h_2\det(A_2 + L_1B_2)/\det(sI-\Sigma)$. It therefore follows that $h_2\det(A_2 + L_1B_2) = \det(sI - \Sigma + N_2L_1)$; hence, $\det(A_2 + L_1B_2)$ is a Hurwitz polynomial. It is now clear that $(A_2 + L_1B_2)^{-1}$ is analytic in $\text{Re}\, s \geq 0$ and $(D_{12}'D_{12})^{\frac{1}{2}}(A_2 + L_1B_2)V_1^{-1}$ is the Wiener–Hopf factor $\Lambda(s)$. Finally,

$$\widetilde{A}_1(s) = A_1(s)\Lambda^{-1}(s) = A_1V_1(L_1B_2 + A_2)^{-1}E_1^{-\frac{1}{2}} = A_2(L_1B_2 + A_2)^{-1}E_1^{-\frac{1}{2}} \tag{5.74}$$

$$\begin{aligned} &= (L_1B_2A_2^{-1} + I)^{-1}E_1^{-\frac{1}{2}} = \left(I + L_1(sI-\Sigma)^{-1}N_2\right)^{-1}E_1^{-\frac{1}{2}} \\ &= \left(I - L_1(sI - \Sigma + N_2L_1)^{-1}N_2\right)E_1^{-\frac{1}{2}}. \end{aligned} \tag{5.75}$$

In a similar way, state-space parameters for $\widetilde{A}$ can be obtained under Assumptions 5.1′ through 5.5′ and the results are presented in the following lemma whose proof is omitted.

Lemma 5.4 *Suppose that*

$$H_2(sI-\Sigma)^{-1}=A_3^{-1}(s)B_3(s) \tag{5.76}$$

denotes a left coprime polynomial fraction description for $H_2(sI-\Sigma)^{-1}$. *Then the Wiener–Hopf factor* $\Omega(s)$ *in (5.33) is given by*

$$\Omega(s)=V_2^{-1}(s)(A_3(s)+B_3(s)L_2)E_2^{\frac{1}{2}}=(A(s)+V_2^{-1}(s)B_3(s)L_2)E_2^{\frac{1}{2}}, \tag{5.77}$$

where E_2 *and* L_2 *are given below in (5.81) and where* $V_2(s)$ *is a stable square polynomial matrix satisfying*

$$A_3=V_2A. \tag{5.78}$$

The state-space parameters of the rational matrix $\widetilde{A}(s)=\Omega^{-1}(s)A(s)$ *are obtained from the formula*

$$\widetilde{A}(s)=\Omega^{-1}(s)A(s)=E_2^{-\frac{1}{2}}\left(I+H_2(sI-\Sigma)^{-1}L_2\right)^{-1} \tag{5.79}$$

$$=E_2^{-\frac{1}{2}}\left(I-H_2(sI-\Sigma+L_2H_2)^{-1}L_2\right), \tag{5.80}$$

where

$$E_2=D_{21}D_{21}',\ L_2=(N_1D_{21}'+M_2H_2')E_2^{-1}. \tag{5.81}$$

The matrix M_2 *is the unique positive semi-definite solution of the ARE*

$$\begin{aligned}(\Sigma-N_1D_{21}'E_2^{-1}H_2)M_2+M_2(\Sigma-N_1D_{21}'E_2^{-1}H_2)'-M_2H_2'E_2^{-1}H_2M_2\\+N_1(I-D_{21}'E_2^{-1}D_{21})N_1'=0\end{aligned} \tag{5.82}$$

with the property that $\Sigma-L_2H_2$ *is stable. The existence of such a solution is assured when* (Σ,H_2) *is detectable and Assumption* 5.5′ *is satisfied (see Example* 5.11 *and its solution).*

Establishing the formula in the following lemma is the last step needed to find a state-space representation for the stable proper rational GBI in (5.42). Since its proof is rather long, it is given in the solution of Example 5.2.

Lemma 5.5

$$\{\Gamma_\phi\}_+ + \{\Lambda A_1^{-1} Y\Omega\}_- = -E_1^{\frac{1}{2}} L_1 (sI - \Sigma)^{-1} L_2 E_2^{\frac{1}{2}}. \tag{5.83}$$

When (Σ, N_2, H_2) is minimal, one can always satisfy (5.63) and (5.76) with $A_2 = A_1$ and $A_3 = A$, respectively. Then, V_1 in Lemma 5.3 and V_2 in Lemma 5.4 are identity matrices. In this case, it can be shown that $\{\Gamma_\phi - \Lambda A_1^{-1} Y\Omega\}_+ = 0$ and $\{\Gamma_\phi\}_+ + \{\Lambda A_1^{-1} Y\Omega\}_- = \{\Lambda A_1^{-1} Y\Omega\}_{sp} = -E_1^{\frac{1}{2}} L_1 (sI - \Sigma)^{-1} L_2 E_2^{\frac{1}{2}}$, where $\{G(s)\}_{sp}$ denotes the strictly proper part of $G(s)$ (see Example 5.3). Moreover, Γ_θ in (5.37) becomes $-\{\Lambda A_1^{-1} Y\Omega\}_\infty$ and the GBI in (5.42) is the same as the one in (2.373) in Sect. 2.7.

Once the state-space parameters for $\widetilde{A}_1(s) = A_1(s)\Lambda^{-1}(s)$, $\widetilde{A}(s) = \Omega^{-1}(s)A(s)$, and $\{\Gamma_\phi\}_+ + \{\Lambda A_1^{-1} Y\Omega\}_-$ are computed, the determination of the state-space parameters for the eight matrices in (5.42) is straightforward. State-space parameters for $\widetilde{B} = \Omega^{-1}B$ and $\widetilde{B}_1 = B_1\Lambda^{-1}$ are easily obtained from the identities $\widetilde{B} = \Omega^{-1}B = -(\Omega^{-1}A)G_{22}$ and $\widetilde{B}_1 = B_1\Lambda^{-1} = -G_{22}(A_1\Lambda^{-1})$. As for $\widetilde{Y}$, $\widetilde{Y}_1$, $\widetilde{X}$, and $\widetilde{X}_1$, the identities in (5.44) through (5.47) are used. Employing straightforward algebra and the formulas in (5.12) through (5.15), one gets the following state-space parameters for the eight matrices in (5.42):

$$\begin{bmatrix} \tilde{A} & \tilde{B} \\ \tilde{Y}_1 & -\tilde{X}_1 \end{bmatrix} = \left[\begin{array}{c|cc} \Sigma - L_2 H_2 & L_2 & N_2 - L_2 D_{22} \\ \hline -E_2^{-\frac{1}{2}} H_2 & E_2^{-\frac{1}{2}} & -E_2^{-\frac{1}{2}} D_{22} \\ -E_1^{\frac{1}{2}} L_1 & 0 & -E_1^{\frac{1}{2}} \end{array}\right], \tag{5.84}$$

$$\begin{bmatrix} \tilde{X} & \tilde{B}_1 \\ \tilde{Y} & -\tilde{A}_1 \end{bmatrix} = \left[\begin{array}{c|cc} \Sigma - N_2 L_1 & -L_2 E_2^{\frac{1}{2}} & N_2 E_1^{-\frac{1}{2}} \\ \hline -(H_2 - D_{22} L_1) & E_2^{\frac{1}{2}} & -D_{22} E_1^{-\frac{1}{2}} \\ L_1 & 0 & -E_1^{-\frac{1}{2}} \end{array}\right]. \tag{5.85}$$

In (5.54), the Wiener–Hopf controllers determined by Corollary 5.1 are given in terms of the homographic transformation defined in (5.18). From an implementation point of view, the linear fractional transformation (LFT) defined in (5.17) is more convenient. Invoking the formula in (5.19) one gets from (5.54) that the LFT form for $C(s)$ is

$$C(s) = G_c(s) \Box Z \tag{5.86}$$

with

$$G_c(s) = \begin{bmatrix} \widetilde{Y}\widetilde{X}^{-1} & \widetilde{A}_1 + \widetilde{Y}\widetilde{X}^{-1}\widetilde{B}_1 \\ \widetilde{X}^{-1} & \widetilde{X}^{-1}\widetilde{B}_1 \end{bmatrix} = \begin{bmatrix} \widetilde{X}_1^{-1}\widetilde{Y}_1 & \widetilde{X}_1^{-1} \\ \widetilde{A} + \widetilde{B}\widetilde{X}_1^{-1}\widetilde{Y}_1 & \widetilde{B}\widetilde{X}_1^{-1} \end{bmatrix}. \tag{5.87}$$

As explained in Sect. 2.7, the controller in (5.86) can be implemented as a tunable realization with a separate block for $Z(s)$ as in Fig. 2.2. It is now possible to state the following theorem.

Theorem 5.1 *The Wiener–Hopf controllers satisfying Corollary 5.1 are obtained in state-space form from (5.86) with*

$$G_c(s)=\left[\begin{array}{c|cc}\Sigma-N_2L_1-L_2H_2+L_2D_{22}L_1 & L_2 & (N_2-L_2D_{22})E_1^{-\frac{1}{2}}\\ \hline -L_1 & 0 & E_1^{-\frac{1}{2}}\\ -E_2^{-\frac{1}{2}}(H_2-D_{22}L_1) & E_2^{-\frac{1}{2}} & -E_2^{-\frac{1}{2}}D_{22}E_1^{-\frac{1}{2}}\end{array}\right]. \tag{5.88}$$

Implementation of $C(s)$ in (5.86) with the realization of $G_c(s)$ in (5.88) is free of unstable hidden poles. The optimal controller minimizing the cost is given by

$$C_{opt}=\tilde{Y}\tilde{X}^{-1}=\left[\begin{array}{c|c}\Sigma-N_2L_1-L_2H_2+L_2D_{22}L_1 & L_2\\ \hline -L_1 & 0\end{array}\right]. \tag{5.89}$$

Proof The state-space parameters for $G_c(s)$ given by (5.88) can be obtained from the state-space parameters in (5.85) by setting

$$V=\begin{bmatrix}V_{11} & V_{12}\\ V_{21} & V_{22}\end{bmatrix}=\begin{bmatrix}\widetilde{A}_1 & \widetilde{Y}\\ -\widetilde{B}_1 & \widetilde{X}\end{bmatrix}=\begin{bmatrix}0 & -I\\ -I & 0\end{bmatrix}\begin{bmatrix}\widetilde{X} & \widetilde{B}_1\\ \widetilde{Y} & -\widetilde{A}_1\end{bmatrix}\begin{bmatrix}0 & -I\\ I & 0\end{bmatrix} \tag{5.90a}$$

in (5.54) and using (5.20) and (5.21). The formula in (5.89) is obtained from (5.86) by taking $Z=0$. In Sect. 2.7, it is shown that the implementation of (5.86) with (5.87) is free of unstable poles iff the realization of (5.88) is free of unstable hidden poles (it is assumed that $Z(s)$ is realized without unstable hidden poles). Hence, it only remains to show that the realization of (5.88) is free of unstable hidden poles. It is first shown that (Σ_c, L_2), where $\Sigma_c=\Sigma-N_2L_1-L_2H_2+L_2D_{22}L_1$, is a stabilizable pair. Since $\Sigma-N_2L_1$ is a stable matrix, the pair $(\Sigma-N_2L_1, L_2)$ is stabilizable, or equivalently, $[sI-\Sigma+N_2L_1 \quad L_2]$ has full row rank in $\mathrm{Re}\,s\geq 0$.

Clearly,

$$[sI-\Sigma+N_2L_1+L_2H_2-L_2D_{22}L_1 \quad L_2]=[sI-\Sigma+N_2L_1 \quad L_2]\begin{bmatrix}I & 0\\ H_2-D_{22}L_1 & I\end{bmatrix}, \tag{5.90b}$$

and it follows that $rank\,[sI-\Sigma_c \quad L_2]=rank[sI-\Sigma+N_2L_1 \quad L_2]$ in $\mathrm{Re}\,s\geq 0$. That is, (Σ_c, L_2) is stabilizable. Similarly, it can be shown that the pair (Σ_c, L_1) is detectable. The condition that (Σ_c, L_2) is stabilizable and (Σ_c, L_1) is detectable assures that $(\Sigma_c, [L_2 \quad (N_2-L_2D_{22})E_1^{-\frac{1}{2}}])$ is stabilizable and $(\Sigma_c, [L_1' \quad (H_2-$

$D_{22}L_1)'E_2^{-\frac{1}{2}}]')$ is detectable, in which case the realization of $G_c(s)$ in (5.88) has no unstable hidden poles (follows from Example (5.9(b)).

An alternative proof that the implementation of $C(s)$ in (5.86) with the realization of $G_c(s)$ in (5.88) is free of unstable hidden poles is given in Example 5.4. The state-space parameters for the controllers $C(s)$ in (5.86) are also determined in the solution of this example.

Remark 5.1 The generalized plant G in Fig. 3.2 is assumed to be compatible; hence, compatibility must be checked before applying the formulas in Corollary 5.1 or Theorem 5.1. Verification of the compatibility of a given generalized plant can be carried out in either the state-space or the frequency domains. A state-space method is now described. A generalized plant G is compatible if the physical system $\tilde{G}$ within it is admissible, $G_{22} = \tilde{G}_{22}$, and the transfer matrices $G_{12}A_1$, AG_{21}, and $G_{11} + G_{12}A_1Y_1G_{21}$ are stable. The verification of compatibility is carried out in two steps. First, the admissibility of the physical system is considered. Suppose that

$$\tilde{G} = \begin{bmatrix} \tilde{G}_{11} & \tilde{G}_{12} \\ \tilde{G}_{21} & \tilde{G}_{22} \end{bmatrix} = \left[\begin{array}{c|cc} \tilde{\Sigma} & \tilde{N}_1 & \tilde{N}_2 \\ \hline \tilde{H}_1 & \tilde{D}_{11} & \tilde{D}_{12} \\ \tilde{H}_2 & \tilde{D}_{21} & \tilde{D}_{22} \end{array}\right] \tag{5.91}$$

denotes an internal description of the physical system block $\tilde{G}(s)$ inside of $G(s)$. Since $\Delta_{\tilde{G}} = \Delta_{\tilde{G}_{22}} = \Delta_{\Sigma} = \det(sI - \Sigma)$ it follows that $\tilde{G}$ is admissible (Definition 2.2) iff $\tilde{G}_{22}$ is admissible. This is the case iff $(\tilde{\Sigma}, \tilde{N}_2)$ is stabilizable and $(\tilde{\Sigma}, \tilde{H}_2)$ is detectable (see Example 5.9(b)). When the physical system is admissible the next step is to check the stabilities of the three transfer matrices (checking of $G_{22} = \tilde{G}_{22}$ is trivial). Suppose that

$$G = \begin{bmatrix} G_{11} & G_{12} \\ G_{21} & G_{22} \end{bmatrix} = \left[\begin{array}{c|cc} \Sigma & N_1 & N_2 \\ \hline H_1 & D_{11} & D_{12} \\ H_2 & D_{21} & D_{22} \end{array}\right] = \left[\begin{array}{c|c} \Sigma & N \\ \hline H & D \end{array}\right] \tag{5.92}$$

is an admissible state-space realization for G which is the case iff (Σ, N_2) is stabilizable and (Σ, H_2) is detectable. Then $\psi_G/\psi_{G_{22}}$ is a stable polynomial and it follows from Lemma 2.3 that $G_{12}A_1$, AG_{21}, and $G_{11} + G_{12}A_1Y_1G_{21}$ are stable.

It is crucial to recognize that there is a basic difference between two state-space descriptions in (5.91) and (5.92). The state-space model in (5.91) is an internal description of the physical system. Thus, all the states of the sub-blocks inside $\tilde{G}$ must be included when the state-space model in (5.91) is found (a systematic way to find a state-space internal description is presented in Example 5.5(a)). In general, the state-space model in (5.91) is not a minimal one. However, the state-space description in (5.92) is a realization of the transfer matrix $G(s)$; hence, it is always possible to find a minimal one. In this regard, the state-space model for the generalized plant in Doyle et al. (1989) is regarded as an internal description and only internal stabilizability and detectability (equivalent to admissibility here) of the generalized plant is imposed. Rigorously speaking, therefore, $\mathcal{H}_2$ design with unstable inputs is not considered in Doyle et al. (1989) because all the requirements for compatibility are not imposed.

A frequency-domain method to check the admissibility of a given physical generalized plant can be found in Example 5.5(b).

Remark 5.2 As explained in Remark 3.2, it is not desirable to include the unstable input signal model G_w into the generalized plant. When W is stable and G_w is strictly acceptable for a compatible G, however, the matrices $WG_{12}A_1$, $AG_{21}G_w$, and $(G_{11}+G_{12}A_1Y_1G_{21})G_w$ are stable. It can then be concluded that the system G_N in (5.57) is compatible. This implies that the strictly acceptable G_w is an unstable input model that can be absorbed into a generalized compatible plant to result in a new compatible system.

5.4 Basic Machinery for Digital Systems

As was done for analog systems, a collection of useful relationships between state-space and frequency-domain representations needed in the sequel are provided in this section for digital systems. A linear time-invariant discrete system described by

$$\hat{x}(k+1)=\Sigma\hat{x}(k)+N\hat{u}(k), \hat{y}(k)=H\hat{x}(k)+D\hat{u}(k) \tag{5.93}$$

is considered where the elements of the vectors $\hat{x}(k)$, $\hat{u}(k)$, and $\hat{y}(k)$ are, respectively the state, the input, and the output variables. Applying the λ-transform to the difference equations in (5.93) yields the polynomial matrix description

$$\left.\begin{aligned}(I-\Sigma\lambda)x(\lambda)&=N\lambda u(\lambda)\\ y(\lambda)&=Hx(\lambda)+Du(\lambda)\end{aligned}\right\}. \tag{5.94}$$

The transfer matrix of the system is $G(\lambda)=H(\lambda^{-1}I-\Sigma)^{-1}N+D$ $=H(I-\Sigma\lambda)^{-1}N\lambda+D$.

Clearly, Eq. (5.94) are a special case of the generalized digital plant equations $T(\lambda)x(\lambda)=U(\lambda)u(\lambda)$, $y(\lambda)=V(\lambda)x(\lambda)+W(\lambda)u(\lambda)$ with system matrix

$$S(\lambda)=\begin{bmatrix}-T(\lambda) & U(\lambda)\\ V(\lambda) & W(\lambda)\end{bmatrix}=\left[\begin{array}{c|cc}-T(\lambda) & U_1(\lambda) & U_2(\lambda)\\ \hline V_1(\lambda) & W_{11}(\lambda) & W_{12}(\lambda)\\ V_2(\lambda) & W_{21}(\lambda) & W_{22}(\lambda)\end{array}\right]. \tag{5.95}$$

In particular, (5.94) corresponds to the case $T(\lambda)=(I-\Sigma\lambda)$, $U(\lambda)=N\lambda$, $V(\lambda)=H$, and $W(\lambda)=D$. Obviously, $T(0)=I$ is nonsingular; hence, the system

(5.94) is nondegenerate and causal (see Sect. A.5). In addition, the characteristic polynomial for the system is $\Delta(\lambda) = \mu \det T(\lambda) = \mu \det(I - \Sigma\lambda)$, where μ is the constant for which $\Delta(\lambda)$ is monic; hence, the system is stable iff the polynomial $\Delta(\lambda)$ is Schur (that is, all zeros of $\Delta(\lambda)$ lie in $|\lambda| > 1$).

In the discrete state-space model, a constant matrix Σ is said to be stable if all its eigenvalues lie in $|\lambda| < 1$ or, equivalently, the polynomial $\det(I - \Sigma\lambda)$ is Schur. A state-space model for $G(\lambda)$ can be conveniently realized from the z- transform version of the transfer matrix given by $\widetilde{G}(z) = G(z^{-1})$ which is obtained from $G(\lambda)$ by setting $\lambda = z^{-1}$. A z-transform rational matrix is state-space realizable iff it is proper; hence, a rational matrix $G(\lambda)$ is state-space realizable iff $G(\lambda)$ is analytic at $\lambda = 0$. That a rational matrix $G(\lambda)$ is not state-space realizable implies that the rational matrix cannot be the transfer matrix of a causal system (see Sect. A.5).

Recognizing the similarity of the digital system described here with the analog case treated in Example 5.9 one gets from Example 5.9(b) that $\Delta(\lambda) = \mu \det T(\lambda) = h_2(\lambda)\psi_{G_{22}}(\lambda)$, where $h_2(\lambda)$ is monic and has no zeros in $\Re$ iff $[T(\lambda) \quad U_2(\lambda)]$ and $[T'(\lambda) \quad V_2'(\lambda)]$ have full row rank in $\Re$.Hence, the system is minimal when $h_2(\lambda) \equiv 1$; this is the case iff $\Re$ is the entire λ-plane or, equivalently, both $[T(\lambda) \quad U_2(\lambda)]$ and $[T'(\lambda) \quad V_2'(\lambda)]$ have full row rank for all λ. When $[T(\lambda) \quad U_2(\lambda)]$ has full row rank for all λ, the pair (T, U_2) is called controllable. When $[T'(\lambda) \quad V_2'(\lambda)]$ has full row rank for all λ, the pair (T, V_2) is called observable. The monic polynomial $h_2(\lambda)$ is Schur iff $\Re$ is the region $|\lambda| \leq 1$ or, equivalently, both $[T(\lambda) \quad U_2(\lambda)]$ and $[T'(\lambda) \quad V_2'(\lambda)]$ have full row rank for all $|\lambda| \leq 1$. When $[T(\lambda) \quad U_2(\lambda)]$ has full row rank for all $|\lambda| \leq 1$, the pair (T, U_2) is called stabilizable. When $[T'(\lambda) \quad V_2'(\lambda)]$ has full row rank for all $|\lambda| \leq 1$, the pair (T, V_2) is called detectable. Clearly, when (T, U_2) is controllable (stabilizable) so is (T, U) and when (T, V_2) is observable (detectable) so is (T, V).

When $T(\lambda) = (I - \Sigma\lambda)$, $U(\lambda) = N\lambda$ is controllable and $T(\lambda) = (I - \Sigma\lambda)$, $V(\lambda) = H$ is observable, the triple (Σ, N, H) is called minimal. In this case, it can be readily shown (see Lemma B.4) that the pairs $I - \Sigma\lambda$, $N\lambda$ and $I - \Sigma\lambda, N$ are left coprime and the pairs $I - \Sigma\lambda, H$ and $I - \Sigma\lambda, H\lambda$ are right coprime. Therefore, when (Σ, N, H) is minimal the characteristic polynomial $\Delta(\lambda) = \mu \det(I - \Sigma\lambda)$ of the system and the characteristic denominator $\psi_G(\lambda)$ of the transfer matrix $G(\lambda) = H(I - \Sigma\lambda)^{-1}N\lambda + D$ are identical. When the triple (Σ, N, H) is not minimal, some poles of $(I - \Sigma\lambda)^{-1}$ are canceled out in the product $H(I - \Sigma\lambda)^{-1}N\lambda$. Hence, the characteristic polynomial $\Delta(\lambda) = \mu \det(I - \Sigma\lambda)$ and the characteristic denominator $\psi_G(\lambda)$ are related by the equation

$$\Delta(\lambda) = \mu \det(I - \Sigma\lambda) = h(\lambda)\psi_G(\lambda), \tag{5.96a}$$

where $h(\lambda)$ is a monic polynomial which accounts for the hidden poles of the system with respect to the transfer matrix $G(\lambda)$. The polynomial $h_2(\lambda)$ accounts for the hidden poles of the system with respect to $G_{22}(\lambda)$ and it is true that $\psi_G =$

$h_1(\lambda)\psi_{G_{22}}$, where $h_1(\lambda)$ is a monic polynomial because $G_{22}(\lambda)$ is a submatrix of $G(\lambda)$. Hence,

$$\Delta(\lambda) = h(\lambda)\psi_G(\lambda) = h(\lambda)h_1(\lambda)\psi_{G_{22}}(\lambda) = h_2(\lambda)\psi_{G_{22}}(\lambda) \Rightarrow h_2(\lambda) = h(\lambda)h_1(\lambda) \tag{5.96b}$$

and $h_2(\lambda)$ contains $h(\lambda)$ as a factor. These observations also follow from the well-known results on equivalence transforms. When (Σ, N, H) is uncontrollable and unobservable, there exists as in the analog case an equivalence transform resulting in

$$HL = [H_1 \quad 0 \mid H_3 \quad 0], \; L^{-1}\Sigma L = \left[\begin{array}{cc|cc} \Sigma_{11} & 0 & \Sigma_{13} & 0 \\ \Sigma_{21} & \Sigma_{22} & \Sigma_{23} & \Sigma_{24} \\ \hline 0 & 0 & \Sigma_{33} & 0 \\ 0 & 0 & \Sigma_{43} & \Sigma_{44} \end{array}\right], \; L^{-1}N = \left[\begin{array}{c} N_1 \\ N_2 \\ \hline 0 \\ 0 \end{array}\right] \tag{5.97}$$

which yields

$$G(\lambda) = H_1(\lambda^{-1}I - \Sigma_{11})^{-1}N_1 + D = H_1(I - \Sigma_{11}\lambda)^{-1}N_1\lambda + D. \tag{5.98}$$

Moreover, the triple (Σ_{11}, N_1, H_1) is controllable and observable and the characteristic polynomial for the system is given by

$$\begin{aligned} \Delta(\lambda) &= \mu \det(I - \Sigma\lambda) \\ &= \mu \det(I - \Sigma_{11}\lambda)\det(I - \Sigma_{22}\lambda)\det(I - \Sigma_{33}\lambda)\det(I - \Sigma_{44}\lambda). \end{aligned} \tag{5.99}$$

Since the triple (Σ_{11}, N_1, H_1) is minimal, it can be concluded from (5.98) that $\psi_G = \mu_G \det(I - \Sigma_{11}\lambda)$ and it follows from (5.96a) and (5.99) that

$$h(\lambda) = \mu_h \det(I - \Sigma_{22}\lambda)\det(I - \Sigma_{33}\lambda)\det(I - \Sigma_{44}\lambda), \tag{5.100}$$

where μ_G and μ_h are the constants for which ψ_G and h are monic. When (Σ, N, H) is controllable and unobservable, or when (Σ, N, H) is uncontrollable and observable, a similar approach can show that the equality in (5.96a) still holds.

When the triple (Σ, N, H) in (5.93) is stabilizable and detectable, the matrices Σ_{22}, Σ_{33}, and Σ_{44} in (5.97) are stable and in this case the polynomial $h(\lambda)$ is Schur. The system in (5.93) is free of unstable hidden poles with respect to the transfer matrix $G(\lambda)$ iff the triple (Σ, N, H) is stabilizable and detectable. Hence, the system is stable iff the transfer matrix $G(\lambda)$ is stable and the triple (Σ, N, H) is stabilizable and detectable. When

$$G(\lambda) = H(I - \Sigma\lambda)^{-1}N\lambda + D = A^{-1}(\lambda)B(\lambda) = B_1(\lambda)A_1^{-1}(\lambda), \tag{5.101}$$

where $A^{-1}B$ and $B_1A_1^{-1}$ denote, respectively, left and right coprime polynomial matrix fraction descriptions for $G(\lambda)$, then

$$\psi_G(\lambda) = \eta \det A(\lambda) = \eta_1 \det A_1(\lambda), \tag{5.102}$$

where η and η_1 are the constants needed for ψ_G to be monic (Lemma B.7). It follows from (5.96a) and (5.102) that

$$\mu \det(I - \Sigma\lambda) = h(\lambda)\psi_G(\lambda) = \eta h(\lambda) \det A(\lambda) = \eta_1 h(\lambda) \det A_1(\lambda). \tag{5.103}$$

The pair (Σ, N) is stabilizable iff $[I - \Sigma\lambda \quad N\lambda]$ or $[I - \Sigma\lambda \quad N]$ has full row rank in $|\lambda| \le 1$. The pair (Σ, H) is detectable iff $[I - \Sigma'\lambda \quad H'\lambda]'$ or $[I - \Sigma'\lambda \quad H']'$ has full column rank in $|\lambda| \le 1$. If (Σ, N_1) or (Σ, N_2) is stabilizable then $(\Sigma, [N_1 \; N_2])$ is stabilizable, and if (Σ, H_1) or (Σ, H_2) is detectable then $(\Sigma, [H_1' \quad H_2']')$ is detectable.

In Sects. 5.4 and 5.5, it is convenient to represent the transfer matrix $G(\lambda)$ associated with the state-space parameters (Σ, N, H, D) as a block matrix distinguished by dashed partition lines. Specifically,

$$G(\lambda) = H(\lambda^{-1}I - \Sigma)^{-1}N + D = H(I - \Sigma\lambda)^{-1}N\lambda + D = \left[\begin{array}{c:c} \Sigma & N \\ \hdashline H & D \end{array}\right]. \tag{5.104}$$

It is not difficult to verify for

$$G(\lambda) = \left[\begin{array}{c:c} \Sigma & N \\ \hdashline H & D \end{array}\right], \; G_1(\lambda) = \left[\begin{array}{c:c} \Sigma_1 & N_1 \\ \hdashline H_1 & D_1 \end{array}\right] \text{ and } G_2(\lambda) = \left[\begin{array}{c:c} \Sigma_2 & N_2 \\ \hdashline H_2 & D_2 \end{array}\right] \tag{5.105}$$

that

$$G(\lambda) = \left[\begin{array}{c:c} \Sigma & N \\ \hdashline H & D \end{array}\right] = \left[\begin{array}{c:c} L^{-1}\Sigma L & L^{-1}N \\ \hdashline HL & D \end{array}\right]; \tag{5.106}$$

$$G^{-1}(\lambda) = \left[\begin{array}{c:c} \Sigma - ND^{-1}H & ND^{-1} \\ \hdashline -D^{-1}H & D^{-1} \end{array}\right] \tag{5.107}$$

where D^{-1} denotes the inverse of D when D is square and nonsingular and the right inverse of D when D has full row rank;

$$G_1(\lambda)G_2(\lambda) = \left[\begin{array}{cc:c} \Sigma_1 & N_1H_2 & N_1D_2 \\ 0 & \Sigma_2 & N_2 \\ \hdashline H_1 & D_1H_2 & D_1D_2 \end{array}\right] = \left[\begin{array}{cc:c} \Sigma_2 & 0 & N_2 \\ N_1H_2 & \Sigma_1 & N_1D_2 \\ \hdashline D_1H_2 & H_1 & D_1D_2 \end{array}\right]; \tag{5.108}$$

$$G_1(\lambda) + G_2(\lambda) = \left[\begin{array}{cc:c} \Sigma_1 & 0 & N_1 \\ 0 & \Sigma_2 & N_2 \\ \hdashline H_1 & H_2 & D_1 + D_2 \end{array}\right]; \tag{5.109}$$

and

$$G_*(\lambda) = \left[\begin{array}{c|c} (\Sigma')^{-1} & (\Sigma')^{-1}H' \\ \hline -N'(\Sigma')^{-1} & D' - N'(\Sigma')^{-1}H' \end{array}\right] \tag{5.110}$$

when Σ is nonsingular. The identity $(\lambda I - \Sigma)^{-1}\lambda - I = (\lambda I - \Sigma)^{-1}\Sigma$ is helpful in the derivation of (5.110).

All definitions and formulas for the LFT and the homographic transformation introduced in Sect. 5.2 for analog systems also apply to digital systems. That is, all the equations from (5.16) to (5.23) with the Laplace variable s replaced by λ are valid for digital systems.

For any constant matrices Σ_a, Σ_b, and M with compatible dimensions, the following equality holds:

$$M - \Sigma_a M \Sigma_b = (\lambda I - \Sigma_a)M(\lambda^{-1}I - \Sigma_b) + \Sigma_a M(\lambda^{-1}I - \Sigma_b) + (\lambda I - \Sigma_a)M\Sigma_b. \tag{5.111}$$

Equation (5.111) also holds when λ is replaced by λ^{-1}. A useful inversion formula is $(N + BMD)^{-1} = N^{-1} - N^{-1}B(DN^{-1}B + M^{-1})^{-1}DN^{-1}$.

5.5 State-Space Representation of the Wiener–Hopf Controllers: The Digital Case

In this section state-space algorithms for the Wiener–Hopf controllers in Chap. 3 are developed for the digital case. In particular, the generalized persistent input standard configuration of Fig. 3.2 with G compatible is considered for the case in which G_w, a weighted $\widetilde{G}_w$, is strictly acceptable.

As explained in Sect. 3.3, when the coprime polynomial fraction descriptions

$$G_{22}(\lambda) = -A^{-1}(\lambda)B(\lambda) = -B_1(\lambda)A_1^{-1}(\lambda) \tag{5.112}$$

associated with the generalized Bezout identity

$$\begin{bmatrix} A & B \\ Y_1 & -X_1 \end{bmatrix}\begin{bmatrix} X & B_1 \\ Y & -A_1 \end{bmatrix} = \begin{bmatrix} I & 0 \\ 0 & I \end{bmatrix} = \begin{bmatrix} X & B_1 \\ Y & -A_1 \end{bmatrix}\begin{bmatrix} A & B \\ Y_1 & -X_1 \end{bmatrix} \tag{5.113}$$

are given, the class of all stabilizing controller transfer matrices is characterized by

$$C(\lambda) = (X_1 - KB)^{-1}(Y_1 + KA) = (Y + A_1K)(X - B_1K)^{-1}, \tag{5.114}$$

where K is a stable rational matrix such that $\det(X - B_1K) \cdot \det(X_1 - KB) \not\equiv 0$. When the system G is compatible, the matrices $G_{12}A_1, AG_{21}$, and

$G_{11} + G_{12}A_1Y_1G_{21}$ are stable. A matrix G_w is strictly acceptable for a compatible G when there exists a stable matrix K such that G_vG_w, $G_\delta G_w$, and G_mG_w are stable where G_δ, G_v, and G_m are the transfer matrices from w to y_δ, v, and y_m, respectively (see Definition 3.3). When G is compatible, a rational matrix G_w is strictly acceptable iff $AG_{21}G_w$ and $(G_{11} + G_{12}A_1Y_1G_{21})G_w$ are stable (see Example 3.21).

The $\mathcal{H}_2$ design problems solved in Chap. 3 with a compatible G in Fig. 3.2 yield all stabilizing controllers that make the performance functional

$$J = \frac{1}{2\pi j} \oint_{|\lambda|=1} \mathrm{Tr}[(WG_\delta G_w)(WG_\delta G_w)_*]\frac{d\lambda}{\lambda} \tag{5.115}$$

finite when the weighting matrix W is stable and G_w is acceptable or strictly acceptable. Here attention is restricted to G_w that are strictly acceptable. The solution for this case can be obtained using the same procedures as employed in Sect. 3.5. In particular, one finds that $\nabla_a = W(X_o\Gamma_w + B_{o1}\Gamma_oX_{g1})B_w$, $\nabla_b = WB_{o1}$, $\nabla_c = B_gB_w$, K_m, and Z in (3.106), (3.107), and (3.117) are instead $\nabla_a = W(G_{11} + G_{12}A_1Y_1G_{21})G_w$, $\nabla_b = WG_{12}A_1$, $\nabla_c = AG_{21}G_w$, $-K$, and $-Z$, respectively. It now easily follows from Theorem 3.2 that the assumptions guaranteeing the existence of a solution and the solution formulas are as follows:

Assumption 5.6 $G(\lambda)$ is compatible and $G_w(\lambda)$ is strictly acceptable.

Assumption 5.7 $\nabla_b(\lambda) = W(\lambda)G_{12}(\lambda)A_1(\lambda)$ has full column rank on $\lambda = e^{j\theta}, \theta \in [0, 2\pi]$.

Assumption 5.8 $\nabla_c(\lambda) = A(\lambda)G_{21}(\lambda)G_w(\lambda)$ has full row rank on $\lambda = e^{j\theta}, \theta \in [0, 2\pi]$.

Theorem 5.2 *Suppose that the matrices Λ and Ω are the square stable Wiener–Hopf spectral factors with stable inverses satisfying*

$$\nabla_{b*}\nabla_b = A_{1*}G_{12*}W_*WG_{12}A_1 = \Lambda_*\Lambda \tag{5.116}$$

and

$$\nabla_c\nabla_{c*} = AG_{21}G_wG_{w*}G_{21*}A_* = \Omega\Omega_*. \tag{5.117}$$

Then all controller transfer matrices C given by (5.114) *for which G_vG_w, $G_\delta G_w$, and G_mG_w are stable and the performance functional J in* (5.115) *is finite are the ones for which*

$$K = \Lambda^{-1}(Z - \{\widetilde{\Gamma}\}_\infty - \{\widetilde{\Gamma}\}_+)\Omega^{-1} \tag{5.118}$$

is chosen with

$$\widetilde{\Gamma} = \Lambda_*^{-1}\nabla_{b*}W(G_{11} + G_{12}A_1Y_1G_{21})G_w\nabla_{c*}\Omega_*^{-1} \tag{5.119}$$

$$= \Gamma_\phi + \Lambda A_1^{-1}Y\Omega, \tag{5.120}$$

where $\Gamma_\phi = \Lambda_*^{-1}\nabla_{b*}WG_{11}G_w\nabla_{c*}\Omega_*^{-1}$ *and* Z *is a stable matrix. The performance functional* J *is minimized when* $Z = 0$.

The stabilizing controllers in (5.114) are described in terms of the polynomial matrices in the generalized Bezout identity (GBI) in (5.113); hence, the same is true for the Wiener–Hopf controllers given by Theorem 5.2. In order to obtain state-space representations for these controllers it is convenient to first establish a stable rational GBI associated with G_{22} using the polynomial GBI in (5.113). Clearly

$$\begin{bmatrix} \Omega^{-1} & 0 \\ -\Gamma_\theta\Omega^{-1} & \Lambda \end{bmatrix}\begin{bmatrix} \Omega & 0 \\ \Lambda^{-1}\Gamma_\theta & \Lambda^{-1} \end{bmatrix} = I = \begin{bmatrix} \Omega & 0 \\ \Lambda^{-1}\Gamma_\theta & \Lambda^{-1} \end{bmatrix}\begin{bmatrix} \Omega^{-1} & 0 \\ -\Gamma_\theta\Omega^{-1} & \Lambda \end{bmatrix}, \tag{5.121}$$

where

$$\Gamma_\theta = \{\widetilde{\Gamma}\}_+ + \{\widetilde{\Gamma}\}_\infty \tag{5.122}$$

$$= \{\Gamma_\phi + \Lambda A_1^{-1}Y\Omega\}_+ + \{\Gamma_\phi + \Lambda A_1^{-1}Y\Omega\}_\infty \tag{5.123}$$

$$= \{\Gamma_\phi\}_+ + \{\Gamma_\phi\}_\infty - \{\Lambda A_1^{-1}Y\Omega\}_- + \Lambda A_1^{-1}Y\Omega. \tag{5.124}$$

Since $\Lambda^{-1}, G_{12}A_1, W, \Omega^{-1}, (G_{11} + G_{12}YAG_{21})G_w$, and $AG_{21}G_w$ are stable, it follows from (5.119) that

$$\widetilde{\Gamma} = \Lambda_*^{-1}A_{1*}G_{12*}W_*W(G_{11} + G_{12}YAG_{21})G_wG_{w*}G_{21*}A_*\Omega_*^{-1} \tag{5.125}$$

is analytic on the unit circle $|\lambda| = 1$; hence, $\{\widetilde{\Gamma}\}_+$ and $\Gamma_\theta = \{\widetilde{\Gamma}\}_+ + \{\widetilde{\Gamma}\}_\infty$ are stable.

Using the equalities in (5.121), one gets from (5.113) that

$$\begin{bmatrix} \Omega^{-1} & 0 \\ -\Gamma_\theta\Omega^{-1} & \Lambda \end{bmatrix}\begin{bmatrix} A & B \\ Y_1 & -X_1 \end{bmatrix}\begin{bmatrix} X & B_1 \\ Y & -A_1 \end{bmatrix}\begin{bmatrix} \Omega & 0 \\ \Lambda^{-1}\Gamma_\theta & \Lambda^{-1} \end{bmatrix} = \begin{bmatrix} I & 0 \\ 0 & I \end{bmatrix} \tag{5.126}$$

and

$$\begin{bmatrix} I & 0 \\ 0 & I \end{bmatrix} = \begin{bmatrix} X & B_1 \\ Y & -A_1 \end{bmatrix} \begin{bmatrix} \Omega & 0 \\ \Lambda^{-1}\Gamma_\theta & \Lambda^{-1} \end{bmatrix} \begin{bmatrix} \Omega^{-1} & 0 \\ -\Gamma_\theta\Omega^{-1} & \Lambda \end{bmatrix} \begin{bmatrix} A & B \\ Y_1 & -X_1 \end{bmatrix} \tag{5.127}$$

which leads to

$$\begin{bmatrix} \widetilde{A} & \widetilde{B} \\ \widetilde{Y}_1 & -\widetilde{X}_1 \end{bmatrix} \begin{bmatrix} \widetilde{X} & \widetilde{B}_1 \\ \widetilde{Y} & -\widetilde{A}_1 \end{bmatrix} = \begin{bmatrix} I & 0 \\ 0 & I \end{bmatrix} = \begin{bmatrix} \widetilde{X} & \widetilde{B}_1 \\ \widetilde{Y} & -\widetilde{A}_1 \end{bmatrix} \begin{bmatrix} \widetilde{A} & \widetilde{B} \\ \widetilde{Y}_1 & -\widetilde{X}_1 \end{bmatrix}, \tag{5.128}$$

where

$$\widetilde{A} = \Omega^{-1}A, \widetilde{B} = \Omega^{-1}B, \widetilde{A}_1 = A_1\Lambda^{-1}, \widetilde{B}_1 = B_1\Lambda^{-1}, \tag{5.129}$$

$$\widetilde{X} = X\Omega + B_1\Lambda^{-1}\Gamma_\theta = A^{-1}\Omega + B_1\Lambda^{-1}(\{\Gamma_\phi\}_+ + \{\Gamma_\phi\}_\infty - \{\Lambda A_1^{-1}Y\Omega\}_-), \tag{5.130}$$

$$\widetilde{Y} = Y\Omega - A_1\Lambda^{-1}\Gamma_\theta = -A_1\Lambda^{-1}(\{\Gamma_\phi\}_+ + \{\Gamma_\phi\}_\infty - \{\Lambda A_1^{-1}Y\Omega\}_-), \tag{5.131}$$

$$\widetilde{X}_1 = \Lambda X_1 + \Gamma_\theta\Omega^{-1}B = \Lambda A_1^{-1} + (\{\Gamma_\phi\}_+ + \{\Gamma_\phi\}_\infty - \{\Lambda A_1^{-1}Y\Omega\}_-)\Omega^{-1}B, \tag{5.132}$$

and

$$\widetilde{Y}_1 = \Lambda Y_1 - \Gamma_\theta\Omega^{-1}A = -(\{\Gamma_\phi\}_+ + \{\Gamma_\phi\}_\infty - \{\Lambda A_1^{-1}Y\Omega\}_-)\Omega^{-1}A. \tag{5.133}$$

It is also clear that

$$G_{22} = -\widetilde{A}^{-1}\widetilde{B} = -\widetilde{B}_1\widetilde{A}_1^{-1}. \tag{5.134}$$

Lemma 5.6 *When Assumptions 5.6, 5.7, and 5.8 are satisfied, then the expression in (5.128) is a stable rational GBI associated with G_{22}.*

Proof By Assumptions 5.6, 5.7, and 5.8, the matrices Λ^{-1} and Ω^{-1} are analytic in $|\lambda| \le 1$. Thus, $\widetilde{X}_1, \widetilde{X}, \widetilde{Y}_1$, and $\widetilde{Y}$ are obviously stable since $X_1, X, Y_1, Y, \Lambda, \Lambda^{-1}, \Omega, \Omega^{-1}$, and Γ_θ are stable.

The eight matrices $\widetilde{A}, \widetilde{B}, \widetilde{X}, \widetilde{Y}, \widetilde{A}_1, \widetilde{B}_1, \widetilde{X}_1$, and $\widetilde{Y}_1$ in (5.129) through (5.133) are stable rational ones; hence, they are analytic at $\lambda = 0$ and have state-space realizations which are established in the sequel.

First, the Wiener–Hopf controllers of Theorem 5.2 are described in terms of the stable rational matrices in (5.128). It is clear from (5.118) and (5.122) that

$$K = \Lambda^{-1}(Z - \{\widetilde{\Gamma}\}_\infty - \{\widetilde{\Gamma}\}_+)\Omega^{-1} = \Lambda^{-1}(Z - \Gamma_\theta)\Omega^{-1}. \tag{5.135}$$

It now follows from (5.114) and (5.135) that

$$C = \left(Y + A_1\Lambda^{-1}(Z - \Gamma_\theta)\Omega^{-1}\right)\left(X - B_1\Lambda^{-1}(Z - \Gamma_\theta)\Omega^{-1}\right)^{-1} \tag{5.136}$$

$$= \left(Y\Omega + A_1\Lambda^{-1}(Z - \Gamma_\theta)\right)\left(X\Omega - B_1\Lambda^{-1}(Z - \Gamma_\theta)\right)^{-1} \tag{5.137}$$

$$= (\widetilde{Y} + \widetilde{A}_1 Z)(\widetilde{X} - \widetilde{B}_1 Z)^{-1} = \begin{bmatrix} \widetilde{A}_1 & \widetilde{Y} \\ -\widetilde{B}_1 & \widetilde{X} \end{bmatrix} \diamond Z. \tag{5.138}$$

Just as in the analog case, an alternative form for the controller matrix in (5.138) is given by

$$C = (\widetilde{Y} + \widetilde{A}_1 Z)(\widetilde{X} - \widetilde{B}_1 Z)^{-1} = (\widetilde{X}_1 - Z\widetilde{B})^{-1}(\widetilde{Y}_1 + Z\widetilde{A}). \tag{5.139}$$

The next step is the derivation of a state-space realization for the controller. Specifically, given

$$G_N = \begin{bmatrix} WG_{11}G_w & WG_{12} \\ G_{21}G_w & G_{22} \end{bmatrix} = \left[\begin{array}{c|cc} \Sigma & N_1 & N_2 \\ \hline H_1 & D_{11} & D_{12} \\ H_2 & D_{21} & D_{22} \end{array}\right] \tag{5.140}$$

the objective is the determination of a state-space realization free of any unstable hidden poles for the controller transfer matrix in (5.138). Moreover, since G is compatible, the response to initial data associated with the physical components within G is stabilized by C. So there is no need to be concerned about this issue in regard to (5.140) although it makes good sense that the state-space model for G_N be consistent with $G_{22} = H_2(I - \Sigma\lambda)^{-1}N_2\lambda + D_{22}$ being stabilized by C. Specifically, with $\Delta = \mu \det(I - \Sigma\lambda)$, it is assumed that $h_2 = \Delta/\psi_{G_{22}}$ is a stable polynomial or, equivalently, (Σ, N_2) is stabilizable and (Σ, H_2) is detectable. This equivalence can be established in the digital case as in the solution of Example 5.9(b) for the analog case. It is shown in the sequel for this case that the derived state-space realization for C is indeed free of unstable hidden poles. In addition, it is established later that the order of the state-space realization for the optimal Wiener–Hopf controller is equal to that of the matrix Σ. Therefore, choosing the lowest order possible for Σ is desirable in this regard. However, the constraint that the controller realization be minimal is not imposed here for more flexibility.

Constraints on the realization in (5.140) imposed by Assumptions 5.6 through 5.8 are summarized in the following lemma.

Lemma 5.7 The *frequency-domain conditions in Assumptions 5.6 through 5.8 impose constraints on the state-space parameters in (5.140). In particular, the following assumptions must be satisfied:*

Assumption 5.6′ (Σ, N_2, H_2) is stabilizable and detectable.

Assumption 5.7′ The matrix $S_{12}(e^{j\theta}) = \begin{bmatrix} \Sigma - e^{j\theta}I & N_2 \\ H_1 & D_{12} \end{bmatrix}$ has full column rank for all $\theta \in [0, 2\pi]$.

Assumption 5.8′ The matrix $S_{21}(e^{j\theta}) = \begin{bmatrix} \Sigma - e^{j\theta}I & N_1 \\ H_2 & D_{21} \end{bmatrix}$ has full row rank for all $\theta \in [0, 2\pi]$.

Proof The need for Assumption 5.6′ has already been discussed. Assumption 5.7′ is considered next. In particular, it is now established under Assumption 5.6′ that Assumption 5.7′ holds iff Assumption 5.7 holds or, equivalently, $\det(A_{1*}G_{12*}W_*WG_{12}A_1) \neq 0$ on $|\lambda| = 1$. One can always write $W(\lambda)G_{12}(\lambda) = B_{p1}(\lambda)A_{p1}^{-1}(\lambda)$, and $G_{22}(\lambda) = -B_1(\lambda)A_1^{-1}(\lambda)$ with A_{p1}, B_{p1} and A_1, B_1 right coprime polynomial matrix pairs. It then clearly follows from Assumption 5.6′ and $\det(I - \Sigma\lambda) = h_{p1}(\lambda)\det A_{p1}(\lambda) = h_{p2}(\lambda)\det A_1(\lambda)$ that the polynomial $h_{p2}(\lambda)$ must be Schur. Moreover, it is easy to confirm that, $\det(A_{1*}G_{12*}W_*WG_{12}A_1) = h_{p1*}\det(B_{p1*}B_{p1})h_{p1}/(h_{p2*}h_{p2})$. Since h_{p2} is Schur, it then follows that $\det(A_{1*}G_{12*}W_*WG_{12}A_1)$ has no zeros on $|\lambda| = 1$ iff both h_{p1} and $\det(B_{p1*}B_{p1})$ have no zeros on $|\lambda| = 1$. Now $[\,(I - \Sigma\lambda) \quad N_2\,]$ has full row rank on $|\lambda| = 1$ because (Σ, N_2) is stabilizable. In this case, h_{p1} has no zeros on $|\lambda| = 1$ iff $[\,(I - \Sigma'\lambda) \quad H_1'\,]$ has full row rank on $|\lambda| = 1$ (follows from the digital version of Example 5.9(b) with $\Re$ the unit circle, $|\lambda| = 1$). Also, $\det(B_{p1*}B_{p1})$ has no zeros on $|\lambda| = 1$ iff $B_{p1}(\lambda)$ has full column rank on $|\lambda| = 1$ or, equivalently, $W(\lambda)G_{12}(\lambda)$ has no transmission zeros on $|\lambda| = 1$. It then follows (from the digital version of Example 5.9(a) with $\Re$ the unit circle) that $\mathrm{diag}\{I, B_{p1}(e^{j\theta})\}$ and $S_{12}(e^{j\theta})$ have the same rank; hence, $S_{12}(e^{j\theta})$ has full column rank for all $0 \leq \theta \leq 2\pi$. That is, under Assumption 5.6′, the Assumptions 5.7 and 5.7′ are equivalent. The proof of the equivalence of Assumption 5.8 and Assumption 5.8′ is similar and is omitted.

In addition to Assumptions 5.6′, 5.7′, and 5.8′, the following assumption is added to assure the existence of solutions for discrete-time algebraic Riccati equations (In Example 5.6, it is shown that this additional assumption is not needed for the frequency-domain solution given by Theorem 5.2).

Assumption 5.9′ D_{12} has full column rank and D_{21} has full row rank.

It is now possible to state a lemma which gives a state-space representation for the stable rational matrix $\widetilde{A}_1$ under the Assumptions 5.6′ through 5.9′.

Lemma 5.8 *Suppose that Assumptions 5.6′ through 5.9′ are satisfied and that*

$$(\lambda^{-1}I - \Sigma)^{-1}N_2 = B_2(\lambda)A_2^{-1}(\lambda) \tag{5.141}$$

denotes a right coprime polynomial fraction description for $(\lambda^{-1}I-\Sigma)^{-1}N_2$. *Then the Wiener–Hopf factor* $\Lambda(\lambda)$ *in (5.116) is given by*

$$\Lambda(\lambda)=E_1^{\frac{1}{2}}(A_2(\lambda)+L_1B_2(\lambda))V_1^{-1}=E_1^{\frac{1}{2}}(A_1(\lambda)+L_1B_2(\lambda)V_1^{-1}(\lambda)), \tag{5.142}$$

where $V_1(\lambda)=A_1^{-1}(\lambda)A_2(\lambda)$ *is a polynomial matrix whose determinant is Schur and* E_1 *and* L_1 *are given below in (5.145). Moreover, state-space parameters for the matrix* $\widetilde{A}_1(\lambda)$ *follow from the formula*

$$\widetilde{A}_1(\lambda)=A_1(\lambda)\Lambda^{-1}(\lambda)=\left(I+L_1(\lambda^{-1}I-\Sigma)^{-1}N_2\right)^{-1}E_1^{-\frac{1}{2}} \tag{5.143}$$

$$=\left(I-L_1(\lambda^{-1}I-\Sigma+N_2L_1)^{-1}N_2\right)E_1^{-\frac{1}{2}}, \tag{5.144}$$

where

$$E_1=D_{12}'D_{12}+N_2'M_1N_2,\ L_1=E_1^{-1}(D_{12}'H_1+H_2'M_1\Sigma). \tag{5.145}$$

With $\Sigma_N=\Sigma-N_2(D_{12}'D_{12})^{-1}D_{12}'H_1$, *the matrix* M_1 *is the unique positive semi-definite solution of the discrete-time algebraic Riccati equation (ARE)*

$$\Sigma_N'M_1(I+N_2(D_{12}'D_{12})^{-1}N_2'M_1)^{-1}\Sigma_N+H_1'(I-D_{12}(D_{12}'D_{12})^{-1}D_{12}')H_1-M_1=0 \tag{5.146}$$

with the property that $(I+N_2(D_{12}'D_{12})^{-1}N_2'M_1)^{-1}\Sigma_N=(I-N_2E_1^{-1}N_2'M_1)\Sigma_N=\Sigma-N_2L_1$ *is stable. The discrete-time ARE in (5.146) simplifies to the equation (see Example 5.7)*

$$\Sigma'M_1\Sigma-L_1'E_1L_1+H_1'H_1-M_1=0. \tag{5.147}$$

Proof Since $G_{22}=H_2(\lambda^{-1}I-\Sigma)^{-1}N_2+D_{22}=-B_1(\lambda)A_1^{-1}(\lambda)$, $(\lambda^{-1}I-\Sigma)^{-1}N_2=B_2(\lambda)A_2^{-1}(\lambda)$, the pair (Σ,N_2) is stabilizable, and the pair (Σ,H_2) is detectable, it follows that $\Delta(\lambda)=\mu\det(I-\Sigma\lambda)=\eta_2h_2(\lambda)\det A_1(\lambda)=\eta_3h_3(\lambda)\det A_2(\lambda)$, where $h_2(s)$ and $h_3(s)$ are monic Schur polynomials and the constants η_2 and η_3 are the ones for which $\Delta(\lambda)$ is monic. In addition, it is clear that $-B_1A_1^{-1}A_2=G_{22}A_2=H_2B_2+D_{22}A_2$ is a polynomial matrix. Hence,

$$V_1(\lambda)=A_1^{-1}A_2=(X_1+Y_1B_1A_1^{-1})A_2=X_1A_2+Y_1B_1A_1^{-1}A_2 \tag{5.148}$$

is a polynomial matrix and $\det V_1 = \det A_2/\det A_1 = (\eta_2 h_2)/(\eta_3 h_3)$ is a Schur polynomial. Moreover, since (Σ, N_2) is stabilizable and Assumption 5.7′ is satisfied, the discrete-time ARE in (5.146) has a unique solution $M_1 \geq 0$ with the property that $(I + N_2(D_{12}'D_{12})^{-1}N_2'M_1)^{-1}\Sigma_N$ is stable (Zhou et al. 1996; Corollary 21.13). A little algebra yields from (5.146) that (see Example 5.7)

$$G_{12*}(\lambda)W_*(\lambda)W(\lambda)G_{12}(\lambda) = \left(L_1(\lambda^{-1}I - \Sigma)^{-1}N_2 + I\right)_* E_1 \left(L_1(\lambda^{-1}I - \Sigma)^{-1}N_2 + I\right) \tag{5.149}$$

$$= (L_1 B_2 A_2^{-1} + I)_* E_1 (L_1 B_2 A_2^{-1} + I) \tag{5.150}$$

and this leads to

$$A_{1*}G_{12*}W_*WG_{12}A_1 = (A_1 + L_1 B_2 V_1^{-1})_* E_1 (A_1 + L_1 B_2 V_1^{-1}) \tag{5.151}$$

$$= V_{1*}^{-1}(A_2 + L_1 B_2)_* E_1 (A_2 + L_1 B_2) V_1^{-1}. \tag{5.152}$$

It is easy to prove that $(A_2 + L_1 B_2)^{-1}$ is analytic in $|\lambda| \leq 1$. In fact, $\det(I + L_1(\lambda^{-1}I - \Sigma)^{-1}N_2) = \det(I + (I - \Sigma\lambda)^{-1}\lambda N_2 L_1) = \det(I - (\Sigma - N_2 L_1)\lambda)/\det(I - \Sigma\lambda)$ and at the same time $\det(I + L_1(\lambda^{-1}I - \Sigma)^{-1}N_2) = \det(I + L_1 B_2 A_2^{-1}) = \det(A_2 + L_1 B_2)h_2/\det(sI - \Sigma)$. Hence, $h_2 \det(A_2 + L_1 B_2) = \det(I - (\Sigma - N_2 L_1)\lambda)$ is Schur and $(A_2 + L_1 B_2)^{-1}$ is analytic in $|\lambda| \leq 1$. Thus, $\Lambda(\lambda) = E_1^{\frac{1}{2}}(A_2 + L_1 B_2)V_1^{-1}$ is a Wiener–Hopf factor. Finally,

$$\begin{aligned}\tilde{A}_1(\lambda) &= A_1(\lambda)\Lambda^{-1}(\lambda) = A_1 V_1 (L_1 B_2 + A_2)^{-1} E_1^{-\frac{1}{2}} = A_2 (L_1 B_2 + A_2)^{-1} E_1^{-\frac{1}{2}} \\ &= (L_1 B_2 A_2^{-1} + I)^{-1} E_1^{-\frac{1}{2}}\end{aligned} \tag{5.153}$$

$$= \left(I + L_1(\lambda^{-1}I - \Sigma)^{-1}N_2\right)^{-1} E_1^{-\frac{1}{2}} = (I - L_1(\lambda^{-1}I - \Sigma + N_2 L_1)^{-1}N_2)E_1^{-\frac{1}{2}}. \tag{5.154}$$

In a similar way, state-space parameters for $\tilde{A}$ can be obtained and the results are presented in the following lemma whose proof is omitted.

Lemma 5.9 *Suppose that Assumptions 5.6′ through 5.9′ are satisfied and that*

$$H_2(\lambda^{-1}I - \Sigma)^{-1} = A_3^{-1}(\lambda)B_3(\lambda) \tag{5.155}$$

denotes a left coprime polynomial fraction description for $H_2(\lambda^{-1}I - \Sigma)^{-1}$. *Then the Wiener–Hopf factor* $\Omega(\lambda)$ *in (5.117) is obtained from*

$$\Omega(\lambda) = V_2^{-1}(\lambda)(A_3(\lambda) + B_3(\lambda)L_2)E_2^{\frac{1}{2}} = (A(\lambda) + V_2^{-1}(\lambda)B_3(\lambda)L_2)E_2^{\frac{1}{2}}, \tag{5.156}$$

where E_2 *and* L_2 *are given below in (5.160) and where* $V_2(\lambda)$ *is a polynomial matrix given by*

$$V_2 = A_3A^{-1} \tag{5.157}$$

which possesses a determinant that is Schur. The state-space parameters of the rational matrix $\widetilde{A}(\lambda)$ *follow from*

$$\widetilde{A}(\lambda) = \Omega^{-1}(\lambda)A(\lambda) = E_2^{-\frac{1}{2}}\left(I + H_2(\lambda^{-1}I - \Sigma)^{-1}L_2\right)^{-1} \tag{5.158}$$

$$= E_2^{-\frac{1}{2}}\left(I - H_2(\lambda^{-1}I - \Sigma + L_2H_2)^{-1}L_2\right), \tag{5.159}$$

where

$$E_2 = D_{21}D_{21}' + H_2M_2H_2', L_2 = (N_1D_{21}' + \Sigma M_2H_2')E_2^{-1}. \tag{5.160}$$

The matrix M_2 *in (5.160) is a solution of the discrete-time ARE*

$$\Sigma_H(I + M_2H_2'(D_{21}D_{21}')^{-1}H_2)^{-1}M_2\Sigma_H' + N_1(I - D_{21}'(D_{21}D_{21}')^{-1}D_{21})N_1' - M_2 = 0, \tag{5.161}$$

where $\Sigma_H = \Sigma - N_1D_{21}'(D_{21}D_{21}')^{-1}H_2$. Specifically, M_2 *is the unique positive semi-definite solution for which* $\Sigma_H(I + M_2H_2'(D_{21}D_{21}')^{-1}H_2)^{-1} = \Sigma - L_2H_2$ *is stable. The discrete-time ARE in (5.161) simplifies to the equation*

$$\Sigma M_2\Sigma' - L_2E_2L_2' + N_1N_1' - M_2 = 0. \tag{5.162}$$

(As is done in Example 5.11, *it can be shown from Corollary 21.13 in Zhou et al. (1996) that when* (Σ, H_2) *is detectable and Assumption 5.8′ is satisfied, the discrete-time ARE in (5.161) has the unique solution* $M_2 \geq 0$ *with the property that* $\Sigma - L_2H_2$ *is stable.)*

The following formula is the final preliminary step needed in order to establish a state-space representation for the stable proper rational GBI in (5.128). Its proof is given in the solution of Example 5.8.

Lemma 5.10

$$\{\Gamma_\phi\}_+ + \{\Gamma_\phi\}_\infty - \{\Lambda A_1^{-1} Y \Omega\}_- = E_1^{\frac{1}{2}}\{L_1(\lambda^{-1}I - \Sigma)^{-1}L_2 + D_\phi\}E_2^{\frac{1}{2}}, \quad (5.163)$$

where

$$D_\phi = (E_1^{-1}(D'_{12}D_{11} + N'_2M_1N_1)D'_{21} + L_1M_2H'_2)E_2^{-1}. \quad (5.164)$$

The state-space parameters for $\widetilde{A}_1(\lambda) = A_1(\lambda)\Lambda^{-1}(\lambda)$, $\widetilde{A}(\lambda) = \Omega^{-1}(\lambda)A(\lambda)$, and $\{\Gamma_\phi\}_+ + \{\Gamma_\phi\}_\infty - \{\Lambda A_1^{-1}Y\Omega\}_-$ enable one to find state-space parameters for the eight matrices in (5.128). State-space parameters for $\widetilde{B} = \Omega^{-1}B$ and $\widetilde{B}_1 = B_1\Lambda^{-1}$ are easily obtained from the identities $\widetilde{B} = \Omega^{-1}B = -(\Omega^{-1}A)G_{22}$ and $\widetilde{B}_1 = B_1\Lambda^{-1} = -G_{22}(A_1\Lambda^{-1})$. As for $\widetilde{Y}$, $\widetilde{Y}_1$, $\widetilde{X}$, and $\widetilde{X}_1$, the identities in (5.130) through (5.133) are used. Employing the formulas in (5.106) through (5.109) then yields after straightforward algebra the state-space parameters for the eight matrices in (5.128):

$$\begin{bmatrix} \widetilde{A} & \widetilde{B} \\ \widetilde{Y}_1 & -\widetilde{X}_1 \end{bmatrix} = \left[\begin{array}{c:cc} \Sigma - L_2H_2 & L_2 & N_2 - L_2D_{22} \\ \hdashline -E_2^{-\frac{1}{2}}H_2 & E_2^{-\frac{1}{2}} & -E_2^{-\frac{1}{2}}D_{22} \\ E_1^{\frac{1}{2}}(D_\phi H_2 - L_1) & -E_1^{\frac{1}{2}}D_\phi & -E_1^{\frac{1}{2}}(I - D_\phi D_{22}) \end{array}\right] \quad (5.165)$$

and

$$\begin{bmatrix} \widetilde{X} & \widetilde{B}_1 \\ \widetilde{Y} & -\widetilde{A}_1 \end{bmatrix} = \left[\begin{array}{c:cc} \Sigma - N_2L_1 & (N_2D_\phi - L_2)E_2^{\frac{1}{2}} & N_2E_1^{-\frac{1}{2}} \\ \hdashline D_{22}L_1 - H_2 & (I - D_{22}D_\phi)E_2^{\frac{1}{2}} & -D_{22}E_1^{-\frac{1}{2}} \\ L_1 & -D_\phi E_2^{\frac{1}{2}} & -E_1^{-\frac{1}{2}} \end{array}\right]. \quad (5.166)$$

As in the analog case, one can again use (5.19) to change the homographic form of the Wiener–Hopf controllers in (5.138) into the LFT form

$$C(\lambda) = G_c(\lambda)\square Z(\lambda) \quad (5.167)$$

with

$$G_c(\lambda) = \begin{bmatrix} \widetilde{Y}\widetilde{X}^{-1} & \widetilde{A}_1 + \widetilde{Y}\widetilde{X}^{-1}\widetilde{B}_1 \\ \widetilde{X}^{-1} & \widetilde{X}^{-1}\widetilde{B}_1 \end{bmatrix} = \begin{bmatrix} \widetilde{X}_1^{-1}\widetilde{Y}_1 & \widetilde{X}_1^{-1} \\ \widetilde{A} + \widetilde{B}\widetilde{X}_1^{-1}\widetilde{Y}_1 & \widetilde{B}\widetilde{X}_1^{-1} \end{bmatrix}. \quad (5.168)$$

The key results needed to prove the following theorem are now available.

Theorem 5.3 *Suppose that Assumptions 5.6′ through 5.9′ are satisfied and suppose that the inverse* $(I - D_{22}D_{\phi})^{-1}$ *exists. Then the state-space Wiener–Hopf controllers yielding finite cost in Theorem 5.2 are given by (5.167) with*

$$G_c(\lambda) = \left[\begin{array}{c|cc} \Sigma_c & N_{c1} & N_{c2} \\ \hline H_{c1} & D_{c11} & D_{c12} \\ H_{c2} & D_{c21} & D_{c22} \end{array}\right] \tag{5.169}$$

where

$$\Sigma_c = \Sigma - N_2L_1 - (L_2 - N_2D_{\phi})(I - D_{22}D_{\phi})^{-1}(H_2 - D_{22}L_1), \tag{5.170}$$

$$N_{c1} = (N_2D_{\phi} - L_2)(I - D_{22}D_{\phi})^{-1}, \tag{5.171}$$

$$N_{c2} = -\left(N_2 + (N_2D_{\phi} - L_2)(I - D_{22}D_{\phi})^{-1}D_{22}\right)E_1^{-\frac{1}{2}}, \tag{5.172}$$

$$H_{c1} = L_1 - D_{\phi}(I - D_{22}D_{\phi})^{-1}(H_2 - D_{22}L_1), \tag{5.173}$$

$$H_{c2} = E_2^{-\frac{1}{2}}(I - D_{22}D_{\phi})^{-1}(H_2 - D_{22}L_1), \tag{5.174}$$

$$D_{c11} = -D_{\phi}(I - D_{22}D_{\phi})^{-1}, D_{c12} = (I - D_{\phi}D_{22})^{-1}E_1^{-\frac{1}{2}}, \tag{5.175}$$

$$D_{c21} = E_2^{-\frac{1}{2}}(I - D_{22}D_{\phi})^{-1}, D_{c22} = -E_2^{-\frac{1}{2}}(I - D_{22}D_{\phi})^{-1}D_{22}E_1^{-\frac{1}{2}}. \tag{5.176}$$

The implementation of $C(\lambda)$ *in (5.167) with the realization of* $G_c(\lambda)$ *in (5.169) is free of unstable hidden poles. The optimal controller minimizing the cost is given by*

$$C_{opt} = \widetilde{Y}\widetilde{X}^{-1} = \left[\begin{array}{cc} \Sigma_c & N_{c1} \\ H_{c1} & D_{c11} \end{array}\right] = G_{c11}. \tag{5.177}$$

Proof The state-space parameters for $G_c(\lambda)$ in (5.169) can be obtained from the state-space parameters in (5.166) by using the digital equivalent of (5.90a) in (5.138) and employing (5.20) and (5.21). The formula in (5.177) is obtained from (5.139) with $Z = 0$ and then making use of (5.168). In Sect. 2.7, it is shown that the implementation of (5.167) with (5.168) is free of unstable poles iff the realization of (5.169) is free of unstable hidden poles (it is assumed that $Z(\lambda)$ is realized without unstable hidden poles). Hence, it only remains to show that the realization of (5.169) is free of unstable hidden poles. It is first shown that (Σ_c, N_{c1}) is a stabilizable pair. Since $\Sigma - N_2L_1$ is a stable matrix, the pair $(\Sigma - N_2L_1, N_{c1})$ is stabilizable, or equivalently, $[\,I - (\Sigma - N_2L_1)\lambda \quad N_{c1}\,]$ has row rank in $|\lambda| \geq 1$. Since

$$[I-\Sigma_c\lambda \quad N_{c1}] = [I-(\Sigma-N_2L_1)\lambda \quad N_{c1}]\begin{bmatrix} I & 0 \\ (H_2-D_{22}L_1)\lambda & I \end{bmatrix}, \tag{5.178}$$

it follows that $rank[I-\Sigma_c\lambda \quad N_{c1}] = rank[I-(\Sigma-N_2L_1)\lambda \quad N_{c1}]$ in $|\lambda|\geq 1$; hence, (Σ_c, N_{c1}) is stabilizable. Next it is shown that (Σ_c, H_{c2}) is a detectable pair. Since $\Sigma-N_2L_1$ is a stable matrix, the pair $(\Sigma-N_2L_1, H_{c2})$ is detectable or, equivalently, $[I-(\Sigma-N_2L_1)'\lambda \quad H'_{c2}]'$ has column rank in $|\lambda|\geq 1$. Since

$$\begin{bmatrix} I-\Sigma_c\lambda \\ H_{c2} \end{bmatrix} = \begin{bmatrix} I & (L_2-N_2D_\phi)E_2^{\frac{1}{2}}\lambda \\ 0 & I \end{bmatrix} = \begin{bmatrix} I-(\Sigma-N_2L_1)\lambda \\ H_{c2} \end{bmatrix}, \tag{5.179}$$

it follows that $rank[I-\Sigma_c'\lambda \quad H'_{c2}]' = rank[I-(\Sigma-N_2L_1)'\lambda \quad H'_{c2}]'$ in $|\lambda|\geq 1$; hence, (Σ_c, H_{c2}) is detectable. The condition that (Σ_c, N_{c1}) is stabilizable and (Σ_c, H_{c2}) is detectable assures that $(\Sigma_c, [N_{c1}\ N_{c2}])$ is stabilizable and $(\Sigma_c, [H'_{c1}\ H'_{c2}]')$ is detectable, which implies that the realization of $G_c(\lambda)$ in (5.169) has no unstable hidden poles (follows from the digital equivalent of Example 5.9(b)).

Remark 5.3 A rational matrix in λ is state-space realizable iff it is analytic at $\lambda=0$. From (5.166),

$$\widetilde{X}(\lambda) = (D_{22}L_1-H_2)(I-(\Sigma-N_2L_1)\lambda)^{-1}(N_2D_\phi-L_2)E_2^{\frac{1}{2}}\lambda + (I-D_{22}D_\phi)E_2^{\frac{1}{2}}. \tag{5.180}$$

Therefore, $\widetilde{X}(0) = (I-D_{22}D_\phi)E_2^{\frac{1}{2}}$ and it is concluded that if $\det(I-D_{22}D_\phi)\neq 0$, $\widetilde{X}^{-1}(\lambda)$ exists and it is analytic at $\lambda=0$. When $\det(I-D_{22}D_\phi)=0$, the matrix $\widetilde{X}^{-1}(\lambda)$, if it exists, is not analytic at $\lambda=0$. In summary, when the transfer matrix $G_N(\lambda)$ has the state-space realization in (5.140), the optimal Wiener–Hopf controller $C_{opt}=\widetilde{Y}\widetilde{X}^{-1}$ is state-space realizable iff $\det(I-D_{22}D_\phi)\neq 0$. A sufficient condition for the existence of $(I-D_{22}D_\phi)^{-1}$ is that $D_{22}=0$; in this case, $G_c(\lambda)$ and $C_{opt}(\lambda)$ simplify to

$$G_c(\lambda) = \left[\begin{array}{c|cc} \Sigma-N_2L_1-L_2H_2+N_2D_\phi H_2 & N_2D_\phi-L_2 & -N_2E_1^{-\frac{1}{2}} \\ \hline L_1-D_\phi H_2 & -D_\phi & E_1^{-\frac{1}{2}} \\ E_2^{-\frac{1}{2}}H_2 & E_2^{-\frac{1}{2}} & 0 \end{array}\right], \tag{5.181}$$

$$C_{opt}(\lambda) = \tilde{Y}\tilde{X}^{-1} = \left[\begin{array}{c|c} \Sigma-N_2L_1-L_2H_2+N_2D_\phi H_2 & N_2D_\phi-L_2 \\ \hline L_1-D_\phi H_2 & -D_\phi \end{array}\right]. \tag{5.182}$$

As for checking compatibility of a given system in the digital state-space domain, Remark 5.1 for the analog case is also valid for the digital case.

5.6 Historical Perspective and Commentary

Compared to the frequency-domain solutions, the state-space algorithms for the $\mathcal{H}_2$ controllers are more efficient in computation, but are limited in the analog case to proper plant systems. Whenever possible, it is useful to solve a feedback control problem in both domains so that more design insight is obtained.

A unifying approach for finding the equivalence of the state-space and frequency-domain solutions for $\mathcal{H}_2$ design problems can be seen in the works of Shaked (1976) and Safonov and Sideris (1985). They show that the solution in one-domain can be derived from the other domain solution by using the equations connecting the spectral factorizations and algebraic Riccati equations. In these earlier works, the 1DOF configuration is considered and some unnecessary assumptions are imposed on the plant transfer matrix. Park and Youla (1992) find the state-space representation of the Wiener–Hopf controllers for the 3-degree-of-freedom model with less restrictive assumptions on the plant than in Shaked (1976) and Safonov and Sideris (1985). State-space algorithms which yield the Wiener–Hopf controllers for the generalized plant model are contained in Park and Choi (1998). In their paper, the derivation of a stable rational generalized Bezout identity (GBI) from a polynomial GBI is developed and used to find the state-space representations for the Wiener–Hopf controllers. In Park, Cho and Kim (2002), the state-space formulas for generalized 2-degree-of-freedom Wiener–Hopf controllers are obtained and the equivalent state-space-domain conditions corresponding to frequency-domain assumptions are found as well. The material in Sect. 5.3 of this chapter is based on Park and Youla (1992), Park and Choi (1998), and Park, Cho and Kim (2002). The assumption that the triple (Σ, N_2, H_2) is minimal in Park and Choi (1998) is loosened here to the one that (Σ, N_2, H_2) is stabilizable and detectable. Derivation of stable rational GBI from polynomial GBI was first introduced by Sugimoto and Yamamoto (1990) and the Wiener–Hopf approach was presented in Park, et al. (1999).

The state-space solution of the $\mathcal{H}_2$ design problem in observer form using two algebraic Riccati equations is contained in Doyle, et al. (1989) for the analog case and in Chen and Francis (1994) for the digital case. The state-space solution reveals the separation property of state estimation and state regulation. Though the state-space formulas in Doyle, et al. (1989) and those presented here look the same, the set-up here is different from that of Doyle, et al. (1989) in that the weighting and input signals are separated from the physical system; thus, the treatment of unstable inputs are clarified which is not the case in Doyle, et al. (1989).

State-space representation of $\mathcal{H}_2$ controllers for acceptable G_w is an open problem. Specifically, state-space algorithms are not yet available for calculating $\mathcal{H}_2$ controllers in the case of 2DOF and 3DOF configurations with unstable inputs including step or ramp signals. When in the analog case, the unstable poles of the reference inputs are limited to the $j\omega$- axis, state-space solutions of the $\mathcal{H}_2$ design problem are possible for the 2DOF case (Park et al. 2002) and for the 3DOF case (Park and Youla 1992).

5.7 Examples

Example 5.1
Establish the equality in (5.70).

Solution
It follows from (5.68) that

$$-\Sigma' M_1 - M_1\Sigma + H_1' D_{12} E_1^{-1} N_2' M_1 + M_1 N_2 E_1^{-1} D_{12}' H_1 + M_1 N_2 E_1^{-1} N_2' M_1 + H_1' D_{12} E_1^{-1} D_{12}' H_1 = H_1' H_1.$$

This simplifies to $-\Sigma' M_1 - M_1\Sigma + L_1' D_{12}' D_{12} L_1 = H_1' H_1$ and adding the equality $-sM_1 + M_1 s = 0$ yields

$$(sI-\Sigma)_* M_1 + M_1(sI-\Sigma) + L_1' D_{12}' D_{12} L_1 = H_1' H_1.$$

Multiplying this equation by $N_2'(sI-\Sigma)_*^{-1}$ on the left and by $(sI-\Sigma)^{-1}N_2$ on the right then gives

$$N_2' M_1 (sI-\Sigma)^{-1} N_2 + N_2'(sI-\Sigma)_*^{-1} M_1 N_2 + N_2'(sI-\Sigma)_*^{-1} L_1' D_{12}' D_{12} L_1 (sI-\Sigma)^{-1} N_2$$
$$= N_2'(sI-\Sigma)_*^{-1} H_1' H_1 (sI-\Sigma)^{-1} N_2.$$

After adding the term $D_{12}' H_1 (sI-\Sigma)^{-1} N_2 + N_2'(sI-\Sigma)_*^{-1} H_1' D_{12} + D_{12}' D_{12}$ to both sides of the above equation one gets

$$E_1 L_1 (sI-\Sigma)^{-1} N_2 + N_2'(sI-\Sigma)_*^{-1} L_1' E_1 + N_2'(sI-\Sigma)_*^{-1} L_1' D_{12}' D_{12} L_1 (sI-\Sigma)^{-1} N_2 + D_{12}' D_{12}$$
$$= N_2'(sI-\Sigma)_*^{-1} H_1' H_1 (sI-\Sigma)^{-1} N_2 + D_{12}' H_1 (sI-\Sigma)^{-1} N_2 + N_2'(sI-\Sigma)_*^{-1} H_1' D_{12} + D_{12}' D_{12},$$

which results in

$$\left(L_1(sI-\Sigma)^{-1}N_2 + I\right)_* D_{12}' D_{12} \left(L_1(sI-\Sigma)^{-1}N_2 + I\right) = G_{12*}(s) W_*(s) W(s) G_{12}(s).$$

Example 5.2
Prove that $\{\Gamma_\phi\}_+ + \{\Lambda A_1^{-1} Y\Omega\}_- = -E_1^{\frac{1}{2}} L_1 (sI-\Sigma)^{-1} L_2 E_2^{\frac{1}{2}}$ in Lemma 5.5.

Solution
Using the equality in (5.65), one obtains

$$\Lambda A_1^{-1} Y\Omega = \widetilde{A}_1^{-1} Y\Omega = E_1^{\frac{1}{2}}\left(I + L_1(sI-\Sigma)^{-1}N_2\right) Y\Omega$$
$$= E_1^{\frac{1}{2}} Y\Omega + E_1^{\frac{1}{2}} L_1 (sI-\Sigma)^{-1} N_2 Y\Omega$$

and this leads to $\{\Lambda A_1^{-1}Y\Omega\}_- = E_1^{\frac{1}{2}}L_1\{(sI-\Sigma)^{-1}N_2Y\Omega\}_-$. Using the equality in (5.77) yields

$$\begin{aligned}\{\Lambda A_1^{-1}Y\Omega\}_- &= E_1^{\frac{1}{2}}L_1\{(sI-\Sigma)^{-1}N_2Y(A+V_2^{-1}B_3L_2)\}_-E_2^{\frac{1}{2}} \\ &= E_1^{\frac{1}{2}}L_1\{(sI-\Sigma)^{-1}N_2YV_2^{-1}B_3\}_-L_2E_2^{\frac{1}{2}}.\end{aligned} \tag{E5.1}$$

The last equality holds since $(sI-\Sigma)^{-1}N_2YA = B_2V_1^{-1}A_1^{-1}YA = B_2V_1^{-1}Y_1$ is analytic in $\mathrm{Re}s \geq 0$. Using the equation

$$-A^{-1}B = H_2(sI-\Sigma)^{-1}N_2 + D_{22} = A_3^{-1}B_3N_2 + D_{22},$$

one gets $B = -(AA_3^{-1}B_3N_2 + AD_{22}) = -(V_2^{-1}B_3N_2 + AD_{22})$. Since $AX+BY=I$, it follows that $AX - (V_2^{-1}B_3N_2 + AD_{22})Y = I$. Multiplying on the left by A^{-1} and on the right by $V_2^{-1}B_3$ then yields

$$\begin{aligned}XV_2^{-1}B_3 - A_3^{-1}B_3N_2YV_2^{-1}B_3 - D_{22}YV_2^{-1}B_3 &= A^{-1}V_2^{-1}B_3 = A_3^{-1}B_3 \\ &= H_2(sI-\Sigma)^{-1}.\end{aligned}$$

Hence, $(X-D_{22}Y)V_2^{-1}B_3 - H_2(sI-\Sigma)^{-1}N_2YV_2^{-1}B_3 = H_2(sI-\Sigma)^{-1}$ and one gets

$$-H_2\{(sI-\Sigma)^{-1}N_2YV_2^{-1}B_3\}_- = H_2\{(sI-\Sigma)^{-1}\}_-. \tag{E5.2}$$

since V_2^{-1} is stable. When (Σ, H_2) is detectable, there is no loss in generality (see Corollary 3.9 in Zhou et al. 1996) to assume that

$$\Sigma = \begin{bmatrix}\Sigma_{11} & 0 \\ \Sigma_{21} & \Sigma_{22}\end{bmatrix}, H_2 = [H_{20} \quad 0], N_2 = \begin{bmatrix}N_{21} \\ N_{22}\end{bmatrix} \tag{E5.3}$$

with the property that Σ_{22} is stable and (Σ_{11}, H_{20}) is an observable pair. Since

$$(sI-\Sigma)^{-1} = \begin{bmatrix}(sI-\Sigma_{11})^{-1} & 0 \\ (sI-\Sigma_{22})^{-1}\Sigma_{21}(sI-\Sigma_{11})^{-1} & (sI-\Sigma_{22})^{-1}\end{bmatrix}, \tag{E5.4}$$

equation (E5.2) becomes

$$-H_{20}\{(sI-\Sigma_{11})^{-1}N_{21}YV_2^{-1}B_3\}_- = H_{20}\{[(sI-\Sigma_{11})^{-1} \quad 0]\}_-. \tag{E5.5}$$

It is established next that $-\{(sI-\Sigma_{11})^{-1}N_{21}YV_2^{-1}B_3\}_- = \{[(sI-\Sigma_{11})^{-1} \quad 0]\}_-$. Using the identity

$$(sI-\Sigma_{11})^{-1}=\sum_{l=1}^{k}\frac{\Sigma_{11}^{l-1}}{s^l}+\frac{\Sigma_{11}^k(sI-\Sigma_{11})^{-1}}{s^k},k\geq 1,$$

one gets

$$\{(sI-\Sigma_{11})^{-1}N_{21}YV_2^{-1}B_3\}_-=\Sigma_{11}^k\left\{\frac{(sI-\Sigma_{11})^{-1}}{s^k}N_{21}YV_2^{-1}B_3\right\}_- \tag{E5.6}$$

since $\left\{\left(\sum_{l=1}^{k}\frac{\Sigma_{11}^{l-1}}{s^l}\right)N_{21}YV_2^{-1}B_3\right\}_-=0$. Similarly it is also true that

$$\{(sI-\Sigma_{11})^{-1}\}_-=\Sigma_{11}^k\left\{\frac{(sI-\Sigma_{11})^{-1}}{s^k}\right\}_-,k\geq 1. \tag{E5.7}$$

Inserting (E5.6) and (E5.7) into (E5.5), one gets

$$H_{20}\Sigma_{11}^k\left\{\frac{(sI-\Sigma_{11})^{-1}}{s^k}N_{21}YV_2^{-1}B_3\right\}_-=-H_{20}\Sigma_{11}^k\left\{\left[\frac{(sI-\Sigma_{11})^{-1}}{s^k}\quad 0\right]\right\}_-,\ k\geq 0$$

so that

$$\left\{\frac{1}{s^k}H_{20}\Sigma_{11}^k\left((sI-\Sigma_{11})^{-1}N_{21}YV_2^{-1}B_3+\left[(sI-\Sigma_{11})^{-1}\quad 0\right]\right)\right\}_-=0,\ k\geq 0.$$

In general, for a rational matrix $P(s)$, $\{P(s)\}_-=0$ implies $\{s^kP(s)\}_-=0$, $k\geq 0$. Hence,

$$H_{20}\Sigma_{11}^k\{\left((sI-\Sigma_{11})^{-1}N_{21}YV_2^{-1}B_3+\left[(sI-\Sigma_{11})^{-1}|0\right]\right)\}_-=0,\ k\geq 0.$$

Since the observability matrix of (Σ_{11},H_{20}) has full column rank, it can be concluded that

$$\{(sI-\Sigma_{11})^{-1}N_{21}YV_2^{-1}B_3\}_-+\{\left[(sI-\Sigma_{11})^{-1}\quad 0\right]\}_-=0.$$

Next it follows from (E5.3) and (E5.4) that

$$(sI-\Sigma)^{-1}N_2YV_2^{-1}B_3=\begin{bmatrix}(sI-\Sigma_{11})^{-1}N_{21}YV_2^{-1}B_3\\(sI-\Sigma_{22})^{-1}\Sigma_{21}(sI-\Sigma_{11})^{-1}N_{21}YV_2^{-1}B_3+(sI-\Sigma_{22})^{-1}N_{22}YV_2^{-1}B_3\end{bmatrix}.$$

Hence, since $(sI - \Sigma_{22})^{-1}$ and V_2^{-1} are stable,

$$\{(sI-\Sigma)^{-1}N_2YV_2^{-1}B_3\}_- = \begin{bmatrix} \{(sI-\Sigma_{11})^{-1}N_{21}YV_2^{-1}B_3\}_- \\ \{(sI-\Sigma_{22})^{-1}\Sigma_{21}(sI-\Sigma_{11})^{-1}N_{21}YV_2^{-1}B_3\}_- \end{bmatrix}$$

$$= \begin{bmatrix} \{(sI-\Sigma_{11})^{-1}N_{21}YV_2^{-1}B_3\}_- \\ \{(sI-\Sigma_{22})^{-1}\Sigma_{21}\{(sI-\Sigma_{11})^{-1}N_{21}YV_2^{-1}B_3\}_-\}_- \end{bmatrix} = \left[\begin{array}{c|c} -\{(sI-\Sigma_{11})^{-1}\}_- & 0 \\ \hline -\{(sI-\Sigma_{22})^{-1}\Sigma_{21}\{(sI-\Sigma_{11})^{-1}\}_-\}_- & 0 \end{array}\right] \quad \text{(E5.8)}$$

$$= -\left\{\begin{bmatrix} (sI-\Sigma_{11})^{-1} & 0 \\ (sI-\Sigma_{22})^{-1}\Sigma_{21}(sI-\Sigma_{11})^{-1} & (sI-\Sigma_{22})^{-1} \end{bmatrix}\right\}_- = -\{(sI-\Sigma)^{-1}\}_- .$$

It therefore follows from (E5.1) and (E5.8) that $\{\Lambda A_1^{-1} Y\Omega\}_- = -E_1^{\frac{1}{2}}L_1\{(sI-\Sigma)^{-1}\}_- L_2 E_2^{\frac{1}{2}}$.

It remains to be shown that

$$\{\Gamma_\phi\}_+ = \{-(WG_{12}\tilde{A}_1)_*(WG_{11}G_w)(\tilde{A}G_{21}G_w)_*\}_+ = -E_1^{\frac{1}{2}}L_1\{(sI-\Sigma)^{-1}\}_+ L_2 E_2^{\frac{1}{2}}.$$

First, state-space parameters for $(WG_{12}\tilde{A}_1)_*$ are obtained from the state-space parameters for WG_{12} and $\tilde{A}_1$ in (5.57) and (5.66). Using (5.14) yields

$$(WG_{12})\tilde{A}_1 = \left[\begin{array}{c|c} \Sigma & N_2 \\ \hline H_1 & D_{12} \end{array}\right]\left[\begin{array}{c|c} \Sigma - N_2L_1 & N_2E_1^{-\frac{1}{2}} \\ \hline -L_1 & E_1^{-\frac{1}{2}} \end{array}\right] = \left[\begin{array}{cc|c} \Sigma & -N_2L_1 & N_2E_1^{-\frac{1}{2}} \\ 0 & \Sigma - N_2L_1 & N_2E_1^{-\frac{1}{2}} \\ \hline H_1 & -D_{12}L_1 & D_{12}E_1^{-\frac{1}{2}} \end{array}\right].$$

Introducing a similarity transformation as in (5.12) with $L = \begin{bmatrix} I & 0 \\ I & I \end{bmatrix}$ and making use of (5.10) then gives

$$(WG_{12})\tilde{A}_1 = \left[\begin{array}{cc|c} \Sigma - N_2L_1 & -N_2L_1 & N_2E_1^{-\frac{1}{2}} \\ 0 & \Sigma & 0 \\ \hline H_1 - D_{12}L_1 & -D_{12}L_1 & D_{12}E_1^{-\frac{1}{2}} \end{array}\right] = \left[\begin{array}{c|c} \Sigma - N_2L_1 & N_2E_1^{-\frac{1}{2}} \\ \hline H_1 - D_{12}L_1 & D_{12}E_1^{-\frac{1}{2}} \end{array}\right].$$

Similarly, the state-space parameters for $\tilde{A}G_{21}G_w$ are obtained from

$$\tilde{A}(G_{21}G_w) = \left[\begin{array}{c|c} \Sigma - L_2H_2 & L_2 \\ \hline -E_2^{-\frac{1}{2}}H_2 & E_2^{-\frac{1}{2}} \end{array}\right]\left[\begin{array}{c|c} \Sigma & N_1 \\ \hline H_2 & D_{21} \end{array}\right] = \left[\begin{array}{cc|c} \Sigma - L_2H_2 & L_2H_2 & L_2D_{21} \\ 0 & \Sigma & N_1 \\ \hline -E_2^{-\frac{1}{2}}H_2 & E_2^{-\frac{1}{2}}H_2 & E_2^{-\frac{1}{2}}D_{21} \end{array}\right]$$

and using the similarity transformation with $L = \begin{bmatrix} I & I \\ 0 & I \end{bmatrix}$ yields

$$\tilde{A}(G_{21}G_w) = \left[\begin{array}{cc|c} \Sigma - L_2H_2 & 0 & L_2D_{21} - N_1 \\ 0 & \Sigma & N_1 \\ \hline -E_2^{-\frac{1}{2}}H_2 & 0 & E_2^{-\frac{1}{2}}D_{21} \end{array}\right] = \left[\begin{array}{c|c} \Sigma - L_2H_2 & N_1 - L_2D_{21} \\ \hline E_2^{-\frac{1}{2}}H_2 & E_2^{-\frac{1}{2}}D_{21} \end{array}\right].$$

Thus,

$$(WG_{12}\tilde{A}_1)_*(WG_{11}G_w)(\tilde{A}G_{21}G_w)_*$$

$$= \left[\begin{array}{c|c} (N_2L_1 - \Sigma)' & (H_1 - D_{12}L_1)' \\ \hline -E_1^{-\frac{1}{2}}N_2' & E_1^{-\frac{1}{2}}D_{12}' \end{array}\right] \left[\begin{array}{c|c} \Sigma & N_1 \\ \hline H_1 & 0 \end{array}\right] \left[\begin{array}{c|c} (L_2H_2 - \Sigma)' & H_2'E_2^{-\frac{1}{2}} \\ \hline (L_2D_{21} - N_1)' & D_{21}'E_2^{-\frac{1}{2}} \end{array}\right]$$

$$= \left[\begin{array}{cc|c} (N_2L_1 - \Sigma)' & (H_1 - D_{12}L_1)'H_1 & 0 \\ 0 & \Sigma & N_1 \\ \hline -E_1^{-\frac{1}{2}}N_2' & E_1^{-\frac{1}{2}}D_{12}'H_1 & 0 \end{array}\right] \left[\begin{array}{c|c} (L_2H_2 - \Sigma)' & H_2'E_2^{-\frac{1}{2}} \\ \hline (L_2D_{21} - N_1)' & D_{21}'E_2^{-\frac{1}{2}} \end{array}\right]$$

$$= \left[\begin{array}{ccc|c} (N_2L_1 - \Sigma)' & (H_1 - D_{12}L_1)'H_1 & 0 & 0 \\ 0 & \Sigma & N_1(L_2D_{21} - N_1)' & N_1D_{21}'E_2^{-\frac{1}{2}} \\ 0 & 0 & (L_2H_2 - \Sigma)' & H_2'E_2^{-\frac{1}{2}} \\ \hline -E_1^{-\frac{1}{2}}N_2' & E_1^{-\frac{1}{2}}D_{12}'H_1 & 0 & 0 \end{array}\right].$$

Introducing a similarity transformation with $L^{-1} = \begin{bmatrix} I & M_1 & 0 \\ 0 & I & M_2 \\ 0 & 0 & I \end{bmatrix}$ and $L = \begin{bmatrix} I & -M_1 & M_1M_2 \\ 0 & I & -M_2 \\ 0 & 0 & I \end{bmatrix}$, one gets

$$(WG_{12}\tilde{A}_1)_*(WG_{11}G_w)(\tilde{A}G_{21}G_w)_* = \left[\begin{array}{ccc|c} \Sigma_l & \Sigma_{12} & \Sigma_{13} & M_1N_1D_{21}'E_2^{-\frac{1}{2}} \\ 0 & \Sigma & \Sigma_{23} & L_2E_2^{\frac{1}{2}} \\ 0 & 0 & \Sigma_h & H_2'E_2^{-\frac{1}{2}} \\ \hline -E_1^{-\frac{1}{2}}N_2' & E_1^{\frac{1}{2}}L_1 & -E_1^{\frac{1}{2}}L_1M_2 & 0 \end{array}\right],$$

where

$$\Sigma_l = (N_2L_1 - \Sigma)', \Sigma_h = (L_2H_2 - \Sigma)'$$

$$\Sigma_{12} = (H_1 - D_{12}L_1)'H_1 + M_1\Sigma - \Sigma_lM_1, \Sigma_{23} = N_1(L_2D_{21} - N_1)' + M_2\Sigma_h - \Sigma M_2$$

$$\Sigma_{13} = \Sigma_lM_1M_2 - \big((H_1 - D_{12}L_1)'H_1 + M_1\Sigma\big)M_2 + M_1N_1(L_2D_{21} - N_1)'.$$

It can be shown that $\Sigma_{12} = 0$ and $\Sigma_{23} = 0$. In fact,

$$\begin{aligned}\Sigma_{12} = &\ (H_1 - D_{12}L_1)'H_1 + M_1\Sigma - \Sigma_l M_1 \\ &= H_1'H_1 - L_1'D_{12}'H_1 + M_1\Sigma + \Sigma' M_1 - L_1'N_2'M_1 \\ &= (\Sigma - N_2E_1^{-1}D_{12}'H_1)'M_1 + M_1(\Sigma - N_2E_1^{-1}D_{12}'H_1) \\ &\quad - M_1N_2E_1^{-1}N_2'M_1 + H_1'(I - D_{12}E_1^{-1}D_{12}')H_1 = 0\end{aligned}$$

follows from (5.68). Similarly,

$$\begin{aligned}-\Sigma_{23} = &- N_1(L_2D_{21} - N_1)' - M_2\Sigma_h + \Sigma M_2 \\ &=(\Sigma - N_1D_{21}'E_2^{-1}H_2)M_2 + M_2(\Sigma - N_1D_{21}'E_2^{-1}H_2)' \\ &\quad - M_2H_2'E_2^{-1}H_2M_2 + N_1(I - D_{21}'E_2^{-1}D_{21})N_1' = 0\end{aligned}$$

follows from (5.82). Hence,

$$(WG_{12}\tilde{A}_1)_*(WG_{11}G_w)(\tilde{A}G_{21}G_w)_* = \left[\begin{array}{ccc:c} \Sigma_l & 0 & \Sigma_{13} & M_1N_1D_{21}'E_2^{-\frac{1}{2}} \\ 0 & \Sigma & 0 & L_2E_2^{\frac{1}{2}} \\ 0 & 0 & \Sigma_h & H_2'E_2^{-\frac{1}{2}} \\ \hdashline -E_1^{-\frac{1}{2}}N_2' & E_1^{\frac{1}{2}}L_1 & -E_1^{\frac{1}{2}}L_1M_2 & 0 \end{array}\right]$$

$$=\begin{bmatrix} -E_1^{-\frac{1}{2}}N_2' & E_1^{\frac{1}{2}}L_1 & -E_1^{\frac{1}{2}}L_1M_2 \end{bmatrix}\begin{bmatrix} (sI-\Sigma_l)^{-1} & 0 & (sI-\Sigma_l)^{-1}\Sigma_{13}(sI-\Sigma_h)^{-1} \\ 0 & (sI-\Sigma)^{-1} & 0 \\ 0 & 0 & (sI-\Sigma_h)^{-1} \end{bmatrix}\begin{bmatrix} M_1N_1D_{21}'E_2^{-\frac{1}{2}} \\ L_2E_2^{\frac{1}{2}} \\ H_2'E_2^{-\frac{1}{2}} \end{bmatrix}$$

$$\begin{aligned}&= -E_1^{-\frac{1}{2}}N_2'(sI-\Sigma_l)^{-1}M_1N_1D_{21}'E_2^{-\frac{1}{2}} + E_1^{\frac{1}{2}}L_1(sI-\Sigma)^{-1}L_2E_2^{\frac{1}{2}} \\ &\quad -E_1^{-\frac{1}{2}}N_2'(sI-\Sigma_l)^{-1}\Sigma_{13}(sI-\Sigma_h)^{-1}H_2'E_2^{-\frac{1}{2}} - E_1^{\frac{1}{2}}L_1M_2(sI-\Sigma_h)^{-1}H_2'E_2^{-\frac{1}{2}}.\end{aligned}$$

Since $(sI - \Sigma_l)^{-1}$ and $(sI - \Sigma_h)^{-1}$ are analytic in $\mathrm{Re}s \leq 0$, it is concluded that

$$\{(WG_{12}\widetilde{A}_1)_*(WG_{11}G_w)(\widetilde{A}G_{21}G_w)_*\}_+ = E_1^{\frac{1}{2}}L_1\{(sI - \Sigma)^{-1}\}_+L_2E_2^{\frac{1}{2}}.$$

Hence,

$$\{\Gamma_\phi\}_+ = \{-(WG_{12}\widetilde{A}_1)_*(WG_{11}G_w)(\widetilde{A}G_{21}G_w)_*\}_+ = -E_1^{\frac{1}{2}}L_1\{(sI - \Sigma)^{-1}\}_+L_2E_2^{\frac{1}{2}}$$

and it follows that

$$\begin{aligned}\{\Gamma_\phi\}_+ + \{\Lambda A_1^{-1} Y\Omega\}_- &= -E_1^{\frac{1}{2}} L_1(\{(sI-\Sigma)^{-1}\}_+ + \{(sI-\Sigma)^{-1}\}_-) L_2 E_2^{\frac{1}{2}} \\ &= -E_1^{\frac{1}{2}} L_1 (sI-\Sigma)^{-1} L_2 E_2^{\frac{1}{2}}.\end{aligned}$$

Example 5.3
Show that when (Σ, N_2, H_2) is minimal,

$$\{\Gamma_\phi\}_+ + \{\Lambda A_1^{-1} Y\Omega\}_- = \{\Lambda A_1^{-1} Y\Omega\}_{sp} = -E_1^{\frac{1}{2}} L_1 (sI-\Sigma)^{-1} L_2 E_2^{\frac{1}{2}}$$

and $\{\Gamma_\phi - \Lambda A_1^{-1} Y\Omega\}_+ = 0$ in Lemma 5.5. The notation $\{G\}_{sp}$ denotes the strictly proper part of G.

Solution
When (Σ, N_2, H_2) is minimal, V_1 in Lemma 5.3 and V_2 in Lemma 5.4 are unimodular matrices and it immediately follows from (5.64) and (5.77) that Λ and Ω are polynomial matrices. Moreover, if $(sI-\Sigma)^{-1}N_2 = \widetilde{B}_2 A_2^{-1}$ is a right coprime matrix fraction description, it follows that $\widetilde{B}_2 A_2^{-1} = \widetilde{B}_2 (A_1 V_1)^{-1} = (\widetilde{B}_2 V_1^{-1}) A_1^{-1} = B_2 A_1^{-1}$ is also a right coprime matrix fraction description. That is, one can always choose a B_2 so that B_2, A_1 is a right coprime pair and $(sI-\Sigma)^{-1}N_2 = B_2 A_1^{-1}$. Similarly, one can always choose a B_3 so that B_3, A is a left coprime pair and $H_2(sI-\Sigma)^{-1} = A^{-1}B_3$. Hence, there is no loss in generality if one sets $A_2 = A_1$ and $A_3 = A$ in (5.63) and (5.76), respectively. It then follows that $V_1 = A_1^{-1}A_2 = I$ and $V_2 = A_3 A^{-1} = I$. Using the equality in (5.65), one now obtains from

$$\begin{aligned}\Lambda A_1^{-1} Y\Omega = \widetilde{A}_1^{-1} Y\Omega &= E_1^{\frac{1}{2}} \left(I + L_1 (sI-\Sigma)^{-1} N_2\right) Y\Omega \\ &= E_1^{\frac{1}{2}} Y\Omega + E_1^{\frac{1}{2}} L_1 (sI-\Sigma)^{-1} N_2 Y\Omega\end{aligned}$$

that $\{\Lambda A_1^{-1} Y\Omega\}_{sp} = E_1^{\frac{1}{2}} L_1 \{(sI-\Sigma)^{-1} N_2 Y\Omega\}_{sp}$. Using the equality in (5.77), then yields

$$\begin{aligned}\{\Lambda A_1^{-1} Y\Omega\}_{sp} &= E_1^{\frac{1}{2}} L_1 \{(sI-\Sigma)^{-1} N_2 Y (A + B_3 L_2)\}_{sp} E_2^{\frac{1}{2}} \\ &= E_1^{\frac{1}{2}} L_1 \{(sI-\Sigma)^{-1} N_2 Y B_3\}_{sp} L_2 E_2^{\frac{1}{2}}.\end{aligned}$$

From the fact that $-A^{-1}B = H_2(sI-\Sigma)^{-1}N_2 + D_{22} = A^{-1}B_3 N_2 + D_{22}$, one gets the equality $B = -(B_3 N_2 + A D_{22})$. Since $AX + BY = I$, it follows that $AX - (B_3 N_2 + A D_{22})Y = I$. Multiplying on the left by A^{-1} and on the right by B_3 then yields $XB_3 - A^{-1}B_3 N_2 Y B_3 - D_{22} Y B_3 = A^{-1} B_3$. Since $A^{-1}B_3 = H_2(sI-\Sigma)^{-1}$ it follows that

$$(X - D_{22}Y)B_3 - H_2(sI - \Sigma)^{-1}N_2YB_3 = H_2(sI - \Sigma)^{-1} \tag{E5.9}$$

and one gets $-H_2\{(sI-\Sigma)^{-1}N_2YB_3\}_{sp} = H_2\{(sI-\Sigma)^{-1}\}_{sp}$. Multiplying on both sides of (E5.9) by s^k and using the identity

$$s^k(sI-\Sigma)^{-1} = \sum_{l=1}^{k} s^{k-l}\Sigma^{l-1} + \Sigma^k(sI-\Sigma)^{-1}, \; k \geq 1,$$

gives $-H_2\Sigma^k\{(sI-\Sigma)^{-1}N_2YB_3\}_{sp} = H_2\Sigma^k\{(sI-\Sigma)^{-1}\}_{sp}$, $k \geq 1$. Since the observability matrix of (Σ, H_2) has full column rank, it can be concluded that $\{(sI-\Sigma)^{-1}N_2YB_3\}_{sp} = -(sI-\Sigma)^{-1}$ and it follows that $\{\Lambda A_1^{-1}Y\Omega\}_{sp} = -E_1^{\frac{1}{2}}L_1 (sI-\Sigma)^{-1}L_2E_2^{\frac{1}{2}}$. Next, it is proven that $\{\Gamma_\phi - \Lambda A_1^{-1}Y\Omega\}_+ = 0$. First it is established that $W(G_{11}+G_{12}YAG_{21})G_w$ is a polynomial matrix. Since $WG_{12} = H_1B_2A_1^{-1}+D_{12}$ and $G_{21}G_w = A^{-1}B_3N_1 + D_{21}$, it follows that

$$\begin{aligned}
W(G_{11}+G_{12}YAG_{21})G_w &= H_1(sI-\Sigma)^{-1}N_1 + (H_1B_2A_1^{-1}+D_{12})Y(B_3N_1+AD_{21})\\
&= H_1(sI-\Sigma)^{-1}N_1 + H_1B_2A_1^{-1}Y(B_3N_1+AD_{21}) + D_{12}Y(B_3N_1+AD_{21})\\
&= H_1\{(sI-\Sigma)^{-1} + B_2A_1^{-1}YB_3\}N_1 + H_1B_2A_1^{-1}YAD_{21} + D_{12}Y(B_3N_1+AD_{21})\\
&= H_1\{(sI-\Sigma)^{-1} + (sI-\Sigma)^{-1}N_2YB_3\}N_1 + H_1B_2Y_1D_{21} + D_{12}Y(B_3N_1+AD_{21}).
\end{aligned}$$

Since $\{(sI-\Sigma)^{-1}N_2YB_3\}_{sp} = -(sI-\Sigma)^{-1}$, the first term of the above equation is a polynomial matrix; hence, $W(G_{11}+G_{12}YAG_{21})G_w$ is a polynomial matrix. It is trivial to show that $WG_{12}A_1$ and $AG_{21}G_w$ are polynomial matrices. Now Λ_*^{-1} and Ω_*^{-1} are analytic in $\mathrm{Re}\, s \leq 0$ and, therefore, it readily follows from

$$\{\Gamma_\phi - \Lambda A_1^{-1}Y\Omega\}_+ = -\{\Lambda_*^{-1}A_{1*}G_{12*}W_*W(G_{11}+G_{12}YAG_{21})G_wG_{w*}G_{21*}A_*\Omega_*^{-1}\}_+ = 0$$

that $\{\Gamma_\phi\}_+ = \{\Lambda A_1^{-1}Y\Omega\}_+$ and $\{\Gamma_\phi\}_+ + \{\Lambda A_1^{-1}Y\Omega\}_- = \{\Lambda A_1^{-1}Y\Omega\}_+ + \{\Lambda A_1^{-1}Y\Omega\}_- = \{\Lambda A_1^{-1}Y\Omega\}_{sp}$.

Example 5.4

Show that the realization of the controller $C(s) = G_c(s)\square Z$ in Theorem 5.1 has no unstable hidden poles by using the state-space description of $C(s) = G_c(s)\square Z$.

Solution

When $Z(s) = H_z(sI-\Sigma_z)^{-1}N_z$ is a minimal realization, the state-space parameters for $C(s)$ are obtained from Theorem 5.1 and (5.23) using $D_\phi = D_2 = 0$, $E_1 = I$, and $E_2 = I$. The result is

$C(s) = G_c(s) \square Z$

$$= \left[\begin{array}{cc|c} \Sigma - N_2L_1 - L_2H_2 + L_2D_{22}L_1 & (N_2 - L_2D_{22})E_1^{-\frac{1}{2}}H_z & L_2 \\ -N_zE_2^{-\frac{1}{2}}(H_2 - D_{22}L_1) & \Sigma_z - N_zE_2^{-\frac{1}{2}}D_{22}E_1^{-\frac{1}{2}}H_z & N_zE_1^{-\frac{1}{2}} \\ \hline -L_1 & E_1^{-\frac{1}{2}}H_z & 0 \end{array}\right] = \left[\begin{array}{c|c} \widetilde{\Sigma}_c & \widetilde{N}_c \\ \hline \widetilde{H}_c & 0 \end{array}\right]$$

Clearly,

$$\left[sI - \widetilde{\Sigma}_c \mid \widetilde{N}_c\right] = \left[\begin{array}{cc|c} sI - \Sigma + N_2L_1 + L_2H_2 - L_2D_{22}L_1 & -(N_2 - L_2D_{22})E_1^{-\frac{1}{2}}H_z & L_2 \\ N_zE_2^{-\frac{1}{2}}(H_2 - D_{22}L_1) & sI - \Sigma_z + N_zE_2^{-\frac{1}{2}}D_{22}E_1^{-\frac{1}{2}}H_z & N_zE_2^{-\frac{1}{2}} \end{array}\right]$$

$$= \left[\begin{array}{cc|c} sI - \Sigma + N_2L_1 & -N_2E_1^{-\frac{1}{2}}H_z & L_2 \\ 0 & sI - \Sigma_z & N_zE_2^{-\frac{1}{2}} \end{array}\right] \left[\begin{array}{cc|c} I & 0 & 0 \\ 0 & I & 0 \\ \hline H_2 - D_{22}L_1 & D_{22}E_1^{-\frac{1}{2}}H_z & I \end{array}\right].$$

The matrices $\Sigma - N_2L_1$ and Σ_z are stable ones; hence, the matrix

$$\begin{bmatrix} sI - \Sigma + N_2L_1 & -N_2E_1^{-\frac{1}{2}}H_z & L_2 \\ 0 & sI - \Sigma_z & N_zE_2^{-\frac{1}{2}} \end{bmatrix}$$

has full row rank in $\mathrm{Re}\,s \geq 0$ and this implies that $[\,sI - \widetilde{\Sigma}_c \quad \widetilde{N}_c\,]$ has full row rank in $\mathrm{Re}\,s \geq 0$. Therefore, the pair $(\widetilde{\Sigma}_c, \widetilde{N}_c)$ is stabilizable. The dual method shows that the pair $(\widetilde{\Sigma}_c, \widetilde{H}_c)$ is detectable and there are no unstable hidden poles in the realization of $C(s) = G_c(s)\square Z$ using Theorem 5.1.

Example 5.5
(Checking of the admissibility condition of a physical generalized plant whose internal structure is specified.) The generalized plant $\widetilde{G}$ in the standard configuration of Fig. 3.1 can be viewed as an interconnection of nondegenerate dynamic subsystems as depicted in Fig. 2.3 of Example 2.1. Each subsystem G_q, $q = 1 \to n$, has the internal state- space description

$$\begin{aligned} \dot{x}_q(t) &= \Sigma_q x_q(t) + N_q u_q(t) \\ y_q(t) &= H_q x_q(t) + D_q u_q(t) \end{aligned}.$$

The connection network is characterized by a constant matrix M according to

$$\begin{bmatrix} u_t \\ y_c \\ y_m \end{bmatrix} = \begin{bmatrix} M_{11} & M_{12} & M_{13} \\ M_{21} & M_{22} & M_{23} \\ M_{31} & M_{32} & M_{33} \end{bmatrix} \begin{bmatrix} e \\ v \\ y_t \end{bmatrix} = M \begin{bmatrix} e \\ v \\ y_t \end{bmatrix},$$

where $u_t = [u_1' u_2' \cdots u_n']'$, and $y_t = [y_1' y_2' \cdots y_n']'$.

(a) (State-space domain approach) Establish that an internal state-space model for the generalized plant is given by

$$\widetilde{G}(s) = \left[\begin{array}{c|cc} \Sigma_t + N_t J_1^{-1} M_{13} H_t & N_t J_1^{-1} M_{11} & N_t J_1^{-1} M_{12} \\ \hline M_{23} J_2^{-1} H_t & M_{21} + M_{23} J_2^{-1} D_t M_{11} & M_{22} + M_{23} J_2^{-1} D_t M_{12} \\ M_{33} J_2^{-1} H_t & M_{31} + M_{33} J_2^{-1} D_t M_{11} & M_{32} + M_{33} J_2^{-1} D_t M_{12} \end{array}\right] = \left[\begin{array}{c|cc} \widetilde{\Sigma} & \widetilde{N}_1 & \widetilde{N}_2 \\ \hline \widetilde{H}_1 & \widetilde{D}_{11} & \widetilde{D}_{12} \\ \widetilde{H}_2 & \widetilde{D}_{21} & \widetilde{D}_{22} \end{array}\right]$$

where

$$\begin{aligned} &\Sigma_t = \text{diag}\{\Sigma_1, \Sigma_2, \cdots, \Sigma_n\}, N_t = \text{diag}\{N_1, N_2, \ldots, N_n\}, \\ &H_t = \text{diag}\{H_1, H_2, \ldots, H_n\}, D_t = \text{diag}\{D_1, D_2, \ldots, D_n\}, \\ &J_1 = I - M_{13} D_t, \text{and } J_2 = I - D_t M_{13}. \end{aligned}$$

Show that the generalized plant is admissible iff the pair $(\widetilde{\Sigma}, \widetilde{N}_2)$ is stabilizable and $(\widetilde{\Sigma}, \widetilde{H}_2)$ is detectable.

(b) (Frequency-domain approach) Let $\widetilde{G}$ denote the transfer matrix of the generalized plant in Fig. 2.3. Establish that $\widetilde{G}$ is obtained by

$$\begin{aligned} \widetilde{G}(s) &= \begin{bmatrix} \widetilde{G}_{11} & \widetilde{G}_{12} \\ \widetilde{G}_{21} & \widetilde{G}_{22} \end{bmatrix} \\ &= \begin{bmatrix} M_{23} \\ M_{33} \end{bmatrix} G_t (I - M_{13} G_t)^{-1} [M_{11} \quad M_{12}] + \begin{bmatrix} M_{21} & M_{22} \\ M_{31} & M_{32} \end{bmatrix}, \end{aligned}$$

where $G_q(s) = H_q(sI - \Sigma_q)^{-1} N_q + D_q$ is the transfer matrix of the $q-th$ sub-block and $G_t = \text{diag}\{G_1, G_2, \ldots, G_n\}$. Let $G_q(s) = B_{q1}(s) A_{q1}^{-1}(s)$, $q = 1 \to n$, denote a right coprime polynomial matrix fraction description and define $A_{t1} = \text{diag}\{A_{q1}, A_{q2}, \ldots, A_{qn}\}$ and $B_{t1} = \text{diag}\{B_{q1}, B_{q2}, \ldots, B_{qn}\}$. Show that the generalized plant is admissible iff each sub-block G_q has no unstable hidden poles and the polynomial $\det(A_{t1} - M_{13} B_{t1}) / \psi_{\widetilde{G}_{22}}$ is Hurwitz. (Similar results can be obtained for the digital systems with minor adjustments.)

Solution

(a) It is convenient to introduce the vector $x_t = [x_1' x_2' \cdots x_n']'$. Then clearly,

$$\dot{x}_t(t) = \Sigma_t x_t(t) + N_t u_t(t) \text{ and } y_t(t) = H_t x_t(t) + D_t u_t(t).$$

Inserting $y_t = H_t x_t + D_t u_t$ into $u_t = [M_{11} \quad M_{12}][e'v']' + M_{13} y_t$ yields

$$u_t = J_1^{-1}[M_{11} \quad M_{12}]\begin{bmatrix} e \\ v \end{bmatrix} + J_1^{-1} M_{13} H_t x_t,$$

where $J_1 = I - M_{13} D_t$. Therefore, it follows that

$$\dot{x}_t(t) = (\Sigma_t + N_t J_1^{-1} M_{13} H_t) x_t + N_t J_1^{-1}[M_{11} \quad M_{12}]\begin{bmatrix} e \\ v \end{bmatrix},$$

which is the state equation of the system $\widetilde{G}$. Since

$$\begin{aligned} y_t &= (I + D_t J_1^{-1} M_{13}) H_t x_t + D_t J_1^{-1}[M_{11} \quad M_{12}]\begin{bmatrix} e \\ v \end{bmatrix} \\ &= J_2^{-1} H_t x_t + J_2^{-1} D_t [M_{11} \quad M_{12}]\begin{bmatrix} e \\ v \end{bmatrix} \end{aligned}$$

with $J_2 = I - D_t M_{13}$, it follows that

$$\begin{aligned} \begin{bmatrix} y_c \\ y_m \end{bmatrix} &= \begin{bmatrix} M_{21} & M_{22} \\ M_{31} & M_{32} \end{bmatrix}\begin{bmatrix} e \\ v \end{bmatrix} + \begin{bmatrix} M_{23} \\ M_{33} \end{bmatrix} y_t \\ &= \begin{bmatrix} M_{23} \\ M_{33} \end{bmatrix} J_2^{-1} H_t x_t + \left(\begin{bmatrix} M_{21} & M_{22} \\ M_{31} & M_{32} \end{bmatrix} + \begin{bmatrix} M_{23} \\ M_{33} \end{bmatrix} J_2^{-1} D_t [M_{11} \quad M_{12}] \right) \begin{bmatrix} e \\ v \end{bmatrix}, \end{aligned}$$

which is the output equation of the system $\widetilde{G}$. These results describe the internal description of the interconnected system $\widetilde{G}(s)$. By Definition 2.2, the physical system $\widetilde{G}$ is admissible when $\Delta_{\widetilde{G}}/\psi_{\widetilde{G}_{22}}$ is a stable polynomial. This is the case iff $(\widetilde{\Sigma}, \widetilde{N}_2)$ is stabilizable and $(\widetilde{\Sigma}, \widetilde{H}_2)$ is detectable (see the solution for Example 5.9(b)).

(b) Inserting $y_t(s) = G_t(s) u_t(s)$ into $u_t = [M_{11} \quad M_{12}][e' \ v']' + M_{13} y_t$ yields $u_t = (I - M_{13} G_t)^{-1}[M_{11} \quad M_{12}][e' \ v']'$ and hence $y_t = G_t u_t = G_t (I - M_{13} G_t)^{-1} [M_{11} \quad M_{12}][e' \ v']'$.

Note that the generalized plant is nondegenerate iff $\det(I - M_{13}G_t) \not\equiv 0$. It now follows that

$$\begin{bmatrix} y_c \\ y_m \end{bmatrix} = \begin{bmatrix} M_{21} & M_{22} \\ M_{31} & M_{32} \end{bmatrix} \begin{bmatrix} e \\ v \end{bmatrix} + \begin{bmatrix} M_{23} \\ M_{33} \end{bmatrix} y_t$$
$$= \left(\begin{bmatrix} M_{23} \\ M_{33} \end{bmatrix} G_t(I - M_{13}G_t)^{-1} [M_{11} \quad M_{12}] + \begin{bmatrix} M_{21} & M_{22} \\ M_{31} & M_{32} \end{bmatrix} \right) \begin{bmatrix} e \\ v \end{bmatrix},$$

which confirms the expression given for the transfer matrix $\widetilde{G}$. Using the internal description obtained in (a), one gets the characteristic polynomial of the generalized plant $\widetilde{G}$ as

$$\begin{aligned}
\Delta_{\widetilde{G}} &= \det(sI - \Sigma_t - N_t J_1^{-1} M_{13} H_t) = \det(sI - \Sigma_t) \det(I - (sI - \Sigma_t)^{-1} N_t J_1^{-1} M_{13} H_t) \\
&= \det(sI - \Sigma_t) \det(I - M_{13} H_t (sI - \Sigma_t)^{-1} N_t J_1^{-1}) \\
&= \det(sI - \Sigma_t) \det(J_1 - M_{13} H_t (sI - \Sigma_t)^{-1} N_t) / \det J_1 \\
&= \det(sI - \Sigma_t) \det(I - M_{13}(D_t + H_t(sI - \Sigma_t)^{-1} N_t)) / \det J_1 = \eta \det(sI - \Sigma_t) \det(I - M_{13} G_t),
\end{aligned}$$

where $\eta = 1/\det J_1$. Since $G_q(s) = B_{q1}(s) A_{q1}^{-1}(s)$ it follows that $G_t(s) = B_{t1}(s) A_{t1}^{-1}(s)$, and

$$\begin{aligned}
\Delta_{\widetilde{G}} &= \eta \det(sI - \Sigma_t) \det(I - M_{13} B_{t1} A_{t1}^{-1}) = \eta (\det(sI - \Sigma_t)/\det A_{t1}) \det(A_{t1} - M_{13} B_{t1}) \\
&= \eta \left(\prod_{q=1}^{n} h_q \right) \det(A_{t1} - M_{13} B_{t1}),
\end{aligned}$$

where the polynomial $h_q(s)$ accounts for the hidden pole of the sub-block $G_q(s)$. By Definition 2.2, the physical system $\widetilde{G}$ is admissible if $\Delta_{\widetilde{G}}/\psi_{\widetilde{G}_{22}}$ is a stable polynomial. Since

$$\Delta_{\widetilde{G}}/\psi_{\widetilde{G}_{22}} = \eta \left(\prod_{q=1}^{n} h_q \right) \det(A_{t1} - M_{13} B_{t1})/\psi_{\widetilde{G}_{22}},$$

this polynomial is Hurwitz iff each sub-block G_q has no unstable hidden poles and the polynomial $\det(A_{t1} - M_{13}B_{t1})/\psi_{\widetilde{G}_{22}}$ is Hurwitz. The fact that $\det(A_{t1} - M_{13}B_{t1})/\psi_{\widetilde{G}_{22}}$ is a polynomial easily follows from the equation

$$\widetilde{G}_{22} = M_{33} G_t (I - M_{13} G_t)^{-1} M_{12} + M_{32} = M_{33} B_{t1} (A_{t1} - M_{13} B_{t1})^{-1} M_{12} + M_{32}$$

for which the following inequality on the McMillan degrees holds:

$$\delta((A_{t1} - M_{13}B_{t1})^{-1}; s_o) \geq \delta(\widetilde{G}_{22}; s_o),$$

where s_o is any finite pole of $\widetilde{G}_{22}$.

Example 5.6
Find the optimal Wiener–Hopf controller using the formula in Theorem 5.2 when W is stable, G_w is strictly acceptable, and (5.140) is given by

$$G_N = \begin{bmatrix} WG_{11}G_w & WG_{12} \\ G_{21}G_w & G_{22} \end{bmatrix} = \begin{bmatrix} \frac{\lambda}{1-\lambda} & \frac{-\lambda}{1-\lambda} \\ \frac{\lambda}{1-\lambda} & \frac{-\lambda}{1-\lambda} \end{bmatrix} = \left[\begin{array}{c|cc} 1 & 1 & -1 \\ \hline 1 & 0 & 0 \\ 1 & 0 & 0 \end{array}\right].$$

(This example shows that the frequency-domain approach in Theorem 5.2 provides a solution while the state-space formula in Theorem 5.3 does not: the state-space parameters D_{12} and D_{21} are zero and Assumption 5.9′ is not met.)

Solution
The eight parameters satisfying the generalized Bezout identity in (5.113) are given by $A = A_1 = 1 - \lambda$, $B = B_1 = \lambda$, $X = X_1 = 1$, and $Y = Y_1 = 1$. In addition, $\nabla_b = WG_{12}A_1 = -\lambda$ and $\nabla_c = AG_{21}G_w = \lambda$; hence, Assumptions 5.7 and 5.8 are satisfied. The Wiener–Hopf spectral factors are therefore $\Lambda = \Omega = 1$ and it can be readily shown that

$$\Gamma_\phi = \frac{-1}{\lambda} + \frac{1}{\lambda - 1} \text{ and } \Lambda A_1^{-1} Y \Omega = \frac{1}{1-\lambda}.$$

Now it follows that

$$\widetilde{\Gamma} = \Gamma_\phi + \Lambda A_1^{-1} Y \Omega = -\frac{1}{\lambda} \text{ and } \{\widetilde{\Gamma}\}_+ + \{\widetilde{\Gamma}\}_\infty = 0.$$

Therefore,

$$K_{opt} = -\Lambda^{-1}(\{\widetilde{\Gamma}\}_\infty + \{\widetilde{\Gamma}\}_+)\Omega^{-1} = 0,$$

and

$$C_{opt}(\lambda) = (X_1 - K_{opt}B)^{-1}(Y_1 + K_{opt}A) = X_1^{-1}Y_1 = 1.$$

Example 5.7
Establish the equality in (5.149).

Solution
It follows from the inversion identity stated at the end of Sect. 5.4 that

$$\begin{aligned}(I+N_2(D_{12}'D_{12})^{-1}N_2'M_1)^{-1} &= I - N_2(N_2'M_1N_2 + D_{12}'D_{12})^{-1}N_2'M_1 \\ &= I - N_2E_1^{-1}N_2'M_1;\end{aligned}$$

Hence, one gets from (5.146) that

$$\Sigma_N'M_1(I - N_2E_1^{-1}N_2'M_1)\Sigma_N + H_1'(I - D_{12}(D_{12}'D_{12})^{-1}D_{12}')H_1 - M_1 = 0.$$

Expanding the above equation yields

$$\begin{aligned}&\Sigma'M_1\Sigma + H_1'D_{12}(D_{12}'D_{12})^{-1}(N_2'M_1N_2 - E_1)E_1^{-1}N_2'M_1\Sigma \\ &+\Sigma'M_1N_2E_1^{-1}(N_2'M_1N_2 - E_1)(D_{12}'D_{12})^{-1}D_{12}'H_1 \\ &-\Sigma'M_1N_2E_1^{-1}N_2'M_1\Sigma + H_1'D_{12}(D_{12}'D_{12})^{-1}((E_1 - N_2'M_1N_2)E_1^{-1}N_2'M_1N_2 \\ &-D_{12}'D_{12})(D_{12}'D_{12})^{-1}D_{12}'H_1 + H_1'H_1 - M_1 = 0\end{aligned}$$

Using the identity $E_1 - N_2'M_1N_2 = D_{12}'D_{12}$ yields

$$\begin{aligned}&\Sigma'M_1\Sigma - H_1'D_{12}E_1^{-1}N_2'M_1\Sigma - \Sigma'M_1N_2E_1^{-1}D_{12}'H_1 - \Sigma'M_1N_2E_1^{-1}N_2'M_1\Sigma \\ &- H_1'D_{12}E_1^{-1}D_{12}'H_1 + H_1'H_1 - M_1 = 0\end{aligned}$$

which simplifies to $\Sigma'M_1\Sigma + H_1'H_1 - M_1 = L_1'E_1L_1$. The identity (5.111) then gives

$$M_1 - \Sigma'M_1\Sigma = (\lambda I - \Sigma')M_1(\lambda^{-1}I - \Sigma) + \Sigma'M_1(\lambda^{-1}I - \Sigma) + (\lambda I - \Sigma')M_1\Sigma$$

and one gets

$$H_1'H_1 - \Sigma'M_1(\lambda^{-1}I - \Sigma) - (\lambda I - \Sigma')M_1\Sigma - (\lambda I - \Sigma')M_1(\lambda^{-1}I - \Sigma) = L_1'E_1L_1.$$

Multiplying this equation by $N_2'(\lambda I - \Sigma')^{-1}$ on the left and by $(\lambda^{-1}I - \Sigma)^{-1}N_2$ on the right leads to

$$N_2'(\lambda I-\Sigma')^{-1}H_1'H_1(\lambda^{-1}I-\Sigma)^{-1}N_2-N_2'(\lambda I-\Sigma')^{-1}\Sigma' M_1N_2$$
$$-N_2'M_1\Sigma(\lambda^{-1}I-\Sigma)^{-1}N_2-N_2'M_1N_2$$
$$=N_2'(\lambda I-\Sigma')^{-1}L_1'E_1L_1(\lambda^{-1}I-\Sigma)^{-1}N_2.$$

Introducing $N_2'M_1\Sigma=E_1L_1-D_{12}'H_1, N_2'M_1N_2=E_1-D_{12}'D_{12}$ which follow from (5.145), then yields

$$N_2'(\lambda I-\Sigma')^{-1}H_1'H_1(\lambda^{-1}I-\Sigma)^{-1}N_2+N_2'(\lambda I-\Sigma')^{-1}H_1'D_{12}+D_{12}'H_1(\lambda^{-1}I-\Sigma)^{-1}N_2+D_{12}'D_{12}$$
$$=N_2'(\lambda I-\Sigma')^{-1}L_1'E_1L_1(\lambda^{-1}I-\Sigma)^{-1}N_2+N_2'(\lambda I-\Sigma')^{-1}L_1'E_1+E_1L_1(\lambda^{-1}I-\Sigma)^{-1}N_2+E_1.$$

Since $WG_{12}=H_1(\lambda^{-1}I-\Sigma)^{-1}N_2+D_{12}$ from (5.14), it is not difficult to confirm that the previous equation is equivalent to

$$(W(\lambda)G_{12}(\lambda))_*W(\lambda)G_{12}(\lambda)=\left(N_2'(\lambda I-\Sigma')^{-1}L_1'+I\right)E_1\left(L_1(\lambda^{-1}I-\Sigma)^{-1}N_2+I\right).$$

Example 5.8
Prove that $\{\Gamma_\phi\}_+ + \{\Gamma_\phi\}_\infty - \{\Lambda A_1^{-1}Y\Omega\}_- = E_1^{\frac{1}{2}}L_1(\lambda^{-1}I-\Sigma)^{-1}L_2E_2^{\frac{1}{2}}$ in Lemma 5.10.

Solution
One gets from (5.143) that

$$\Lambda A_1^{-1}Y\Omega=\widetilde{A}_1^{-1}Y\Omega=E_1^{\frac{1}{2}}\left(I+L_1(\lambda^{-1}I-\Sigma)^{-1}N_2\right)Y\Omega$$
$$=E_1^{\frac{1}{2}}Y\Omega+E_1^{\frac{1}{2}}L_1(\lambda^{-1}I-\Sigma)^{-1}N_2Y\Omega$$

and this leads to $\{\Lambda A_1^{-1}Y\Omega\}_-=E_1^{\frac{1}{2}}L_1\{(\lambda^{-1}I-\Sigma)^{-1}N_2Y\Omega\}_-$. Employing (5.156) then gives

$$\begin{aligned}\{\Lambda A_1^{-1}Y\Omega\}_- &= E_1^{\frac{1}{2}}L_1\{(\lambda^{-1}I-\Sigma)^{-1}N_2Y(A+V_2^{-1}B_3L_2)\}_-E_2^{\frac{1}{2}}\\ &= E_1^{\frac{1}{2}}L_1\{(\lambda^{-1}I-\Sigma)^{-1}N_2YV_2^{-1}B_3\}_-L_2E_2^{\frac{1}{2}}\end{aligned}\qquad\text{(E5.10)}$$

which follows from the fact that $(\lambda^{-1}I-\Sigma)^{-1}N_2YA=B_2V_1^{-1}A_1^{-1}YA=B_2V_1^{-1}Y_1$ is analytic in $|\lambda|<1$. Since $-A^{-1}B=H_2(\lambda^{-1}I-\Sigma)^{-1}N_2+D_{22}=A_3^{-1}B_3N_2+D_{22}$, it is clear that

$$B = -(AA_3^{-1}B_3N_2 + AD_{22}) = -(V_2^{-1}B_3N_2 + AD_{22}).$$

From $AX + BY = I$, moreover, one gets $AX - (V_2^{-1}B_3N_2 + AD_{22})Y = I$. Multiplying by A^{-1} on the left and by $V_2^{-1}B_3$ on the right then yields $XV_2^{-1}B_3 - A_3^{-1}B_3N_2YV_2^{-1}B_3 - D_{22}YV_2^{-1}B_3 = A^{-1}V_2^{-1}B_3$. Since $A^{-1}V_2^{-1}B_3 = A_3^{-1}B_3 = H_2(\lambda^{-1}I - \Sigma)^{-1}$ it follows that

$$(X - D_{22}Y)V_2^{-1}B_3 - H_2(\lambda^{-1}I - \Sigma)^{-1}N_2YV_2^{-1}B_3 = H_2(\lambda^{-1}I - \Sigma)^{-1}.$$

Now V_2^{-1} is stable; hence,

$$-H_2\{(\lambda^{-1}I - \Sigma)^{-1}N_2YV_2^{-1}B_3\}_- = H_2\{(\lambda^{-1}I - \Sigma)^{-1}\}_-. \tag{E5.11}$$

When (Σ, H_2) is a detectable pair there is no loss in generality in assuming

$$\Sigma = \begin{bmatrix} \Sigma_{11} & 0 \\ \Sigma_{21} & \Sigma_{22} \end{bmatrix}, H_2 = [H_{20} \quad 0], N_2 = \begin{bmatrix} N_{21} \\ N_{22} \end{bmatrix}, \tag{E5.12}$$

where Σ_{22} is a stable matrix and (Σ_{11}, H_{20}) is an observable pair. In this case,

$$(\lambda^{-1}I - \Sigma)^{-1} = \begin{bmatrix} (\lambda^{-1}I - \Sigma_{11})^{-1} & 0 \\ (\lambda^{-1}I - \Sigma_{22})^{-1}\Sigma_{21}(\lambda^{-1}I - \Sigma_{11})^{-1} & (\lambda^{-1}I - \Sigma_{22})^{-1} \end{bmatrix} \tag{E5.13}$$

and equation (E5.11) becomes

$$-H_{20}\{(\lambda^{-1}I - \Sigma_{11})^{-1}N_{21}YV_2^{-1}B_3\}_- = H_{20}\{[(\lambda^{-1}I - \Sigma_{11})^{-1} \ 0]\}_-. \tag{E5.14}$$

It is established next that $-\{(\lambda^{-1}I - \Sigma_{11})^{-1}N_{21}YV_2^{-1}B_3\}_- = \{[(\lambda^{-1}I - \Sigma_{11})^{-1} \ 0]\}_-$. Using the identity

$$(\lambda^{-1}I - \Sigma_{11})^{-1} = \sum_{l=1}^{k} \lambda^l \Sigma_{11}^{l-1} + \lambda^k \Sigma_{11}^k (\lambda^{-1}I - \Sigma_{11})^{-1}, \ k \geq 1$$

and the fact that $\left\{\left(\sum_{l=1}^{k} \lambda^l \Sigma_{11}^{l-1}\right) N_{21}YV_2^{-1}B_3\right\}_- = 0$ one gets

$$\{(\lambda^{-1}I - \Sigma_{11})^{-1}N_{21}YV_2^{-1}B_3\}_- = \Sigma_{11}^k\{\lambda^k(\lambda^{-1}I - \Sigma_{11})^{-1}N_{21}YV_2^{-1}B_3\}_-. \tag{E5.15}$$

Similarly it is also true that

$$\{(\lambda^{-1}I - \Sigma_{11})^{-1}\}_- = \Sigma_{11}^k\{\lambda^k(\lambda^{-1}I - \Sigma_{11})^{-1}\}_-, k \geq 1. \tag{E5.16}$$

Inserting (E5.15) and (E5.16) into (E5.14) then yields ($k = 0$ corresponds to (E5.14))

$$H_{20}\Sigma_{11}^{k}\{\lambda^{k}(\lambda^{-1}I-\Sigma_{11})^{-1}N_{21}YV_{2}^{-1}B_{3}\}_{-} = -H_{20}\Sigma_{11}^{k}\{[\lambda^{k}(\lambda^{-1}I-\Sigma_{11})^{-1} \quad 0]\}_{-},\ k\geq 0.$$

Thus, $\{\lambda^{k}\Phi(\lambda)\}_{-} = 0, k\geq 0$, where

$$\Phi(\lambda) = H_{20}\Sigma_{11}^{k}\left((\lambda^{-1}I-\Sigma_{11})^{-1}N_{21}YV_{2}^{-1}B_{3} + [(\lambda^{-1}I-\Sigma_{11})^{-1} \quad 0]\right).$$

In general, for a rational matrix $\Phi(\lambda)$, $\{\lambda^{k}\Phi(\lambda)\}_{-} = 0$, $k\geq 0$, implies $\Phi(\lambda)$ is analytic in $|\lambda|<1$ except possibility at the point $\lambda = 0$. Since $(\lambda^{-1}I-\Sigma_{11})^{-1}$ is analytic at $\lambda = 0$, $\Phi(\lambda)$ is analytic at $\lambda = 0$; hence, $\{\Phi(\lambda)\}_{-} = 0$. This leads to

$$H_{20}\Sigma_{11}^{k}\left\{\left((\lambda^{-1}I-\Sigma_{11})^{-1}N_{21}YV_{2}^{-1}B_{3} + [(\lambda^{-1}I-\Sigma_{11})^{-1} \quad 0]\right)\right\}_{-} = 0,\ k\geq 0. \tag{E5.17a}$$

Since the observability matrix for (Σ_{11}, H_{20}) has full column rank, it follows that

$$\{(\lambda^{-1}I-\Sigma_{11})^{-1}N_{21}YV_{2}^{-1}B_{3}\}_{-} + \{[(\lambda^{-1}I-\Sigma_{11})^{-1} \quad 0]\}_{-} = 0.$$

In addition, (E5.12) and (E5.13) yield

$$(\lambda^{-1}I-\Sigma)^{-1}N_{2}YV_{2}^{-1}B_{3} = \begin{bmatrix} (\lambda^{-1}I-\Sigma_{11})^{-1}N_{21}YV_{2}^{-1}B_{3} \\ (\lambda^{-1}I-\Sigma_{22})^{-1}\Sigma_{21}(\lambda^{-1}I-\Sigma_{11})^{-1}N_{21}YV_{2}^{-1}B_{3} + (\lambda^{-1}I-\Sigma_{22})^{-1}N_{22}YV_{2}^{-1}B_{3} \end{bmatrix}.$$

Since $(\lambda^{-1}I-\Sigma_{22})^{-1}$ and V_{2}^{-1} are stable, it can now be concluded with the aid of (E5.17a) that

$$\begin{aligned}\{(\lambda^{-1}I-\Sigma)^{-1}N_{2}YV_{2}^{-1}B_{3}\}_{-} &= \begin{bmatrix} \{(\lambda^{-1}I-\Sigma_{11})^{-1}N_{21}YV_{2}^{-1}B_{3}\}_{-} \\ \{(\lambda^{-1}I-\Sigma_{22})^{-1}\Sigma_{21}(\lambda^{-1}I-\Sigma_{11})^{-1}N_{21}YV_{2}^{-1}B_{3}\}_{-} \end{bmatrix} \\ &= \begin{bmatrix} \{(\lambda^{-1}I-\Sigma_{11})^{-1}N_{21}YV_{2}^{-1}B_{3}\}_{-} \\ \{(\lambda^{-1}I-\Sigma_{22})^{-1}\Sigma_{21}\{(\lambda^{-1}I-\Sigma_{11})^{-1}N_{21}YV_{2}^{-1}B_{3}\}_{-}\}_{-} \end{bmatrix} \\ &= \left[\begin{array}{c|c} -\{(\lambda^{-1}I-\Sigma_{11})^{-1}\}_{-} & 0 \\ \hline -\{(\lambda^{-1}I-\Sigma_{22})^{-1}\Sigma_{21}\{(\lambda^{-1}I-\Sigma_{11})^{-1}\}_{-}\}_{-} & 0 \end{array}\right]\end{aligned}$$

$$= -\left\{ \begin{bmatrix} (\lambda^{-1}I - \Sigma_{11})^{-1} & 0 \\ (\lambda^{-1}I - \Sigma_{22})^{-1}\Sigma_{21}(\lambda^{-1}I - \Sigma_{11})^{-1} & (\lambda^{-1}I - \Sigma_{22})^{-1} \end{bmatrix} \right\}_{-} = -\{(\lambda^{-1}I - \Sigma)^{-1}\}_{-}. \tag{E5.17b}$$

Hence, one gets from (E5.10) and (E5.17b) that $\{\Lambda A_1^{-1} Y \Omega\}_{-} = -E_1^{\frac{1}{2}} L_1 \{(\lambda^{-1}I - \Sigma)^{-1}\}_{-} L_2 E_2^{\frac{1}{2}}$.

It is established next that

$$\begin{aligned} \{\Gamma_\phi\}_{+,\infty} &= \{(WG_{12}\widetilde{A}_1)_*(WG_{11}G_w)(\widetilde{A}G_{21}G_w)_*\}_{+,\infty} \\ &= E_1^{\frac{1}{2}} L_1 \{(\lambda^{-1}I - \Sigma)^{-1}\}_{+,\infty} L_2 E_2^{\frac{1}{2}}, \end{aligned}$$

where $\{\cdot\}_{+,\infty} = \{\cdot\}_{+} + \{\cdot\}_{\infty}$. First, state-space parameters for $(WG_{12}\widetilde{A}_1)_*$ are sought. From the state-space parameters for WG_{12} and $\widetilde{A}_1$ in (5.140) and (5.144), it follows that

$$(WG_{12})\tilde{A}_1 = \left[\begin{array}{cc|c} \Sigma & -N_2 L_1 & N_2 E_1^{-\frac{1}{2}} \\ 0 & \Sigma - N_2 L_1 & N_2 E_1^{-\frac{1}{2}} \\ \hline H_1 & -D_{12} L_1 & D_{12} E_1^{-\frac{1}{2}} \end{array}\right] = \left[\begin{array}{c|c} \Sigma - N_2 L_1 & N_2 E_1^{-\frac{1}{2}} \\ \hline H_1 - D_{12} L_1 & D_{12} E_1^{-\frac{1}{2}} \end{array}\right].$$

Similarly, the state-space parameters for $\widetilde{A}G_{21}G_w$ are given by

$$\tilde{A}(G_{21}G_w) = \left[\begin{array}{cc|c} \Sigma - L_2 H_2 & L_2 H_2 & L_2 D_{21} \\ 0 & \Sigma & N_1 \\ \hline -E_2^{-\frac{1}{2}} H_2 & E_2^{-\frac{1}{2}} H_2 & E_2^{-\frac{1}{2}} D_{21} \end{array}\right] = \left[\begin{array}{c|c} \Sigma - L_2 H_2 & N_1 - L_2 D_{21} \\ \hline E_2^{-\frac{1}{2}} H_2 & E_2^{-\frac{1}{2}} D_{21} \end{array}\right].$$

Hence,

$$\begin{aligned} (WG_{12}\widetilde{A}_1)_*(WG_{11}G_w) &= (H_{1l}(\lambda^{-1}I - \Sigma_n')^{-1} N_2 E_1^{-\frac{1}{2}} + D_{12} E_1^{-\frac{1}{2}})_*(H_1(\lambda^{-1}I - \Sigma)^{-1} N_1 + D_{11}) \\ &= E_1^{-\frac{1}{2}}(N_2'(\lambda I - \Sigma_n)^{-1} H_{1l}' H_1 (\lambda^{-1}I - \Sigma)^{-1} N_1 + N_2'(\lambda I - \Sigma_n)^{-1} H_{1l}' D_{11} \\ &\quad + D_{12}' H_1 (\lambda^{-1}I - \Sigma)^{-1} N_1 + D_{12}' D_{11}), \end{aligned}$$

where $\Sigma_n' = \Sigma - N_2 L_1$ and $H_{1l} = H_1 - D_{12} L_1$. The Eq. (5.147) changes to $H_{1l}' H_1 = -\Sigma_n M_1 \Sigma + M_1$ since

$$\begin{aligned} &\Sigma' M_1 \Sigma - L_1' E_1 L_1 + H_1' H_1 - M_1 \\ &= \Sigma' M_1 \Sigma - L_1'(D_{12}' H_1 + N_2' M_1 \Sigma) + H_1' H_1 - M_1 \\ &= \Sigma' M_1 \Sigma - L_1' N_2' M_1 \Sigma - L_1' D_{12}' H_1 + H_1' H_1 - M_1 \\ &= (\Sigma' - L_1' N_2') M_1 \Sigma + (H_1' - L_1' D_{12}') H_1 - M_1. \end{aligned}$$

Using the identity

$$M_1 - \Sigma_n M_1 \Sigma = (\lambda I - \Sigma_n) M_1 (\lambda^{-1} I - \Sigma) + \Sigma_n M_1 (\lambda^{-1} I - \Sigma) + (\lambda I - \Sigma_n) M_1 \Sigma,$$

one gets

$$H'_{1l} H_1 = (\lambda I - \Sigma_n) M_1 \Sigma + \Sigma_n M_1 (\lambda^{-1} I - \Sigma) + (\lambda I - \Sigma_n) M_1 (\lambda^{-1} I - \Sigma).$$

Multiplying by $N'_2 (\lambda I - \Sigma_n)^{-1}$ on the left and by $(\lambda^{-1} I - \Sigma)^{-1} N_1$ on the right then yields

$$\begin{aligned} N'_2 (\lambda I - \Sigma_n)^{-1} H'_{1l} H_1 (\lambda^{-1} I - \Sigma)^{-1} N_1 &= N'_2 M_1 \Sigma (\lambda^{-1} I - \Sigma)^{-1} N_1 \\ &+ N'_2 (\lambda I - \Sigma_n)^{-1} \Sigma_n M_1 N_1 + N'_2 M_1 N_1. \end{aligned}$$

Therefore, it follows that

$$(WG_{12}\widetilde{A}_1)_* (WG_{11} G_w) = E_1^{-\frac{1}{2}} (\Phi_n(\lambda) + E_1 L_1 (\lambda^{-1} I - \Sigma)^{-1} N_1 + D_a),$$

where

$$\Phi_n(\lambda) = N'_2 (\lambda I - \Sigma_n)^{-1} (H'_{1l} D_{11} + \Sigma_n M_1 N_1) \text{ and } D_a = D'_{12} D_{11} + N'_2 M_1 N_1.$$

Since Σ_n is a stable matrix, $\Phi_n(\lambda)$ is analytic in $|\lambda| \geq 1$ and strictly proper. Next, define $\Sigma'_h = \Sigma - L_2 H_2$ and $N_{1l} = N_1 - L_2 D_{21}$. Since $\widetilde{A} G_{21} G_w = E_2^{-\frac{1}{2}} H_2 (\lambda^{-1} I - \Sigma'_h)^{-1} N_{1l} + E_2^{-\frac{1}{2}} D_{21}$ and Σ_h is stable, the matrix $(\widetilde{A} G_{21} G_w)_* = (N'_{1l} (\lambda I - \Sigma_h)^{-1} H'_2 + D'_{21}) E_2^{-\frac{1}{2}}$ is analytic in $|\lambda| \geq 1$ and proper. Hence, $\Phi_n (\widetilde{A} G_{21} G_w)_*$ has no poles in $|\lambda| \geq 1$, $\{\Phi_n (\widetilde{A} G_{21} G_w)_*\}_{+,\infty} = 0$ and it follows from

$$\begin{aligned} \Gamma_\phi &= (WG_{12}\widetilde{A}_1)_* (WG_{11} G_w)(\widetilde{A} G_{21} G_w)_* \\ &= E_1^{-\frac{1}{2}} (\Phi_n(\lambda) + E_1 L_1 (\lambda^{-1} I - \Sigma)^{-1} N_1 + D_a)(\widetilde{A} G_{21} G_w)_* \end{aligned}$$

that

$$\begin{aligned} \{\Sigma_\phi\}_{+,\infty} &= E_1^{-\frac{1}{2}} \{(E_1 L_1 (\lambda^{-1} I - \Sigma)^{-1} N_1 + D_a)(\widetilde{A} G_{21} G_w)_*\}_{+,\infty} \\ &= E_1^{-\frac{1}{2}} \{D_a D'_{21} + E_1 L_1 (\lambda^{-1} I - \Sigma)^{-1} N_1 N'_{1l} (\lambda I - \Sigma_h)^{-1} H'_2 \\ &+ E_1 L_1 (\lambda^{-1} I - \Sigma)^{-1} N_1 D'_{21}\}_{+,\infty} E_2^{-\frac{1}{2}}. \end{aligned} \tag{E5.18}$$

The Eq. (5.162) is changed to $N_1 N'_{1l} = M_2 - \Sigma M_2 \Sigma_h$ since

$$\begin{aligned}\Sigma M'_2 \Sigma' - L_2 E_2 L'_2 + N_1 N'_1 - M_2 &= \Sigma M_2 \Sigma' - (N_1 D'_{21} + \Sigma M_2 H'_2) L'_2 + N_1 N'_1 - M_2 \\ &= \Sigma M_2 \Sigma' - \Sigma M_2 H'_2 L'_2 + N_1 N'_1 - N_1 D'_{21} L'_2 - M_2 \\ &= \Sigma M_2 (\Sigma' - H'_2 L'_2) + N_1 (N'_1 - D'_{21} L'_2) - M_2.\end{aligned}$$

Using the identity

$$M_2 - \Sigma M_2 \Sigma_h = (\lambda^{-1} I - \Sigma) M_2 (\lambda I - \Sigma_h) + \Sigma M_2 (\lambda I - \Sigma_h) + (\lambda^{-1} I - \Sigma) M_2 \Sigma_h,$$

one gets

$$N_1 N'_{1l} = (\lambda^{-1} I - \Sigma) M_2 (\lambda I - \Sigma_h) + \Sigma M_2 (\lambda I - \Sigma_h) + (\lambda^{-1} I - \Sigma) M_2 \Sigma_h.$$

Thus,

$$\begin{aligned}&E_1 L_1 (\lambda^{-1} I - \Sigma)^{-1} N_1 N'_{1l} (\lambda I - \Sigma_h)^{-1} H'_2 \\ &= E_1 L_1 M_2 H'_2 + E_1 L_1 (\lambda^{-1} I - \Sigma)^{-1} \Sigma M_2 H'_2 + E_1 L_1 M_2 \Sigma_h (\lambda I - \Sigma_h)^{-1} H'_2.\end{aligned}$$

Inserting this equality into (E5.18) and noticing that $(\lambda I - \Sigma_h)^{-1}$ is analytic in $|\lambda| \geq 1$ yields

$$\begin{aligned}\{\Gamma_\phi\}_{+,\infty} &= E_1^{-\frac{1}{2}} \{D_a D'_{21} + E_1 L_1 M_2 H'_2 + E_1 L_1 (\lambda^{-1} I - \Sigma)^{-1} (\Sigma M_2 H'_2 + N_1 D'_{21})\}_{+,\infty} E_2^{-\frac{1}{2}} \\ &= E_1^{\frac{1}{2}} L_1 \{(\lambda^{-1} I - \Sigma)^{-1} L_2\}_{+,\infty} E_2^{\frac{1}{2}} + E_1^{\frac{1}{2}} D_\phi E_2^{\frac{1}{2}},\end{aligned}$$

where $D_\phi = \left(E_1^{-1} (D'_{12} D_{11} + N'_2 M_1 N_1) D'_{21} + L_1 M_2 H'_2\right) E_2^{-1}$. Now it easily follows from (E5.10) and (E5.17b) that

$$\begin{aligned}&\{\Gamma_\phi\}_{+,\infty} - \{\Lambda A_1^{-1} Y \Omega\}_- = E_1^{\frac{1}{2}} L_1 (\{(\lambda^{-1} I - \Sigma)^{-1}\}_{+,\infty} + \{(\lambda^{-1} I - \Sigma)^{-1}\}_-) L_2 E_2^{\frac{1}{2}} + E_1^{\frac{1}{2}} D_\phi E_2^{\frac{1}{2}} \\ &= E_1^{\frac{1}{2}} L_1 (\lambda^{-1} I - \Sigma)^{-1} L_2 E_2^{\frac{1}{2}} + E_1^{\frac{1}{2}} D_\phi E_2^{\frac{1}{2}}.\end{aligned}$$

Example 5.9

The system matrix introduced in (5.9) is a special case of the more general one given by

$$S(s) = \begin{bmatrix} -T(s) & U(s) \\ V(s) & W(s) \end{bmatrix} = \left[\begin{array}{c|cc} -T(s) & U_1(s) & U_2(s) \\ V_1(s) & W_{11}(s) & W_{12}(s) \\ V_2(s) & W_{21}(s) & W_{22}(s) \end{array}\right]$$

which is associated with a generalized plant having the polynomial matrix description (PMD) of (2.10) through (2.12). All the submatrices in $S(s)$ are polynomial matrices and attention is restricted to nondegenerate systems ($\det T(s) \not\equiv 0$).

Clearly, (5.9) corresponds to the special case $T = sI - \Sigma$, $U = N$, $V = H$, and $W = D$, where Σ, N, H, and D are constant matrices. By analogy then it is natural to call (T, U) stabilizable when $[\,T(s) \quad U(s)\,]$ has full row rank in Re$s \geq 0$ and (T, V) detectable when $[\,T'(s) \quad V'(s)\,]'$ has full column rank in Re$s \geq 0$. Connected with the PMD are two rational transfer matrices given by

$$G(s) = V(s)T^{-1}(s)U(s) + W(s) = B_1(s)A_1^{-1}(s)$$

and

$$G_{22}(s) = V_2(s)T^{-1}(s)U_2(s) + W_{22}(s) = \mathcal{B}_1(s)\mathcal{A}_1^{-1}(s),$$

where $B_1(s)A_1^{-1}(s)$ and $\mathcal{B}_1(s)\mathcal{A}_1^{-1}(s)$ are right coprime polynomial matrix fraction descriptions for $G(s)$ and $G_{22}(s)$, respectively. From Sect. 2.2, $\Delta(s) = \mu \det T(s)$ is the characteristic polynomial for the generalized plant and $\psi_{G_{22}}(s)$ is the characteristic denominator for $G_{22}(s)$. Moreover, $h_2(s) = \Delta/\psi_{G_{22}}$ is a polynomial and when it is stable $G(s)$ is called admissible (Definition 2.2). Clearly, $\Delta(s)$ and $\psi_{G_{22}}(s)$ for $G_{22}(s)$ are the same as for $G(s)$; hence, $G(s)$ is admissible iff $G_{22}(s)$ is free of unstable hidden poles or, equivalently, $G_{22}(s)$ is admissible.

(a) When both $[\,T(s) \quad U(s)\,]$ and $[\,T'(s) \quad V'(s)\,]$ have full row rank in some region $\Re$ of the s-plane, establish that

$$\text{rank } S(s) = \text{rank}(\text{diag}\{I, B_1(s)\}, s \in \Re.$$

(b) Clearly, $\Delta = \mu \det T = h_2(s)\psi_{G_{22}}(s)$, where $h_2(s)$ and $\psi_{G_{22}}(s) = \eta_1 \det \mathcal{A}_1(s)$ are monic polynomials. Establish that $h_2(s)$ has no zeros in $\Re$ iff $[\,T(s) \quad U_2(s)\,]$ and $[\,T'(s) \quad V_2'(s)\,]$ have full row rank in $\Re$. When $\Re$ is the half-plane Re$s \geq 0$, it then follows that G is admissible iff (T, U_2) is stabilizable and (T, V_2) is detectable.

(c) When in (a) $\Re$ is the entire s-plane, establish that $S(s)$ and diag$\{I, B_1(s)\}$ have the same Smith form (see Section B.4).

(d) Apply the result in (c) to the state-space system matrix in (5.9) to establish as stated in the paragraph containing Eq. (5.9) that the state-space system matrix $S(s)$ has the same Smith form as diag$\{I, B_1(s)\}$ when (Σ, N, H) is minimal or, equivalently, the system is controllable and observable.

Solution

(a) Since $\det T(s) \not\equiv 0$, both $[\,T(s) \quad U(s)\,]$ and $[\,T'(s) \quad V'(s)\,]$ have full normal row rank which equals n the number of rows of the square matrix $T(s)$. It then follows from the Smith form (see Section B.4) that $[\,-T(s) \quad U(s)\,] = M(s)[\,E(s) \quad 0\,]L(s)$, where $M(s)$ and $L(s)$ are unimodular polynomial matrices and

$$E(s) = \text{diag}\{e_1(s), e_2(s), \ldots, e_n(s)\}.$$

The polynomials $e_i(s)$, $i = 1 \to n$, are the invariant polynomials for $[-T(s) \quad U(s)]$. Moreover, $R(s) = L^{-1}(s)$ is a unimodular polynomial matrix and one has

$$LR = \begin{bmatrix} L_{11} & L_{12} \\ L_{21} & L_{22} \end{bmatrix} \begin{bmatrix} R_{11} & R_{12} \\ R_{21} & R_{22} \end{bmatrix} = \begin{bmatrix} I & 0 \\ 0 & I \end{bmatrix}.$$

Hence, $L_{11}R_{11} + L_{12}R_{21} = I$ and L_{11}, L_{12} is a left coprime pair. Clearly,

$$[-T \quad U] = [ME \quad 0] \begin{bmatrix} L_{11} & L_{12} \\ L_{21} & L_{22} \end{bmatrix} = [MEL_{11} \quad MEL_{12}],$$

$$\begin{bmatrix} -T \\ V \end{bmatrix} = \begin{bmatrix} MEL_{11} \\ V \end{bmatrix} = \begin{bmatrix} ME & 0 \\ 0 & I \end{bmatrix} \begin{bmatrix} L_{11} \\ V \end{bmatrix}, \text{and } S = \begin{bmatrix} ME & 0 \\ 0 & I \end{bmatrix} \begin{bmatrix} L_{11} & L_{12} \\ V & W \end{bmatrix}.$$

Now except at the zeros of the invariant polynomials, ME is nonsingular. Thus, the normal row rank of $[L'_{11} \quad V']$ is n, the normal row rank of $[-T' \quad V']$. So one can again invoke the Smith form to get

$$\begin{bmatrix} L_{11} \\ V \end{bmatrix} = P \begin{bmatrix} \widetilde{E} \\ 0 \end{bmatrix} Q = \begin{bmatrix} P_{11} & P_{12} \\ P_{21} & P_{22} \end{bmatrix} \begin{bmatrix} \widetilde{E}Q \\ 0 \end{bmatrix} = \begin{bmatrix} P_{11}\widetilde{E}Q \\ P_{21}\widetilde{E}Q \end{bmatrix},$$

where P and Q are unimodular polynomial matrices; P_{11}, P_{21} is a right coprime pair, and $\widetilde{E}$ is a diagonal matrix whose elements are the invariant polynomials for $[L'_{11} \quad V']$. Thus,

$$S = \begin{bmatrix} ME & 0 \\ 0 & I \end{bmatrix} \begin{bmatrix} P_{11} & L_{12} \\ P_{21} & W \end{bmatrix} \begin{bmatrix} \widetilde{E}Q & 0 \\ 0 & I \end{bmatrix}.$$

Moreover, from

$$L_{11}R_{11} + L_{12}R_{21} = P_{11}(\widetilde{E}QR_{11}) + L_{12}R_{21} = P_{11}X_c + L_{12}Y_c = I$$

it follows that P_{11}, L_{12} is a left coprime pair. In addition, one can always write $P_{11}^{-1}L_{12} = B_{c1}A_{c1}^{-1}$, where B_{c1}, A_{c1} is a right coprime polynomial matrix pair for which there exists a polynomial matrix pair $X_{c1}(s), Y_{c1}(s)$ such that $X_{c1}A_{c1} + Y_{c1}B_{c1} = I$ and $\det P_{11} = \eta \det A_{c1}$, where η is a nonzero constant. It is now easy to see that

$$\widetilde{S} = \begin{bmatrix} P_{11} & L_{12} \\ P_{21} & W \end{bmatrix}, M_1 = \begin{bmatrix} X_c & B_{c1} \\ Y_c & -A_{c1} \end{bmatrix} \Rightarrow \widetilde{S}M_1 = \begin{bmatrix} P_{11} & L_{12} \\ P_{21} & W \end{bmatrix} \begin{bmatrix} X_c & B_{c1} \\ Y_c & -A_{c1} \end{bmatrix}$$
$$= \begin{bmatrix} I & 0 \\ Z_1 & Z_2 \end{bmatrix},$$

where Z_1 and Z_2 are polynomial matrices. In particular, $Z_1 = P_{21}X_c + WY_c$ and

$$Z_2 = P_{21}B_{c1} - WA_{c1} = (P_{21}B_{c1}A_{c1}^{-1} - W)A_{c1} = (P_{21}P_{11}^{-1}L_{12} - W)A_{c1} = -GA_{c1}$$

follows from

$$G = VT^{-1}U + W = (P_{21}\widetilde{E}Q)(-MEP_{11}\widetilde{E}Q)^{-1}(MEL_{12}) + W = -P_{21}P_{11}^{-1}L_{12} + W.$$

It is also true that

$$M_2M_1 = \begin{bmatrix} P_{11} & L_{12} \\ Y_{c1} & -X_{c1} \end{bmatrix} \begin{bmatrix} X_c & B_{c1} \\ Y_c & -A_{c1} \end{bmatrix} = \begin{bmatrix} I & 0 \\ (Y_{c1}X_c - X_{c1}Y_c) & I \end{bmatrix}.$$

Therefore, $\det(M_2M_1) = (\det M_1)(\det M_2) = 1$ and the polynomial matrices M_1 and M_2 are unimodular. Clearly, $GA_{c1} = B_1A_1^{-1}A_{c1} = -Z_2$ is a polynomial matrix. Moreover, there exists a polynomial matrix pair X_1, Y_1 such that $X_1A_1 + Y_1B_1 = I$. Hence,

$$K = A_1^{-1}A_{c1} = (X_1A_1 + Y_1B_1)A_1^{-1}A_{c1} = X_1A_{c1} + Y_1GA_{c1}$$

is a polynomial matrix. Thus,

$$M_3 = \begin{bmatrix} I & 0 \\ Z_1 & -I \end{bmatrix}, M_4 = \begin{bmatrix} I & 0 \\ 0 & K \end{bmatrix} \Rightarrow M_3\widetilde{S}M_1 = \begin{bmatrix} I & 0 \\ 0 & GA_{c1} \end{bmatrix} = \begin{bmatrix} I & 0 \\ 0 & B_1 \end{bmatrix} M_4.$$

Obviously, M_3 is unimodular and it is established next that M_4 is also. Since P_{21} and W are polynomial matrices, $\delta(G; s_o) \leq \delta(P_{11}^{-1}L_{12}; s_o)$ for any finite pole s_o of $G(s)$. Moreover, there exists a polynomial matrix pair W_1, W_2 such that $W_1P_{11} + W_2P_{21} = I$ because the pair P_{11}, P_{21} is right coprime. Thus,

$$(W_1P_{11} + W_2P_{21})P_{11}^{-1}L_{12} = W_1L_{12} + W_2P_{21}P_{11}^{-1}L_{12} = P_{11}^{-1}L_{12}$$

and one gets

$$\delta(P_{11}^{-1}L_{12}; s_o) \leq \delta(P_{21}P_{11}^{-1}L_{12}; s_o) = \delta((W - G); s_o) = \delta(G; s_o).$$

Hence, $\delta(G;s_o) = \delta(P_{11}^{-1}L_{12};s_o) = \delta(P_{11}^{-1};s_o)$ since P_{11}, L_{12} is a left coprime pair. It immediately follows that

$$\psi_G(s) = \eta_1 \det A_1 = \psi_{P_{11}^{-1}}(s) = \eta_2 \det P_{11} = \eta_3 \det A_{c1},$$

where the constants η_1, η_2, η_3 are introduced to accommodate the fact that the characteristic denominators ψ_G and $\psi_{P_{11}^{-1}}$ are monic polynomials. Clearly,

$$\det M_4 = \det K = \det(A_1^{-1}A_{c1}) = \det A_{c1}/\det A_1 = \eta_1/\eta_3 = \eta$$

is a nonzero constant and M_4 is indeed unimodular. It is also evident that

$$S = \begin{bmatrix} ME & 0 \\ 0 & I \end{bmatrix}\begin{bmatrix} P_{11} & L_{12} \\ P_{21} & W \end{bmatrix}\begin{bmatrix} \widetilde{E}Q & 0 \\ 0 & I \end{bmatrix} = \begin{bmatrix} ME & 0 \\ 0 & I \end{bmatrix}\widetilde{S}\begin{bmatrix} \widetilde{E}Q & 0 \\ 0 & I \end{bmatrix}$$
$$= \Psi(s)\text{diag}\{I, B_1(s)\}\Phi(s),$$

where

$$\Psi(s) = \begin{bmatrix} ME & 0 \\ 0 & I \end{bmatrix}M_3^{-1}, \Phi(s) = M_4M_1^{-1}\begin{bmatrix} \widetilde{E}Q & 0 \\ 0 & I \end{bmatrix}.$$

Moreover, when $[T \quad U]$ and $[T' \quad V']$ have full row rank in $\Re$, there can be no point in $\Re$ that is a zero of any of the invariant polynomials. In this case, E and $\widetilde{E}$ are nonsingular for every $s \in \Re$ and, therefore, so are Ψ and Φ. It immediately follows that

$$\text{rank}S(s) = \text{rank}(\text{diag}\{I, B_1(s)\}), s \in \Re.$$

(b) It is not difficult to confirm that the solution in (a) still holds with U, V, W, and G replaced with U_2, V_2, W_{22}, and G_{22}. Now $T = -MEL_{11} = -MEP_{11}\widetilde{E}Q$ again, but

$$\det T = (-1)^n(\det M)(\det E)(\det P_{11})(\det \widetilde{E})(\det Q) = \eta h_2(s)\psi_{G_{22}}(s),$$

where $\eta = (-1)^n(\det M)(\det Q)/\eta_2$, $h_2(s) = (\det E)(\det \widetilde{E})$, and $\psi_{G_{22}}(s) = \eta_1 \det \mathcal{A}_1 = \eta_2 \det P_{11}$ Clearly, $E(s)$ is nonsingular in $\Re$ iff $[T \quad U_2]$. has full row rank in $\Re$. Moreover, when $E(s)$ is nonsingular in $\Re$, then $\widetilde{E}(s)$is nonsingular in $\Re$ iff $[T' \quad V_2']$ has full row rank in $\Re$. Hence, the polynomial $h_2(s)$has no zeros in $\Re$ iff $[T \quad U_2]$ and $[T' \quad V_2']$ have full row rank in $\Re$ Clearly, when $\Re$ is the half-plane Re$s \geq 0$, G is admissible iff (T, U_2) is stabilizable and (T, V_2) is detectable.

(c) When $\Re$ is the entire $s-$ plane, $E = \widetilde{E} = I$; hence, Ψ and Φ are unimodular. It immediately follows that $S(s)$ and $\operatorname{diag}\{I, B_1(s)\}$ are equivalent matrices and, therefore, have the same Smith form.

(d) When the system associated with (5.9) is controllable and observable, then $[T \quad U] = [(sI - \Sigma) \quad N]$ and $[T' \quad V'] = [(sI - \Sigma') \quad H']$ have full row rank for all s. That is, $\Re$ is the entire $s-$ plane and it follows from (c) that $S(s)$ and $\operatorname{diag}\{I, B_1(s)\}$ have the same Smith form.

Example 5.10

(a) For the system shown in Fig. 5.1, establish that

$$\begin{bmatrix} y_1 \\ y_2 \end{bmatrix} = U\begin{bmatrix} x_1 \\ x_2 \end{bmatrix} = \begin{bmatrix} V_{12}V_{22}^{-1} & (V_{11} - V_{12}V_{22}^{-1}V_{21}) \\ V_{22}^{-1} & -V_{22}^{-1}V_{21} \end{bmatrix}\begin{bmatrix} x_1 \\ x_2 \end{bmatrix}$$

and

$$\begin{bmatrix} y_1 \\ w \end{bmatrix} = \begin{bmatrix} V_{11} & V_{12} \\ V_{21} & V_{22} \end{bmatrix}\begin{bmatrix} x_2 \\ y_2 \end{bmatrix} = V\begin{bmatrix} x_2 \\ y_2 \end{bmatrix}.$$

When the state-space parameters for V are the ones given in (5.20) show that a set of state-space parameters for U is given by (5.21).

(b) For the system shown in Fig. 5.2 below establish that $y = T_{yx}x$, where

$$T_{yx} = U\Box\phi = \begin{bmatrix} U_{11} & U_{12} \\ U_{21} & U_{22} \end{bmatrix}\Box\phi.$$

Then prove for the state-space descriptions in (5.22) that one gets the state-space descriptions for $U(s)\Box\phi(s)$ in (5.23).

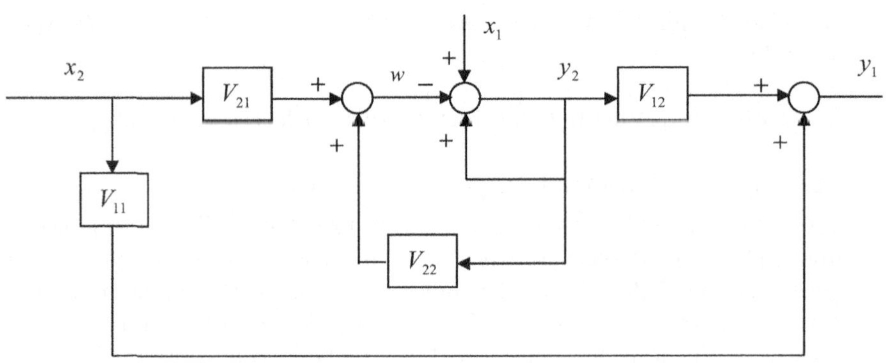

Fig. 5.1 System for Example 5.10(a)

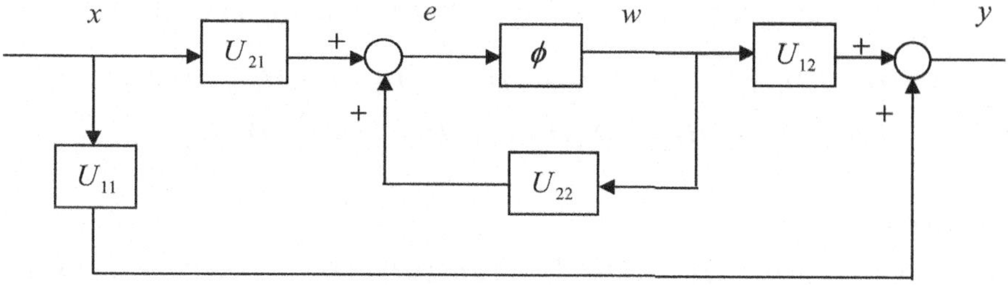

Fig. 5.2 System for Example 5.10(b)

Solution

(a) From Fig. 5.1, $y_2 = x_1 - (V_{21}x_2 + V_{22}y_2) + y_2$ which yields $y_2 = V_{22}^{-1}x_1 - V_{22}^{-1}V_{21}x_2$. It then follows that $y_1 = V_{12}y_2 + V_{11}x_2 = V_{12}V_{22}^{-1}x_1 + (V_{11} - V_{12}V_{22}^{-1}V_{21})x_2$. Thus,

$$\begin{bmatrix} y_1 \\ y_2 \end{bmatrix} = U\begin{bmatrix} x_1 \\ x_2 \end{bmatrix} = \begin{bmatrix} V_{12}V_{22}^{-1} & (V_{11} - V_{12}V_{22}^{-1}V_{21}) \\ V_{22}^{-1} & -V_{22}^{-1}V_{21} \end{bmatrix}\begin{bmatrix} x_1 \\ x_2 \end{bmatrix}$$

and from $y_1 = V_{11}x_2 + V_{12}y_2$, $w = V_{21}x_2 + V_{22}y_2$ one gets

$$\begin{bmatrix} y_1 \\ w \end{bmatrix} = \begin{bmatrix} V_{11} & V_{12} \\ V_{21} & V_{22} \end{bmatrix}\begin{bmatrix} x_2 \\ y_2 \end{bmatrix} = V\begin{bmatrix} x_2 \\ y_2 \end{bmatrix}.$$

In accordance with (5.20), a state-space realization for V is given by

$$\begin{aligned} \dot{\hat{z}} &= \Sigma\hat{z} + N_1\hat{x}_2 + N_2\hat{y}_2 \\ \hat{y}_1 &= H_1\hat{z} + D_{11}\hat{x}_2 + D_{12}\hat{y}_2 \\ \hat{w} &= H_2\hat{z} + D_{21}\hat{x}_2 + D_{22}\hat{y}_2. \end{aligned}$$

From Fig. 5.1, the constraint $\hat{y}_2 = \hat{x}_1 - \hat{w} + \hat{y}_2$ leads to $\hat{w} = \hat{x}_1$. Thus, from the equation above for $\hat{w}$, one gets $D_{22}\hat{y}_2 = \hat{x}_1 - H_2\hat{z} - D_{21}\hat{x}_2$ or, since D_{22} non-singular is assumed,

$$\hat{y}_2 = -D_{22}^{-1}H_2\hat{z} + D_{22}^{-1}\hat{x}_1 - D_{22}^{-1}D_{21}\hat{x}_2.$$

Hence,

$$\dot{\hat{z}} = (\Sigma - N_2D_{22}^{-1}H_2)\hat{z} + N_2D_{22}^{-1}\hat{x}_1 - (N_2D_{22}^{-1}D_{21} - N_1)\hat{x}_2$$

which agrees with the first block row of the final matrix in (5.21). Moreover,

$$\begin{aligned}\hat{y}_1 &= H_1\hat{z} + D_{11}\hat{x}_2 + D_{12}\hat{y}_2 = H_1\hat{z} + D_{11}\hat{x}_2 + D_{12}(-D_{22}^{-1}H_2\hat{z} + D_{22}^{-1}\hat{x}_1 - D_{22}^{-1}D_{21}\hat{x}_2)\\ &= (H_1 - D_{12}D_{22}^{-1}H_2)\hat{z} + D_{12}D_{22}^{-1}\hat{x}_1 + (D_{11} - D_{12}D_{22}^{-1}D_{21})\hat{x}_2.\end{aligned}$$

That is,

$$\begin{bmatrix}\hat{y}_1\\ \hat{y}_2\end{bmatrix} = \begin{bmatrix}(H_1 - D_{12}D_{22}^{-1}H_2)\\ -D_{22}^{-1}H_2\end{bmatrix}\hat{z} + \begin{bmatrix}D_{12}D_{22}^{-1} & (D_{11} - D_{12}D_{22}^{-1}D_{21})\\ D_{22}^{-1} & -D_{22}^{-1}D_{21}\end{bmatrix}\begin{bmatrix}\hat{x}_1\\ \hat{x}_2\end{bmatrix}$$

which agrees with the second block row of the final matrix in (5.21).

(b) It is clear from Fig. 5.2 that $w = \phi e$ and

$$\begin{bmatrix}y\\ e\end{bmatrix} = \begin{bmatrix}U_{11} & U_{12}\\ U_{21} & U_{22}\end{bmatrix}\begin{bmatrix}x\\ w\end{bmatrix} = U\begin{bmatrix}x\\ w\end{bmatrix}.$$

Hence, $e = U_{21}x + U_{22}w = U_{21}x + U_{22}\phi e$ which yields $e = (I - U_{22}\phi)^{-1}U_{21}x$. Thus,

$$y = U_{11}x + U_{12}\phi e = [U_{11} + U_{12}\phi(I - U_{22}\phi)^{-1}U_{21}]x = (U(s)\Box\phi(s))x.$$

A state-space description for $T_{yx} = U(s)\Box\phi(s)$ in terms of the state-space descriptions given in (5.22) for $U(s)$ and $\phi(s)$ is now shown to be the one given by (5.23). It easily follows that

$$\begin{aligned}\dot{\hat{x}}_u &= \Sigma\hat{x}_u + N_1\hat{x} + N_2\hat{w}\\ \hat{y} &= H_1\hat{x}_u + D_{11}\hat{x} + D_{12}\hat{w}\\ \hat{e} &= H_2\hat{x}_u + D_{21}\hat{x} + D_{22}\hat{w}\end{aligned}$$

is the realization for $U(s)$ and

$$\begin{aligned}\dot{\hat{x}}_\phi &= \Sigma_\phi\hat{x}_\phi + N_\phi\hat{e}\\ \hat{w} &= H_\phi\hat{x}_\phi + D_\phi\hat{e}\end{aligned}$$

is the realization for $\phi(s)$. Using the equation for $\hat{w}$ in the equation for $\hat{e}$ yields

$$\hat{e} = H_2\hat{x}_u + D_{21}\hat{x} + D_{22}\hat{w} = H_2\hat{x}_u + D_{21}\hat{x} + D_{22}(H_\phi\hat{x}_\phi + D_\phi\hat{e}).$$

It is now convenient to introduce $\hat{x}_\xi = [\hat{x}_u' \quad \hat{x}_\phi']'$, $E_1 = (I - D_\phi D_{22})$, and $E_2 = (I - D_{22}D_\phi)$. Then it easily follows that

$$\hat{e} = H_e\hat{x}_\xi + E_2^{-1}D_{21}\hat{x},\ H_e = E_2^{-1}[H_2 \quad D_{22}H_\phi].$$

Using this equation for $\hat{e}$ in the equation for $\hat{w}$ gives

$$\begin{aligned}\hat{w} &= H_\phi\hat{x}_\phi + D_\phi\hat{e} = [0 \quad H_\phi]\hat{x}_\xi + D_\phi H_e\hat{x}_\xi + D_\phi E_2^{-1}D_{21}\hat{x}\\ &= H_w\hat{x}_\xi + D_\phi E_2^{-1}D_{21}\hat{x},\end{aligned}$$

where $H_w = [0 \quad H_\phi] + D_\phi H_e = [0 \quad H_\phi] + D_\phi E_2^{-1}[H_2 \quad D_{22}H_\phi]$. That is,

$$\begin{aligned}H_w &= [D_\phi E_2^{-1}H_2 \quad (I + D_\phi E_2^{-1}D_{22})\,H_\phi] = [D_\phi E_2^{-1}H_2 \quad (I + E_1^{-1}D_\phi D_{22})\,H_\phi]\\ &= [D_\phi E_2^{-1}H_2 \quad E_1^{-1}\,H_\phi]\end{aligned}$$

since $E_1D_\phi = D_\phi E_2$. It is now possible to obtain

$$\begin{aligned}\hat{y} &= H_1\hat{x}_u + D_{11}\hat{x} + D_{12}\hat{w} = [H_1 \quad 0]\hat{x}_\xi + D_{11}\hat{x} + D_{12}(H_w\hat{x}_\xi + D_\phi E_2^{-1}D_{21}\hat{x})\\ &= ([H_1 \quad 0] + D_{12}[D_\phi E_2^{-1}H_2 \quad E_1^{-1}\,H_\phi])\hat{x}_\xi + (D_{11} + D_{12}D_\phi E_2^{-1}D_{21})\hat{x}\\ &= [(H_1 + D_{12}D_\phi E_2^{-1}H_2) \quad D_{12}E_1^{-1}\,H_\phi]\hat{x}_\xi + (D_{11} + D_{12}D_\phi E_2^{-1}D_{21})\hat{x}\end{aligned}$$

which confirms the second block row of (5.23). Finally,

$$\begin{aligned}\dot{\hat{x}}_\xi = \begin{bmatrix}\dot{\hat{x}}_u\\ \dot{\hat{x}}_\phi\end{bmatrix} &= \begin{bmatrix}\Sigma & 0\\ 0 & \Sigma_\phi\end{bmatrix}\begin{bmatrix}\hat{x}_u\\ \hat{x}_\phi\end{bmatrix} + \begin{bmatrix}N_1 & N_2\\ 0 & 0\end{bmatrix}\begin{bmatrix}\hat{x}\\ \hat{w}\end{bmatrix} + \begin{bmatrix}0 & 0\\ 0 & N_\phi\end{bmatrix}\begin{bmatrix}\hat{x}\\ \hat{e}\end{bmatrix}\\ &= \begin{bmatrix}\Sigma & 0\\ 0 & \Sigma_\phi\end{bmatrix}\begin{bmatrix}\hat{x}_u\\ \hat{x}_\phi\end{bmatrix} + \begin{bmatrix}N_1\\ 0\end{bmatrix}\hat{x} + \begin{bmatrix}N_2\\ 0\end{bmatrix}\hat{w} + \begin{bmatrix}0\\ N_\phi\end{bmatrix}\hat{e}\end{aligned}$$

When the expressions for $\hat{e}$ and $\hat{w}$ derived above are used, one gets

$$\dot{\hat{x}}_\xi = \begin{bmatrix}(\Sigma + N_2D_\phi E_2^{-1}H_2) & N_2E_1^{-1}H_\phi\\ N_\phi E_2^{-1}H_2 & (\Sigma_\phi + N_\phi E_2^{-1}D_{22}H_\phi)\end{bmatrix}\hat{x}_\xi + \begin{bmatrix}N_1 + N_2D_\phi E_2^{-1}D_{21}\\ N_\phi E_2^{-1}D_{21}\end{bmatrix}\hat{x}$$

Since $E_1D_\phi = D_\phi E_2$, it is clear that the first block row of (5.23) is confirmed.

Example 5.11
Derive the dual statement of the following fact that can be obtained from Corollary 13.10 and Theorem 13.5 of Zhou et al. (1966): If (Σ, N_2) is stabilizable, D_{12} has full column rank, and the matrix

$$S_{12}(j\omega) = \begin{bmatrix}\Sigma - j\omega I & N_2\\ H_1 & D_{12}\end{bmatrix}$$

has full column rank for all ω, then the algebraic Riccati equation

$$(\Sigma - N_2E_1^{-1}D_{12}'H_1)'M_1 + M_1(\Sigma - N_2E_1^{-1}D_{12}'H_1) - M_1N_2E_1^{-1}N_2'M_1 \\ + H_1'(I - D_{12}E_1^{-1}D_{12}')H_1 = 0$$

has a unique solution $M_1 \geq 0$ in which $E_1 = D_{12}'D_{12}$ and $L_1 = E_1^{-1}(D_{12}'H_1 + N_2'M_1)$. Furthermore, the matrix $\Sigma - N_2L_1$ is stable.

Solution

Transposing the matrices in this example yields that if (Σ', N_2') is detectable, D_{12}' has full row rank, and the matrix

$$S_{12}'(j\omega) = \begin{bmatrix} \Sigma' - j\omega I & H_1' \\ N_2' & D_{12}' \end{bmatrix}$$

has full row rank for all ω, then the algebraic Riccati equation

$$(\Sigma' - H_1'D_{12}E_1^{-1}N_2')'M_1' + M_1'(\Sigma - N_2E_1^{-1}D_{12}'H_1) - M_1'N_2E_1^{-1}N_2'M_1' \\ + H_1'(I - D_{12}E_1^{-1}D_{12}')H_1 = 0$$

has a unique solution $M_1' \geq 0$ and the matrix $\Sigma' - L_1'N_2' = \Sigma' - (H_1'D_{12} + M_1'N_2)E_1^{-1}N_2'$ is stable. Making substitutions of $\Sigma' = F$, $H_1' = N_1$, $N_2' = H_2$, $D_{12}' = D_{21}$, and $M_1' = M_2$ yields the dual fact that if (F, H_2) is detectable, D_{21} has full row rank, and the matrix

$$S_{21}(j\omega) = \begin{bmatrix} F - j\omega I & N_1 \\ H_2 & D_{21} \end{bmatrix}$$

has full row rank for all ω, then the algebraic Riccati equation

$$(F - N_1D_{21}'E_2^{-1}H_2)M_2 + M_2(F - N_1D_{21}'E_2^{-1}H_2)' - M_2H_2'E_2^{-1}H_2M_2 + N_1(I \\ - D_{21}'E_1^{-1}D_{21})N_1' \\ = 0$$

has a unique solution $M_2 \geq 0$ and the matrix $F - L_2H_2$ is stable, where $E_2 = D_{21}D_{21}'$ and $L_2 = (N_1D_{21}' + M_2H_2')E_2^{-1}$.

Appendix A
Complex Functions, Transforms, Parseval's Formula, and Stochastic Processes

A.1 Overview

The essential elements of complex function theory, transforms, and stochastic processes that play an important role in Wiener–Hopf design of multivariable control systems are summarized here for easy review. For more rigor and the omitted proofs mathematical sources on these subjects should be consulted. The primary references for the abbreviated treatment given in this Appendix are Churchill (1948) and Papoulis (1962, 1965, 1977).

A.2 Complex Function Theory

Complex function theory involves complex functions $f(p)$ of the complex variable p. A single-valued function $f(p)$ is called analytic at a finite point p_o in the complex $p-$plane when it has a derivative at all points in the neighborhood $|p - p_o| < \varepsilon$, for some $\varepsilon > 0$. When a function is analytic at some point in every neighborhood of p_o, but not analytic at p_o itself, then p_o is called a singular point or a singularity of the function. When p_o is a singular point of $f(p)$ and $f(p)$ is analytic everywhere else in some neighborhood of p_o, then p_o is called an isolated singular point.

For the case in which $f(p)$ is a matrix of complex functions, the derivative of $f(p)$ is the matrix that results when each element of $f(p)$ is replaced with its derivative. A function or matrix is analytic in a region when it is or all its elements are analytic at every point in that region. A function or matrix is said to be analytic everywhere when it is analytic at all finite points of the complex p-plane.

J. J. Bongiorno Jr. and K. Park, *Design of Linear Multivariable Feedback Control Systems*, https://doi.org/10.1007/978-3-030-44356-6

Attention is restricted in this book to rational functions $f(p) = n(p)/d(p)$ expressed as the ratio of relatively prime polynomials $n(p)$ and $d(p)$. Such functions can be represented as the sum of a polynomial and a strictly proper rational function using the division algorithm. The strictly proper part, in turn, can be expanded in partial fractions. It then easily follows that the derivative of $f(p)$ exists for all finite p except those points where $d(p)$ has zeros. That is, a rational function $f(p)$ is analytic everywhere except for those singular points which are the zeros of its denominator. These isolated singularities are the finite poles of $f(p)$. When the degree of $n(p)$ exceeds the degree of $d(p)$, then $|f(p)| \to \infty$, $|p| \to \infty$ and $f(p)$ is said to have a pole at infinity. The poles of a rational matrix are the poles of its elements (also see Sect. B.6).

One of the most important results in the entire theory of analytic functions is the ***Cauchy integral theorem***. This theorem states that when a single-valued function $f(p)$ is analytic within and on a closed curve C, then

$$\int_C f(p)dp = 0. \tag{A.1}$$

The Cauchy integral theorem is easily extended to multiply connected regions. In particular, when a region R of the complex plane is bounded by a finite number of disconnected closed curves $C_1, C_2, \ldots, C_n$ and the single-valued function $f(p)$ is analytic at every point of the region R and its boundary, then

$$\int_{C_1} f(p)dp + \int_{C_2} f(p)dp + \cdots + \int_{C_n} f(p)dp = 0 \tag{A.2}$$

provided the direction of integration around each closed curve is such that R is on the same side of all boundary lines.

For the case in which $f(p)$ is a matrix of complex functions, the integral of $f(p)$ is the matrix that results when each element of $f(p)$ is replaced with its integral. Hence, (A.1) and (A.2) hold when $f(p)$ is a matrix whose elements are single-valued functions that are analytic at every point of the region R and its boundary.

When there exists an integer m such that $f_k(p) = (p - p_k)^m f(p)$ is analytic at p_k and $f_k(p_k) \neq 0$, then p_k is called a pole of the function $f(p)$ of order m. The residue of $f(p)$ at the pole p_k is the quantity

$$R_k = \frac{f_k^{(m-1)}(p_k)}{(m-1)!} = \frac{1}{(m-1)!} \frac{d^{(m-1)} f_k(p)}{dp^{(m-1)}}\bigg|_{p=p_k}. \tag{A.3}$$

When $f(p)$ is analytic on and within a closed curve C except at a finite number of singular points $p_1, p_2, \cdots, p_m$ inside C and $R_1, R_2, \ldots, R_m$ denote the residues of $f(p)$ at these points, then

$$\int_{C} f(p)\,dp = 2\pi j \sum_{k=1}^{m} R_k \tag{A.4}$$

when the direction of integration along C is such that the region R bounded by C is on the left-hand side. Equation (A.4) is referred to as the ***Cauchy residue theorem*** and it extends in an obvious and natural way to the case in which $f(p)$ is a matrix.

The Cauchy residue theorem can often be used effectively to evaluate certain line integrals such as

$$J = \int_{-j\infty}^{j\infty} f(p)dp = \lim_{R\to\infty} \int_{-jR}^{jR} f(p)dp \tag{A.5}$$

when $f(p)$ is either a function or a matrix analytic on the imaginary axis and satisfies $f(p) \leq 0(p^{-2})$. This integral can be computed from

$$J_C = \int_{C_R} f(p)\,dp = \int_{-jR}^{jR} f(p)dp + \int_{\Gamma_R} f(p)\,dp, R \to \infty, \tag{A.6}$$

where Γ_R is the semicircular contour $p = Re^{j\theta}$, $\pi/2 < \theta < 3\pi/2$ traversed in a counter-clockwise direction from $\pi/2$ to $3\pi/2$. Clearly, when $f(p)$ only has singularities that are poles and none of these singularities lie on the imaginary axis,

$$J_C = 2\pi j \sum_{k=1}^{m} R_k, \tag{A.7}$$

where the R_k are the residues in the poles of $f(p)$ that lie in $\mathrm{Re}\, p < 0$ (i.e., inside C_R, $R \to \infty$). When

$$\lim_{R\to\infty} \int_{\Gamma_R} f(p)\,dp = 0, \tag{A.8}$$

it follows that

$$J = \int_{-j\infty}^{j\infty} f(p)dp = 2\pi j \sum_{k=1}^{m} R_k = J_C. \tag{A.9}$$

That (A.8) is true when $f(p) \leq 0(p^{-2})$ is now established. It suffices to treat the case in which $f(p)$ is scalar since the same arguments can be used for each element of $f(p)$ when it is a matrix. Substituting $p = Re^{j\theta}$ into the integral in (A.8) yields

$$\left| \int_{\Gamma_R} f(p)dp \right| = \left| \int_{\frac{\pi}{2}}^{\frac{3\pi}{2}} f(Re^{j\theta})jRe^{j\theta}d\theta \right| \leq \int_{\frac{\pi}{2}}^{\frac{3\pi}{2}} |f(Re^{j\theta})|R\,d\theta. \tag{A.10}$$

On account of $f(p) \leq 0(p^{-2})$, $\lim_{p\to\infty} pf(p) = 0$; hence,

$$\lim_{R\to\infty} |f(Re^{j\theta})|R = 0 \tag{A.11}$$

and (A.8) follows immediately from (A.10).

When Γ_R is chosen instead as the semicircular contour $p = Re^{j\theta}$, $-\pi/2 < \theta < \pi/2$, and traversed in a clockwise direction, similar reasoning as above with $f(p) \leq 0(p^{-2})$ and analytic on the imaginary axis leads to

$$J = \int_{-j\infty}^{j\infty} f(p)dp = -2\pi j \sum_{k=1}^{m} R_k, \tag{A.12}$$

where the R_k are now the residues in the poles of $f(p)$ that lie in Re $p > 0$: the negative sign is needed because C_R is now traversed in a clockwise direction.

The results obtained in (A.9) and (A.12) can be compactly stated as follows. With $f(p) \leq 0(p^{-2})$ and analytic on the imaginary axis,

$$J = \int_{-j\infty}^{j\infty} f(p)\,dp = \int_C f(p)\,dp \tag{A.13}$$

provided the contour C includes the imaginary axis and is closed by an infinitely large semicircular arc into either the left or right half-planes. Of course, when $f(p)$ is in addition analytic in either Re $p < 0$ or Re $p > 0$, then the residues on the right-hand side of either (A.9) or (A.12), respectively, are zero and one gets $J = 0$. This also follows immediately from (A.13) when the contour C is closed into the half-plane in which $f(p)$ is analytic. Equation (3.168) easily follows from (A.13) with p replaced by s and f replaced by M when $M(s)$ is a real rational matrix.

A.3 Laplace Transforms and Parseval's Formula

The ***unilateral Laplace transform*** of a given function $\hat{f}(t)$ specified for $t \geq 0$ is defined by

$$f(s) = \int_0^\infty \hat{f}(t)e^{-st}dt = \mathcal{L}_I\{\hat{f}(t)\} \tag{A.14}$$

for those values of the complex variable $s = \sigma + j\omega$ for which the integral converges. It can be shown that if the integral exists for some value of s, then there exists an abscissa of convergence such that the integral converges for Re $s > \sigma_c$ and diverges for Re $s < \sigma_c$. Moreover, $f(s)$ is an analytic function in the domain of its existence, the half-plane Re $s > \sigma_c$.

A sufficient condition for the existence of the integral in (A.14) is that $\hat{f}(t)$ be piecewise continuous and $|\hat{f}(t)| \leq \alpha e^{\beta t}$. For it then follows that

$$|f(s)| \leq \int_0^\infty |\hat{f}(t)e^{-st}|\,dt \leq \int_0^\infty \alpha e^{\beta t}e^{-\sigma t}dt = \frac{\alpha}{\sigma - \beta}, \sigma > \beta. \tag{A.15}$$

Moreover, $\lim_{\sigma \to \infty} |f(s)| = 0$. Hence, $f(s)$ exists in this case for Re $s > \beta$ and when it is a rational function it is strictly proper. Conversely, when $f(s)$ exists and is strictly proper, it can be established that wherever $\hat{f}(t)$ is continuous

$$\left.\begin{matrix} \hat{f}(t), & t > 0 \\ 0, & t < 0 \end{matrix}\right\} = \frac{1}{2\pi j}\int_{\sigma - j\infty}^{\sigma + j\infty} f(s)e^{st}\,ds \tag{A.16}$$

provided $\sigma > \sigma_c$. The inversion integral in (A.16) is called the Bronwich integral and the line of integration is called the Bronwich line. It gives at a point t_o, where $\hat{f}(t)$ has a discontinuity the mean value $[\hat{f}(t_o^+) + \hat{f}(t_o^-)]/2$. It is significant that the Bronwich integral returns a causal function $\hat{f}(t)$: i.e., $\hat{f}(t) = 0$ for $t < 0$. Specifically, when $f(s)$ is strictly proper and the Laplace transform of a causal time function, then (A.16) gives that causal time function provided the integral is evaluated along the Bronwich line which lies to the right of all singularities of $f(s)$. Clearly, all statements made above regarding the unilateral Laplace transform extend in straightforward fashion to the case in which $\hat{f}(t)$ is a matrix of time functions.

It is important to mention that the Bronwich integral in (A.16) can be evaluated as a closed-contour integral when $\hat{f}(t)$ is causal and $f(s)$ is strictly proper and rational. In particular, one can always choose the Bronwich line to be in the right half-plane and to the right of all poles of $f(s)$. Then for $t<0$, the closed contour can be taken to be the Bronwich line closed by an infinite semicircular arc into the right half-plane: the integral along the semicircular arc is zero because $ds = jRe^{j\theta}d\theta$ and

$$f(s)e^{st}R|_{s=Re^{j\theta}} = f(Re^{j\theta})e^{tR\cos\theta}e^{jtR\sin\theta}R \to 0, R \to \infty \tag{A.17a}$$

when $\cos\theta > 0$, $t<0$, and $f(s)$ is strictly proper. Since $f(s)$ is analytic in its domain of existence, it immediately follows on account of the Cauchy integral theorem that the closed-contour integral and, therefore, the Bronwich integral is zero for $t<0$ in agreement with (A.16).

For $t > 0$, the same Bronwich line can be chosen. But now the contour is closed with a semicircular arc $s = Re^{j\theta}$, $\pi/2<\theta<3\pi/2$ and two horizontal line segments, $s = \sigma + jR$ and $s = \sigma - jR$ with $0 \leq \sigma \leq \sigma_b$, where σ_b is the point where the Bronwich line intersects the real axis. Again $R \to \infty$ is invoked. Now $t\cos\theta<0$ and so (A.17a) still holds; hence, the contribution to the integral along the semicircular arc is zero. The contribution to the integral along the horizontal line segments is also zero because $ds = d(\sigma + jR) = d\sigma$ and

$$f(s)e^{st}|_{s=\sigma+jR} = f(\sigma + jR)e^{\sigma t}e^{jRt} \to 0, R \to \infty \tag{A.17b}$$

when σt is finite and $f(s)$ is strictly proper.

The unilateral Laplace transform is well suited for dealing with transient responses of systems to initial conditions and to causal inputs such as steps and ramps. When the response to a noncausal input is desired, the ***bilateral Laplace transform***

$$f(s) = \int_{-\infty}^{\infty} \hat{f}(t)e^{-st}\,dt = \mathcal{L}_{II}\{\hat{f}(t)\} \tag{A.18}$$

plays a role. The bilateral Laplace transform also comes into play for stochastic signals. When analog signals are the stochastic wide-sense stationary ones described in Sect. A.4, they are characterized by noncausal autocorrelation functions whose bilateral Laplace transforms play an important role.

Clearly, (A.18) is equivalent to

$$f(s) = \int_{-\infty}^{0} \hat{f}_1(-t)e^{-st}\,dt + \int_{0}^{\infty} \hat{f}_2(t)e^{-st}\,dt, \tag{A.19}$$

where

$$\hat{f}_1(-t) = \begin{cases} \hat{f}(t), & t<0 \\ 0, & t>0 \end{cases} \tag{A.20}$$

and

$$\hat{f}_2(t) = \begin{cases} \hat{f}(t), & t>0 \\ 0, & t<0 \end{cases}. \tag{A.21}$$

That is,

$$f(s) = \int_{0}^{\infty} \hat{f}_1(t)e^{st}dt + \int_{0}^{\infty} \hat{f}_2(t)e^{-st}dt \tag{A.22}$$

or

$$f(s) = f_1(-s) + f_2(s), \tag{A.23}$$

where

$$f_1(s) = \int_{0}^{\infty} \hat{f}_1(t)e^{-st}dt = \mathcal{L}_I\{\hat{f}_1(t)\} = \mathcal{L}_I\{\hat{f}(-t)\} \tag{A.24}$$

and

$$f_2(s) = \int_{0}^{\infty} \hat{f}_2(t)e^{-st}dt = \mathcal{L}_I\{\hat{f}_2(t)\} = \mathcal{L}_I\{\hat{f}(t)\}. \tag{A.25}$$

Thus, when σ_1 is the abscissa of convergence for $f_1(s)$ and when σ_2 is the abscissa of convergence for $f_2(s)$, then $f_1(-s)$ is analytic in Re $s < -\sigma_1$ and $f_2(s)$ is analytic in Re $s > \sigma_2$; hence, $f(s)$ is analytic in the strip of convergence $\sigma_2 < \mathrm{Re}\, s < -\sigma_1$

provided $\sigma_2 < -\sigma_1$. This strip is the domain of existence for $f(s)$ as the bilateral Laplace transform of $\hat{f}(t)$. In this case, it can be shown that

$$\hat{f}(t) = \frac{1}{2\pi j} \int_{\sigma - j\infty}^{\sigma + j\infty} f(s) e^{st} \, ds \tag{A.26}$$

provided the line of integration lies in the strip of convergence or $\sigma_2 < \sigma < -\sigma_1$. It is important to note that when $f_1(s)$ and $f_2(s)$ are stable unilateral transforms, then σ_1 and σ_2 are both negative and the $j\omega$-axis of the complex s-plane lies in the strip of convergence. Moreover, when $f_1(s)$ and $f_2(s)$ are strictly proper and rational it is clear from (A.23) that $f(s)$ is also.

A quantity which is often of interest in control system analysis and design is the integral-square-value

$$J_{isv} = \int_0^\infty \hat{f}^*(t) \hat{f}(t) dt = \sum_{i=1}^n \int_0^\infty \left| \hat{f}_i(t) \right|^2 dt \tag{A.27}$$

of a causal vector function $\hat{f}(t)$ whose elements satisfy $|\hat{f}_i(t)| \leq \alpha e^{-\beta t}$, $\beta > 0$. In this case, the integrals in (A.27) are finite. Moreover, $f(s)$ is strictly proper and analytic in Re $s > -\beta$. Hence, the $j\omega$-axis can be selected for the Bronwich line. It then follows from (A.16) and (A.27) that

$$J_{isv} = \int_0^\infty \hat{f}^*(t) \left[\frac{1}{2\pi j} \int_{-j\infty}^{j\infty} f(s) e^{st} ds \right] dt. \tag{A.28}$$

Interchanging the order of integration which is permissible for the functions $\hat{f}_i(t)$ under consideration here then yields

$$\begin{aligned} J_{isv} &= \frac{1}{2\pi j} \int_{-j\infty}^{j\infty} \left[\int_0^\infty \hat{f}^*(t) e^{st} \, dt \right] f(s) ds \\ &= \frac{1}{2\pi j} \int_{-j\infty}^{j\infty} \left[\int_0^\infty \hat{f}(t) e^{\bar{s}t} \, dt \right]^* f(s) ds \\ &= \frac{1}{2\pi j} \int_{-j\infty}^{j\infty} f^*(-\bar{s}) f(s) \, ds = \frac{1}{2\pi j} \int_{-j\infty}^{j\infty} f_*(s) f(s) ds. \end{aligned} \tag{A.29}$$

That is

$$J_{isv} = \int_{0}^{\infty} \hat{f}^{*}(t)\hat{f}(t)dt = \frac{1}{2\pi j}\int_{-j\infty}^{j\infty} f_{*}(s)f(s)ds \tag{A.30}$$

which is Parseval's formula. Equation (A.30) is the basis for (1.7) since when $\hat{f}(t)$ is real one has $\hat{f}^{*}(t) = \hat{f}'(t)$ and so

$$f_{*}(s) = f^{*}(-\bar{s}) = \left[\int_{0}^{\infty} \hat{f}(t)e^{\bar{s}t}\,dt\right]^{*} = \int_{0}^{\infty} \hat{f}'(t)e^{st}\,dt = f'(-s). \tag{A.31}$$

A.4 Analog Stochastic Processes and Mean-Square-Value

Any complex or real vector function of time $\hat{f}(t)$ whose elements are random functions of time characterized by appropriate probability density functions is called an analog stochastic process. Interest is restricted here to such processes that are ergodic, have constant mean $\hat{m}$, and are second-order stationary: i.e., the mean

$$\hat{m} = <\hat{f}(t)> = \lim_{T\to\infty}\frac{1}{2T}\int_{-T}^{T} \hat{f}(t)dt \tag{A.32a}$$

and the autocorrelation matrix

$$\hat{\Phi}_{ff}(\tau) = <\hat{f}(t+\tau)\hat{f}^{*}(t)> = \lim_{T\to\infty}\frac{1}{2T}\int_{-T}^{T} \hat{f}(t+\tau)\hat{f}^{*}(t)dt \tag{A.32b}$$

do not depend on the time variable t. It is because of ergodicity that the ensemble averages are equal to the time averages. Such processes are said to be wide-sense stationary and possess the covariance matrix

$$\hat{C}_{ff}(\tau) = <[\hat{f}(t+\tau) - \hat{m}][\hat{f}(t) - \hat{m}]^{*}> = \hat{\Phi}_{ff}(\tau) - \hat{m}\hat{m}^{*}. \tag{A.33}$$

Since any nonzero mean $\hat{m}$ can be absorbed into the shape-deterministic component of any input signals, there is no loss in generality in restricting attention here to cases in which $\hat{m} = 0$; hence, $\hat{\Phi}_{ff}(\tau)$ is both the autocorrelation and covariance matrix for the zero-mean process $\hat{f}(t)$. When $\hat{g}(t)$ is a similar stochastic process, the cross-correlation matrix

$$\hat{\Phi}_{fg}(\tau) = <\hat{f}(t+\tau)\hat{g}^*(t)> \tag{A.34}$$

is also often of interest. When $\hat{\Phi}_{fg}(\tau) \equiv 0, \hat{f}(t)$ and $\hat{g}(t)$ are said to be uncorrelated. When

$$\hat{\Phi}_{fg}(\tau) = <\hat{f}(t+\tau)> \, <\hat{g}^*(t)> \, = [<\hat{f}(t+\tau)>][<\hat{g}(t)>]^* \tag{A.35}$$

then $\hat{f}(t)$ and $\hat{g}(t)$ are said to be independent. Clearly, when $\hat{f}(t)$ and $\hat{g}(t)$ are independent and either one has zero mean, then they are uncorrelated.

There are a number of important properties that are possessed by autocorrelation matrices. First,

$$\hat{\Phi}^*_{ff}(\tau) = <\hat{f}(t)\hat{f}^*(t+\tau)> \, = <\hat{f}(t-\tau)\hat{f}^*(t)> \, = \hat{\Phi}_{ff}(-\tau) \tag{A.36}$$

because t can be replaced with $t-\tau$ on account of the process being stationary. An immediate consequence of (A.36) is that autocorrelation matrices are hermitian at $\tau = 0$: $\hat{\Phi}^*_{ff}(0) = \hat{\Phi}_{ff}(0)$. Moreover, for any constant vector $a \neq 0$,

$$a^*\hat{\Phi}_{ff}(0)a = <a^*\hat{f}(t)\hat{f}^*(t)a> \, = <|\hat{f}^*(t)a|^2> \tag{A.37}$$

is nonnegative. This establishes the second important property that $\Phi_{ff}(0)$ is a hermitian nonnegative-definite matrix or, equivalently,

$$\hat{\Phi}_{ff}(0) = \hat{\Phi}^*_{ff}(0) \geq 0. \tag{A.38}$$

Finally, it is evident that

$$\mathrm{Tr}\,[\hat{\Phi}_{ff}(0)] = <\hat{f}^*(t)\hat{f}(t)> \, = \sum_{i=1}^{n} <|\hat{f}_i(t)|^2> \tag{A.39}$$

is the mean norm-squared value or simply the mean-square-value of $\hat{f}(t)$ for any t.

The bilateral Laplace transform of the autocorrelation matrix is also useful in control system applications and is given by

$$\Phi_{ff}(s) = \mathcal{L}_{II}\{\hat{\Phi}_{ff}(\tau)\} = \int_{-\infty}^{\infty} \hat{\Phi}_{ff}(\tau)e^{-s\tau}d\tau. \tag{A.40}$$

When $\hat{\phi}(\tau)$ denotes an element of $\hat{\Phi}_{ff}(\tau)$ and all such elements satisfy $|\hat{\phi}(\tau)| \le Me^{-\beta|\tau|}$, $\beta > 0$, then the existence of $\Phi_{ff}(s)$ with a strip of convergence that includes the $j\omega$-axis is assured as follows. Clearly, by analogy with (A.23),

$$\phi(s) = \phi_1(-s) + \phi_2(s), \tag{A.41}$$

where the unilateral Laplace transforms

$$\phi_1(s) = \mathcal{L}_I\{\hat{\phi}(-\tau)\}, \phi_2(s) = \mathcal{L}_I\{\hat{\phi}(\tau)\} \tag{A.42}$$

exist, are analytic in Re $s > -\beta$, and are strictly proper when $\phi_1(s)$ and $\phi_2(s)$ are rational. Hence, $\phi(s)$ exists, is analytic in the strip $-\beta <$Re $s < \beta$, and is strictly proper when $\phi_1(s)$ and $\phi_2(s)$ are rational. In this case one can use

$$\hat{\Phi}_{ff}(\tau) = \frac{1}{2\pi j} \int_{-j\infty}^{j\infty} \Phi_{ff}(s)e^{s\tau}ds \tag{A.43}$$

for the inversion integral with the $j\omega$-axis as the Bronwich line. Moreover, the integral in (A.43) can be evaluated as a closed-contour integral: for $\tau > 0$, the path of integration is closed with an infinite-radius semicircle into the left half-plane and for $\tau < 0$, the path of integration is closed with an infinite-radius semicircle into the right half-plane.

One easily gets from (A.36) and (A.40) that

$$\Phi_{ff*}(s) = \Phi^*_{ff}(-\bar{s}) = \int_{-\infty}^{\infty} \hat{\Phi}^*_{ff}(\tau)e^{s\tau}d\tau = \int_{-\infty}^{\infty} \hat{\Phi}_{ff}(-\tau)e^{s\tau}d\tau = \int_{-\infty}^{\infty} \hat{\Phi}_{ff}(\tau)e^{-s\tau}d\tau$$
$$= \Phi_{ff}(s). \tag{A.44a}$$

Hence, $\Phi_{ff}(j\omega) = \Phi^*(j\omega)$ is an hermitian matrix. Moreover, it easily follows from (A.40) when $\hat{\Phi}_{ff}(\tau)$ is a real matrix that $\overline{\Phi_{ff}}(s) = \Phi_{ff}(\bar{s})$; thus, in this case $\Phi_{ff*}(s) = \Phi'_{ff}(-s) = \Phi_{ff}(s)$ and $\Phi_{ff}(s)$ is parahermitian. Equations (A.32b) and (A.40) also yield that

$$\begin{aligned}\Phi_{ff}(s) &= \lim_{T\to\infty}\frac{1}{2T}\int_{-T}^{T}\int_{-\infty}^{\infty}\hat{f}(t+\tau)\hat{f}^*(t)e^{-s\tau}d\tau dt \\ &= \lim_{T\to\infty}\frac{1}{2T}\int_{-\infty}^{\infty}\int_{-\infty}^{\infty}\hat{f}_T(t+\tau)\hat{f}_T^*(t)e^{-s\tau}d\tau dt \qquad \text{(A.44b)} \\ &= \lim_{T\to\infty}\frac{1}{2T}\int_{-\infty}^{\infty}f_T(s)e^{st}\hat{f}_T^*(t)dt = \lim_{T\to\infty}\frac{1}{2T}f_T(s)f_T^*(-\bar{s}),\end{aligned}$$

where $f_T(s) = \mathcal{L}_{II}\{\hat{f}_T(t)\}$ and

$$\hat{f}_T(t) = \begin{cases}\hat{f}(t), & |t| \le T \\ 0, & |t| > T\end{cases}. \qquad \text{(A.44c)}$$

Therefore,

$$a^*\Phi_{ff}(j\omega)a = \lim_{T\to\infty}\frac{1}{2T}|f_T^*(j\omega)a|^2 \ge 0 \qquad \text{(A.44d)}$$

for any constant vector a and $\Phi_{ff}(j\omega) \ge 0$. That is, $\Phi_{ff}(j\omega)$ is a nonnegative-definite hermitian matrix. It also follows from (A.44b) that $\Phi_{ff}(j\omega) \to 0,\ \omega \to \infty$ iff

$$\lim_{T\to\infty}\frac{f_T(j\omega)}{\sqrt{T}} \to 0,\ \omega \to \infty \qquad \text{(A.44e)}$$

or $\Phi_{ff}(j\omega) \le 0(\omega^{-2})$. Hence, when $\Phi_{ff}(s)$ is strictly proper and rational one has in fact that $\Phi_{ff}(s) \le 0(s^{-2})$.

It easily follows from (A.39) and (A.43) that the mean-square-value of $\hat{f}(t)$ is given by

$$<\hat{f}^*(t)\hat{f}(t)> = \text{Tr}\,[\hat{\Phi}_{ff}(0)] - \frac{1}{2\pi j}\int_{-j\infty}^{j\infty}\text{Tr}\,[\Phi_{ff}(s)]\,ds \qquad \text{(A.45)}$$

which is the basis for (1.9) when $\hat{f}(t)$ is real. It is of interest to note that were the $\hat{f}_i(t)$ real voltages across one-ohm resistors, then $<\hat{f}^*(t)\hat{f}(t)>$ would be the total average power dissipated in the resistors. Since this average power can be obtained from the integral in (A.45), $\Phi_{ff}(s)$ is called the power spectral density matrix for the stochastic process $\hat{f}(t)$.

An issue of interest in connection with analog stochastic processes is the relationship between the power spectral density matrix of the output from a linear causal time-invariant stable multivariable system possessing a rational transfer matrix $G(s) = \mathcal{L}_I\{\hat{G}(t)\}$ to the power spectral density matrix of its input in the steady-state (i.e., after all transients and all contributions from initial conditions have disappeared). For systems of this type with output $\hat{y}_o(t)$ and input $\hat{x}(t)$ one gets

$$\hat{y}_o(t) = \hat{y}_i(t, t_o) + \int_{t_o}^{t} \hat{G}(t-\tau)\hat{x}(\tau)d\tau, \tag{A.46}$$

where $\hat{y}_i(t, t_o)$ is the contribution from the initial conditions at t_o and $\hat{G}(t)$ is the causal inverse unilateral Laplace transform of $G(s)$. When the system is stable, $G(s)$ is stable and

$$\lim_{t_o \to -\infty} \hat{y}_i(t, t_o) = 0. \tag{A.47}$$

Since $\hat{G}(t)$ is causal, $\hat{G}(t-\tau) = 0,\ \tau > t$ and one gets for the steady-state system response

$$\hat{y}(t) = \lim_{t_o \to -\infty} \hat{y}_o(t) = \int_{-\infty}^{\infty} \hat{G}(t-\tau)\hat{x}(\tau)\,d\tau = \int_{-\infty}^{\infty} \hat{G}(\tau)\hat{x}(t-\tau)\,d\tau. \tag{A.48}$$

Then

$$\hat{y}(t+\tau)\hat{y}^*(t) = \int_{-\infty}^{\infty}\int_{-\infty}^{\infty} \hat{G}(\tau_1)\hat{x}(t+\tau-\tau_1)\hat{x}^*(t-\tau_2)\hat{G}^*(\tau_2)d\tau_1\,d\tau_2 \tag{A.49}$$

and taking ensemble averages on both sides of this equation yields

$$\hat{\Phi}_{yy}(\tau) = \int_{-\infty}^{\infty}\int_{-\infty}^{\infty} \hat{G}(\tau_1)\hat{\Phi}_{xx}(\tau+\tau_2-\tau_1)\hat{G}^*(\tau_2)d\tau_1\,d\tau_2. \tag{A.50}$$

Since

$$\mathcal{L}_{II}\{\hat{\Phi}_{xx}(\tau+\tau_2-\tau_1)\} = \mathcal{L}_{II}\{\hat{\Phi}_{xx}(\tau)\}e^{s(\tau_2-\tau_1)} = \Phi_{xx}(s)e^{-s\tau_1}e^{s\tau_2} \tag{A.51}$$

and since

$$G_*(s) = G^*(-\bar{s}) = [\int_{-\infty}^{\infty} \hat{G}(\tau_2)e^{\bar{s}\tau_2}\,d\tau_2]^* = \int_{-\infty}^{\infty} \hat{G}^*(\tau_2)e^{s\tau_2}\,d\tau_2 \tag{A.52}$$

because $\hat{G}(t) = 0,\ t < 0$, it follows from the bilateral Laplace transform of (A.50) that

$$\Phi_{yy}(s) = \int_{-\infty}^{\infty}\int_{-\infty}^{\infty} \hat{G}(\tau_1)e^{-s\tau_1}\Phi_{xx}(s)\hat{G}^*(\tau_2)e^{s\tau_2}d\tau_1\,d\tau_2 = G(s)\Phi_{xx}(s)G_*(s). \tag{A.53}$$

In arriving at (A.53) the integrals beginning with (A.46) are assumed to exist and the order of integration for multiple integrals is assumed to be interchangeable. This is an acceptable assumption for the systems and signals of interest here.

Clearly, when $G(s)$ is a stable proper matrix, when $\Phi_{xx}(s)$ is analytic on the $j\omega$-axis, and when $\Phi_{xx}(s) \leq 0(s^{-2})$, it follows that $\Phi_{yy}(s)$ is analytic on the $j\omega$-axis and

$$\Phi_{yy}(s) = G(s)\Phi_{xx}(s)G_*(s) \leq 0(s^{-2}). \tag{A.54}$$

In this case, the mean-square-value of $\hat{y}(t)$ is given by

$$<\hat{y}^*(t)\hat{y}(t)> \; = \mathrm{Tr}\,[\hat{\Phi}_{yy}(0)] = \frac{1}{2\pi j}\int_{-j\infty}^{j\infty} \mathrm{Tr}\,[G(s)\Phi_{xx}(s)G_*(s)]\,ds, \tag{A.55}$$

which is the basis for (3.141). It is also convenient at times to consider cases in which $\Phi_{xx}(s) = \Sigma$, a constant diagonal matrix with nonnegative elements. That is, cases in which $\hat{\Phi}_{xx}(\tau) = \Sigma\hat{\delta}(\tau)$, where $\hat{\delta}(\tau)$ is the unit impulse function. Such processes are called "white-noise" processes and in these cases the mean-square-value of $\hat{y}(t)$ is still given by (A.55) when $G(s)$ is stable and strictly proper.

An important property of the cross-correlation matrix for stationary processes is

$$\begin{aligned}\hat{\Phi}_{fg}(\tau) = \; &<\hat{f}(t+\tau)\hat{g}^*(t)> \; = \; <\hat{f}(t)\hat{g}^*(t-\tau)> \; = [<\hat{g}(t-\tau)\hat{f}^*(t)>]^* \\ &= \hat{\Phi}^*_{gf}(-\tau).\end{aligned} \tag{A.56}$$

Hence,

$$\Phi_{fg}(s) = \mathcal{L}_{II}\{\hat{\Phi}^*_{gf}(-\tau)\} = [\int_{-\infty}^{\infty} \hat{\Phi}_{gf}(-\tau)e^{-\bar{s}\tau}\,d\tau]^* = [\int_{-\infty}^{\infty} \hat{\Phi}_{gf}(\tau)e^{\bar{s}\tau}\,d\tau]^* = \Phi_{gf*}(s) \tag{A.57}$$

which is in agreement with (A.44a) when $\hat{g}(t) = \hat{f}(t)$. The matrix $\Phi_{fg}(s)$ is called the cross-power spectral density matrix for the processes $\hat{f}(t)$ and $\hat{g}(t)$. Clearly, when these processes are uncorrelated, $\hat{\Phi}_{fg}(\tau) = 0$ and $\Phi_{fg}(s) = 0$. In this regard, it is of interest to establish the power spectral density matrix $\Phi_{yy}(s)$ for the output $\hat{y}(t)$ from the system described by

$$y(s) = \sum_{i=1}^{n} G_i(s)x_i(s), \tag{A.58}$$

where the $G_i(s)$ are stable proper transfer matrices and the $\hat{x}_i(t)$ are independent zero-mean stationary stochastic exogenous inputs to the system. In this case, the inputs are uncorrelated;

hence,

$$\hat{\Phi}_{x_ix_j}(\tau) = 0,\ i \neq j. \tag{A.59}$$

Thus, with

$$\hat{x}(t) = [\hat{x}_1'(t) \quad \hat{x}_2'(t) \quad \cdots \quad \hat{x}_n'(t)]' \tag{A.60}$$

one gets

$$\hat{\Phi}_{xx}(\tau) = \ <\hat{x}(t+\tau)\hat{x}^*(t)> \ = \text{diag}\ \{\hat{\Phi}_{x_1x_1}(\tau), \hat{\Phi}_{x_2x_2}(\tau), \ldots, \hat{\Phi}_{x_nx_n}(\tau)\} \tag{A.61}$$

and

$$\Phi_{xx}(s) = \text{diag}\ \{\hat{\Phi}_{x_1x_1}(s), \hat{\Phi}_{x_2x_2}(s), \ldots, \hat{\Phi}_{x_nx_n}(s)\}. \tag{A.62}$$

Clearly,

$$x(s) = [x_1'(s) \quad \hat{x}_2'(s) \quad \cdots \quad \hat{x}_n'(s)]' \tag{A.63}$$

and with

$$G(s) = [\,G_1(s) \quad G_2(s) \quad \cdots \quad G_n(s)\,] \tag{A.64}$$

one has that (A.58) is equivalent to $y(s) = G(s)x(s)$. It immediately follows from (A.53) that

$$\Phi_{yy}(s) = G(s)\Phi_{xx}(s)G_*(s) = \sum_{i=1}^{n} G_i(s)\Phi_{x_ix_i}(s)G_{i*}(s). \tag{A.65}$$

That is, when (A.59) holds, the power spectral density matrix for $y(s)$ is the sum of the power spectral density matrices for the outputs $y_i(s) = G_i(s)x_i(s)$ from each of the subsystems $G_i(s)$.

A.5 λ–Transforms and Parseval's Formula

The ***unilateral λ–transform*** of a given sequence $\hat{f}(k)$ specified for $k = 0, 1, 2, \ldots$ is defined by

$$f(\lambda) = \sum_{k=0}^{\infty} \hat{f}(k)\lambda^k = \lambda_I\{\hat{f}(k)\} \tag{A.66}$$

for those values of the complex variable $\lambda = re^{-j\theta}$ for which the sum converges. It can also be shown that when the sum converges for some λ then it converges for all λ in an open circular region $|\lambda| < r_c$, where the radius of convergence r_c depends on the sequence $\hat{f}(k)$. Moreover, $f(\lambda)$ is an analytic function in its domain of existence, $|\lambda| < r_c$. Clearly, this definition extends naturally to the case in which $\hat{f}(k)$ is a matrix whose elements are sequences: $f(\lambda)$ is then the matrix whose elements are the λ - transforms of the elements of $\hat{f}(k)$.

A sufficient condition for the existence of the sum in (A.66) is $|\hat{f}(k)| \le \alpha\beta^k$. For it then follows that

$$|f(\lambda)| \le \alpha \sum_{k=0}^{\infty} (\beta|\lambda|)^k = \frac{\alpha}{1-\beta|\lambda|}, \quad |\lambda| < \frac{1}{\beta}. \tag{A.67}$$

When $f(\lambda)$ exists, it can be established that

$$\left.\begin{matrix} \hat{f}(k), & k \ge 0 \\ 0, & k < 0 \end{matrix}\right\} = \frac{1}{2\pi j} \oint_{|\lambda|=r} \lambda^{-k} f(\lambda) \frac{d\lambda}{\lambda} \tag{A.68}$$

provided $0<r<r_c$ and the circular contour of integration is traversed in a counter-clockwise direction. So when $\hat{f}(k)$ is a causal sequence (i.e., $\hat{f}(k)=0,\ k<0$), it is given by the inversion integral (A.68). That (A.68) is correct for $k<0$ follows immediately from the Cauchy integral theorem since $f(\lambda)$ is analytic on and within $|\lambda|=r<r_c$: in this case, the integrand of (A.68) is analytic on and within the closed-contour of integration for $k\leq -1$ because $\lambda^{-k-1}=\lambda^{|k|-1}$ has no pole at $\lambda=0$. So given that $f(\lambda)$ is the λ - transform of a causal sequence, one can recover the sequence from the inversion integral by choosing any r such that all poles of $f(\lambda)$ lie outside the circle $|\lambda|=r$. It is important to emphasize that this can be done iff $f(\lambda)$ is analytic at $\lambda=0$ which is the fundamental restriction on $f(\lambda)$ for it to have a causal inverse.

It is important to recognize that the Cauchy residue theorem can be used to evaluate the integral in (A.68) for $k\geq 0$ when $f(\lambda)$ is strictly proper. In particular,

$$\int_C \lambda^{-k}f(\lambda)\frac{d\lambda}{\lambda}=\int_{C_1}\lambda^{-k}f(\lambda)\frac{d\lambda}{\lambda}+\int_{C_2}\lambda^{-k}f(\lambda)\frac{d\lambda}{\lambda} \tag{A.69}$$

is considered where C_1 is the circle $|\lambda|=r<r_c$ traversed in a counter-clockwise direction and C_2 is the circle $|\lambda|=R>r$ traversed in a clockwise direction. That is, the integral on the left-hand side of (A.69) is the integral over the boundary of a ring with inner radius r and outer radius R. When $f(\lambda)$ is strictly proper, it then follows for $k\geq 0$ that

$$\int_{C_2}\lambda^{-k}f(\lambda)\frac{d\lambda}{\lambda}=-\frac{1}{R^k}\int_0^{2\pi}e^{jk\theta}f(Re^{-j\theta})jd\theta\to 0,\ R\to\infty. \tag{A.70}$$

Hence, in this case

$$\hat{f}(k)=\frac{1}{2\pi j}\int_{C_1}\lambda^{-k}f(\lambda)\frac{d\lambda}{\lambda}=\lim_{R\to\infty}\frac{1}{2\pi j}\int_C\lambda^{-k}f(\lambda)\frac{d\lambda}{\lambda}=-\sum_{i=1}^{m}R_i, \tag{A.71}$$

where the R_i are the residues in $\lambda^{-(k+1)}f(\lambda)$ at the poles of $f(\lambda)$. the negative sign is required in (A.71) because the region enclosed by the boundary C is on the right-hand side as C is traversed.

The unilateral λ-transform is well suited for dealing with causal sequences. For noncausal sequences, the ***bilateral λ-transform***

$$f(\lambda)=\sum_{k=-\infty}^{\infty}\hat{f}(k)\lambda^k=\lambda_{II}\{\hat{f}(k)\} \tag{A.72}$$

plays a role. The bilateral λ-transform also comes into play for wide-sense stationary stochastic sequences of the kind described in Sect. A.6 which are characterized by noncausal autocorrelation and cross-correlation functions.

Clearly, (A.72) is equivalent to

$$f(\lambda) = \sum_{k=-\infty}^{0} \hat{f}_1(-k)\lambda^k + \sum_{k=0}^{\infty} \hat{f}_2(k)\lambda^k - \hat{f}(0), \tag{A.73}$$

where

$$\hat{f}_1(-k) = \begin{cases} \hat{f}(k), & k \leq 0 \\ 0, & k > 0 \end{cases} \tag{A.74}$$

and

$$\hat{f}_2(k) = \begin{cases} \hat{f}(k), & k \geq 0 \\ 0, & k < 0 \end{cases}. \tag{A.75}$$

That is,

$$f(\lambda) = \sum_{k=0}^{\infty} \hat{f}_1(k)\lambda^{-k} + \sum_{k=0}^{\infty} \hat{f}_2(k)\lambda^k - \hat{f}(0) \tag{A.76}$$

or

$$f(\lambda) = f_1(\lambda^{-1}) + f_2(\lambda) - \hat{f}(0), \tag{A.77}$$

where

$$f_1(\lambda) = \sum_{k=0}^{\infty} \hat{f}_1(k)\lambda^k = \lambda_I\{\hat{f}_1(k)\} = \lambda_I\{\hat{f}(-k)\} \tag{A.78}$$

and

$$f_2(\lambda) = \sum_{k=0}^{\infty} \hat{f}_2(k)\lambda^k = \lambda_I\{\hat{f}_2(k)\} = \lambda_I\{\hat{f}(k)\}. \tag{A.79}$$

Thus, when r_{c1} is the radius of convergence for $f_1(\lambda)$ and when r_{c2} is the radius of convergence for $f_2(\lambda)$, then $f_1(\lambda^{-1})$ is analytic in $|\lambda| > r_{c1}^{-1}$ and $f_2(\lambda)$ is analytic in $|\lambda| < r_{c2}$; hence, $f(\lambda)$ is analytic in the ring of convergence $r_{c1}^{-1} < |\lambda| < r_{c2}$ provided $r_{c1}^{-1} < r_{c2}$. This ring is the domain of existence for $f(\lambda)$ as the bilateral λ-transform of $\hat{f}(k)$. In this case, it can be shown that

$$\hat{f}(k) = \frac{1}{2\pi j} \oint_{|\lambda|=r} \lambda^{-k} f(\lambda) \frac{d\lambda}{\lambda} \tag{A.80}$$

provided the circle of integration lies in the ring of convergence (i.e., $r_{c1}^{-1} < r < r_{c2}$). It is important to note that when $f_1(\lambda)$ and $f_2(\lambda)$ are stable unilateral λ-transforms, then r_{c1} and r_{c2} are both greater than unity and the unit circle of the complex λ-plane lies in the ring of convergence.

A quantity which is often of interest in control system analysis and design is the sum-square-value

$$J_{ssv} = \sum_{k=0}^{\infty} \hat{f}^*(k)\hat{f}(k) = \sum_{i=1}^{n}\sum_{k=0}^{\infty} |\hat{f}_i(k)|^2 \tag{A.81}$$

of a causal vector function $\hat{f}(k)$ whose elements satisfy $|\hat{f}_i(k)| \le \alpha\beta^k$, $0 < \beta < 1$. In this case, the infinite sums in (A.81) are finite. Moreover, $f(\lambda)$ is analytic in $|\lambda| < r_c$, $r_c = \beta^{-1} > 1$ and therefore stable. Hence, the unit circle (i.e., $r = 1$) can be selected for the integral in (A.68). It then follows that

$$J_{ssv} = \sum_{k=0}^{\infty} \hat{f}^*(k) [\frac{1}{2\pi j} \oint_{|\lambda|=1} \lambda^{-k} f(\lambda) \frac{d\lambda}{\lambda}]. \tag{A.82}$$

Interchanging the order of integration and summation which is permissible for the sequences $\hat{f}_i(k)$ under consideration here then yields

$$\begin{aligned} J_{ssv} &= \frac{1}{2\pi j} \oint_{|\lambda|=1} \left[\sum_{k=0}^{\infty} \hat{f}^*(k)\lambda^{-k} \right] f(\lambda) \frac{d\lambda}{\lambda} \\ &= \frac{1}{2\pi j} \oint_{|\lambda|=1} \left[\sum_{k=0}^{\infty} \hat{f}(k)(\bar{\lambda})^{-k} \right]^* f(\lambda) \frac{d\lambda}{\lambda} \\ &= \frac{1}{2\pi j} \oint_{|\lambda|=1} f^*(\bar{\lambda}^{-1}) f(\lambda) \frac{d\lambda}{\lambda} = \frac{1}{2\pi j} \oint_{|\lambda|=1} f_*(\lambda) f(\lambda) \frac{d\lambda}{\lambda}. \end{aligned} \tag{A.83}$$

That is

$$J_{ssv} = \sum_{k=0}^{\infty} \hat{f}^*(k)\hat{f}(k) = \frac{1}{2\pi j} \oint_{|\lambda|=1} f_*(\lambda) f(\lambda) \frac{d\lambda}{\lambda} \tag{A.84}$$

which is Parseval's formula for the digital case. Equation (A.84) is the basis for (1.6) since when $\hat{f}(k)$ is real one has $\hat{f}^*(k) = \hat{f}'(k)$ and so

$$f_*(\lambda) = f^*(\bar{\lambda}^{-1}) = \left[\sum_{k=0}^{\infty} \hat{f}(k)(\bar{\lambda})^{-k} \right]^* = \sum_{k=0}^{\infty} \hat{f}'(k)(\lambda)^{-k} = f'(\lambda^{-1}). \tag{A.85}$$

A.6 Digital Stochastic Processes and Mean-Square-Value

Any complex or real vector sequence $\hat{f}(k)$ whose elements are random functions of the discrete variable k characterized by appropriate probability density functions is called a digital stochastic process. Interest is restricted here to such processes that are ergodic, have constant mean $\hat{m}$, and are second-order stationary: i.e., the mean

$$\hat{m} = <\hat{f}(k)> = \lim_{N\to\infty} \frac{1}{2N+1} \sum_{k=-N}^{N} \hat{f}(k) \tag{A.86a}$$

and the autocorrelation matrix

$$\hat{\Phi}_{ff}(l) = <\hat{f}(k+l)\hat{f}^*(k)> = \lim_{N\to\infty} \frac{1}{2N+1} \sum_{k=-N}^{N} \hat{f}(k+l)\hat{f}^*(k) \tag{A.86b}$$

do not depend on the discrete variable k. It is because of ergodicity that the ensemble averages are equal to the time averages. Such processes are said to be wide-sense stationary and possess the covariance matrix

$$\hat{C}_{ff}(l) = <[\hat{f}(k+l) - \hat{m}][\hat{f}(k) - \hat{m}]^*> = \hat{\Phi}_{ff}(l) - \hat{m}\hat{m}^*. \tag{A.87}$$

Since any nonzero mean $\hat{m}$ can be absorbed into the shape-deterministic component of any input signals, there is no loss in generality in restricting attention here to cases in which $\hat{m} = 0$; hence, $\hat{\Phi}_{ff}(l)$ is both the autocorrelation and covariance matrix for the zero-mean process $\hat{f}(k)$. When $\hat{g}(k)$ is a similar stochastic process, the cross-correlation matrix

$$\hat{\Phi}_{fg}(l) = <\hat{f}(k+l)\hat{g}^*(k)> \tag{A.88}$$

is also often of interest. When $\hat{\Phi}_{fg}(l) \equiv 0, \hat{f}(k)$ and $\hat{g}(k)$ are said to be uncorrelated. When

$$\hat{\Phi}_{fg}(l) = <\hat{f}(k+l)> \; <\hat{g}^*(k)> \; = [<\hat{f}(k+l)>][<\hat{g}(k)>]^* \tag{A.89}$$

then $\hat{f}(k)$ and $\hat{g}(k)$ are said to be independent. Clearly, when $\hat{f}(k)$ and $\hat{g}(k)$ are independent and either one has zero mean, then they are uncorrelated.

There are a number of important properties that are possessed by autocorrelation matrices. First,

$$\hat{\Phi}^*_{ff}(l) = <\hat{f}(k)\hat{f}^*(k+l)> \; = \; <\hat{f}(k-l)\hat{f}^*(k)> \; = \hat{\Phi}_{ff}(-l) \tag{A.90}$$

because k can be replaced with $k-l$ on account of the process being stationary. An immediate consequence of (A.90) is that autocorrelation matrices are hermitian at $l=0$: $\hat{\Phi}^*_{ff}(0) = \hat{\Phi}_{ff}(0)$. Moreover, for any constant vector $a \neq 0$,

$$a^*\hat{\Phi}_{ff}(0)a = <a^*\hat{f}(k)\hat{f}^*(k)a> \; = \; <|\hat{f}^*(k)a|^2> \tag{A.91}$$

is nonnegative. This establishes the second important property that $\Phi_{ff}(0)$ is a hermitian nonnegative-definite matrix or, equivalently,

$$\hat{\Phi}_{ff}(0) = \hat{\Phi}^*_{ff}(0) \geq 0. \tag{A.92}$$

Finally, it is evident that

$$\text{Tr}\,[\hat{\Phi}_{ff}(0)] = <\hat{f}^*(k)\hat{f}(k)> \; = \sum_{i=1}^{n} <|\hat{f}_i(k)|^2> \tag{A.93}$$

is the mean norm-squared value or simply the mean-square-value of $\hat{f}(k)$ for any k.

The bilateral λ-transform of the autocorrelation matrix is also useful in control system applications and is given by

$$\Phi_{ff}(\lambda) = \lambda_{II}\{\hat{\Phi}_{ff}(l)\} = \sum_{l=-\infty}^{\infty} \hat{\Phi}_{ff}(l)\lambda^l. \tag{A.94}$$

When $\hat{\phi}(l)$ denotes an element of $\hat{\Phi}_{ff}(l)$ and all such elements satisfy $|\hat{\phi}(l)| \leq \alpha\beta^{|l|}$, $0<\beta<1$, then the existence of $\Phi_{ff}(\lambda)$ with a ring of convergence that includes the unit circle is assured as follows. Clearly, by analogy with (A.77),

$$\phi(\lambda) = \phi_1(-\lambda) + \phi_2(\lambda) - \hat{\phi}(0), \tag{A.95}$$

where the unilateral λ-transforms

$$\phi_1(\lambda) = \lambda_I\{\hat{\phi}(-l)\}, \phi_2(\lambda) = \lambda_I\{\hat{\phi}(l)\} \tag{A.96}$$

exist and are analytic in $|\lambda| < \beta^{-1}$. Hence, $\phi(\lambda)$ exists and is analytic in the ring $\beta < |\lambda| < \beta^{-1}$. In this case one can use

$$\hat{\Phi}_{ff}(l) = \frac{1}{2\pi j} \oint_{|\lambda|=1} \lambda^{-l}\Phi_{ff}(\lambda)\frac{d\lambda}{\lambda} \tag{A.97}$$

for the inversion integral with the path of integration being the unit circle.

One easily gets from (A.90) and (A.94) that

$$\Phi_{ff*}(\lambda) = \Phi^*_{ff}(\bar{\lambda}^{-1}) = \sum_{l=-\infty}^{\infty} \hat{\Phi}^*_{ff}(l)\lambda^{-l} = \sum_{l=-\infty}^{\infty} \hat{\Phi}_{ff}(-l)\lambda^{-l} = \sum_{l=-\infty}^{\infty} \hat{\Phi}_{ff}(l)\lambda^{l} = \Phi_{ff}(\lambda). \tag{A.98}$$

Hence, $\Phi_{ff}(e^{-j\theta}) = \Phi^*(e^{-j\theta})$ is an hermitian matrix. Moreover, it easily follows from (A.94) when $\hat{\Phi}_{ff}(l)$ is a real matrix that $\overline{\Phi}_{ff}(\lambda) = \Phi_{ff}(\bar{\lambda})$; thus, in this case $\Phi_{ff*}(\lambda) = \Phi'_{ff}(\lambda^{-1}) = \Phi_{ff}(\lambda)$ and $\Phi_{ff}(\lambda)$ is parahermitian. Equations (A.86b) and (A.94) also yield that

$$\begin{aligned} \Phi_{ff}(\lambda) &= \lim_{N\to\infty} \frac{1}{2N+1} \sum_{k=-N}^{N} \sum_{l=-\infty}^{\infty} \hat{f}(k+l)\hat{f}^*(k)\lambda^l \\ &- \lim_{N\to\infty} \frac{1}{2N+1} \sum_{k=-\infty}^{\infty} \sum_{l=-\infty}^{\infty} \hat{f}_N(k+l)\hat{f}^*_N(k)\lambda^l \\ &= \lim_{N\to\infty} \frac{1}{2N+1} \sum_{k=-\infty}^{\infty} f_N(\lambda)\lambda^k\hat{f}^*_N(k) = \lim_{N\to\infty} \frac{1}{2N+1} f_N(\lambda)f^*_N(-\bar{\lambda}^{-1}), \end{aligned} \tag{A.99}$$

where $f_N(\lambda) = \lambda_{II}\{\hat{f}_N(k)\}$ and

$$\hat{f}_N(k) = \begin{cases} \hat{f}(k), & |k| \le N \\ 0, & |k| > N \end{cases}. \tag{A.100}$$

Therefore,

$$a^* \Phi_{ff}(e^{-j\theta}) a = \lim_{N \to \infty} \frac{1}{2N+1} |f_N^*(e^{-j\theta}) a|^2 \ge 0 \tag{A.101}$$

for any constant vector a and $\Phi_{ff}(e^{-j\theta}) \ge 0$. That is, $\Phi_{ff}(\lambda)$ is a nonnegative-definite hermitian matrix on the unit circle $|\lambda| = |e^{-j\theta}| = 1$.

It easily follows from (A.93) and (A.97) with $r = 1$ that the mean-square-value of $\hat{f}(k)$ is given by

$$<\hat{f}^*(k)\hat{f}(k)> \; = \text{Tr}\;[\hat{\Phi}_{ff}(0)] = \frac{1}{2\pi j} \oint_{|\lambda|=1} \text{Tr}\;[\Phi_{ff}(\lambda)] \frac{d\lambda}{\lambda} \tag{A.102}$$

which is the basis for (1.8) when $\hat{f}(k)$ is real. It is of interest to note that were the $\hat{f}_i(k)$ real voltages across one-ohm resistors, then $<\hat{f}^*(k)\hat{f}(k)>$ would be the total average power dissipated in the resistors. Since this average power can be obtained from the integral in (A.102), $\Phi_{ff}(\lambda)$ is called the power spectral density matrix for the stochastic process $\hat{f}(k)$.

An issue of interest in connection with digital stochastic processes is the relationship between the power spectral density matrix of the output from a linear causal time-invariant stable multivariable system possessing a rational transfer matrix $G(\lambda) = \lambda_I\{\hat{G}(k)\}$ to the power spectral density matrix of its input in the steady-state (i.e., after all transients and all contributions from initial data strings have disappeared). For systems of this type with output $\hat{y}_o(k)$ and input $\hat{x}(k)$ one gets

$$\hat{y}_o(k) = \hat{y}_i(k, k_o) + \sum_{l=k_o}^{k} \hat{G}(k - l)\hat{x}(l), \tag{A.103}$$

where $\hat{y}_i(k, k_o)$ is the contribution from the initial data at k_o and $\hat{G}(k)$ is the causal inverse unilateral λ-transform of $G(\lambda)$. When the system is stable, $G(\lambda)$ is stable and

$$\lim_{k_o \to -\infty} \hat{y}_i(k, k_o) = 0. \tag{A.104}$$

Since $\hat{G}(k)$ is causal, $\hat{G}(k-l)=0,\ l>k$ and one gets for the steady-state system response

$$\hat{y}(k)=\lim_{k_o\to-\infty}\hat{y}_o(k)=\sum_{l=-\infty}^{\infty}\hat{G}(k-l)\hat{x}(l)=\sum_{l=-\infty}^{\infty}\hat{G}(l)\hat{x}(k-l). \tag{A.105}$$

Then

$$\hat{y}(k+l)\hat{y}^*(k)=\sum_{l_1=-\infty}^{\infty}\sum_{l_2=-\infty}^{\infty}\hat{G}(l_1)\hat{x}(k+l-l_1)\hat{x}^*(k-l_2)\hat{G}^*(l_2) \tag{A.106}$$

and taking ensemble averages on both sides of this equation yields

$$\hat{\Phi}_{yy}(l)=\sum_{l_1=-\infty}^{\infty}\sum_{l_2=-\infty}^{\infty}\hat{G}(l_1)\Phi_{xx}(l+l_2-l_1)\hat{G}^*(l_2). \tag{A.107}$$

Since

$$\lambda_{II}\{\hat{\Phi}_{xx}(l+l_2-l_1)\}=\lambda_{II}\{\hat{\Phi}_{xx}(l)\}\lambda^{(l_1-l_2)}=\Phi_{xx}(\lambda)\lambda^{l_1}\lambda^{-l_2} \tag{A.108}$$

and since

$$G_*(\lambda)=G^*(\bar{\lambda}^{-1})=\left[\sum_{l_2=-\infty}^{\infty}\hat{G}(l_2)\bar{\lambda}^{-l_2}\right]^*=\sum_{l_2=-\infty}^{\infty}\hat{G}^*(l_2)\lambda^{-l_2} \tag{A.109}$$

because $\hat{G}(k)=0,\ k<0$, it follows from the bilateral λ-transform of (A.107) that

$$\Phi_{yy}(\lambda)=\sum_{l_1=-\infty}^{\infty}\sum_{l_2=-\infty}^{\infty}\hat{G}(l_1)\lambda^{l_1}\Phi_{xx}(\lambda)\hat{G}^*(l_2)\lambda^{-l_2}=G(\lambda)\Phi_{xx}(\lambda)G_*(\lambda). \tag{A.110}$$

In arriving at (A.110) the sums beginning with (A.103) are assumed to exist and the order of multiple sums is assumed to be interchangeable. This is an acceptable assumption for the systems and signals of interest here.

Clearly, when $G(\lambda)$ is a stable matrix and $\Phi_{xx}(\lambda)$ is analytic on the unit circle, it follows that $\Phi_{yy}(\lambda)$ is analytic on the unit circle. In this case, the mean-square-value of $\hat{y}(k)$ easily follows from (A.102) and is given by

$$<\hat{y}^*(k)\hat{y}(k)>\ =\text{Tr}\,[\hat{\Phi}_{yy}(0)]=\frac{1}{2\pi j}\oint_{|\lambda|=1}\text{Tr}\,[G(\lambda)\Phi_{xx}(\lambda)G_*(\lambda)]\frac{d\lambda}{\lambda}, \tag{A.111}$$

which is the basis for (3.102). It is also convenient at times to consider cases in which $\Phi_{xx}(\lambda) = \Sigma$, a constant diagonal matrix with nonnegative elements. That is, cases in which $\hat{\Phi}_{xx}(l) = \Sigma\hat{\delta}(l)$, where $\hat{\delta}(l)$ is the unit pulse function: $\hat{\delta}(0) = 1$ and $\hat{\delta}(k) = 0,\ k \neq 0$. Such processes are called "white-noise" processes and in these cases the mean-square-value of $\hat{y}(k)$ is given by (A.111) when $G(\lambda)$ is stable.

An important property of the cross-correlation matrix for stationary digital processes is

$$\begin{aligned}\hat{\Phi}_{fg}(l) &= <\hat{f}(k+l)\hat{g}^*(k)> = <\hat{f}(k)\hat{g}^*(k-l)> = [<\hat{g}(k-l)\hat{f}^*(k)>]^* \\ &= \hat{\Phi}^*_{gf}(-l).\end{aligned} \tag{A.112}$$

Hence,

$$\Phi_{fg}(\lambda) = \lambda_{II}\{\hat{\Phi}^*_{gf}(-l)\} = [\sum_{l=-\infty}^{\infty} \hat{\Phi}_{gf}(-l)\bar{\lambda}^{l}]^* = [\sum_{l=-\infty}^{\infty} \hat{\Phi}_{gf}(l)\bar{\lambda}^{-l}]^* = \Phi_{gf*}(\lambda) \tag{A.113}$$

which is in agreement with (A.98) when $\hat{g}(k) = \hat{f}(k)$. The matrix $\Phi_{fg}(\lambda)$ is called the cross-power spectral density matrix for the processes $\hat{f}(k)$ and $\hat{g}(k)$. Clearly, when these processes are uncorrelated, $\hat{\Phi}_{fg}(l) = 0$ and $\Phi_{fg}(\lambda) = 0$. In this regard it is of interest to establish the power spectral density matrix $\Phi_{yy}(\lambda)$ for the output $\hat{y}(k)$ from the system described by

$$y(\lambda) = \sum_{i=1}^{n} G_i(\lambda)x_i(\lambda), \tag{A.114}$$

where the $G_i(\lambda)$ are stable transfer matrices and the $\hat{x}_i(k)$ are independent zero-mean stationary stochastic exogenous inputs to the system. In this case, the inputs are uncorrelated; hence,

$$\hat{\Phi}_{x_ix_j}(l) = 0,\ i \neq j. \tag{A.115}$$

Thus, with

$$\hat{x}(k) = [\hat{x}'_1(k)\quad \hat{x}'_2(k)\quad \cdots\quad \hat{x}'_n(k)]' \tag{A.116}$$

one gets

$$\hat{\Phi}_{xx}(l) = <\hat{x}(k+l)\hat{x}^*(k)> = \text{diag } \{\hat{\Phi}_{x_1x_1}(l), \hat{\Phi}_{x_2x_2}(l), \ldots, \hat{\Phi}_{x_nx_n}(l)\} \quad \text{(A.117)}$$

and

$$\Phi_{xx}(\lambda) = \text{diag } \{\hat{\Phi}_{x_1x_1}(\lambda), \hat{\Phi}_{x_2x_2}(\lambda), \ldots, \hat{\Phi}_{x_nx_n}(\lambda)\}. \quad \text{(A.118)}$$

Clearly,

$$x(\lambda) = [x_1'(\lambda) \quad \hat{x}_2'(\lambda) \quad \cdots \quad \hat{x}_n'(\lambda)]' \quad \text{(A.119)}$$

and with

$$G(\lambda) = [G_1(\lambda) \quad G_2(\lambda) \quad \cdots \quad G_n(\lambda)] \quad \text{(A.120)}$$

one has that (A.114) is equivalent to $y(\lambda) = G(\lambda)x(\lambda)$. It immediately follows from (A.110) that

$$\Phi_{yy}(\lambda) = G(\lambda)\Phi_{xx}(\lambda)G_*(\lambda) = \sum_{i=1}^{n} G_i(\lambda)\Phi_{x_ix_i}(\lambda)G_{i*}(\lambda). \quad \text{(A.121)}$$

That is, when (A.115) holds, the power spectral density matrix for $y(\lambda)$ is the sum of the power spectral density matrices for the outputs $y_i(\lambda) = G_i(\lambda)x_i(\lambda)$ from each of the subsystems $G_i(\lambda)$.

Appendix B
Coprime Polynomial Matrix Fraction Descriptions for Rational Matrices

B.1 Overview

The mathematical infrastructure that is the basis for the coprime polynomial matrix fraction description of rational matrices is reviewed in this appendix. Specifically, the McMillan form (also called the Smith–Mcmillan form) for a rational matrix and its McMillan degree are described and several properties of the McMillan degree are stated. In addition, the relationship between McMillan forms and coprime polynomial matrix fraction descriptions is presented. Omitted proofs can be found either in Rosenbrock (1970), Kailath (1980), Chen (1984), Maciejowski (1989), Antsaklis and Michel (2006), or Youla (2015). A number of other results that are made use of repeatedly are highlighted as well.

B.2 Notation and Basic Concepts

A rectangular array of elements with m rows and n columns is called an $m \times n$ *matrix*. When the symbol for the element in the i-row, j-column is a_{ij}, the symbol used for the matrix is A. Specifically, the matrix A is depicted as

$$A = [a_{ij}] = \begin{bmatrix} a_{11} & a_{12} & \cdots & a_{1n} \\ a_{21} & a_{22} & \cdots & a_{2n} \\ \vdots & \vdots & \cdots & \vdots \\ a_{m1} & a_{m2} & \cdots & a_{mn} \end{bmatrix}. \tag{B.1}$$

J. J. Bongiorno Jr. and K. Park, *Design of Linear Multivariable Feedback Control Systems*, https://doi.org/10.1007/978-3-030-44356-6

Typical examples are matrices whose elements are real numbers, complex numbers, polynomials, or ratios of polynomials (i.e., rational functions). When the elements a_{ij} are polynomials, the matrix is called a *polynomial matrix*. When the elements a_{ij} are rational functions, the matrix is called a *rational matrix*. The notation $a(p)|b(p)$ is used for two rational functions $a(p)$ and $b(p)$ to denote that the ratio $b(p)/a(p)$ is a polynomial. It is said in this case that $a(p)$ *divides* $b(p)$. When $a(p)$ and $b(p)$ are both polynomials, $a(p)|b(p)$ is equivalent to the statement that $a(p)$ divides $b(p)$ without remainder. The *least common multiple* (*lcm*) of a set of polynomials $a_1(p)$, $a_2(p), \ldots, a_k(p)$ is the monic polynomial of least degree which is divisible without remainder by every polynomial in the set. The *greatest common divisor* (*gcd*) of a set of polynomials $a_1(p)$, $a_2(p)$, ..., $a_k(p)$ is the monic polynomial of highest degree which divides every polynomial in the set without remainder. Two polynomials are said to be *coprime* when their *gcd* is one (equivalently, they have no common zeros).

The determinant of a square $r \times r$ submatrix of A formed from the elements in the rows $i_1, i_2, \ldots, i_r$ and the columns $j_1, j_2, \ldots, j_r$ with $1 \le i_1 < i_2 < \cdots < i_r \le m$ and $1 \le j_1 < j_2 < \cdots < j_r \le n$ is represented by

$$A^{i_1,i_2,\ldots,i_r}_{j_1,j_2,\ldots,j_r} = \det \begin{bmatrix} a_{i_1j_1} & a_{i_1j_2} & \cdots & a_{i_1j_r} \\ a_{i_2j_1} & a_{i_2j_2} & \cdots & a_{i_2j_r} \\ \vdots & \vdots & \cdots & \vdots \\ a_{i_rj_1} & a_{i_rj_2} & \cdots & a_{i_rj_r} \end{bmatrix}. \tag{B.2}$$

This determinant is called a *minor* of A of order r. The *rank* of a polynomial or rational matrix $A(p)$ is r if there is a minor of order r which is not zero for at least one point in the complex p-plane while all minors of order $r+1$ are identically zero. Clearly, rank $(A) \le \min(m, n)$. When rank $(A) = \min(m, n)$, A is said to have *full rank*. For a square matrix A of full rank, $\det A \ne 0$ and A is called a *nonsingular* matrix.

B.3 Binet–Cauchy Formula and Related Results

The Binet–Cauchy formula (Gantmacher 1960) is an expression for a minor of order r of a matrix product. Specifically, when $C = AB$, where A is an $l \times n$ matrix and B is an $n \times m$ matrix, then with $r \le \min\,(l, m)$

$$C^{i_1,i_2,\ldots,i_r}_{j_1,j_2,\ldots,j_r} = \begin{cases} \sum\limits_{\{k_i\}} A^{i_1,i_2,\ldots,i_r}_{k_1,k_2,\ldots,k_r} B^{k_1,k_2,\ldots,k_r}_{j_1,j_2,\ldots,j_r}, & r \le n \\ 0, & r > n \end{cases} \tag{B.3}$$

where the sum is taken over all integer sets $\{k_i\} = \{k_1, k_2, \ldots, k_r\}$ satisfying

$$1 \le k_1 < k_2 < \cdots < k_r \le n. \tag{B.4}$$

When $l = m = r = n$, one gets from (B.3) that

$$\det C = (\det A)(\det B), \tag{B.5}$$

the formula for the determinant of the product of two square matrices. It also readily follows from (B.3) that

$$\text{rank } (C) \le \min \,[\text{rank } (A), \text{rank } (B)]. \tag{B.6}$$

Another important result is obtained from the two identities

$$\begin{bmatrix} I & 0 \\ -N & I \end{bmatrix} \begin{bmatrix} I & M \\ N & I \end{bmatrix} = \begin{bmatrix} I & M \\ 0 & (I - NM) \end{bmatrix} \tag{B.7}$$

and

$$\begin{bmatrix} I & -M \\ 0 & I \end{bmatrix} \begin{bmatrix} I & M \\ N & I \end{bmatrix} = \begin{bmatrix} (I - MN) & 0 \\ N & I \end{bmatrix}, \tag{B.8}$$

where M is an $m \times n$ matrix and N is an $n \times m$ matrix. Applying (B.5) to both of these identities yields

$$\det \begin{bmatrix} I & M \\ N & I \end{bmatrix} = \det(I - NM) = \det(I - MN). \tag{B.9}$$

B.4 Smith Form of a Polynomial Matrix

A square polynomial matrix with nonzero constant determinant is called *unimodular*. It is important to recognize that if A is unimodular, then $A^{-1} = (\text{adj } A)/(\det A)$ is a polynomial matrix. Moreover, from $A^{-1}A = I$ and (B.5) one gets $\det A^{-1} = (\det A)^{-1}$ is a nonzero constant. That is, the inverse of a unimodular matrix is also unimodular. Two polynomial matrices $A(p)$ and $B(p)$ are called *equivalent* when there exist unimodular polynomial matrices $U(p)$ and $V(p)$ such that $B(p) = U(p)A(p)V(p)$.

For a polynomial matrix $A(p)$ of rank r, the ratios defined by

$$e_1(p) = \frac{D_1(p)}{D_0(p)},\ e_2(p) = \frac{D_2(p)}{D_1(p)},\ \cdots, e_r(p) = \frac{D_r(p)}{D_{r-1}(p)}, \tag{B.10}$$

where $D_0(p) \equiv 1$ and where for $k = 1 \to r$ the polynomial $D_k(p)$ is the *gcd* of all minors of $A(p)$ of order k play an important role. It can be established that each ratio is a monic polynomial and that $e_{k-1}|e_k$ for $k = 2 \to r$. Clearly, $D_k = e_1 e_2 \cdots e_k$. The polynomials $e_k(p)$, $k = 1 \to r$, are called the invariant polynomials for $A(p)$.

Theorem B.1 *There exist unimodular matrices $U(p)$, $V(p)$ for a polynomial matrix $A(p)$ of rank r such that*

$$S(p) = U(p)A(p)V(p) = \begin{bmatrix} Q(p) & 0 \\ 0 & 0 \end{bmatrix}, \tag{B.11}$$

where

$$Q(p) = \text{diag}\ \{e_1(p), e_2(p), \ldots, e_r(p)\} \tag{B.12}$$

and the $e_k(p)$ are the invariant polynomials for $A(p)$ defined in (B.10). When $A(p)$ is $m \times n$ and $r < \min(m, n)$, the null matrix in the second block row, second block column of $S(p)$ is $(m - r) \times (n - r)$. When $m = n = r$, $S(p) = Q(p)$; when $r = m$, $S(p) = [Q(p) \quad 0]$; and when $r = n$, $S(p) = [Q'(p) \quad 0']'$. The $m \times n$ matrix $S(p)$ is equivalent to $A(p)$ and is called the Smith form for $A(p)$.

Remark B.1 The unimodular matrices $U(p)$ and $V(p)$ in (B.11) correspond, respectively, to performing a sequence of elementary row and column operations on $A(p)$ of the following type: (*i*) multiplication of any row(column) by a nonzero constant, (*ii*) the addition to any row(column) of another row(column) which has been multiplied by a polynomial, and (*iii*) the interchange of any two rows (columns). Moreover, when $A(p)$ is a real polynomial matrix, then so is $Q(p)$ and one can choose real unimodular matrices $U(p)$ and $V(p)$.

Remark B.2 It is evident from (B.11) that $UA = SV^{-1} = [Q_r' \quad 0]'$, where $Q_r = [Q \quad 0]V^{-1}$. Similarly, $AV = U^{-1}S = [Q_c \quad 0]$, where $Q_c = U^{-1}[Q' \quad 0]'$.

B.5 The McMillan Form of a Rational Matrix

For any rational matrix $G(p) = [g_{ij}(p)]$, one can always write $g_{ij}(p) = n_{ij}(p)/d_{ij}(p)$, where $n_{ij}(p)$, $d_{ij}(p)$ are coprime and $d_{ij}(p)$ is monic. The *lcm* of the $d_{ij}(p)$ is called the *least common denominator* (*lcd*) of $G(p)$. Clearly, with $d(p)$ denoting the *lcd* of

$G(p)$, it follows that $N(p) = d(p)G(p)$ is a polynomial matrix. Hence, there exist unimodular matrices $U(p)$ and $V(p)$ such that

$$U(p)G(p)V(p) = \frac{U(p)N(p)V(p)}{d(p)} = \frac{S(p)}{d(p)} = \begin{bmatrix} H(p) & 0 \\ 0 & 0 \end{bmatrix} = M(p), \tag{B.13}$$

where $S(p)$ is the Smith form for $N(p)$. That is,

$$H(p) = \frac{Q(p)}{d(p)} = \text{diag}\left\{\frac{e_1(p)}{d(p)}, \frac{e_2(p)}{d(p)}, \ldots, \frac{e_r(p)}{d(p)}\right\}, \tag{B.14}$$

where r is the rank of $N(p)$ and the $e_k(p)$, $k = 1 \to r$, are the invariant polynomials for $N(p)$. The matrix $M(p)$ is called the *Smith-McMillan form* or for brevity simply the *McMillan form* for $G(p)$.

It is important to recognize that the $e_k(p)$ are not usually all coprime to $d(p)$. So it is convenient to introduce

$$\frac{e_k(p)}{d(p)} = \frac{\varepsilon_k(p)}{\psi_k(p)} = h_k(p), \tag{B.15}$$

where $\varepsilon_k(p)$ and $\psi_k(p)$ are the monic coprime polynomials that remain after cancelation of all common factors in the ratio $e_k(p)/d(p)$. Obviously, $h_{k-1}|h_k$ since

$$\frac{h_k}{h_{k-1}} = \left(\frac{e_k}{d}\right)\left(\frac{d}{e_{k-1}}\right) = \left(\frac{\varepsilon_k}{\psi_k}\right)\left(\frac{\psi_{k-1}}{\varepsilon_{k-1}}\right) = \frac{e_k}{e_{k-1}} \tag{B.16}$$

is a polynomial because $e_{k-1}|e_k$. Moreover, no zero of ψ_{k-1} can cancel a zero of ε_{k-1}; so all zeros of ε_{k-1} must be canceled by zeros of ε_k otherwise (B.16) is not a polynomial. That is, $\varepsilon_{k-1}|\varepsilon_k$ is true. Similarly, no zero of ε_k can cancel a zero of ψ_k; so all zeros of ψ_k must be canceled by zeros of ψ_{k-1} otherwise (B.16) cannot be a polynomial. That is, $\psi_k|\psi_{k-1}$ is true.

Another important observation to make is that

$$\text{rank }(G) = \text{rank }(M) = \text{rank }(H) = \text{rank }(Q) = r. \tag{B.17}$$

The first equality is a consequence of the following considerations. From (B.6) and (B.13), one gets rank $(M) \leq$ rank (G) since the unimodular matrices U and V have full rank. Similarly, from $G = \tilde{U}M\tilde{V}$, where $\tilde{U} = U^{-1}$ and $\tilde{V} = V^{-1}$ are unimodular, one gets rank $(G) \leq$ rank (M). Hence, rank (M) = rank (G). Obviously, the order of the largest nonzero minor of M is the same as the order of the largest nonzero minor of H. So rank $(M) =$ rank (H) and the rest easily follows. Specifically,

Theorem B.2 *There exist unimodular matrices $U(p), V(p)$ for a rational matrix $G(p)$ of rank r such that*

$$M(p) = U(p)G(p)V(p) = \begin{bmatrix} H(p) & 0 \\ 0 & 0 \end{bmatrix}, \tag{B.18}$$

where

$$H(p) = \text{diag}\ \{h_1(p), h_2(p), \ldots, h_r(p)\} = \text{diag}\ \{\frac{\varepsilon_1(p)}{\psi_1(p)}, \frac{\varepsilon_2(p)}{\psi_2(p)}, \ldots, \frac{\varepsilon_r(p)}{\psi_r(p)}\} \tag{B.19}$$

is such that $\varepsilon_k(p)$ and $\psi_k(p)$ are coprime for each $k = 1 \to r$; $\varepsilon_1|\varepsilon_2|\cdots|\varepsilon_r$; and *$\psi_r|\psi_{r-1}|\cdots|\psi_1$. When $G(p)$ is $m \times n$ and $r < \min(m, n)$, the null matrix in the second block row, second block column of $M(p)$ is $(m-r) \times (n-r)$. When $m = n = r$, $M(p) = H(p)$; when $r = m$, $M(p) = [H(p) \quad 0]$; and when $r = n$, $M(p) = [H'(p) \quad 0']'$. The $m \times n$ matrix $M(p)$ is called the McMillan form for $G(p)$.*

Remark B.3 It is evident from Remark B.1 that when $G(p)$ is a real rational matrix, then $H(p)$ is a real rational matrix and one can choose real unimodular matrices $U(p)$ and $V(p)$ in (B.18).

The rational functions $h_k(p) = \varepsilon_k(p)/\psi_k(p)$, $k = 1 \to r$, can be determined directly from the minors of the rational matrix $G(p)$. In particular,

$$\tilde{D}_k = \prod_{l=1}^{k} h_l = h_1 h_2 \cdots h_k, \quad 1 \le k \le r \tag{B.20}$$

divides all minors of $G(p)$ of order k and all other rational divisors of these minors divides $\tilde{D}_k$. Moreover, on account of the monic character of the polynomials $e_k(p)$ and $d(p)$ in (B.15), the $h_k(p)$ are ratios of monic polynomials; hence, the same is true of the $\tilde{D}_k(p)$. The rational function $\tilde{D}_k(p)$ when expressed as the ratio of coprime polynomials is therefore called the greatest common divisor (*gcd*) of the minors of $G(p)$ of order k. Clearly, $h_1 = \tilde{D}_1$ and

$$h_k = \frac{h_1 h_2 \cdots h_{k-1} h_k}{h_1 h_2 \cdots h_{k-1}} = \frac{\tilde{D}_k}{\tilde{D}_{k-1}}, \quad k = 2 \to r. \tag{B.21}$$

A simple formula is now given for the determination of the $\tilde{D}_k$ from the set of all minors of G of order k. When for each $f_l(p) = n_l(p)/d_l(p)$, $l = 1 \to \mu$, the polynomials $n_l(p)$ and $d_l(p)$ are coprime and the complete set represents all the nonzero minors of G of order k, then

$$\begin{aligned}\tilde{D}_k(p) &= \frac{N(p)}{D(p)} \\ &= \frac{gcd \text{ of}\{n_1(p), n_2(p), \ldots, n_\mu(p)\}}{lcm \text{ of } \{d_1(p), d_2(p), \ldots, d_\mu(p)\}}.\end{aligned} \tag{B.22}$$

Example B.1 The methodology described above is applied to the real rational $2x2$ matrix

$$G(p) = \begin{bmatrix} \frac{p-1}{p+1} & \frac{1}{p} \\ p & 1 \end{bmatrix} \tag{B.23}$$

in this example to find its McMillan form. The *gcd* of all the numerators of all the minors of order one is the *gcd* of all the numerators of all the elements of $G(p)$ or unity. The *lcm* of all the denominators of all the minors of order one is the *lcm* of all the denominators of all the elements of $G(p)$ or $p(p+1)$. Hence, $h_1 = \tilde{D}_1 = 1/[p(p+1)]$. There is only one minor of order two and it is given by $\det G(p) = -2/(p+1)$. Hence, $h_1 h_2 = \tilde{D}_2 = 1/(p+1)$. It then follows that $h_2 = \tilde{D}_2/h_1 = 1/[h_1(p+1)] = p$. Clearly, rank of $G(p)$ is two. So

$$M(p) = H(p) = \text{diag } \{h_1(p), h_2(p)\} = \text{diag } \{\frac{1}{p(p+1)}, p\} \tag{B.24}$$

It is easy to see that $\varepsilon_1(p) = 1$, $\varepsilon_2(p) = p$, and $\varepsilon_1|\varepsilon_2$. Also, $\psi_1(p) = p(p+1)$, $\psi_2(p) = 1$, and $\psi_2|\psi_1$.

B.6 Characteristic Denominator, Poles, and McMillan Degree of a Rational Matrix

The *characteristic denominator* of a rational matrix $G(p)$ is the *lcm* of the denominator of all minors expressed as ratios of coprime polynomials and is denoted by $\psi_G(p)$. A rational matrix $G(p)$ possesses a *pole* at $p = p_o$ (finite or infinite) if any element of $G(p)$ does. The *order* of the pole at p_o is the largest multiplicity it possesses as a pole of any element of $G(p)$. The *McMillan degree of p_o as a pole of* $G(p)$, denoted by $\delta(G; p_o)$, is the largest multiplicity it possesses as a pole of any minor of $G(p)$. When p_o is not a pole of any minor of $G(p)$, its multiplicity as a pole of every minor of $G(p)$ is zero; hence, $\delta(G; p_o) = 0$ in this case. The *McMillan degree of* $G(p)$ denoted by $\delta(G)$ is the sum of the degrees of the distinct poles of $G(p)$. Specifically, when p_i, $i = 1 \rightarrow n$, are the finite distinct poles of $G(p)$ then

$$\delta(G) = \delta(G;\infty) + \sum_{i=1}^{n} \delta(G;p_i) = \delta(G;\infty) + \nu(G), \tag{B.25}$$

where $\nu(G) = \delta(G) - \delta(G;\infty)$ is called the *finite McMillan degree of* $G(p)$. When $G(p)$ has no pole at infinity, $\delta(G;\infty) = 0$ and $\delta(G) = \nu(G)$. It should be evident that

$$\psi_G(p) = \prod_{i=1}^{n} (p - p_i)^{\delta_i}, \tag{B.26}$$

where $\delta_i = \delta(G;p_i)$, $i = 1 \rightarrow n$. Since a polynomial is a rational function with only poles at infinity having a multiplicity equal to the degree of the polynomial, it follows that

$$\nu(G) = \sum_{i=1}^{n} \delta_i = \delta(\psi_G). \tag{B.27}$$

For Example B.1, the distinct poles are at $p_1 = 0$, $p_2 = -1$, and ∞ and each has McMillan degree of one. So $\delta(G) = 3$, $\nu(G) = 2$, and $\psi_G = p(p+1)$. For this example it is also true that $\psi_G = \psi_1\psi_2$.

The usual definition of the degree of a rational function is the larger of the degrees of the numerator and denominator polynomials. That is, the degree of a rational function is the sum of the multiplicities of all its poles including any at infinity. So the definition of the McMillan degree of a rational matrix given in Youla (2015) and used here is a natural extension of the definition for the degree of a rational function. It leads to similar properties, and treats poles at infinity in the same way as any other poles. The latter point distinguishes the definition for McMillan degree used here from the equivalent ones given in Rosenbrock (1970) and Kailath (1980) which treat the issue of poles at infinity in an indirect way.

B.7 Some Properties of the McMillan Degree

In this section, several properties of the McMillan degree that are particularly important in connection with control systems are stated. Additional properties are covered in Examples B.6 and B.7.

Property B.1 If $G(p) = G_1(p)G_2(p)$ and p_o is a pole of $G_1(p)$ and/or $G_2(p)$, then $\delta(G;p_o) \leq \delta(G_1;p_o) + \delta(G_2;p_o)$ and $\delta(G) \leq \delta(G_1) + \delta(G_2)$.

Property B.2 If $G(p) = G_1(p) + G_2(p)$ and p_o is a pole of $G_1(p)$ and/or $G_2(p)$, then $\delta(G\,; p_o) \leq \delta(G_1; p_o) + \delta(G_2; p_o)$ and $\delta(G) \leq \delta(G_1) + \delta(G_2)$.

Property B.3 If $G(p) = G_1(p) + G_2(p)$, $G_1(p)$ and $G_2(p)$ have no common poles, and p_o is a pole of $G(p)$, then $\delta(G; p_o) = \delta(G_1; p_o)$ when p_o is a pole of $G_1(p)$, $\delta(G; p_o) = \delta(G_2; p_o)$ when p_o is a pole of $G_2(p)$, and $\delta(G) = \delta(G_1) + \delta(G_2)$.

Property B.4 If $G_S(p)$ is a submatrix of $G(p)$ and p_o is a pole of $G_S(p)$, then $\delta(G_S; p_o) \leq \delta(G; p_o)$ and $\delta(G_S) \leq \delta(G)$.

The properties described above are employed in the following two examples to generate additional results that are useful.

Example B.2 The matrix

$$G_o(p) = \begin{bmatrix} G_S(p) & 0 \\ 0 & 0 \end{bmatrix} + C = G(p) + C \tag{B.28}$$

is considered with C a constant matrix. Since C has no poles, G and C have no common poles and $\delta(C) = 0$. It then follows from Property B.3 for any pole p_o that $\delta(G_o; p_o) = \delta(G; p_o)$. Now G_S is a submatrix of G; hence, $\delta(G_S; p_o) \leq \delta(G; p_o)$ according to Property B.4. Moreover,

$$G = \begin{bmatrix} I \\ 0 \end{bmatrix} G_S [I \quad 0]. \tag{B.29}$$

Thus, from Property B.1,

$$\delta(G; p_o) \leq \delta(\begin{bmatrix} I \\ 0 \end{bmatrix}) + \delta(G_S; p_o) + \delta([I \quad 0]) = \delta(G_S; p_o) \tag{B.30}$$

and it follows that

$$\delta(G_o; p_o) = \delta(G; p_o) = \delta(G_S; p_o) \Rightarrow \delta(G_o) = \delta(G) = \delta(G_S). \tag{B.31}$$

Example B.3 The case $G_1(p) = \Gamma(p)G_2(p)$ with both Γ and Γ^{-1} stable is considered. When p_o is a finite point in the unstable region of the p-plane, $\delta(\Gamma; p_o) = 0$ and it follows from Property B.1 that $\delta(G_1; p_o) \leq \delta(G_2; p_o)$. On the other hand, from $G_2(p) = \Gamma^{-1}(p)G_1(p)$ and $\delta(\Gamma^{-1}; p_o) = 0$ it follows that $\delta(G_2; p_o) \leq \delta(G_1; p_o)$. Hence, $\delta(G_1; p_o) = \delta(G_2; p_o)$.

When Γ in Example B.3 is a constant nonsingular matrix, then $\delta(\Gamma; p_o) = \delta(\Gamma^{-1}; p_o) = 0$ for all p_o finite or infinite. So the same reasoning as in the example yields $\delta(G_1; p_o) = \delta(G_2; p_o)$ for any pole p_o in this case. That is,

Property B.5 If $G_1(p) = \Gamma G_2(p)$ and Γ is a constant nonsingular matrix, then $\delta(G_1; p_o) = \delta(G_2; p_o)$ for any pole p_o and $\delta(G_1) = \delta(G_2)$.

An additional property that is important in network synthesis where for the same circuitry the driving point admittance and the driving point impedance, which are the inverses of one another, should have the same McMillan degree is

Property B.6 For a square nonsingular rational matrix $G(p)$ it is true that $\delta(G) = \delta(G^{-1})$.

B.8 Relationship Between the Characteristic Denominator and the McMillan Form

It is established in the sequel that the finite McMillan degree and the characteristic denominator of a rational matrix G are connected with the polynomial

$$\psi(p) = \prod_{k=1}^{r} \psi_k(p) \tag{B.32}$$

associated with the McMillan form for G. A preliminary step in obtaining the connection is the following lemma.

Lemma B.1 *When $M(p)$ is the McMillan form for $G(p)$ given by (B.18), then $\psi_G(p) = \psi_M(p) = \psi_H(p)$.*

Proof Clearly, $G = \tilde{U} M \tilde{V}$, where $\tilde{U} = U^{-1}$ and $\tilde{V} = V^{-1}$ are unimodular polynomial matrices. Since $\tilde{U}$ and $\tilde{V}$ are polynomial matrices, they have no finite poles. It then follows that G can only have finite poles where M does and $\delta(\tilde{U}; p_o) = \delta(\tilde{V}; p_o) = 0$ for any such finite pole p_o of M. Hence,

$$\delta(G; p_o) \le \delta(\tilde{U}; p_o) + \delta(M; p_o) + \delta(\tilde{V}; p_o) = \delta(M; p_o). \tag{B.33}$$

Conversely, from $M = UGV$ one gets with similar reasoning that $\delta(M; p_o) \le \delta(G; p_o)$. Hence, $\delta(M; p_o) = \delta(G; p_o)$. Moreover, every nonzero minor of M is a nonzero minor of H and conversely. Thus,

$$\delta(G; p_o) = \delta(M; p_o) = \delta(H; p_o) \tag{B.34}$$

and the three matrices G, M, and H have the same finite distinct poles with the same respective McMillan degrees. It immediately follows that $\psi_G(p) = \psi_M(p) = \psi_H(p)$.

Remark B.4 It is readily evident from (B.27) and Lemma B.1 that $\nu(G) = \nu(M) = \nu(H)$.

Lemma B.2 *The polynomials* $\psi(p)$ *and* $\psi_G(p)$ *are identical.*

Proof From Lemma B.1, p_i is a finite pole of G iff it is a finite pole of H with the same McMillan degree, δ_i. Since $\psi_r|\psi_{r-1}|\cdots|\psi_1$, it follows when p_i is a pole of G that for some integer k, $1 \le k \le r$, p_i is a zero of ψ_l, $l = 1 \to k$, with respective multiplicities α_l, $l = 1 \to k$, that satisfy $\alpha_1 \ge \alpha_2 \ge \cdots \ge \alpha_k \ge 1$. Moreover, when $k < r$, then $\psi_l(p_i) \ne 0$ for $l = (k+1) \to r$. Clearly, p_i is a zero of $\psi(p)$ with multiplicity $\alpha = \alpha_1 + \alpha_2 + \cdots + \alpha_k$. In addition, since $\varepsilon_1|\varepsilon_2|\cdots|\varepsilon_k$ and ε_k and ψ_k are coprime, it follows that $\varepsilon_l(p_i) \ne 0$ for each $l = 1 \to k$. Hence, p_i is a pole of

$$H^{1,2,\ldots,k}_{1,2,\ldots,k} = \prod_{l=1}^{k} h_l = \prod_{l=1}^{k} \frac{\varepsilon_l}{\psi_l} \tag{B.35}$$

with multiplicity α and this is the largest multiplicity this pole has as a pole of any principal minor of H, the only nonzero minors of H. Thus,

$$\alpha = \delta(H; p_i) = \delta_i \tag{B.36}$$

and each finite distinct pole p_i of G with McMillan degree δ_i is a zero of $\psi(p)$ of multiplicity δ_i. Moreover, there are no other zeros of $\psi(p)$ since any zero of $\psi(p)$ is a zero of at least one of the $\psi_l(p)$ and, therefore, a pole of H which must be a pole of G. It then follows from (B.31) that

$$\psi(p) = \prod_{i=1}^{n} (p - p_i)^{\delta_i} = \psi_G(p), \tag{B.37}$$

where p_i, $i = 1 \to n$, are the finite distinct poles of $G(p)$.

B.9 Coprime Polynomials and the Matrix Resultant

Given the coefficients of two polynomials one can of course decide whether or not they are coprime by finding all their zeros. A more direct approach which leads to additional insights is now described. The two polynomials considered are

$$\begin{aligned} a(p) &= p^n + a_1 p^{n-1} + a_2 p^{n-2} + \cdots + a_n, \\ b(p) &= b_0 p^n + b_1 p^{n-1} + b_2 p^{n-2} + \cdots + b_n \end{aligned} \tag{B.38}$$

with $n \ge 1$. When one of the polynomials is lower in degree than the other, it would be identified with $b(p)$ by setting $b_i = 0$, $i = 0 \to k$, where $n - k - 1 = \delta(b)$. Associated with this pair of polynomials is a *Sylvester matrix*

$$R = \begin{bmatrix} b_n & b_{n-1} & b_{n-2} & \cdots & b_2 & b_1 & b_0 & 0 & 0 & \cdots & 0 & 0 \\ 0 & b_n & b_{n-1} & \cdots & b_3 & b_2 & b_1 & b_0 & 0 & \cdots & 0 & 0 \\ \vdots & \vdots & \vdots & \cdots & \vdots & \vdots & \vdots & \vdots & \vdots & \cdots & \vdots & \vdots \\ 0 & 0 & 0 & \cdots & 0 & b_n & b_{n-1} & b_{n-2} & b_{n-3} & \cdots & b_1 & b_0 \\ a_n & a_{n-1} & a_{n-2} & \cdots & a_2 & a_1 & 1 & 0 & 0 & \cdots & 0 & 0 \\ 0 & a_n & a_{n-1} & \cdots & a_3 & a_2 & a_1 & 1 & 0 & \cdots & 0 & 0 \\ \vdots & \vdots & \vdots & \cdots & \vdots & \vdots & \vdots & \vdots & \vdots & \cdots & \vdots & \vdots \\ 0 & 0 & 0 & \cdots & 0 & a_n & a_{n-1} & a_{n-2} & a_{n-3} & \cdots & a_1 & 1 \end{bmatrix} \tag{B.39}$$

whose determinant is called the *resultant* for $a(p)$ and $b(p)$. Moreover,

$$R\xi(p) = [\,b(p) \quad pb(p) \quad \cdots \quad p^{n-1}b(p) \quad a(p) \quad pa(p) \quad \cdots \quad p^{n-1}a(p)]', \tag{B.40}$$

where

$$\xi(p) = [\,1 \quad p \quad \cdots \quad p^n \quad p^{n+1} \quad p^{n+2} \quad \cdots \quad p^{2n-1}\,]'. \tag{B.41}$$

Theorem B.3 *The nonzero polynomials $a(p)$ and $b(p)$ with $n \geq 1$ are coprime iff the $2n \times 2n$ coefficient matrix R is nonsingular; i.e., iff* $\det R \neq 0$.

Corollary B.1 *When $a(p)$ and $b(p)$ are coprime polynomials and $n \geq 1$, there exist two polynomials $x(p)$ and $y(p)$ such that $\delta(x) < \delta(b)$, $\delta(y) < \delta(a)$, and*

$$a(p)x(p) + b(p)y(p) = 1. \tag{B.42}$$

Proof When (B.40) is multiplied on the left by $\Gamma = [\gamma_{ij}] = R^{-1}$, whose existence is guaranteed because $a(p)$ and $b(p)$ are coprime polynomials, and then the elements in the first row on each side of the equation obtained are equated, one gets (B.42) where

$$y(p) = \gamma_{11} + \gamma_{12}p + \cdots + \gamma_{1n}p^{n-1} \tag{B.43}$$

and

$$x(p) = \gamma_{1,n+1} + \gamma_{1,n+2}p + \cdots + \gamma_{1,2n}p^{n-1}. \tag{B.44}$$

Clearly, $\delta(y) \leq n-1 < \delta(a)$ since $\delta(a) = n$. Moreover, since $a(p)x(p) + b(p)y(p) = 1$, one also has for $n \geq 1$ that $\delta(ax) = \delta(by)$. Thus,

$$\delta(x) = \delta(b) + \delta(y) - \delta(a) < \delta(b) \tag{B.45}$$

follows from the fact that $\delta(y) < \delta(a)$ and this completes the proof of the corollary.

Corollary B.2 *There exist polynomials $x(p)$ and $y(p)$ so that $a(p)x(p)+b(p)y(p)=1$ iff the polynomials $a(p)$ and $b(p)$ are coprime.*

Proof That the polynomials $x(p)$ and $y(p)$ exist when $n \geq 1$ and $a(p)$ and $b(p)$ are coprime follows immediately from Corollary B.1. When $n=0$, both $a(p)$ and $b(p)$ are constants and if one is nonzero it is still true that constant polynomials $x(p)$ and $y(p)$ exist for which (B.42) is satisfied. That polynomials $x(p)$ and $y(p)$ do not exist when $a(p)$ and $b(p)$ are not coprime follows from the fact that $a(p)=b(p)=0$ for some p precludes $a(p)x(p)+b(p)y(p)=1$ for all p.

Remark B.5 It is important to note that when $a(p)$ and $b(p)$ are real polynomials, their matrix resultant R is real. Hence, $\Gamma = R^{-1}$ is real and the polynomials $x(p)$ and $y(p)$ are real.

B.10 Coprime Polynomial Matrices

The distinguishing feature of two coprime polynomials is that their *gcd* is one: the only common polynomial divider of both polynomials is a nonzero constant. This notion can be extended to polynomial matrices; however, because matrix multiplication is not commutative in general, left and right divisors of polynomial matrices have to be considered. The appropriate generalization is embodied in the following definition.

Definition B.1 Two polynomial matrices $A(p), B(p)$ with the same number of rows are *left coprime* if every square common left polynomial matrix divisor of both $A(p)$ and $B(p)$ is unimodular. Specifically, if $A(p)=Q(p)\tilde{A}(p)$ and $B(p)=Q(p)\tilde{B}(p)$, where $Q(p)$, $\tilde{A}(p)$, and $\tilde{B}(p)$ are polynomial matrices and $Q(p)$ is square, then $\det Q(p)$ is a nonzero constant. Two polynomial matrices $A_1(p), B_1(p)$ with the same number of columns are *right coprime* if every square common right polynomial matrix divisor of both $A_1(p)$ and $B_1(p)$ is unimodular. Specifically, if $A_1(p)=\tilde{A}_1(p)Q_1(p)$ and $B_1(p)=\tilde{B}_1(p)Q_1(p)$, where $Q_1(p)$, $\tilde{A}_1(p)$, and $\tilde{B}_1(p)$ are polynomial matrices and $Q_1(p)$ is square, then $\det Q_1(p)$ is a nonzero constant.

There are a number of alternative and more practical ways to determine whether or not two polynomial matrices are coprime other than checking that all common polynomial matrix divisors are unimodular. Two important alternatives are provided by the following lemmas.

Lemma B.3

(a) *Two polynomial matrices $A(p), B(p)$ with the same number of rows are left coprime iff there exists polynomial matrices $X(p), Y(p)$ so that*

$$A(p)X(p) + B(p)Y(p) = I. \tag{B.46}$$

(b) *Two polynomial matrices $A_1(p), B_1(p)$ with the same number of columns are right coprime iff there exists polynomial matrices $X_1(p), Y_1(p)$ so that*

$$X_1(p)A_1(p) + Y_1(p)B_1(p) = I. \tag{B.47}$$

Lemma B.4

(a) *Two polynomial matrices $A(p), B(p)$ with the same number of rows r are left coprime iff for all p*

$$\text{rank}\ [A(p) \quad B(p)] = r. \tag{B.48}$$

(b) *Two polynomial matrices $A_1(p), B_1(p)$ with the same number of columns c are right coprime iff for all p*

$$\text{rank}\ [A_1'(p) \quad B_1'(p)]' = c. \tag{B.49}$$

B.11 Coprime Polynomial Matrix Fraction Descriptions

The reduction of rational transfer functions to ratios of coprime polynomials or, equivalently, coprime polynomial fractions can be extended to rational transfer matrices. Again, because matrix multiplication is not commutative in general, both left coprime and right coprime polynomial matrix fraction descriptions play a role.

Definition B.2 When $G(p) = A^{-1}(p)B(p)$, where $G(p)$ is a rational matrix and $A(p), B(p)$ is a left coprime polynomial matrix pair, $A^{-1}(p)B(p)$ is called a *left coprime polynomial matrix fraction description for* $G(p)$. When $G(p) = B_1(p)A_1^{-1}(p)$, where $G(p)$ is a rational matrix and $A_1(p), B_1(p)$ is a right coprime polynomial matrix pair, $B_1(p)A_1^{-1}(p)$ is called a *right coprime polynomial matrix fraction description for* $G(p)$.

Lemma B.5 *Any rational matrix $G(p)$ always possesses a left and a right coprime polynomial matrix fraction description. Specifically, one can always write*

$$G(p) = A^{-1}(p)B(p) = B_1(p)A_1^{-1}(p), \tag{B.50a}$$

where $A(p), B(p)$ is a left coprime polynomial matrix pair and $A_1(p), B_1(p)$ is a right coprime polynomial matrix pair. Moreover, there exist polynomial matrices $X(p), X_1(p)$, $Y(p)$, and $Y_1(p)$ so that the generalized Bezout identity (Kailath 1980*)*

$$\begin{bmatrix} A & B \\ Y_1 & -X_1 \end{bmatrix}\begin{bmatrix} X & B_1 \\ Y & -A_1 \end{bmatrix} = \begin{bmatrix} X & B_1 \\ Y & -A_1 \end{bmatrix}\begin{bmatrix} A & B \\ Y_1 & -X_1 \end{bmatrix} = \begin{bmatrix} I & 0 \\ 0 & I \end{bmatrix} \tag{B.50b}$$

is satisfied.

Proof The matrix $H(p)$ in (B.19) can be written as

$$H(p) = \mathrm{E}(p)\Psi^{-1}(p) = \Psi^{-1}(p)\mathrm{E}(p), \tag{B.51}$$

where

$$\mathrm{E}(p) = \mathrm{diag}\{\varepsilon_1(p), \varepsilon(p)_2, \ldots, \varepsilon_r(p)\}, \Psi(p) = \mathrm{diag}\{\psi_1(p), \psi(p)_2, \ldots, \psi_r(p)\}. \tag{B.52}$$

Then the McMillan form for $G(p)$ in (B.18) yields

$$\begin{aligned} G = U^{-1}\begin{bmatrix} H & 0 \\ 0 & 0 \end{bmatrix}V^{-1} &= U^{-1}\begin{bmatrix} \Psi^{-1} & 0 \\ 0 & I \end{bmatrix}UU^{-1}\begin{bmatrix} \mathrm{E} & 0 \\ 0 & 0 \end{bmatrix}V^{-1} \\ &= U^{-1}\begin{bmatrix} \mathrm{E} & 0 \\ 0 & 0 \end{bmatrix}V^{-1}V\begin{bmatrix} \Psi^{-1} & 0 \\ 0 & I \end{bmatrix}V^{-1} \end{aligned} \tag{B.53}$$

and it is natural to make the identifications

$$A = U^{-1}\begin{bmatrix} \Psi & 0 \\ 0 & I \end{bmatrix}U, A_1 = V\begin{bmatrix} \Psi & 0 \\ 0 & I \end{bmatrix}V^{-1} \tag{B.54}$$

and

$$B = B_1 = U^{-1}\begin{bmatrix} \mathrm{E} & 0 \\ 0 & 0 \end{bmatrix}V^{-1} \tag{B.55}$$

to get (B.50a). It remains to establish that the pair $A(p)$, $B(p)$ is left coprime and the pair $A_1(p)$, $B_1(p)$ is right coprime. For the former pair it suffices to show that there exist polynomial matrices $X(p)$, $Y(p)$ such that (B.46) is satisfied or

$$U^{-1}\begin{bmatrix} \Psi & 0 \\ 0 & I \end{bmatrix}UX + U^{-1}\begin{bmatrix} \mathrm{E} & 0 \\ 0 & 0 \end{bmatrix}V^{-1}Y = I. \tag{B.56}$$

For the latter pair it suffices to show that there exist polynomial matrices $X_1(p)$, $Y_1(p)$ such that (B.47) is satisfied or

$$X_1 V\begin{bmatrix} \Psi & 0 \\ 0 & I \end{bmatrix} V^{-1} + Y_1 U^{-1}\begin{bmatrix} \mathrm{E} & 0 \\ 0 & 0 \end{bmatrix} V^{-1} = I. \tag{B.57}$$

The choices

$$X = U^{-1}\begin{bmatrix} \tilde{X} & 0 \\ 0 & I \end{bmatrix} U, X_1 = V\begin{bmatrix} \tilde{X} & 0 \\ 0 & I \end{bmatrix} V^{-1}, Y = Y_1 = V\begin{bmatrix} \tilde{Y} & 0 \\ 0 & 0 \end{bmatrix} U, \tag{B.58}$$

where

$$\tilde{X} = \mathrm{diag}\{x_1, x_2, \ldots, x_r\}, \tilde{Y} = \mathrm{diag}\{y_1, y_2, \ldots, y_r\}, \tag{B.59}$$

reduces the problem of finding polynomial matrices $X(p)$, $Y(p)$ that satisfy (B.56) and polynomial matrices $X_1(p)$, $Y_1(p)$ that satisfy (B.57) to the problem of finding for $k = 1 \rightarrow r$ polynomial pairs $x_k(p)$, $y_k(p)$ for which $\Psi\tilde{X} + \mathrm{E}\tilde{Y} = I$ or, equivalently,

$$\psi_k(p)x_k(p) + \varepsilon_k(p)y_k(p) = 1, \; k = 1 \rightarrow r. \tag{B.60}$$

As noted in Sect. B.9, this can always be done since the polynomials in each pair $\varepsilon_k(p)$, $\psi_k(p)$ in the McMillan form for $G(p)$ are coprime. Finally, it is a straightforward process to confirm that the choices made in (B.54), (B.55), and (B.58) which are identical with the ones made in Appendix A of Youla et al. (1976b) satisfy (B.50b) when (B.60) is satisfied. Actually, it is evident that the first matrix product in (B.50b) is an identity matrix since $Y_1 X - X_1 Y = 0$ for (B.58). Then each matrix in the first matrix product is the inverse of the other; hence, the product is commutative and the second matrix product follows. Equation (B.50b) is a compact expression for the many relationships that hold among the submatrices and is in agreement with those in Eq. (4) of Antsaklis (1979).

Remark B.6 It is impossible to satisfy (B.60) for all p with $x_k(p) \equiv 0$ when $\varepsilon_k(p)$ is not a constant. So in these cases $x_k(p) \not\equiv 0$. When the monic polynomial $\varepsilon_k(p)$ is a constant, it is unity and (B.60) is satisfied for any $x_k(p) \not\equiv 0$ with $y_k(p) = 1 - \psi_k(p)x_k(p)$. So it is always possible to satisfy the r equations in (B.60) with $x_k(p) \not\equiv 0$, $k = 1 \rightarrow r$. That is, it is always possible to choose $X(p)$ and $X_1(p)$ so that

$$\det X = \det X_1 = \det\tilde{X} = \prod_{k=1}^{r} x_k(p) \not\equiv 0. \tag{B.61}$$

Remark B.7 In light of Remarks B.1, B.3, and B.5, it should be evident that when $G(p)$ is a real rational matrix then the formulas (B.54) and (B.55) give real polynomial matrices. The same is true for (B.58) through (B.60).

Example B.4 As an illustrative example, left and right coprime polynomial matrix fraction descriptions for the $G(p)$ of Example B.1 are now determined. From (B.23),

$$N = p(p+1)G(p) = \begin{bmatrix} p(p-1) & (p+1) \\ p^2(p+1) & p(p+1) \end{bmatrix} \tag{B.62}$$

and the unimodular matrices U and V for which UNV gives the Smith form for N are needed so that formulas (B.54) and (B.55) can be used. The first step is to move the element of lowest degree into the first row, first column:

$$NV_1 = \begin{bmatrix} p(p-1) & (p+1) \\ p^2(p+1) & p(p+1) \end{bmatrix} \begin{bmatrix} 0 & 1 \\ 1 & 0 \end{bmatrix} = \begin{bmatrix} (p+1) & p(p-1) \\ p(p+1) & p^2(p+1) \end{bmatrix}. \tag{B.63}$$

Clearly, a zero can be created in the second row, first column by subtracting p times the first row from the second row:

$$U_1(NV_1) = \begin{bmatrix} 1 & 0 \\ -p & 1 \end{bmatrix} \begin{bmatrix} (p+1) & p(p-1) \\ p(p+1) & p^2(p+1) \end{bmatrix} = \begin{bmatrix} (p+1) & p(p-1) \\ 0 & 2p^2 \end{bmatrix}. \tag{B.64}$$

Next polynomial division gives $p(p-1) = (p+1)(p-2) + 2$. So multiplying the first column by $(p-2)$ and subtracting the result from the second column gives

$$(U_1NV_1)V_2 = \begin{bmatrix} (p+1) & p(p-1) \\ 0 & 2p^2 \end{bmatrix} \begin{bmatrix} 1 & -(p-2) \\ 0 & 1 \end{bmatrix} = \begin{bmatrix} (p+1) & 2 \\ 0 & 2p^2 \end{bmatrix}. \tag{B.65}$$

Now the element of lowest degree again has to be moved to the first row, first column:

$$(U_1NV_1V_2)V_3 = \begin{bmatrix} (p+1) & 2 \\ 0 & 2p^2 \end{bmatrix} \begin{bmatrix} 0 & 1 \\ 1/2 & 0 \end{bmatrix} = \begin{bmatrix} 1 & (p+1) \\ p^2 & 0 \end{bmatrix}. \tag{B.66}$$

It should now be evident that

$$\begin{aligned} U_2(U_1NV_1V_2V_3)V_4 &= \begin{bmatrix} 1 & 0 \\ p^2 & -1 \end{bmatrix} \begin{bmatrix} 1 & (p+1) \\ p^2 & 0 \end{bmatrix}. \begin{bmatrix} 1 & -(p+1) \\ 0 & 1 \end{bmatrix} \\ &= \begin{bmatrix} 1 & 0 \\ 0 & p^2(p+1) \end{bmatrix}. \end{aligned} \tag{B.67}$$

is the Smith form for N. So

$$U = U_2U_1 = \begin{bmatrix} 1 & 0 \\ p^2 & -1 \end{bmatrix}\begin{bmatrix} 1 & 0 \\ -p & 1 \end{bmatrix} = \begin{bmatrix} 1 & 0 \\ p(p+1) & -1 \end{bmatrix} \tag{B.68}$$

and

$$\begin{aligned} V = V_1V_2V_3V_4 &= \begin{bmatrix} 0 & 1 \\ 1 & 0 \end{bmatrix}\begin{bmatrix} 1 & -(p-2) \\ 0 & 1 \end{bmatrix}\begin{bmatrix} 0 & 1 \\ 1/2 & 0 \end{bmatrix}\begin{bmatrix} 1 & -(p+1) \\ 0 & 1 \end{bmatrix} \\ &= \frac{1}{2}\begin{bmatrix} 1 & -(p+1) \\ -(p-2) & p(p-1) \end{bmatrix}. \end{aligned} \tag{B.69}$$

It is easy to verify that

$$M = UGV = \frac{1}{p(p+1)}UNV = \frac{1}{p(p+1)}\,\text{diag}\,\{1, p^2(p+1)\} = \text{diag}\{\frac{1}{p(p+1)}, p\} \tag{B.70}$$

in agreement with (B.24). Finally, since $G(p)$ is 2×2 and its rank is 2, (B.54) reduces to

$$A(p) = U^{-1}\Psi U = \begin{bmatrix} 1 & 0 \\ p(p+1) & -1 \end{bmatrix}\begin{bmatrix} p(p+1) & 0 \\ 0 & 1 \end{bmatrix}\begin{bmatrix} 1 & 0 \\ p(p+1) & -1 \end{bmatrix} \tag{B.71}$$

or

$$A(p) = \begin{bmatrix} p(p+1) & 0 \\ p(p+1)(p^2+p-1) & 1 \end{bmatrix} \tag{B.72}$$

and

$$\begin{aligned} A_1(p) &= V\Psi V^{-1} \\ &= \frac{1}{2}\begin{bmatrix} 1 & -(p+1) \\ -(p-2) & p(p-1) \end{bmatrix}\begin{bmatrix} p(p+1) & 0 \\ 0 & 1 \end{bmatrix}\begin{bmatrix} p(p-1) & (p+1) \\ (p-2) & 1 \end{bmatrix} \end{aligned} \tag{B.73}$$

or

$$A_1(p) = \frac{1}{2}\begin{bmatrix} (p+1)(p^3-p^2-p+2) & (p+1)(p^2+p-1) \\ -p(p-1)(p-2)(p^2+p-1) & -p(p^3-4p-1) \end{bmatrix}. \tag{B.74}$$

Equation (B.55) reduces to

$$B(p) = B_1(p) = U^{-1}\mathrm{E}V^{-1} = \begin{bmatrix} 1 & 0 \\ p(p+1) & -1 \end{bmatrix} \begin{bmatrix} 1 & 0 \\ 0 & p \end{bmatrix} \begin{bmatrix} p(p-1) & (p+1) \\ (p-2) & 1 \end{bmatrix} \tag{B.75}$$

or

$$B(p) = B_1(p) = \begin{bmatrix} p(p-1) & (p+1) \\ p(p^3-2p+2) & p^2(p+2) \end{bmatrix}. \tag{B.76}$$

It is of interest and instructive to verify that the final results obtained in this example for $A(p)$, $A_1(p)$, and $B(p) = B_1(p)$ are indeed such that $A(p), B(p)$, and $A_1(p), B_1(p)$ are left coprime and right coprime polynomial matrix pairs, respectively. This is easily accomplished after it is recognized that (B.76) yields

$$\det B(p) = \det B_1(p) = p \det \begin{bmatrix} (p-1) & (p+1) \\ (p^3-2p+2) & p^2(p+2) \end{bmatrix} = -2p \tag{B.77}$$

which is nonzero for all p except $p = 0$. So rank $[A(p) \quad B(p)] = 2$ for all $p \neq 0$. Since the rank of

$$[A(0) \quad B(0)] = \begin{bmatrix} 0 & 0 & 0 & 1 \\ 0 & 1 & 0 & 0 \end{bmatrix} \tag{B.78}$$

is also two, rank $[A(p) \quad B(p)] = 2$ for all p and it follows from Lemma B.4 that $A(p), B(p)$ is left coprime. Similarly, rank $[A_1'(p) \quad B_1'(p)] = 2$ for all $p \neq 0$. Since the rank of

$$[A_1'(0) \quad B_1'(0)] = \begin{bmatrix} 0 & 1 & 0 & 0 \\ 0 & -1/2 & 1 & 0 \end{bmatrix} \tag{B.79}$$

is also two, rank $[A_1'(p) \quad B_1'(p)] = 2$ for all p and it follows from Lemma B.4 that $A_1(p), B_1(p)$ is right coprime.

Finally, the polynomial matrices chosen in (B.58) are now evaluated for this example. First, the pair $x_1(p), y_1(p)$ is determined such that $x_1(p)$ is a nonzero polynomial and in accordance with (B.60)

$$\psi_1(p)x_1(p) + \varepsilon_1(p)y_1(p) = p(p+1)x_1(p) + y_1(p) = 1. \tag{B.80}$$

Clearly,

$$x_1(p) = 1, y_1(p) = 1 - p(p+1) = -(p^2 + p - 1) \tag{B.81}$$

are acceptable choices. Next, the pair $x_2(p), y_2(p)$ is determined such that $x_2(p)$ is a nonzero polynomial and in accordance with (B.60)

$$\psi_2(p)x_2(p) + \varepsilon_2(p)y_2(p) = x_2(p) + py_2(p) = 1. \tag{B.82}$$

Clearly,

$$y_2(p) = 1, x_2(p) = -(p-1) \tag{B.83}$$

are acceptable choices and one has

$$X(p) = U^{-1}\tilde{X}U = \begin{bmatrix} 1 & 0 \\ p(p+1) & -1 \end{bmatrix} \begin{bmatrix} 1 & 0 \\ 0 & -(p-1) \end{bmatrix} \begin{bmatrix} 1 & 0 \\ p(p+1) & -1 \end{bmatrix}, \tag{B.84}$$

$$\begin{aligned} X_1(p) &= V\tilde{X}V^{-1} \\ &= \frac{1}{2} \begin{bmatrix} 1 & -(p+1) \\ -(p-2) & p(p-1) \end{bmatrix} \begin{bmatrix} 1 & 0 \\ 0 & -(p-1) \end{bmatrix} \begin{bmatrix} p(p-1) & (p+1) \\ (p-2) & 1 \end{bmatrix}, \end{aligned} \tag{B.85}$$

and

$$\begin{aligned} Y(p) &= Y_1(p) = V\tilde{Y}U \\ &= \frac{1}{2} \begin{bmatrix} 1 & -(p+1) \\ -(p-2) & p(p-1) \end{bmatrix} \begin{bmatrix} -(p^2+p-1) & 0 \\ 0 & 1 \end{bmatrix} \begin{bmatrix} 1 & 0 \\ p(p+1) & -1 \end{bmatrix} \end{aligned} \tag{B.86}$$

which yields

$$X(p) = \begin{bmatrix} 1 & 0 \\ p^2(p+1) & -(p-1) \end{bmatrix}, \tag{B.87}$$

$$X_1(p) = \frac{1}{2} \begin{bmatrix} (p-1)(p^2-2) & p(p+1) \\ -p^2(p-1)(p-2) & -(p^3-p^2-2) \end{bmatrix}, \tag{B.88}$$

and

$$Y(p) = Y_1(p) = \frac{1}{2}\begin{bmatrix} -(p^3+3p^2+2p-1) & (p+1) \\ (p^4+p^3-2p^2-3p+2) & -p(p-1) \end{bmatrix}. \tag{B.89}$$

Coprime polynomial matrix fraction descriptions of the form (B.50a) are not unique. Indeed, one has the following lemma.

Lemma B.6 *(a) When $A^{-1}(p)B(p)$ is a left coprime polynomial matrix fraction description for $G(p)$, then $\tilde{A}^{-1}(p)\tilde{B}(p)$ is another iff $\tilde{A}(p) = N(p)A(p)$ and $\tilde{B}(p) = N(p)B(p)$, where $N(p)$ is a unimodular matrix. (b) When $B_1(p)A_1^{-1}(p)$ is a right coprime polynomial matrix fraction description for $G(p)$, then $\tilde{B}_1(p)\tilde{A}_1^{-1}(p)$ is another iff $\tilde{A}_1(p) = A_1(p)N_1(p)$ and $\tilde{B}_1(p) = B_1(p)N_1(p)$, where $N_1(p)$ is a unimodular matrix.*

Proof When $G = A^{-1}B = \tilde{A}^{-1}\tilde{B}$ and $AX + BY = I$, then $\tilde{B} = \tilde{A}A^{-1}B = NB$, where

$$N = \tilde{A}A^{-1} = \tilde{A}A^{-1}(AX + BY) = \tilde{A}X + \tilde{B}Y \tag{B.90}$$

is a polynomial matrix. So if $\tilde{A}\tilde{X} + \tilde{B}\tilde{Y} = I$, it follows from $\tilde{A} = NA$, $\tilde{B} = NB$ that

$$\det(\tilde{A}\tilde{X} + \tilde{B}\tilde{Y}) = (\det N)\det(A\tilde{X} + B\tilde{Y}) = 1 \tag{B.91}$$

or $\det N$ is a nonzero constant. That is, $\tilde{A} = NA$, $\tilde{B} = NB$, where N is unimodular. Conversely, when $\tilde{A} = NA$, $\tilde{B} = NB$, $AX + BY = I$, and N is unimodular, then $\tilde{A}^{-1}\tilde{B} = A^{-1}B$ and since

$$N(AX + BY)N^{-1} = \tilde{A}\tilde{X} + \tilde{B}\tilde{Y} = I, \tag{B.92}$$

where $\tilde{X} = XN^{-1}$ and $\tilde{Y} = YN^{-1}$ are polynomial matrices the pair $\tilde{A}, \tilde{B}$ is left coprime from Lemma B.3(a). The modifications needed in this proof of part (a) of the theorem to get a proof for part (b) are straightforward (see Example B.8).

The characteristic denominator of a rational matrix is simply related to its coprime polynomial matrix fraction descriptions. Indeed, one has the following lemma.

Lemma B.7 *When $A^{-1}(p)B(p)$ and $B_1(p)A_1^{-1}(p)$ are left and right coprime polynomial matrix fraction descriptions for $G(p)$, then*

$$\psi_G(p) = \eta \det A(p) = \eta_1 \det A_1(p), \tag{B.93}$$

where the constants η and η_1 make their associated polynomials monic.

Proof The specific choices for the A and A_1 matrices given in (B.54) are designated here as A_o and A_{1o}, respectively. Then any other choice is given by $A = NA_o$ and $A_1 = A_{1o}N_1$, where N and N_1 are unimodular. It follows from Lemma B.2 that

$$\det A_o(p) = \det A_{1o}(p) = \det \Psi(p) = \prod_{k=1}^{r} \psi_k(p) = \psi_G(p). \tag{B.94}$$

That is,

$$\psi_G(p) = \eta \det A(p) = \eta_1 \det A_1(p), \tag{B.95}$$

where

$$\eta = (\det N)^{-1}, \eta_1 = (\det N_1)^{-1} \tag{B.96}$$

are the constants which render $\psi_G(p)$ in (B.95) monic.

Remark B.8 Clearly, (B.48) is satisfied for all p when $B(p) = I$. Hence, the characteristic denominator for $G(p) = A^{-1}(p)B(p) = A^{-1}(p)$ is $\psi_G(p) = \psi_{A^{-1}}(p) = \eta \det A(p)$. That is, the characteristic denominator of the inverse of a square nonsingular polynomial matrix is to within a multiplicative constant the determinant of the polynomial matrix.

An important issue that arises in connection with unstable coprime polynomial matrix fraction descriptions are the conditions under which multiplication by a stable rational matrix yields a stable product. In this regard, the next lemma provides a frequently used tool in Wiener–Hopf design of multivariable control systems.

Lemma B.8 *(a) When $M(p)$ is a stable rational matrix and $A(p)^{-1}B(p)$ is a left coprime polynomial matrix fraction description, then $MA^{-1}B$ is stable iff MA^{-1} is stable. (b) When $M_1(p)$ is a stable rational matrix and $B_1(p)A_1^{-1}(p)$ is a right coprime polynomial matrix fraction description, then $B_1A_1^{-1}M_1$ is stable iff $A_1^{-1}M_1$ is stable.*

Proof Cleary, MA^{-1} stable is sufficient for $MA^{-1}B$ stable. On the other hand, from $AX + BY = I$ one gets

$$MA^{-1} = M(X + A^{-1}BY) = MX + MA^{-1}BY \tag{B.97}$$

must be stable when $MA^{-1}B$ is stable. That is, MA^{-1} stable is necessary. The modifications needed in this proof of part (a) of the lemma to get a proof for part (b) are straightforward (see Example B.9).

B.12 Computational Issues

It is evident from Example B.4 that the calculations in more complicated cases will be quite tedious to carry out by hand. Fortunately, a number of algorithms suitable for implementation on a digital computer are available. Some of the early contributions in this regard are Kučera (1979), Chang and Pearson (1982), and Chang (1982). The latter two built on the work of Forney (1975), Bitmead et al. (1978), and Kung and Kailath (1980). Also of interest are Lai (1989) and Yamada et al. (1995). More recently, algorithms developed by Henrion and Šebek (1998a, b, c, 1999a, b, 2000) and Hromcik and Šebek (2001) have appeared. Most importantly, software is available commercially in the Polynomial Toolbox for MATLAB® which, among other things, can be used to find coprime polynomial matrix fraction descriptions and solutions to the generalized Bezout equation (B.50b). For the latter, the technique used in Example 2.1 to choose the M matrix so that (2.241) is satisfied and in the Solution of Example 2.3(b) to choose the $\tilde{M}$ matrix so that (2.62) is satisfied can be employed. MATLAB® is a trademark of the MathWorks, Inc. (www.mathworks.com) and the Polynomial Toolbox for MATLAB® can be obtained from Polyx, Ltd. (www.polyx.cz).

B.13 Additional Examples

Example B.5 Prove that the polynomials $x(p)$ and $y(p)$ in Corollary B.1 are unique.

Solution When $x+\Delta x, y+\Delta y$ is any other solution pair, then $\delta(\Delta x)<\delta(b)$ and $\delta(\Delta y)<\delta(a)$ is necessary. Also, $a\Delta x+b\Delta y=0$. Hence, $\Delta x=-b\Delta y/a$ must be a polynomial. Since $a(p)$ and $b(p)$ are coprime, it follows that a must divide Δy or $\Delta y=ma$, where m is a polynomial. Then $\Delta x=-bm$. Hence, $m=0$ is necessary; otherwise, $\delta(\Delta x)\geq\delta(b)$, $\delta(\Delta y)\geq\delta(a)$ and the requirement that $\delta(\Delta x)<\delta(b)$, $\delta(\Delta y)<\delta(a)$ is violated.

Example B.6 For the block diagonal matrix $G(p)=\text{diag}\ \{G_1(p), G_2(p),\cdots, G_k(p)\}$, where the $G_l(p)$, $l=1\rightarrow k$ are rational matrices establish that $\delta(G)=\delta(G_1)+\delta(G_2)+\cdots+\delta(G_k)$.

Solution Clearly, it suffices to establish the result for $G(p)=\text{diag}\ \{G_1(p), G_2(p)\}$ since the general result would then follow immediately by induction. Let $M_1(p)$ be one of the minors of $G_1(p)$ which possesses the pole p_o with maximum multiplicity $m_1=\delta(G_1;p_o)$. Let $M_2(p)$ be one of the minors of $G_2(p)$ which possesses the same pole p_o with maximum multiplicity $m_2=\delta(G_2;p_o)$. Then $M(p)=M_1(p)M_2(p)$ is a minor of $G(p)$ which possesses the pole p_o with multiplicity $m=m_1+m_2$. Hence, $\delta(G;p_o)\geq\delta(G_1;p_o)+\delta(G_2;p_o)$ in this case. The result still holds if p_o is a pole of $G_1(p)$, but not a pole of $G_2(p)$: in this case, $\delta(G_2;p_o)=0$ and $M_1(p)$ is still a minor of $G(p)$. Likewise, if p_o is a pole of $G_2(p)$, but not a pole of $G_1(p)$, $\delta(G_2;p_o)=0$

and $M_2(p)$ is still a minor of $G(p)$. Now p_o is a pole of $G(p)$ iff it is a pole of $G_1(p)$ and/or $G_2(p)$. So summing up over all the poles of $G(p)$ it immediately follows that $\delta(G) \geq \delta(G_1) + \delta(G_2)$. On the other hand, from

$$G = \begin{bmatrix} G_1 & 0 \\ 0 & 0 \end{bmatrix} + \begin{bmatrix} 0 & 0 \\ 0 & G_2 \end{bmatrix}$$

it follows from Property B.2 and Example B.2 that $\delta(G) \leq \delta(G_1) + \delta(G_2)$. Hence, $\delta(G) = \delta(G_1) + \delta(G_2)$.

Example B.7 When $G(p) = g(p)K$, where $g(p)$ is a rational function and K is a constant matrix, prove that $\delta(G) = (\text{rank } K)\,\delta(g)$.

Solution Every nonzero minor of $G(p)$ of order r is given by $cg^r(p)$, c a nonzero constant. The largest possible value for r is the rank of K. Hence, any pole p_o of $g(p)$ with multiplicity m_o is a pole of $G(p)$ with McMillan degree of (rank $K)m_o$. Summing these degrees over all the poles of $g(p)$ gives $\delta(G) = (\text{rank } K)\delta(g)$.

Example B.8 Prove that when $B_1(p)A_1^{-1}(p)$ is a right coprime polynomial matrix fraction description for $G(p)$, then $\tilde{B}_1(p)\tilde{A}_1^{-1}(p)$ is another iff $\tilde{A}_1(p) = A_1(p)N_1(p)$ and $\tilde{B}_1(p) = B_1(p)N_1(p)$, where $N_1(p)$ is a unimodular matrix. This is part (b) of Lemma B.6.

Solution When $G = B_1A_1^{-1} = \tilde{B}_1\tilde{A}_1^{-1}$ and $X_1A_1 + Y_1B_1 = I$, then $\tilde{B}_1 = B_1A_1^{-1}\tilde{A}_1 = B_1N_1$, where

$$N_1 = A_1^{-1}\tilde{A}_1 = (X_1A_1 + Y_1B_1)A_1^{-1}\tilde{A}_1 = (X_1\tilde{A}_1 + Y_1\tilde{B}_1)$$

is a polynomial matrix. So if $\tilde{X}_1\tilde{A}_1 + \tilde{Y}_1\tilde{B}_1 = I$, it follows from $\tilde{A}_1 = A_1N_1$, $\tilde{B}_1 = B_1N_1$ that

$$\det(\tilde{X}_1\tilde{A}_1 + \tilde{Y}_1\tilde{B}_1) = (\det N_1)\det(\tilde{X}_1A_1 + \tilde{Y}_1B_1) = 1$$

or $\det N_1$ is a nonzero constant. That is, $\tilde{A}_1 = A_1N_1$, $\tilde{B}_1 = B_1N_1$, where N_1 is unimodular. Conversely, when $\tilde{A}_1 = A_1N_1$, $\tilde{B}_1 = B_1N_1$, $X_1A_1 + Y_1B_1 = I$, and N_1 is unimodular, then $\tilde{B}_1\tilde{A}_1^{-1} = B_1A_1^{-1}$ and since

$$N_1^{-1}(X_1A_1 + Y_1B_1)N_1 = \tilde{X}_1\tilde{A}_1 + \tilde{Y}_1\tilde{B}_1 = I,$$

where $\tilde{X}_1 = N_1^{-1}X_1$ and $\tilde{Y}_1 = N_1^{-1}Y_1$ are polynomial matrices, the pair $\tilde{A}_1, \tilde{B}_1$ is right coprime from Lemma B.3(b).

Example B.9 Prove that when $M_1(p)$ is a stable rational matrix and $B_1(p)A_1^{-1}(p)$ is a right coprime polynomial matrix fraction description, then $B_1A_1^{-1}M_1$ is stable iff $A_1^{-1}M_1$ is stable. This is part (b) of Lemma B.8.

Solution Clearly, $A_1^{-1}M_1$ stable is sufficient for $B_1A_1^{-1}M_1$ stable. On the other hand, from $X_1A_1 + Y_1B_1 = I$ one gets

$$A_1^{-1}M_1 = (X_1 + Y_1B_1A_1^{-1})M_1 = X_1M_1 + Y_1B_1A_1^{-1}M_1$$

must be stable when $B_1A_1^{-1}M_1$ is stable. That is, $A_1^{-1}M_1$ stable is necessary.

Appendix C
Spectral Factorization of Rational Parahermitian-Positive Matrices

C.1 Overview

The spectral factorization of real rational parahermitian-positive matrices is clearly an important step in the Wiener–Hopf design approach covered in this book. In this appendix the key results connected with such a spectral factorization are provided for easy reference; proofs for the analog case can be found in Youla (1961, 2015). A transformation is used to obtain the desired results for the digital case. In particular, conditions are stated under which it is possible to obtain a factorization into the product of two matrices—one of which together with its inverse is quasi-stable. A quasi-stable matrix is one that is stable except possibly for poles on $s = j\omega$ in the analog case or on $|\lambda| = 1$ in the digital case. When the quasi-stable factor is on the left, it is called a left Wiener–Hopf factor and it may only possess a left inverse. When the quasi-stable factor is on the right, it is called a right Wiener–Hopf factor and it may only possess a right inverse. When the matrix to be factored is nonsingular on $s = j\omega$, ω finite, in the analog case or on $|\lambda| = 1$ in the digital case and is good in addition, then the quasi-stable factors are in fact stable and stable inverses exist for them.

C.2 Basic Definitions and Preliminary Results

As already noted in Sect. 1.4, a real rational matrix $G(s)$ satisfying $G_*(s) = G(s)$, where $G_*(s) = G^*(-\bar{s}) = G'(-s)$, is called a *parahermitian* matrix. Clearly, $G(j\omega) = G^*(j\omega)$ is Hermitian in this case. When $G(j\omega)$ is also nonnegative definite, then $G(s)$ is called *parahermitian-positive*. The notation $G \geq 0$ ($G > 0$) is used to denote a nonnegative-definite matrix (positive-definite matrix).

J. J. Bongiorno Jr. and K. Park, *Design of Linear Multivariable Feedback Control Systems*, https://doi.org/10.1007/978-3-030-44356-6

A real rational matrix $G(s)$ not necessarily square is called *paraunitary* when $G_*(s)G(s) = I$ or $G(s)G_*(s) = I$ or both. When $G(s)$ is a square paraunitary matrix, it follows that $G_*(s) = G'(-s) = G^{-1}(s)$ and for $s = j\omega$, $G_*(j\omega) = G^*(j\omega) = G^{-1}(j\omega)$ or $G(j\omega)$ is unitary in the usual sense. A paraunitary matrix is said to be *regular* when it is analytic in $\mathrm{Re}\, s > 0$.

Lemma C.1 *A paraunitary matrix $G(s)$ and its inverse (left or right) are analytic on $s = j\omega$, infinity included.*

Proof Either $G^*(j\omega)G(j\omega) = I$ or $G(j\omega)G^*(j\omega) = I$. In the former case, it follows from setting each diagonal element of the product matrix on the left-hand side of the equation equal to the unity diagonal element on the right-hand side of the equation that the sum of the squared magnitudes of all the elements in any column of $G(j\omega)$ must be unity; hence, $|g_{ij}(j\omega)| \le 1$ for any element. Thus, $G(j\omega)$ and its left inverse $G^*(j\omega)$ are finite for all ω. In the latter case, similar reasoning would lead to the fact that the sum of the squared magnitudes of all the elements in any row of $G(j\omega)$ must be unity; now $G(j\omega)$ and its right-inverse $G^*(j\omega)$ are finite for all ω. When $G(s)$ is square, it immediately follows from $G^*(j\omega) = G^{-1}(j\omega)$ that $G^{-1}(j\omega)$ is finite for all ω.

C.3 Main Results

Theorem C.1 *A real, rational, $n \times n$, parahermitian-positive matrix $G(s)$ of rank r possesses a spectral factorization*

$$G(s) = \Omega(s)\Omega_*(s), \tag{C.1}$$

where $\Omega(s)$ is an $n \times r$ real rational matrix (left Wiener–Hopf factor) that is analytic in $\mathrm{Re}\, s > 0$ *and also possesses a left inverse $\Omega^{-1}(s)$ analytic in* $\mathrm{Re}\, s > 0$. *Moreover, $\Omega(s)$ is unique up to multiplication on the right by an arbitrary real constant orthogonal $r \times r$ matrix. When $G(s)$ is good and rank $G(j\omega) = r$ for all finite ω, $\Omega(s)$ is analytic in* $\mathrm{Re}\, s > 0$ *and one can construct a left inverse $\Omega^{-1}(s)$ also analytic in* $\mathrm{Re}\, s > 0$.

Remark C.1 When $n = r$ in Theorem C.1, $\Omega(s)$ is square and $\Omega^{-1}(s)$ is both a left and right inverse of $\Omega(s)$.

Remark C.2 When $G'(s) = \Omega(s)\Omega_*(s)$ and $\Omega(s)$ is a left Wiener–Hopf factor, then $G(s) = \Omega'_*(s)\Omega'(s)$ and $\Omega'(s)$ is a right Wiener–Hopf factor for $G(s)$. That is, for any rational parahermitian-positive matrix it is always possible to write $G(s) = \Omega_l(s)\Omega_{l*}(s) = \Omega_{r*}(s)\Omega_r(s)$, where $\Omega_l(s)$ and $\Omega_r(s)$ are left and right Wiener–Hopf factors, respectively.

C.4 Digital Systems

In digital systems the spectral factorization of a real rational parahermitian-positive $n \times n$ matrix $G(\lambda) = G_*(\lambda) = G^*(1/\bar{\lambda}) = G'(\lambda^{-1})$, $G(e^{-j\theta}) \geq 0$ is of interest. Specifically, when $G(\lambda)$ is of rank r, a factorization $G(\lambda) = \Omega(\lambda)\Omega_*(\lambda)$ is desired in which the $n \times r$ matrix $\Omega(\lambda)$ is analytic together with its left inverse in $|\lambda| < 1$. It is established in the sequel that this is always possible.

Consider the transformation

$$\lambda = \frac{1-s}{1+s} \tag{C.2}$$

which maps the interior of the unit circle in the complex λ-plane into Re $s > 0$ in the s - plane. Then

$$\tilde{G}(s) = G\left(\frac{1-s}{1+s}\right) = G(\lambda) = G^*(1/\bar{\lambda}) = G^*\left(\frac{1+\bar{s}}{1-\bar{s}}\right) = \tilde{G}_*(s) \tag{C.3}$$

and $\tilde{G}(j\omega) = G(e^{-j\theta}) \geq 0$. That is, $\tilde{G}(s)$ is parahermitian-positive and of rank r. It then follows in accordance with Theorem C.1 that one can construct an $n \times r$ left Wiener–Hopf factor $\tilde{\Omega}(s)$ so that

$$\tilde{G}(s) = G\left(\frac{1-s}{1+s}\right) = \tilde{\Omega}(s)\tilde{\Omega}_*(s) \tag{C.4}$$

and

$$\Omega(\lambda) = \tilde{\Omega}\left(\frac{1-\lambda}{1+\lambda}\right) \tag{C.5}$$

together with its left inverse are analytic in $|\lambda| < 1$. Substituting $s = (1-\lambda)/(1+\lambda)$ into (C.4) then gives the desired result $G(\lambda) = \Omega(\lambda)\Omega_*(\lambda)$ since

$$\tilde{\Omega}_*\left(\frac{1-\lambda}{1+\lambda}\right) = \tilde{\Omega}^*\left(-\frac{1-\bar{\lambda}}{1+\bar{\lambda}}\right) = \Omega^*(1/\bar{\lambda}) = \Omega_*(\lambda). \tag{C.6}$$

It also follows from the fact that $\tilde{\Omega}(s)$ is unique to within multiplication on the right by an arbitrary real constant orthogonal $r \times r$ matrix that the same is true of $\Omega(\lambda)$.

Clearly, one can write $\Omega(\lambda) = N(\lambda)/g(\lambda)$, where $g(\lambda)$ is the *lcd* of $\Omega(\lambda)$ and $N(\lambda)$ is a $n \times r$ polynomial matrix. Now rank (Ω) =rank $(N) = r$: otherwise rank $(G) < r$, a contradiction. So there exists a unimodular matrix $U(\lambda)$ which reduces $N(\lambda)$ to its canonical form:

$$U(\lambda)N(\lambda) = \begin{bmatrix} N_o(\lambda) \\ 0 \end{bmatrix}, \tag{C.7}$$

where $N_o(\lambda)$ is a full rank $r \times r$ matrix. Thus,

$$U(\lambda)\Omega(\lambda) = \frac{U(\lambda)N(\lambda)}{g(\lambda)} = \begin{bmatrix} N_o(\lambda)/g(\lambda) \\ 0 \end{bmatrix} = \begin{bmatrix} \Omega_o(\lambda) \\ 0 \end{bmatrix}, \tag{C.8}$$

where $\Omega_o(\lambda)$ is analytic in $|\lambda| < 1$ because $U(\lambda)\Omega(\lambda)$ is. Moreover, $\Omega_o(\lambda)$ is a full rank $r \times r$ matrix which has an inverse. Suppose $M(\lambda)$ is a left inverse of $\Omega(\lambda)$. Then from

$$M(\lambda)\Omega(\lambda) = M(\lambda)U^{-1}(\lambda)\begin{bmatrix} \Omega_o(\lambda) \\ 0 \end{bmatrix} = I \tag{C.9}$$

one gets that $M(\lambda)$ is a left inverse iff

$$M(\lambda)U^{-1}(\lambda) = [\Omega_o^{-1}(\lambda) \quad M_o(\lambda)] \tag{C.10}$$

or

$$M(\lambda) = [\Omega_o^{-1}(\lambda) \quad M_o(\lambda)]U(\lambda) \tag{C.11}$$

It readily follows from (C.10) and (C.11) that a left inverse $M(\lambda) = \Omega^{-1}(\lambda)$ analytic in $|\lambda| < 1$ exists iff $\Omega_o^{-1}(\lambda)$ is analytic in $|\lambda| < 1$: necessity follows from (C.10) and sufficiency follows from (C.11) with the choice $M_o(\lambda) = 0$. That is,

$$\Omega(\lambda) = U^{-1}(\lambda)\begin{bmatrix} \Omega_o(\lambda) \\ 0 \end{bmatrix}, \tag{C.12}$$

where $\Omega_o(\lambda)$ and $\Omega_o^{-1}(\lambda)$ are analytic in $|\lambda| < 1$ and

$$\Omega^{-1}(\lambda) = [\Omega_o^{-1}(\lambda) \quad 0]U(\lambda) \tag{C.13}$$

is a left inverse analytic in $|\lambda| < 1$.

It is often the case that in addition to being parahermitian-positive and of rank r, the matrix $G(\lambda)$ is good and rank $[G(\lambda)] = r$ for all $|\lambda| = 1$. When $G(\lambda)$ is good,

$$\mathrm{Tr}\,[G(e^{-j\theta})] = \mathrm{Tr}\,[\Omega(e^{-j\theta})\Omega^*(e^{-j\theta})] = \sum_{i=1}^{n}\sum_{j=1}^{r}|\Omega_{i,j}(e^{-j\theta})|^2 \tag{C.14}$$

is finite for all θ. Hence, the elements $\Omega_{i,j}(e^{-j\theta})$ of $\Omega(e^{-j\theta})$ must be finite for all θ. That is, $\Omega(\lambda)$ and, therefore, $\Omega_o(\lambda)$ must be good in addition to being analytic in $|\lambda|<1$. So $\Omega(\lambda)$ and $\Omega_o(\lambda)$ are analytic in $|\lambda|\leq 1$ when $G(\lambda)$ is good.

It should be clear that rank $[G(\lambda)] = r$ for all $|\lambda| = 1$ iff rank (Ω) =rank $(\Omega_o) = r$ for all $|\lambda| = 1$. Thus, det$\Omega_o \neq 0$ on $|\lambda| = 1$ and $\Omega_o^{-1} =$ (adj Ω_o)/(det Ω_o) is good in addition to being analytic in $|\lambda|<1$. That is, $\Omega_o^{-1}(\lambda)$ and $\Omega^{-1}(\lambda) = [\Omega_o^{-1}(\lambda) \quad 0]U(\lambda)$ are analytic in $|\lambda|\leq 1$. It is now possible to state without further proof

Theorem C.2 *A real, rational, $n \times n$, parahermitian-positive matrix $G(\lambda)$ of rank r possesses a spectral factorization $G(\lambda) = \Omega(\lambda)\Omega_*(\lambda)$, where $\Omega(\lambda)$ is an $n \times r$ real rational left Wiener–Hopf factor that is analytic in $|\lambda|<1$ and possesses a left inverse $\Omega^{-1}(\lambda)$ analytic in $|\lambda|<1$. Moreover, $\Omega(\lambda)$ is unique to within multiplication on the right by an arbitrary real constant orthogonal $r \times r$ matrix. When $G(\lambda)$ is good and rank $G(\lambda) = r$ for all $|\lambda| = 1$, $\Omega(\lambda)$ is analytic in $|\lambda|\leq 1$ and one can construct a left inverse $\Omega^{-1}(\lambda)$ also analytic in $|\lambda|\leq 1$.*

Remark C.3 It should be evident that Remarks C.1 and C.2 hold here as well with s replaced by λ.

C.5 Historical Perspective and Discussion

The existence of Wiener–Hopf factors in the most general case (i.e., including the case $G(s)$ less than full rank) for rational parahermitian-positive matrices is proven in Youla (1961). This was the first paper to provide a comprehensive treatment of the spectral factorization of rational matrices and the first to show that the Wiener–Hopf factor is a polynomial matrix when $G(s)$ is a polynomial matrix. The constructive proof given involves finding the McMillan form for $G(s)$. An alternative approach which does not require the McMillan form and relies instead on a symmetric factor extraction methodology has a conceptual simplicity which is appealing; it is described in Davis (1963) and Callier (1985). But this methodology is restricted to the case in which $G(s)$ has full rank and as with the methodology in Youla (1961) is not well suited for calculations on a digital computer.

A methodology for factoring proper rational matrices which reduces the problem to finding the solution of an associated matrix Riccati equation is described in Anderson (1967). However, partial fraction expansions are needed, state-space realizations have to be determined, and a transformation to Jordan canonical form is employed in the solution of the Riccati equation. So this methodology is also not well suited for calculations on a digital computer and, moreover, does not lend itself to the case in which $G(s)$ has $j\omega$-axis poles.

The first successful computer-oriented factorization algorithm for a full-rank polynomial matrix $A(s)$ satisfying $A(j\omega) \geq 0$ is described in Tuel (1968). As pointed out there it can be applied to a full-rank rational matrix $G(s)$ as well. In contrast with Anderson (1967), the Tuel algorithm makes a more direct connection between spectral factorization and a solution of a nonlinear difference equation of the Riccati type. Moreover, the solution of the Riccati equation is obtained by iteration which is shown to converge for the case $A(j\omega) > 0$. For test cases in which $A(j\omega)$ was singular for some $j\omega$, the iteration was observed to converge more slowly. An additional advantage of the Tuel algorithm is that the spectral factorization is accomplished without the need to spectrally factor scalar polynomials. Similar computer-oriented algorithms of this kind are described in Park and Youla (1992), Chap. 4 of Aliev and Larin (1998), Oară and Varga (2000), Aliev and Larin (2003), and Gu et al. (2006) under a variety of different assumptions on the matrix to be factored regarding its structure, rank, and poles and zeros on $s = j\omega$ in the analog case and on $|\lambda| = 1$ in the digital case. Useful works in connection with finding the needed solutions of the Riccati equations are Youla and Paterno (1971) and Laub (1979).

In Ježek and Kučera (1985), an algorithm for spectral factorization of polynomial matrices that are positive-definite on $s = j\omega$ in the analog case and on $|\lambda| = 1$ in the digital case is described. The methodology is based on Newton's method and at each iteration of the algorithm full use is made of the structure of the equations to be solved with an increase in efficiency as a result. Just as with the Tuel (1968) algorithm, the methodology appears to work even when the matrix to be factored has zeros of its determinant at points on the stability boundary. However, convergence again slows down and is degraded.

In Youla and Kazanjian (1978), Wiener–Hopf factorization of an arbitrary integrable full-rank periodic nonnegative-definite Hermitian matrix $K(\theta)$ is described for the case in which the Paley–Wiener criterion

$$\int_{-\pi}^{\pi} \ln \det K(\theta) d\theta > -\infty \tag{C.15}$$

is satisfied. The algorithm given begins with the coefficient matrices in the Fourier series expansion of $K(\theta)$. It can be applied to rational matrices using (C.2) and $\lambda = e^{-j\theta}$ and factoring the associated $K(\theta)$ obtained; however, the matrix $K(\theta)$ is not restricted to the boundary value of some rational matrix and the algorithm is more general in this regard. Moreover, a connection is made between spectral factorization and two sets of matrix orthogonal polynomials generated by the weight $K(\theta)$ that provides an effective alternative scheme for the construction of the desired Wiener–Hopf factor. In addition, poles and zeros on $s = j\omega$ in the analog case and on $|\lambda| = 1$ in the digital case can be handled as long as (C.15) holds although convergence is slower in these cases.

A summary and discussion of spectral factorization methods including ones other than those already mentioned here is presented in Sayed and Kailath (2001). It is stated there that spectral factorization via a Riccati equation turns out to be very effective which contrasts with the claim in Ježek and Kučera (1985) that their algorithm based on Newton's method is more efficient. In any case, it is fortunate that there are several methodologies available for calculating Wiener–Hopf spectral factors on a digital computer. No doubt in difficult cases where there are poles and/or zeros close to the stability boundary it may be necessary to try more than one methodology in order to obtain a satisfactory result. Worthy of a first try is the software available commercially in the Polynomial Toolbox for MATLAB® which, among other things, can be used to find Wiener–Hopf spectral factors. MATLAB® is a trademark of the MathWorks, Inc. (www.mathworks.com) and the Polynomial Toolbox for MATLAB® can be obtained from Polyx, Ltd. (www.polyx.cz).

References

F.A. Aliev, V.B. Larin, *Optimization of Linear Control Systems: Analytical Methods and Computational Algorithms* (Gordon and Breach, The Netherlands, 1998)

F.A. Aliev, V.B. Larin, Special cases in optimization problems for stationary linear closed-loop systems. Int. Appl. Mech. **39**(3), 251–273 (2003)

F.A. Aliev, V.B. Larin, Comments on "Optimizing simultaneously over the numerator and denominator polynomials in the Youla-Kučera parameterization". IEEE Trans. Automat. Control **52**, 763 (2007)

F.A. Aliev, V.B. Larin, Parameterization of sets of stabilizing controllers in mechanical systems. Int. Appl. Mech. **44**(6), 599–618 (2008)

F.A. Aliev, V.B. Larin, Comment on 'Persistent inputs and the standard $\mathcal{H}_2$ multivariable control problem' by K. Park, J.J. Bongiorno Jr. Int. J. Control **83**(6), 1296–1298 (2010)

B.D.O. Anderson, An algebraic solution to the spectral factorization problem. IEEE Trans. Automat. Control **AC-12**, 410–414 (1967)

B.D.O. Anderson, J.B. Moore, *Optimal Control: Linear Quadratic Methods* (Prentice Hall, New Jersey, 1989)

P.J. Antsaklis, Some relations satisfied by prime polynomial matrices and their role in linear multivariable system theory. IEEE Trans. Automat. Control **AC-24**, 611–616 (1979)

P.J. Antsaklis, Proper stable transfer matrix factorizations and internal system descriptions. IEEE Trans. Automat. Control **AC-31**, 634–638 (1986)

P. Antsaklis, A.N. Michel, *Linear Systems* (Birkhaüser, Boston, 2006)

K.J. Åström, *Introduction to Stochastic Control Theory* (Academic Press, New York, 1970)

K.J. Åström, K.H. Johansson, Q.G. Wang, Design of decoupled PI controllers for two-by-two systems. IEE Proc.-Control Theory Appl. **149**(1), 74–81 (2002)

R.R. Bitmead, S.Y. Kung, B.D.O. Anderson, T. Kailath, Greatest common divisor via generalized Sylvester and Bezout matrices. IEEE Trans. Automat. Control **AC-23**, 1043–1047 (1978)

J.J. Bongiorno Jr., Minimum sensitivity design of linear multivariable feedback control systems by matrix spectral factorization. IEEE Trans. Automat. Control **AC-14**, 665–673 (1969)

J.J. Bongiorno Jr., F. Khorrami, T.C. Lin, A Wiener–Hopf approach to trade-offs between stability margin and performance in two- and three-degree-of-freedom multivariable control systems. Int. J. Control **68**(6), 1337–1366 (1997)

J.J. Bongiorno Jr., K. Park, Response to comment on 'Persistent inputs and the standard $\mathcal{H}_2$ multivariable control problem' by K. Park, J.J. Bongiorno Jr. Int. J. Control **83**(6), 1299–1302 (2010)

J. J. Bongiorno Jr. and K. Park, *Design of Linear Multivariable Feedback Control Systems*, https://doi.org/10.1007/978-3-030-44356-6

J.J. Bongiorno Jr., D.C. Youla, Wiener–Hopf design of optimal decoupling one-degree-of-freedom controllers for plants with rectangular transfer matrices. Int. J. Control **74**(14), 1393–1411 (2001)

S.P. Boyd, C.H. Barratt, *Linear Controller Design: Limits of Performance* (Prentice Hall, Englewood Cliffs, NJ, 1991)

S.P. Boyd, L. ElGhaoui, E. Feron, V. Balakrishnan, *Linear Matrix Inequalities in Control Theory* (SIAM Studies in Applied Mathematics, SIAM, Philadelphia, PA, 1994)

J.W. Brewer, Kronecker products and matrix calculus in system theory. IEEE Trans. Circuits Syst. **CAS-25**, 772–781 (1978)

T.S. Brinsmead, G.C. Goodwin, Cheap decoupled control. Automatica **37**, 1465–1471 (2001)

F.M. Callier, On polynomial matrix spectral factorization by symmetric extraction, IEEE Trans. Automat. Control **AC-30**, 453–464 (1985)

F.M. Callier, C.A. Desoer, *Multivariable Feedback Systems* (Springer, Berlin, 1982)

B.-C. Chang, J.B. Pearson, *Algorithms for the Solution of Polynomial Equations Arising in Multivariable Control Theory*. Houston, Texas: Technical Report 8208 (Department of Electrical Engineering, Rice University, 1982)

B.-C. Chang, *Programs for Solving Polynomial Equations*. Houston, Texas: Technical Report 8209 (Department of Electrical Engineering, Rice University, 1982)

S.S.L. Chang, *Synthesis of Optimal Control Systems* (McGraw-Hill, New York, 1961)

B. M. Chen, *Robust and $\mathcal{H}_\infty$-Control*. (Springer, London, 2000)

C.-T. Chen, *Linear System Theory and Design* (Holt, Rinehart, and Winston, New York, 1984)

T. Chen, B. Francis, *Optimal Sampled-Data Control Systems*. Version 4.4 (1994)

T. Chen, B. Francis, *Optimal Sampled-Data Control Systems* (Springer, London, 1995)

L. Cheng, J.B. Pearson, Synthesis of linear multivariable regulators. IEEE Trans. Automat. Control **AC-26**, 194–202 (1981)

R.V. Churchill, *Introduction to Complex Variables and Applications* (McGraw-Hill, New York, 1948)

G.O. Corrêa, M.A. Da Silveira, On the design of servomechanisms via $\mathcal{H}_2$-optimization. Int. J. Control **61**(2), 475–491 (1995)

G.O. Corrêa, T.M. Soares, D.M. Sales, Optimal tracking under decoupling constraints. Int. J. Control **74**(1), 10–27 (2001)

M.A. Da Silveira, G.O. Corrêa, $\mathcal{H}_2$-optimal control of linear systems with tracking/disturbance rejection constraints. Int. J. Control **55**(5), 1115–1139 (1992)

M.C. Davis, Factoring the spectral matrix. IEEE Trans. Automat Control **AC-8**, 296–305 (1963)

C.A. Desoer, A.N. Gündes, Decoupling linear multiinput multioutput plants by dynamic output feedback: an algebraic theory. IEEE Trans. Automat. Control **31**, 744–750 (1986)

C.A. Desoer, C.L. Gustafson, Algebraic theory of linear multivariable feedback systems, IEEE Trans. Automat. Control **AC-29**, 909–917 (1984)

C.A. Desoer, R.-W. Liu, J. Murray, R. Saeks, Feedback system design: the fractional representation approach to analysis and synthesis. IEEE Trans. Automat. Control **AC-25**, 399–412 (1980)

J.C. Doyle, K. Glover, P.P. Khargonekar, B.A. Francis, State-space solutions to standard $\mathcal{H}_2$ and $\mathcal{H}_\infty$ control problems. IEEE Trans. Automat. Control **AC-34**, 831–847 (1989)

P.L. Falb, W.A. Wolovich, Decoupling in design and synthesis of multivariable control systems. IEEE Trans. Automat. Control **AC-12**, 651–669 (1967)

G.D. Forney Jr., Minimal bases of rational vector spaces with applications to multivariable linear systems. SIAM J. Control **13**(3), 493–520 (1975)

F.R. Gantmacher, *The Theory of Matrices*, vol. I (Chelsea, New York, 1960)

G.I. Gómez, G.C. Goodwin, An algebraic approach to decoupling in linear multivariable systems. Int. J. Control **73**(7), 582–599 (2000)

M.J. Grimble, Two-degrees of freedom feedback and feedforward optimal control of multivariable stochastic systems. Automatica **24**, 809–817 (1988)

M.J. Grimble, *Robust Industrial Control Systems: Optimal Design Approach for Polynomial Systems* (Wiley, England, 2006)

G. Gu, X.-R. Cao, H. Badr, Generalized LQR control and Kalman filtering with relations to computations of inner-outer and spectral factorizations. IEEE Trans. Automat. Control **51**, 595–605 (2006)

A.N. Gündes, C.A. Desoer, *Algebraic Theory of Linear Feedback Systems with Full and Decentralized Compensators* (Springer-Verlag, Berlin, 1990)

M.L.J. Hautus, M. Heyman, Linear feedback decoupling-transfer function analysis. IEEE Trans. Automat. Control **AC-28**, 823–832 (1983)

D. Henrion, M. Šebek, Numerical methods for polynomial rank evaluation, in *Proceedings of the IFAC Conference on System Structure and Control* (Nantes, France, 1998a)

D. Henrion, M. Šebek, Symmetric matrix polynomial equation: Interpolation results. Automatica **34**, 811–824 (1998b)

D. Henrion, M. Šebek, Efficient numerical method for the discrete-time symmetric matrix polynomial equation, in *IEE Proceedings, Control Theory and Applications*, vol. 145 (1998c), pp. 443–448

D. Henrion, M. Šebek, Reliable numerical methods for polynomial matrix triangulariza-tion. IEEE Trans. Automat. Control **44**, 497–508 (1999a)

D. Henrion, M. Šebek, Improved polynomial matrix determinant computation. IEEE Trans. Circuits Syst. Part I: Fund. Theory Appl. **46**, 1307–1308 (1999b)

D. Henrion, M. Šebek, An algorithm for polynomial matrix factor extraction. Int. J. Control **73**(8), 686–695 (2000)

B.L. Ho, R.E. Kalman, Effective construction of linear state-variable models from input-output data, in *Third Allerton Conference on Circuit and System Theory* (University of Illinois, Monticello, 1965), pp. 449–459

R.A. Horn, C.R. Johnson, *Matrix Analysis* (Cambridge University Press, New York, 1985)

R.A. Horn, C.R. Johnson, *Topics in Matrix Analysis* (Cambridge University Press, New York, 1991)

I.M. Horowitz, *Synthesis of Feedback Systems* (Academic, New York, 1963)

J.W. Howze, S.P. Bhattacharyya, Robust tracking, error feedback, and two-degree-of-freedom controllers. IEEE Trans. Automat. Control **AC-42**, 980–983 (1997)

M. Hromcik, M. Šebek, Fast Fourier transform and linear polynomial matrix equations, in *First IFAC/IEEE Symposium on System Structure and Control 2001 [CD-ROM]* (Pergamon Press, Oxford, 2001)

K.J. Hunt, M. Šebek, V. Kučera, Polynomial solution of the standard multivariable $\mathcal{H}_2$-optimal control problem. IEEE Trans. Automat. Control **AC-39**, 1502–1507 (1994)

J. Ježek, V. Kučera, Efficient algorithm for matrix spectral factorization. Automatica **21**, 663–669 (1985)

T. Kailath, *Linear Systems* (Prentice-Hall, New Jersey, 1980)

R.E. Kalman, Irreducible realizations and the degree of a matrix of rational functions, Technical Report No. 64-5 (Research Institute for Advanced Studies, Baltimore, Maryland, 1964)

R.E. Kalman, Irreducible realizations and the degree of a rational matrix. SIAM J. **13**, 520–544 (1965)

R.J. Kavanagh, Noninteracting controls in linear multivariable systems. AIEE Trans. Appl. Ind. **76**, 95–100 (1957)

P.P. Khargonekar, M.A. Rotea, Multiple objective optimal control of linear systems: the quadratic norm case. IEEE Trans. Automat. Control **36**, 14–24 (1991)

S. Kung, T. Kailath, Fast projection methods for minimal design problems in linear system theory. Automatica **16**, 399–403 (1980)

V. Kučera, Closed-loop stability of discrete linear single-variable systems. Kybernetika **10**(2), 146–171 (1974)

V. Kučera, *Discrete Linear Control: The Polynomial Equation Approach* (Wiley, New York, 1979)

V. Kučera, *Analysis and Design of Discrete Linear Control Systems* (Prentice Ha ll International, Hertfordshire, England, 1991)

V. Kučera, Author's reply. IEEE Trans. Automat. Control **52**, 763 (2007)

V. Kučera, Decoupling Optimal Controllers, in *Proceedings of the 18th International Conference on Process Control* (Tatranská Lomnica, Slovakia, 2011), pp. 400–407

H. Kwakernaak, R. Sivan, *Linear Optimal Control Systems* (Wiley, New York, 1972)

Y.-S. Lai, An algorithm for solving the matrix polynomial equation B(s)D(s) + A(s)N(s) = H(s). IEEE Trans. Circuits Syst. **36**, 1087–1089 (1989)

V.B. Larin, K.I. Naumenko, V.N. Suntsev, *Spectral Methods for Synthesis of Linear Systems with Feedback* (Naukova Dumka, Kiev, USSR, 1971) [Russian]

V.B. Larin, K.I. Naumenko, V.N. Suntsev, Synthesis of optimal linear stabilization systems, Cybernetics and Control Theory. Soviet Phys.-Doklady **17**(5), 438–440 (1972) (English, Russian original); translation from Doklady Akademii SSSR **204**(2), 306–308 (1972)

V.B. Larin, K.I. Naumenko, V.N. Suntsev, *Synthesis of Optimal Linear Systems with Feedback* (Naukova Dumka, Kiev, USSR, 1973). [Russian]

H.-P. Lee, J.J. Bongiorno Jr., Wiener–Hopf design of optimal decoupled multivariable feedback control systems. IEEE Trans. Automat. Control **AC-38**, 1838–1843 (1993a)

H.-P. Lee, J.J. Bongiorno Jr., Wiener–Hopf design of optimal decoupling controllers for plants with non-square transfer matrices. Int. J. Control **58**(6), 1227–1246 (1993b)

C.A. Lin, Necessary and sufficient conditions for existence of decoupling controllers. IEEE Trans. Automat. Control **AC-42**, 1157–1161 (1997)

C.-A. Lin, T.-F. Hsieh, Stabilization, parameterization, and decoupling controller design for linear multivariable plants. IEEE Trans. Automat. Control **38**, 620–625 (1993)

A. Linnemann, Q.-Q. Wang, Block decoupling with stability by unity output feedback- solution and performance limitations. Automatica **29**, 735–744 (1993)

J.M. Maciejowski, *Multivariable Feedback Design* (Addison-Wesley, New York, 1989)

B.S. Morgan, The synthesis of linear multivariable systems by state-variable feedback. IEEE Trans. Automat. Control **AC-9**, 405–411 (1964)

A.S. Morse, W.A. Wonham, Decoupling and pole assignment by dynamic compensation. SIAM J. Control **8**, 317–337 (1970)

C.N. Nett, Algebraic aspects of linear control system stability. IEEE Trans. Automat. Control **AC-31**, 941–949 (1986)

G.C. Newton Jr., L.A. Gould, J.F. Kaiser, *Analytical Design of Linear Feedback Controls* (Wiley, New York, 1957)

C. Oară, A. Varga, Computation of general inner-outer and spectral factorizations. IEEE Trans. Automat. Control **45**, 2307–2325 (2000)

A. Papoulis, *The Fourier Integral and Its Applications* (McGraw-Hill, New York, 1962)

A. Papoulis, *Probability, Random Variables, and Stochastic Processes* (McGraw-Hill, New York, 1965)

A. Papoulis, *Signal Ananlysis* (McGraw-Hill, New York, 1977)

K. Park, $\mathcal{H}_2$ design of one-degree-of-freedom decoupling controllers for square plants. Int. J. Control **81**(9), 1343–1351 (2008)

K. Park, $\mathcal{H}_2$ design of decoupling controllers based on directional interpolations, in *Joint 48th IEEE Conference on Decision and Control and 28th Chinese Control Conference* (Shanghai, P.R. China, 2009)

K. Park, Parameterization of decoupling controllers in the generalized plant. IEEE Trans. Automat. Control **AC-57**, 1067–1070 (2012)

K. Park, J.J. Bongiorno Jr, A general theory for the Wiener–Hopf design of multivariable control systems. IEEE Trans. Automat. Control **AC-34**, 619–626 (1989)

K. Park, J.J. Bongiorno Jr., Wiener–Hopf design of servo-regulator-type multivariable control systems including feedforward compensation. Int. J. Control **52**(5), 1189–1216 (1990)

K. Park, J.J. Bongiorno Jr., Persistent inputs and the standard $\mathcal{H}_2$ multivariable control problem. Int. J. Control **82**(11), 2002–2012 (2009)

K. Park, Y.-S. Cho, J.-S. Lee, K. Kang, Wiener–Hopf approach to derivation of rational doubly coprime factorizations. Automatica **35**, 725–728 (1999)

K. Park, Y.-S. Cho, J.G. Kim, Wiener–Hopf design of the general two-degree-of-freedom controllers and the connection to state-space formulas. Int. J. Control **75**(18), 1435–1448 (2002)

K. Park, G.-H. Choi, State space representation of the general Wiener–Hopf controllers. J. Control Autom. Syst. Eng. **4**(3), 301–307 (1998)

K. Park, G.-H. Choi, T.-Y. Kuc, Wiener–Hopf design of the optimal decoupling control system with state-space formulas. Automatica **38**, 319–326 (2002)

K. Park, D.C. Youla, Numerical calculation of the optimal three-degree-of-freedom Wiener–Hopf controller. Int. J. Control **56**(1), 227–244 (1992)

L. Pernebo, An algebraic theory for the design of controllers for linear mutivariable systems-Part I: Structure matrices and feedforward design. IEEE Trans. Automat. Control **AC-26**, 171–182 (1981a)

L. Pernebo, An algebraic theory for the design of controllers for linear mutivariable systems-Part II: Feedback realizations and feedback design. IEEE Trans. Automat. Control **AC-26**, 183–194 (1981b)

K.Yu. Polyakov, Polynomial design of optimal sampled-data tracking systems I. Quadratic optimization. Autom. Remote Control **62**(2), 304–316 (2001)

H.H. Rosenbrock, *State Space and Multivariable Theory* (Wiley, New York, 1970)

E.N. Rosenwasser, B.P. Lampe, *Multivariable Computer-Controlled Systems: A Transfer Function Approach* (Springer, Berlin, 2006)

A. Saberi, P. Sannuti, B.M. Chen, *H_2 Optimal Control* (Prentice Hall International, Hertfordshire, England, 1995)

R. Saeks, J. Murray, Feedback system design: the tracking and disturbance rejection problems. IEEE Trans. Automat. Control **AC-26**, 203–217 (1981)

M.G. Safonov, B.S. Chen, Multivariable stability margin optimization with decoupling and output regulation. Proc. Instit. Electric. Eng. **129**, 276–282 (1982)

M.G. Safonov, A. Sideris, Unification of Wiener–Hopf and state space approaches to quadratic optimal control, in *Digital Techniques in Simulation, Communications and Control*, ed. by S. G. Tzafestas (Elsevier, New York, 1985)

A.H. Sayed, T. Kailath, A survey of spectral factorization methods. Numer. Linear Algebra Appl. **8**(6/7), 467–496 (2001)

R.E. Skelton, T. Iwasaki, Increased roles of linear algebra in control education. IEEE Control Syst. Mag. **15**, 76–90 (1995)

U. Shaked, A general transfer function approach to the steady-state linear quadratic Gaussian stochastic control problem. Int. J. Control **24**(6), 771–800 (1976)

A. Sideris, $\mathcal{H}_\infty$ optimal control as a weighted Wiener–Hopf problem. IEEE Trans. Automat. Control **35**, 361–366 (1990)

J. Stefanovski, Simplified formula for a controller in optimal control problems. SIAM J. Control Optim. **45**(6), 2011–2034 (2007)

V. Strejc, The general theory of autonomity and invariance of linear systems of control. Acta Technica **5**, 235–258 (1960)

K. Sugimoto, Y. Yamamoto, A polynomial matrix method for computing stable rational doubly coprime factorizations. Syst. Control Lett. **14**, 267–273 (1990)

W.G. Tuel Jr., Computer algorithm for spectral factorization of rational matrices. IBM J. Res. Develop. **12**, 163–170 (1968)

A.I.G. Vardulakis, Internal stabilization and decoupling in linear multivariable systems by unity output feedback compensation. IEEE Trans. Automat. Control **32**, 735–739 (1987)

M. Vidyasagar, *Control System Synthesis: A Factorization Approach* (MIT Press, Cambridge, MA, 1985)

I.N. Voznesenskij, A control system with many outputs (in Russian). Automat. i Telemech. **4**, 7–38 (1936)

Q.-G. Wang, Decoupling with internal stability for unity output feedback systems. Automatica **28**, 411–415 (1992)

Q.-G. Wang, *Decoupling Control*. Lecture Notes in Control and Information sciences (Springer, Berlin, 2003)

J. E. Weston, J.J. Bongiorno Jr., Extension of analytical design techniques to multivariable feedback control systems. IEEE Trans. Automat. Control **AC-17**, 613–620 (1972)

N. Wiener, *Extrapolation, Interpolation, and Smoothing of Stationary Time Series* (Technology Press, Cambridge, MA, 1949)

W.A. Wolovich, *Linear Multivariable Systems* (Springer, New York, 1974)

L. Xie, D. Xue, S.K. Tso, On the two-degree-of-freedom Wiener–Hopf optimal design with tracking and disturbance rejection constraints. Automatica **36**, 1897–1904 (2000)

M. Yamada, P.Z. Zun, Y. Funahashi, On solving diophantine equations by real matrix manipulation. IEEE Trans. Automat. Control **40**, 118–122 (1995)

D.C. Youla, On the factorization of rational matrices. IRE Trans. Inf. Theory **IT-7**, 172–189 (1961)

D.C. Youla, *Course Notes for Linear Systems (EL610)* (Polytechnic University, Brooklyn, New York, 1981)

D.C. Youla, *Lectures on Passive Multiport Synthesis* (Polytechnic University, Brooklyn, New York, 1985)

D.C. Youla, *Theory and Synthesis of Linear Passive Time-Invariant Networks, Cambridge* (Cambridge University Press, England, 2015)

D.C. Youla, J.J. Bongiorno Jr., A feedback theory of two-degree-of-freedom optimal Wiener–Hopf design. IEEE Trans. Automat. Control **AC-30**, 652–665 (1985)

D.C. Youla, J.J. Bongiorno Jr., Wiener–Hopf design of optimal one-degree-of-freedom decoupling controllers. Int. J. Control **73**(18), 1657–1670 (2000)

D.C. Youla, J.J. Bongiorno Jr., H.A. Jabr, Modern Wiener–Hopf design of optimal controllers-part I: the single-input-output case. IEEE Trans. Automat. Control **AC-21**, 3–13 (1976a)

D.C. Youla, J.J. Bongiorno Jr., C.N. Lu, Single-loop feedback-stabilization of linear multivariable plants. Automatica **10**, 159–173 (1974)

D.C. Youla, H.A. Jabr, J.J. Bongiorno Jr., Modern Wiener–Hopf design of optimal controllers-part II: the multivariable case. IEEE Trans. Automat. Control **AC-21**, 319–338 (1976b)

D.C. Youla, N. Kazanjian, Bauer-type factorization of positive matrices and the theory of matrix polynomials on the unit circle. IEEE Trans. Circuits Syst. **CAS-25**, 57–69 (1978)

D.C. Youla, P. Paterno, *An APL/360 package for the linear regulator, quadratic-cost Riccati equation*. Brooklyn, NewYork: Technical Report PIBMRI-71-085 (Polytechnic Institute of Brooklyn, 1971)

D.C. Youla, P. Tissi, *An explicit formula for the degree of a rational matrix*. Brooklyn, New York: Technical Report PIBMRI-1273-65 (Polytechnic Institute of Brooklyn, 1965)

D.C. Youla, P. Tissi, n-port synthesis via reactance extraction – part I. 1966 IEEE Int. Conv. Record **14**(pt. 7), 183–208 (1966)

K. Zhou, J.C. Doyle, K. Glover, *Robust and Optimal Control* (Prentice-Hall, Upper Saddle River, New Jersey, 1996)

Index

J. J. Bongiorno Jr. and K. Park, *Design of Linear Multivariable Feedback Control Systems*, https://doi.org/10.1007/978-3-030-44356-6

The manufacturer's authorised representative in the EU is Springer Nature Customer Service Centre GmbH, Europaplatz 3, 69115 Heidelberg, Germany. If you have any concerns regarding our products, please contact ProductSafety@springernature.com

Printed and bound by CPI Group (UK) Ltd, Croydon, CR0 4YY
22/09/2026
02220985-0010